Lecture Notes in Electrical Engineering

Volume 511

** **Indexing: The books of this series are submitted to ISI Proceedings, EI-Compendex, SCOPUS, MetaPress, Springerlink** **

Lecture Notes in Electrical Engineering (LNEE) is a book series which reports the latest research and developments in Electrical Engineering, namely:

- Communication, Networks, and Information Theory
- Computer Engineering
- Signal, Image, Speech and Information Processing
- Circuits and Systems
- Bioengineering
- Engineering

The audience for the books in LNEE consists of advanced level students, researchers, and industry professionals working at the forefront of their fields. Much like Springer's other Lecture Notes series, LNEE will be distributed through Springer's print and electronic publishing channels.

For general information about this series, comments or suggestions, please use the contact address under "service for this series".

To submit a proposal or request further information, please contact the appropriate Springer Publishing Editors:

Asia:

China, *Jessie Guo, Assistant Editor* (jessie.guo@springer.com) (Engineering)

India, *Swati Meherishi, Senior Editor* (swati.meherishi@springer.com) (Engineering)

Japan, *Takeyuki Yonezawa, Editorial Director* (takeyuki.yonezawa@springer.com) (Physical Sciences & Engineering)

South Korea, *Smith (Ahram) Chae, Associate Editor* (smith.chae@springer.com) (Physical Sciences & Engineering)

Southeast Asia, *Ramesh Premnath, Editor* (ramesh.premnath@springer.com) (Electrical Engineering)

South Asia, *Aninda Bose, Editor* (aninda.bose@springer.com) (Electrical Engineering)

Europe:

Leontina Di Cecco, Editor (Leontina.dicecco@springer.com)

(Applied Sciences and Engineering; Bio-Inspired Robotics, Medical Robotics, Bioengineering; Computational Methods & Models in Science, Medicine and Technology; Soft Computing; Philosophy of Modern Science and Technologies; Mechanical Engineering; Ocean and Naval Engineering; Water Management & Technology)

(christoph.baumann@springer.com)

(Heat and Mass Transfer, Signal Processing and Telecommunications, and Solid and Fluid Mechanics, and Engineering Materials)

North America:

Michael Luby, Editor (michael.luby@springer.com) (Mechanics; Materials)

More information about this series at http://www.springer.com/series/7818

Vijay Nath · Jyotsna Kumar Mandal
Editors

Nanoelectronics, Circuits and Communication Systems

Proceeding of NCCS 2017

Springer

Editors
Vijay Nath
Department of Electronics and Communication Engineering
Birla Institute of Technology, Mesra
Ranchi, Jharkhand
India

Jyotsna Kumar Mandal
Department of Computer Science and Engineering
University of Kalyani
Kalyani
India

ISSN 1876-1100 ISSN 1876-1119 (electronic)
Lecture Notes in Electrical Engineering
ISBN 978-981-13-0775-1 ISBN 978-981-13-0776-8 (eBook)
https://doi.org/10.1007/978-981-13-0776-8

Library of Congress Control Number: 2018942516

Printed on acid-free paper

This Springer imprint is published by the registered company Springer Nature Singapore Pte Ltd.
The registered company address is: 152 Beach Road, #21-01/04 Gateway East, Singapore 189721, Singapore

Preface

Nowadays the development of electronic goods provides major benefits to the world market in terms of high-level digital portability, connectivity, safety, and security. Designs of complex integrated circuits (ICs) are possible due to Electronic Design Automation (EDA) software such as Cadence, Mentor Graphics, Synopsis, Xilinx, and Active HDL. Presently, three types of advanced techniques play a major role in IC design and manufacture, such as the design of full custom ICs, semi-custom ICs, and Application Specific Integrated Circuit (ASIC). Up to 2015, more than 1 billion transistors existed in desktop computers. For their complete design they require more than 500 rules which are statistical in nature. To meet the demand of the market for sophisticated systems, both designers and engineers are developing automated EDA software tools for efficient, bulk design. These tools consider the length of transistors in the nanometer range. These range-designed ICs require good support in terms of nano-materials and their related chemicals. Testing of ICs is also a major issue in sophisticated systems. Testing should occur at each step of design, from the Simulation Program of Integrated Circuit Emphasis (SPICE) level, through the schematics and layout levels, to fabrication. As levels of design increase, so does the testing cost of ICs. Without efficient software support, testing of ICs is difficult.

System on Chip (SoC) provides an excellent platform where analog and digital IC layouts can come together and be fabricated on a single wafer. Because of this, the cost of fabrication per IC and chip, accommodated on a single wafer, decreases and the product quality increases. Nanotechnology has allowed the integration of electronic devices, chips, circuits, and systems. The nanoscale dimensions of nano-electronics components of systems having giga-scale complexity can be measured on a chip. Nanotechnology improves the capability of electronic components by reducing the size of the transistors in ICs, increasing the density of memory chips, reducing power consumption, and improving display screens, e.g., their thickness. Today, in global automation, control, and functional environments, the Internet of Things (IoT) plays a major role. Nowadays peoples are leading toward a cashless world. The computing systems driving the cashless world are reliable, robust, correct, and highly accurate. IoT system may be adopted for smart education systems to make a global education hub. It is also adopted in global factory operation, control, and monitoring, for perfect and optimized products. NCCS–2017 provides an excellent forum for young researchers, engineers, and professors to work together and share knowledge. It also generates ideas about how to work in electronic media both safely and securely. Manufacturing companies/industries and universities make significant contributions to the development of their countries. However, they are facing several challenges such as rapid product development, flexibility, Low to medium volume, transportation, low cost, etc. Many advanced and unconventional technologies, tools, and software programs are being developed worldwide to face up to these challenges. Among these technologies, IC design, manufacturing, and the IoT have become more popular due to their ability in terms of precision. For research, development, sharing knowledge, and exchanging ideas in current trends, the Third International Conference on Nanoelectronics, Circuits and Communication Systems (NCCS-2017) was organized by the Indian Society of Very Large Scale Integration (VLSI) Education (ISVE) in Ranchi and the Institution of Electronics and Telecommunication Engineers (IETE), also in Ranchi, at the Advanced Regional Telecom Training Centre (ARTTC) Bharat Sanchar Nigam Limited (BSNL) near the Jumar River, Hazaribag Road, Ranchi, from 11–12 November 2017. This conference covered advancement in Micro Electro Mechanical System (MEMS) and nanoelectronics, wireless communication, optical communication, instrumentation, signal processing, the IoT, image processing, bioengineering, green energy, hybrid vehicles, environmental science, weather forecasting, cloud computing, renewable energy, Radio Frequency ID (RFID), CMOS sensors, actuators, transducers, telemetry systems, embedded systems, sensor network applications in mines, etc. Additionally, there is a demand for manpower to drive electronic system design and manufacturing. Because populations are increasing, society has a duty to provide all types of facilities to new users in order that they can live their lives efficiently. This is only possible with the help of research and development. As people move into the digital era, knowledge of safety and security is very important, something which is driven in part by technical meetings, conferences, workshops, and seminars on current trends in technology. NCCS-2017 represents a forum for young researches to develop their knowledge in current trends and provide solutions for future technologies demanded by the world market.

At this conference around 300 papers were received, of which 58 were peer blind reviewed and registered. Presented papers were accepted for publication in the conference proceedings of the Springer Scopus book series, *Lecture Notes in Electrical Engineering (LNEE)*. In addition, 11 outstanding papers were selected for SCI Journals like *Microsystem Technologies* and *IETE Technical Review; IETE Technical Research*, etc. Expert evaluators also guided to authors for extended version of the research articles. They suggest continue on same track and update their articles 70–80% with new titles, contents and results and submit to listed Science Citation Index (SCI) and Scopus journals for publications. All these articles were blindly peer reviewed by at least three reviewers with their detailed comments being passed to the authors for final decisions. During presentation of authors, six expert committee members evaluated their work. They also guided to authors for IPR/patents and new innovative project for funding in their area of research. NCCS–2017 organized by the ISVE and IETE in Ranchi at ARTTC BSNL provided a unique platform to young researchers, scientists, engineers, and professors to present their work worldwide. An expert committee guided authors to improve their articles, and provided advice about how to get work published in high-impact journals and book series. This conference was totally different to other conferences and symposia. This conference provided very good support to young researchers in terms of learning (by expert lectures, suggestions & comments) and updating their knowledge. This forum provided outcome-based learning and research strategies.

Consent for the conference was received from the Honorable Governor of Jharkhand, Smt. Draupadi Murmu, and President of the IETE, New Delhi, Dr. K. T. V. Reddy. Their message was published in the conference souvenir booklet along with other messages from dignitaries, wishing success for the two-day event. In the inaugural session of the conference on dais present as the Chief Guest Dr. Gopal Pathak, Vice-chancellor of Jharkhand Technical University, Ranchi; Guest of Honor Dr. K. K. Thakur, CGMT BSNL, Ranchi; Guest of Honor Dr. A. A. Khan, Former Vice-chancellor of Ranchi University, Ranchi; Guest of Honor, Dr. R. K. Singh, Former Chairman of the IETE, Ranchi; Guest of Honor Sh. Sanjay Kumar Jha, Past Chairman of the IETE, Ranchi, and Chief Executive Engineer of the Government of Jharkhand; Guest of Honor Dr. P. R. Thakura, Professor at the BIT Mesra and Executive Member of the ISVE and IETE, Ranchi; Guest of Honor, Sh. Ajay Kumar, Chairman of the IETE, Ranchi, and AGM (Admin) ARTTC BSNL, Ranchi; Dr. Anand Kumar Thakur, Faculty at SSMC RU and Organizing Secretary of NCCS-2017; Dr. Raj Kumar Singh, Faculty at RLSY RU and Convenor of NCCS-2017, Ranchi; Dr. Vijay Nath, Faculty at BIT, Mesra, and General Chair of NCCS-2017; and Keynote Speaker Dr. S. Jit, Professor at the Indian Institute of Technology, BHU (Uttar Pradesh). The conference began with a welcome address by Sh. Ajay Kumar, Chairman of the IETE Ranchi Centre with technical details about the conference being delivered by Dr. Vijay Nath, General Chair of the conference. A keynote address was given by Dr. S. Jit, Professor at the Indian Institute of Technology, BHU (Uttar Pradesh), entitled "Nanoelectronics and its future prospective." The chief guest at this function, Dr. Gopal Pathak, demonstrated the impact of research in new technology and its applications in society. He also shared his research and development views. He explained that without quality of research, technical education cannot grow, because day by day new technologies are emerging. He also demonstrated that through this conference in nearby Indian Institute of Technology (IITs), National Institute of Technology (NITs), Birla Institute of Technology (BITs), and other parallel technical universities, researchers were following trend by publishing their articles in high-quality journals. All of the other dignitaries also gave views about the growth of society, quality of publications, research innovations, and challenges.

The first technical session started with an enlightening lecture by Dr. S. Jit on the topic of "Nanotechnology and its future perspective." After his lecture he took charge as the chairman of the first technical session. M. Selvi et al. presented their ideas on the Chapter "Classification of Medical Dataset Along with Topic Modeling Using LDA." Manish Mohan Baral and Amitabh Verma demonstrated Chapter "Cloud-Based Intelligent System for Supply Chain Management: A Future Roadmap for SCM Technologies." Shipra and Mahesh Chandra explained the Chapter "Effect of Processing Combined MFCC and DSCC Features with QCN for Hindi Vowel Classification in Noisy Environments." Sanjiv Kumar Srivastava et al. demonstrated Chapter "The Impact of Knowledge Management and Data Mining on CRM in the Service Industry." R. S. Nancy Noella and J. Priyadarshini defined Chapter "Efficient Computer-Aided Diagnosis of Alzheimer's Disease and Parkinson's Disease—A Survey." Brij Mohan Prasad and P. R. Thakura explained the Chapter "Design and Analysis of Dedicated Power Converter for Hybrid Electric Vehicles." Keshav Sinha and Partha Paul demonstrated Chapter "An Underground Mine Safety of Personnel's Using IoT." Ravi Prakash et al. demonstrated the Chapter "Implementation of Trapdoor Functionality to Two-Layer Encryption and Decryption by Using RSA-AES Cryptography Algorithms." Bathula Siva Kumar Reddy explained the Chapter "Experimental Validation of Spectrum Sensing Techniques Using Software-Defined Radio." Shaik Qadeer et al. demonstrated Chapter "Smart Switch for Power Saving." Debdatta Kandar and Babu Sena Paul defined Chapter "SDN-Based Programmable RSU Selection Method in Heterogeneous Networks." K. Damayanti et al. demonstrated the Chapter "Design and Implementation of an Energy-Efficient, Low-Cost Robotic Solar Tracker." Arindam Banerjee et al. explained the Chapter "Fast Squaring Technique for Radix Vicinity Numbers for Radix $2^n \pm M$ with Reduced Computational Complexity." Annu Priya and Sudip Kumar Sahana presented Chapter "A Survey on Multiprocessor Scheduling Using Evolutionary Technique." Minal Padlia and Jankiballabh Sharma defined Chapter "Fractional Sobel Filter Based Brain Tumor Detection and Segmentation Using Statistical Features and SVM." Anupam Kumar and Manoj Kumar demonstrated Chapter "Cladding Mode Analysis of Photonic Crystal Fiber Using Scalar Effective Index Model." M. J. Abinash and V. Vasudevan defined Chapter "A Hybrid Forward Selection Based LASSO Technique for Liver Cancer Classification." Smita Pallavi et al. explained the Chapter "Feature Subset Selection Using IULDA Model for Prediction." Md Sajid et al. defined the Chapter "Effect of RC Surge Suppressor in Reduction of Over Voltages at Motor Terminal Caused by PWM-Based Inverter." Nalini Singh and Satchidananda Dehuri explained the Chapter "Usage of Deep Learning in Epileptic Seizure Detection Through EEG Signal." Annapurna Mishra and Satchidananda Dehuri defined Chapter "An Experimental Study of Filter Bank Approach and Biogeography-Based Optimized ANN in Fingerprint Classification." Sandeep Dabhade et al. demonstrated Chapter "Performance and Capacity Testing of Medium and Large Managed Plane and Control Plane Optical Networks." Prabhat Kumar Ranjan and P. R. Thakura demonstrated Chapter "Analysis of Single-Stage Three-Phase DC–AC Boost Inverter for Distributed Generation System." Avanish Kumar and P. R. Thakura defined a Chapter "Close Loop Speed Controller for Brushless DC Motor for Hybrid Electric Vehicles." Ankita and Sudip Kumar Sahana defined Chapter "A Survey on Grid Schedulers." M. K. Mandal and A. K. Das demonstrated Chapter "Chaos-Based Colour Image Encryption Using Microcontroller ATMEGA 32." Rahul Priyadarshi et al. defined Chapter "An Enhanced GEAR Protocol for Wireless Sensor Networks." C. Kavitha and M. Ganesh Madhan demonstrated Chapter "A Novel Equivalent Circuit Approach for Modal Analysis of MEMS Cantilever Beam." M. Karuppasamy and S. P. Balakannan explained Chapter "Energy-Efficient Data Delivery in Green Cloud Networks." Shalini Mahato and Sanchita Paul defined Chapter "Electroencephalogram (EEG) Signal Analysis for Diagnosis of Major Depressive Disorder (MDD): A Review." K. Murali Krishna and M. Ganesh Madhan defined the Chapter "Numerical Simulation of High-Temperature VCSEL Operation and Its Impact on Digital Optical Link Performance." Jatindeep Singh et al. illustrated a Chapter "Smart Activity Sequence Generator in Wearable IoT. " Shamama Anwar et al. defined Chapter "Hand Gesture Recognition: A Survey." Qaiser Razi and Vijay Nath demonstrated the Chapter "Design of Smart Embedded System for Agricultural Update Using Internet of Things." Sabiha Fatma and Vijay Nath defined the Chapter "Study and Design of Smart Embedded System for Train Track Monitoring Using IoTs." Abhinav Kumar and Vijay Nath explained the Chapter "Study and Design of Smart Embedded System for Smart City Using Internet of Things." Satvika Anand and Vijay Nath demonstrated the Chapter "Study and Design of Smart Embedded System for Remote Health Monitoring Using Internet of Things." Chandan Kumar and Vijay Nath illustrated the Chapter "Design of Smart Embedded System for Auto Toll Billing System Using IoTs." Pratik Mondal and Susanta Kumar Parui explained the Chapter "Two Step Coupled Gap Resonator and Its Application as Bandpass Filter." Ashwani Sharma et al. defined the Chapter "Performance Comparison of DCF and FBG as Dispersion Compensation Techniques at 100 Gbps Over 120 km Using SMF." Neha Nidhi et al. explained Chapter "Different Aspects of Smart Grid: An overview." Sumit Srivastava et al. defined Chapter "Robust Voiceprint Based Audio Watermarking Using Wavelet Transform." Neha Nidhi et al. demonstrated Chapter "A High-Performance Energy-Efficient 75.17 dB Two-Stage Operational Amplifier." Utkarsh Raj et al. explained the Chapter "Automated Toll Plaza Using Barcode-Laser Scanning Technology." Trisha Ghosh et al. defined the Chapter "MIMO Wideband Antenna Technique in DGS for Enhanced Wireless Communication." Jyoti et al. explained the Chapter "Designing of FIR Filter Using FPGA: A Review." Shantanu Chaudhary et al. defined the Chapter "Design of All-Terrain Rover Quadcopter for Military Engineering Services." Dipti Kumari and Kumar Rajnish explained Chapter "A Systematic Approach Towards Development of Universal Software Fault Prediction Model Using Object-Oriented Design Measurement." Priyanka Parihar et al. defined the Chapter "6T SRAM Cell Design and Investigation for Ultra-Low-Power Application." Rohit Mohan et al. defined the Chapter "Design of Robot Monitoring System for Aviation." E. V. V. Hari Charan et al. demonstrated the Chapter "Electronic Toll Collection System Using Barcode Technology." Vidushi Goel et al. explained the Chapter "Design of Smartphone Controlled Robot Using Bluetooth." Paritosh Kumar Sinha et al. defined the Chapter "Design of Earthquake Indicator System Using ATmega328p and ADXL335 for Disaster Management." Deril Raju et al. defined the Chapter "Study and Design of Smart Embedded System for Aviation System: A Review." Deepak Prasad et al.

explained the Chapter "Study and Design of Smart Industry: A Review." S. Sanjay Kumar et al. explained the Chapter "Design of Smart Security Systems for Home Automation." Vidushi Goel et al. defined the Chapter "Auto-Train Track Fault Detection System." Sonali B. Wankhede demonstrated the "Study of Network-Based DoS Attacks."

At this conference authors were invited to submit original papers with selected, quality papers being presented. Authors described their articles very well. Authors and editors took the utmost care in presenting information and acknowledging original sources whenever necessary. The editors would like to express their gratitude toward the authors, organizers of the IC-NCCS-2017, and staff of Springer (India) for publishing these conference proceedings. Readers are requested to provide their valuable feedback on the quality of presentation and inadvertent errors or omission of information. We expect that the book will be welcomed by students as well as practicing engineers, researchers, and professors.

Ranchi, India Vijay Nath

Editorial Acknowledgements

We extend our thanks to all the authors for contributing to this book/proceedings by sharing their valuable research findings. We would especially like to thank a number of reviewers for promptly reviewing the papers submitted to the conference. We are grateful to the volunteers, invited speakers, session chairs, sponsors, sub-committee members, members of the international advisory committee and the national advisory committee, members of the technical program committee, members of the joint secretary, and members of scientific advisory committee, for the part they all played in making the conference a success. The editors express their heartfelt gratitude toward Dr. K. T. V. Reddy, President of the IETE New Delhi; Dr. A. K. S. Chandelle, Immediate Past President of the IETE New Delhi; Smt. Srimati Dagur, Former President of the IETE New Delhi; Sh. Sanjay Kumar Jha, Immediate Past Chairman of the IETE Ranchi and Executive Engineer Government of Jharkhand; Prof. Bernd Michel, Micro Materials Centre (MMC) Berlin Germany; Prof. Bharath Bhushan, Ohio Eminent Scholar and The Horward D. Winbigler Professor and Director of the NBLL at Ohio State University Columbus Ohio U.S.A.; Prof. P. S. Neelakanta (C. Engg.), Fellow at the IEE Florida Atlantic University (FAU) U.S.A.; Sh. Prasad Vijay Bhushan Pandey, DTO Term Cell1 BSNL Ranchi and Chairman of the ISVE Ranchi; Prof. A. A. Khan, former Vice-chancellor of Ranchi University; Prof. M. K. Mishra, Vice-chancellor of BIT Mesra; Prof. Gopal Pathak, Vice-chancellor of Jharkhand Technical University Ranchi; Dr. K. K. Thakur, CGMT BSNL Ranchi; Prof. R. K. Pandey, Vice-chancellor of Ranchi University; Prof. P. K. Barhai, former Vice-chancellor of BIT Mesra; Sh. R. Mishra, former CMD-HEC Ranchi; Dr. M. Chakraborty, Professor at IIT Kharagpur; Dr. Ramgopal Rao, Professor at IIT Bombay, and Director at IIT Delhi; Dr. P. Chakraborty, Professor at IIT BHU; Dr. S. Jit, Professor at IIT BHU; Dr. J. K. Mandal, Professor at Kalyani University; Dr. Abhijit Biswas, Professor at Kolkata University; Dr. Subir Kumar Sarkar, Professor at Jadavpur University; Dr. Gaurav Trivedi, Associate Professor at IIT Guwahati; Dr. Y. S. Chauhan, Associate Professor at IIT Kanpur; Dr. B. K. Kaushik, Professor at IIT Roorkee; Dr. Shree Prakash Tiwari, Faculty at IIT Jodhpur; Dr. P. Kumar, Associate Professor at IIT Patna; Dr. M. Bhaskar, Professor at NIT

Tirchy; Dr. Adesh Kumar, Faculty at UPES University, Dehradun; Dr. Manish Kumar Associate Professor at MMMUT Gorakhpur; Dr. Manish ishra, Associate Professor at DDU University Gorakhpur; Dr. Umesh Yadav, Professor at DDU University Gorakhpur; Dr. J. K. Mandal, Professor at Kalyani University; Prof. D. Acharjee, President at ISTM Kolkata; Dr. N. Gupta, Professor at BIT Mesra, Ranchi; Dr. Vibha Rani Gupta, Professor at BIT Mesra; Dr. S. Pal, Professor at BIT Mesra, Dr. B. K. Mishra, Principal Jumeritelaya Government of Jharkhand; Dr. V. K. Jha, BIT Mesra; Sh. Ajay Kumar, AGM (administration) ARTTC BSNL Ranchi, and Chairman of IETE Ranchi; Dr. P. R. Thakura, Executive Member of IETE and ISVE, Ranchi and Professor at BIT Mesra Ranchi; Dr. M. Chandra, Dr. M. Chandra, Executive Member of IETE Ranchi, and Professor at BIT Mesra Ranchi; Dr. S. K. Ghorai, Executive Member of IETE Ranchi and Professor at BIT Mesra Ranchi; Dr. B. Chakraborty, Executive Member of IETE Ranchi, and Executive Engineer at Mecon Ranchi; Dr. S. Chakraborty, Executive Member of IETE Ranchi and Professor at BIT Mesra Ranchi; Dr. S. S. Solanki, Professor at BIT Mesra Ranchi; Dr. S. Pal, Professor at BIT Mesra Ranchi; Dr. S. Kumar, Executive Member of IETE Ranchi and Associate Professor at BIT Mesra Ranchi; Dr. B. K. Bhattacharya, Professor at NIT Agartala; Dr. Anand Kumar Thakur, Treasurer of IETE Ranchi & Faculty SSMC and Director FM Ranchi University; Dr. Raj Kumar Singh, Executive Member of IETE Ranchi and Faculty at RLSYC Ranchi University and Coordinator of UGC Refresher Course Ranchi University; Dr. R. K. Lal, Associate Professor at BIT Mesra Ranchi; Dr. Sudip Sahana ; Dr. P. Pal; Dr. Amritanjali; Dr. Rishi Sharma; Dr. K. K. Senapati; Dr. S. S. Sahu; Dr. M. K. Mukul; Dr. K. Bose, BIT Mesra; Smt. Saroj, Treasurer of ISVE Ranchi; Prof. Jyoti Singh Joint Secretary of ISVE Ranchi; Prof. A. K. Pandey, Secretary of ISVE Ranchi; Sh. Suraj Kumar Saw; Sh. Subro Chakraborty; Sh. Dipayan Ghosh; Sh. Ramkrishna Kundu, Executive Member of ISVE Ranchi; Sh. Deepak Prasad; Sh. Sumit Singh; Sh. H. Kar; Sh. Rajanish Yadav; and Sh. Anup Tirkey member of ISVE Ranchi for their endless support, encouragement, and motivation to organize such a prestigious event which paved the way for publishing the Proceedings of the Third International Nanoelectronics, Circuits and Communication Systems (NCCS-2017) conference. At last we express our sincere gratitude toward the staff members at Springer (India) who helped publish this book.

Contents

Classification of Medical Dataset Along with Topic Modeling Using LDA 1
M. Selvi, K. Thangaramya, M. S. Saranya, K. Kulothungan, S. Ganapathy and A. Kannan

Cloud-Based Intelligent System for Supply Chain Management: A Future Roadmap for SCM Technologies 13
Manish Mohan Baral and Amitabh Verma

Effect of Processing Combined MFCC and DSCC Features with QCN for Hindi Vowel Classification in Noisy Environments 25
Shipra and Mahesh Chandra

The Impact of Knowledge Management and Data Mining on CRM in the Service Industry 37
Sanjiv Kumar Srivastava, Bibhas Chandra and Praveen Srivastava

Efficient Computer-Aided Diagnosis of Alzheimer's Disease and Parkinson's Disease—A Survey 53
R. S. Nancy Noella and J. Priyadarshini

Design and Analysis of Dedicated Power Converter for Hybrid Electric Vehicles 65
Brij Mohan Prasad and P. R. Thakura

An Underground Mine Safety of Personnel's Using IoT 77
Keshav Sinha and Partha Paul

Implementation of Trapdoor Functionality to Two-Layer Encryption and Decryption by Using RSA-AES Cryptography Algorithms 89
Ravi Prakash, Premkumar Chithaluru, Deepak Sharma and P. Srikanth

Experimental Validation of Spectrum Sensing Techniques Using Software-Defined Radio 97
Bathula Siva Kumar Reddy

Smart Switch for Power Saving 105
Shaik Qadeer, Ajaz Fatima, Asfia Aleem and Amreen Begum

SDN-Based Programmable RSU Selection Method in Heterogeneous Networks 115
Debdatta Kandar and Babu Sena Paul

Design and Implementation of an Energy-Efficient, Low-Cost Robotic Solar Tracker 127
K. Damayanti, T. Sunil Reddy, B. M. Reddy, Avireni Srinivasulu and SM-IEEE

Fast Squaring Technique for Radix Vicinity Numbers for Radix $2^n \pm M$ with Reduced Computational Complexity 139
Arindam Banerjee, Arpan Deyasi, Swapan Bhattacharyya and Angsuman Sarkar

A Survey on Multiprocessor Scheduling Using Evolutionary Technique 149
Annu Priya and Sudip Kumar Sahana

Fractional Sobel Filter Based Brain Tumor Detection and Segmentation Using Statistical Features and SVM 161
Minal Padlia and Jankiballabh Sharma

Cladding Mode Analysis of Photonic Crystal Fiber Using Scalar Effective Index Model 177
Anupam Kumar and Manoj Kumar

A Hybrid Forward Selection Based LASSO Technique for Liver Cancer Classification 185
M. J. Abinash and V. Vasudevan

Feature Subset Selection Using IULDA Model for Prediction 195
Smita Pallavi, Akshay Kumar and Utkarsh Mohan

Effect of RC Surge Suppressor in Reduction of Over Voltages at Motor Terminal Caused by PWM-Based Inverter 209
Md Sajid, Amer Ali Khan, M. Suryakalavathi and B. P. Singh

Usage of Deep Learning in Epileptic Seizure Detection Through EEG Signal 219
Nalini Singh and Satchidananda Dehuri

An Experimental Study of Filter Bank Approach and Biogeography-Based Optimized ANN in Fingerprint Classification 229
Annapurna Mishra and Satchidananda Dehuri

Performance and Capacity Testing of Medium and Large Managed Plane and Control Plane Optical Networks 239
Sandeep Dabhade, Sumit Kumar, Shishir Kumar and K. B. Sivasubramanian

Analysis of Single-Stage Three-Phase DC–AC Boost Inverter for Distributed Generation System 245
Prabhat Kumar Ranjan and P. R. Thakura

Close Loop Speed Controller for Brushless DC Motor for Hybrid Electric Vehicles 255
Avanish Kumar and P. R. Thakura

A Survey on Grid Schedulers 269
Ankita and Sudip Kumar Sahana

Chaos-Based Colour Image Encryption Using Microcontroller ATMEGA 32 281
M. K. Mandal and A. K. Das

An Enhanced GEAR Protocol for Wireless Sensor Networks 289
Rahul Priyadarshi, Surender Kumar Soni and Prashant Sharma

A Novel Equivalent Circuit Approach for Modal Analysis of MEMS Cantilever Beam 299
C. Kavitha and M. Ganesh Madhan

Energy-Efficient Data Delivery in Green Cloud Networks 313
M. Karuppasamy and S. P. Balakannan

Electroencephalogram (EEG) Signal Analysis for Diagnosis of Major Depressive Disorder (MDD): A Review 323
Shalini Mahato and Sanchita Paul

Numerical Simulation of High-Temperature VCSEL Operation and Its Impact on Digital Optical Link Performance 337
K. Murali Krishna and M. Ganesh Madhan

Smart Activity Sequence Generator in Wearable IoT 353
Jatindeep Singh, Punit Mishra, Satyajit Mohapatra, Hari Shanker Gupta and Nihar Mohapatra

Hand Gesture Recognition: A Survey 365
Shamama Anwar, Subham Kumar Sinha, Snehanshu Vivek and Vishal Ashank

Design of Smart Embedded System for Agricultural Update Using Internet of Things 373
Qaiser Razi and Vijay Nath

Study and Design of Smart Embedded System for Train Track Monitoring Using IoTs 383
Sabiha Fatma and Vijay Nath

Study and Design of Smart Embedded System for Smart City Using Internet of Things 397
Abhinav Kumar and Vijay Nath

Study and Design of Smart Embedded System for Remote Health Monitoring Using Internet of Things 409
Satvika Anand and Vijay Nath

Design of Smart Embedded System for Auto Toll Billing System Using IoTs 415
Chandan Kumar and Vijay Nath

Two Step Coupled Gap Resonator and Its Application as Bandpass Filter 425
Pratik Mondal and Susanta Kumar Parui

Performance Comparison of DCF and FBG as Dispersion Compensation Techniques at 100 Gbps Over 120 km Using SMF 435
Ashwani Sharma, Inder Singh, Suman Bhattacharya and Shalini Sharma

Different Aspects of Smart Grid: An Overview 451
Neha Nidhi, Deepak Prasad and Vijay Nath

Robust Voiceprint Based Audio Watermarking Using Wavelet Transform 457
Sumit Srivastava, Mahesh Chandra and G. Sahoo

A High-Performance Energy-Efficient 75.17 dB Two-Stage Operational Amplifier 469
Neha Nidhi, Deepak Prasad and Vijay Nath

Automated Toll Plaza Using Barcode-Laser Scanning Technology 475
Utkarsh Raj, Neha Nidhi and Vijay Nath

MIMO Wideband Antenna Technique in DGS for Enhanced Wireless Communication 483
Trisha Ghosh, Sneha Tiwari and J. Sahay

Designing of FIR Filter Using FPGA: A Review 493
Jyoti, Adesh Kumar and Anil Sangwan

Design of All-Terrain Rover Quadcopter for Military Engineering Services 507
Shantanu Chaudhary, Arka Prava, Neha Nidhi and Vijay Nath

A Systematic Approach Towards Development of Universal Software Fault Prediction Model Using Object-Oriented Design Measurement 515
Dipti Kumari and Kumar Rajnish

6T SRAM Cell Design and Investigation for Ultra-Low-Power Application 527
Priyanka Parihar, Neha Gupta, Vaibhav Neema and Praveen Singh

Design of Robot Monitoring System for Aviation 535
Rohit Mohan, Akash Keneth Suraj, Sakshi Agarawal, Sananaya Majumdar and Vijay Nath

Electronic Toll Collection System Using Barcode Technology 549
E. V. V. Hari Charan, Indrajit Pal, Akash Sinha, Raj Kamal Roye Baro and Vijay Nath

Design of Smartphone Controlled Robot Using Bluetooth 557
Vidushi Goel, Riya, Pinki Kumari, Prachi Shikha, Tanushree, Deepak Prasad and Vijay Nath

Design of Earthquake Indicator System Using ATmega328p and ADXL335 for Disaster Management 565
Paritosh Kumar Sinha, Satyam Saraiyan, Momojit Ghosh and Vijay Nath

Study and Design of Smart Embedded System for Aviation System: A Review 573
Deril Raju, Lalitha Eleswarapu, Rohith Saiv and Vijay Nath

Study and Design of Smart Industry: A Review 591
Divyesh Kumar Maurya, Ankush Kumar, Suraj Kaunoujiya, Deepak Prasad and Vijay Nath

Design of Smart Security Systems for Home Automation 599
S. Sanjay Kumar, Ayushman Khalkho, Sparsh Agarwal, Suraj Prakash, Deepak Prasad and Vijay Nath

Auto-Train Track Fault Detection System 605
Vidushi Goel, Shubham Kumar, Aditya Muralidharan, Naveen Markham, Deepak Prasad and Vijay Nath

Study of Network-Based DoS Attacks 611
Sonali B. Wankhede

Author Index 617

About the Editors

Vijay Nath received his bachelor's degree in Physics and master's degree in Electronics from DDU Gorakhpur University, India, in 1998 and 2001, respectively. He received PGDCN from MMM Engineering College Gorakhpur (GM) in 1999 and Ph.D. in VLSI Design & Technology from Dr. RML Avadh University, Faizabad, in association with CEERI, Pilani, in 2008. He served as faculty in the Department of Electronics, DDU Gorakhpur University, Gorakhpur (2002–2006). In 2006, he joined as faculty in the Department of Electronics and Communication Engineering, Birla Institute of Technology, Mesra, Ranchi. Currently, he is Professor In-charge of Embedded System Design Lab of the Department of ECE, Member of BIT Brand Management, Assistant Examination Controller of BIT, Mesra, Ranchi. His research interests include microelectronics, low power VLSI design, temperature sensors, ASICs design, SoC, FPGA-based wireless system design, signal conditioning, real embedded systems designs, smart cardiac pacemakers, etc. He has completed three projects funded by DRDO, MHRD, ISRO, Government of India. He has 16 years of teaching and research experience, guided two Ph.D. scholars, developed pedagogy course on VLSI design, edited four books as volume editor of Springer, published 120 research articles in international journals and conferences, and is a member of several respected professional and academic bodies including IETE, ISVE, and IEEE.

Jyotsna Kumar Mandal received his M.Tech. from the Department of Computer Science, University of Calcutta, India, and Ph.D. in the field of Data Compression and Error Correction Techniques from Jadavpur University. He is Professor of Computer Science and Engineering, Director of IQAC, and Former Dean of FETM, University of Kalyani. He has 29 years of teaching and research experience and is currently working in the fields of network security, steganography, image processing, and wireless and sensor networks. He has guided 21 Ph.D. scholars, published six books and more than 380 papers, including 154 publications in various international journals, and edited 27 volumes as volume editor for ScienceDirect, Springer, CSI, etc.

Classification of Medical Dataset Along with Topic Modeling Using LDA

M. Selvi, K. Thangaramya, M. S. Saranya, K. Kulothungan, S. Ganapathy and A. Kannan

Abstract Nowadays, medical applications need a lot of storage for storing and providing access to the medical information seekers. Moreover in medical applications, information grows tremendously and hence they must be stored using a suitable storage structure so that it is possible to retrieve them faster from the text corpus in which the medical information is stored. The existing methods for storage and retrieval do not focus on classified organization. However, classified data storage will facilitate fast retrieval. Therefore, a new Latent Dirichlet Allocation (LDA) based topic modeling approach is proposed in this paper which uses temporal rules for effective manipulation of stored data. Therefore, a temporal rule based classification algorithm is proposed in this work by combining Naïve Bayes Classifier with LDA and temporal rules to store the data more efficiently and it helps to retrieve the documents faster. From the experiments conducted in this work by storing and retrieving medical data in a corpus, it is proved that the proposed model is more efficient with respect to classification accuracy leading to organized storage and fast retrieval.

Keywords Topic modeling · Latent Dirichlet Allocation (LDA)
Word cloud · Topic of words · Naive Bayes classification

M. Selvi (✉) · K. Thangaramya · M. S. Saranya · K. Kulothungan · A. Kannan
Department of IST, Anna University, CEG Campus, Chennai, India
e-mail: arpudhaselvi123@gmail.com

K. Thangaramya
e-mail: thangaramya112@gmail.com

M. S. Saranya
e-mail: saranyasivaraman5@gmail.com

K. Kulothungan
e-mail: kulo@auist.net

A. Kannan
e-mail: kannan@annauniv.edu

S. Ganapathy
School of CSE, VIT University, Chennai Campus, Chennai, India
e-mail: sganapathy@vit.ac.in

V. Nath and J. K. Mandal (eds.), *Nanoelectronics, Circuits and Communication Systems*, Lecture Notes in Electrical Engineering 511,
https://doi.org/10.1007/978-981-13-0776-8_1

1 Introduction

In this work, an intelligent medical data mining system has been developed for effective storage and retrieval of medical data with relevancy. For this purpose, classification algorithms from machine learning are used to perform effective decisions. From University of California, Irvine (UCI) repository breast and liver datasets are collected. Data preprocessing is done on the real benchmark life sciences dataset to remove missing values. The preprocessed data is given as a input to various classification algorithms namely Navie Bayes (NB), Support Vector Machine (SVM), Differential Topic Modeling (DTM) [1] which uses Latent Dirichlet Allocation (LDA) with SVM classification algorithm and finally with a combination of Naïve Bayes and temporal constraints with LDA (NBTC) which is the newly designed technique which is proposed in this paper to mine medical data storage system. To provide a system with user-friendly environment, a user interface with a facility to output clearly is provided in this system which displays the mined results to the user with graphical output.

Topic modeling is considered in this work for organizing, analyzing, understanding, and summarizing the textual documents related to medical applications. For this purpose, the topic modeling technique namely Latent Dirichlet Allocation (LDA) algorithm is used in this work for providing relevant list of topics by mainly focusing the retrieval of textual documents. In this model, each textual document is viewed as a mixture of topics. Moreover, LDA algorithm assigns topics of words with certain probabilities. In topic modeling, distribution of topics is combined to form documents and distribution of words is viewed as topics. R tool has been employed in this work for implementation of the proposed NBTC classifier which generates word cloud with weight of the words. This word cloud is composed of images which contain words occurred in particular documents. Word cloud is used to find the most occurred words in given corpus. High-frequency words are highlighted in bold and bigger using word cloud. Using this, the relevant topics are retrieved from the text corpus and are given to the user. From the experiment conducted in this work, it is proved that the proposed model provides most relevant information to medical information seekers.

The rest of the paper is organized as follows: Sect. 2 deals with the Literature Survey, Sect. 3 explains about the proposed Methodology, Sect. 4 describes the details of the proposed work, Sect. 5 explains about the results and discussions and Sect. 6 provides conclusions on this work and suggests some suitable future works.

2 Literature Survey

There are many works on topic modeling, classification [2–4], and information retrieval which are available in the literature [5–7]. Among them, Saranya et al. [5] proposed a new approach for effective storage and retrieval of medical data by using intelligent techniques. Farid et al. [8] proposed an adaptive rule based classifier for mining and extracting the relevant knowledge from biological datasets for providing efficient diagnosis for health seekers. Nahato et al. [9] proposed an effective classification model for classifying clinical datasets. This model enables classifier to predict presence or absence of a disorder from the clinical datasets. Back Propagation Neural Network (BPNN) is used in the classification model. Farid et al. [10] used Naive Bayesian is one of the popular algorithms in data mining and its main advantage is that the algorithm needs only one scan for training data and also handles missing values. Lawrence et al. [11] compared traditional Classification Tree Analysis (CTA) results to Stochastic Gradient Boosting (SGB) for remote sensing based datasets. Farid and Rahman [12] proposed a method for weights assigned to each training instance in the training data that improves classification accuracy of the decision tree.

Liu and Yu [13] proposed a feature selection algorithm for classification and clustering. Feature selection algorithm is integrated with meta data to provide detailed description about classification system. Alghamdi and Alfalqi [14] explained four methods for analyzing the huge volume of unclassified text. Methodology includes Latent Semantic Analysis (LSA), Probabilistic Latent Semantic Analysis (PLSA), and Latent Dirichlet Allocation (LDA). Zeng et al. [15] used Belief Propagations (BP) in Latent Dirichlet allocation (LDA) for achieving better speed and accuracy in classifying large-scale document datasets. Author-Topic Models (ATM) using belief propagations (BP). Chen et al. [1] proposed a transformed Pitman-Yor Process (TPYP) to compare vocabulary similarities and dissimilarities between topics of two different document collections by using differential topic model. Cheng et al. [16] proposed a Biterm Topic Model (BTM) for finding biterms that occur in the entire corpus. This will make the topic modeling effectively.

3 Methodology

Figure 1 shows the overall view of the machine learning based probabilistic classification approach for effective storage of medical records. From University of California, Irvine (UCI) repository breast and liver datasets are collected. Dataset preprocessing is done on the real benchmark life sciences dataset to remove missing values. Preprocessed data can be feed into Naive Bayes algorithm. Classified results are obtained using Naive Bayes algorithm. Knowledge discovery will be intimated whether it is true positive, true negative, false positive, and false negative values based on the confusion matrix which allows visualizing the algorithm. Then the formula will be executed and will be displayed in graph mode.

In this work, LDA is used for topic modeling. The main focus of the LDA model [6, 7] is the distribution of words representing topics. In this model, mixture of topics is viewed as documents and Bag of Words concept is also used in the modeling.

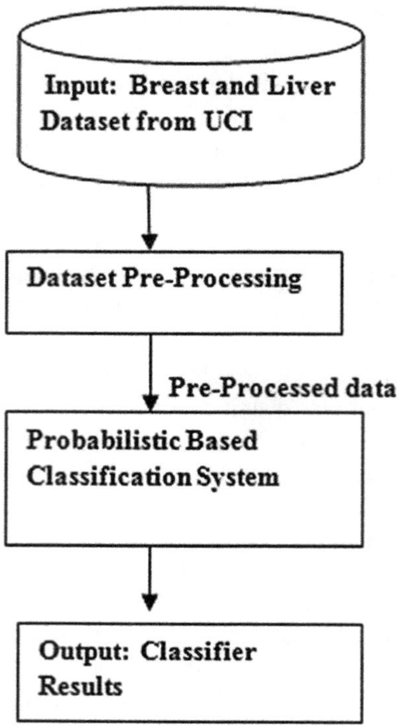

Fig. 1 Probabilistic based classification system

4 Proposed Work

The proposed scheme consists of medical related topics which are collected and stored in medical database. Using Latent Dirichlet Allocation (LDA) based topic modeling used to provide quick access to humans to visualize the topics which are mainly discussed in documents.

4.1 Document Topic Modeling

The most common and popular method for topic modeling is Latent Dirichlet Allocation (LDA). LDA is an algorithm which is used to generate topics with related keywords from the entire document. LDA discovers the topical patterns from the documents. From a collection of documents, we represent the frequent keywords and assign those keywords to a particular topic. The topic modeling process developed in this work is shown in Fig. 2.

Topic modeling is efficiently used in this work to analyze large volumes of text. In this topic-based probabilistic generative model, words are distributed over topics and topics are distributed over documents. In Natural Language Processing,

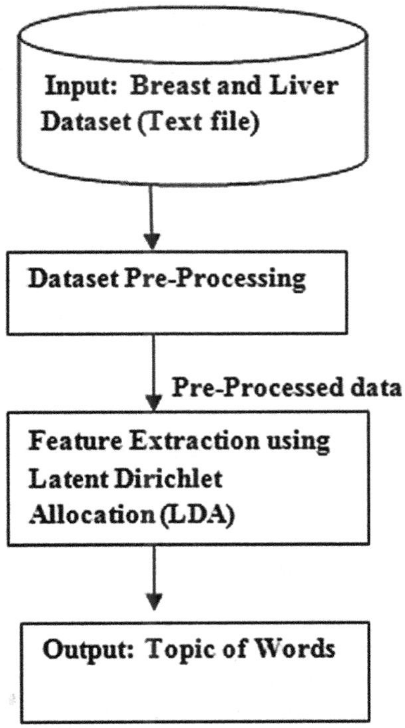

Fig. 2 Topic modeling

LDA plays an important role. Topic model is an unsupervised learning technique which extracts the content from the raw and unlabeled data. Topics are generated along with the most likely and frequently occurring words. In this work, feature extraction is performed using LDA with temporal rules and the topics are classified using the existing classification algorithms namely SVM, SVM with LDA, NB and the proposed NBTC classifier which uses NB with LDA and temporal constraints. It is proved through experiments that the proposed NBTC provides better classification accuracy when it is compared with the existing classification algorithms discussed in this paper.

4.2 *Word Cloud Generation*

A word cloud is a text mining method that allows us to highlight the most frequently used keywords in a paragraph of texts. It is also referred to as a text cloud or tag cloud. The procedure of creating word cloud is very simple in R software if you know the different steps to execute. A text mining package (tm) and word cloud generator package (word cloud) are available in R for helping us to analyze texts and to quickly visualize the keywords as a word cloud. Table 1 shows the distribution of data in University of California, Irvine (UCI) Repository.

Table 1 Dataset collection from UCI repository

No.	Dataset	Instance	No. of Attributes	Attribute type	Classes
1.	Breast cancer	286	9	Nominal	2
2.	Liver disorders	345	7	Numeric	2

Table 2 Confusion Matrix

Current = Predicted class	True class	False class
True class	True positive (TP)	False positive (FP)
False class	True negative (TN)	False negative (TN)

Table 2 shows the confusion matrix for the use of evaluation of the proposed model using the UCI repository.

5 Results and Discussions

From University of California, Irvine (UCI) repository breast and liver datasets are collected for classification. Datasets consist of breast cancer and liver cancer text file for experimenting topic modeling and word cloud generation. Figure 3 shows a sample word cloud generation for breast cancer dataset.

This word cloud has been generated by running the entire corpus of preprocessed breast cancer corpus. Moreover, the word cloud indicates the weights and frequently occurred words in the document. In this model, breast is the most frequently occurring word as shown in Fig. 4.

Frequently occurring words have been plotted by using Word Plot in R tool. It can be performed by setting the minimum frequency limit as 100 so the words which are repeated more than 100 times are listed. Breast is the medical term that occurs frequently in the entire corpus.

Figure 5 shows the list of topic generated from the breast cancer dataset. This dataset consists of breast cancer text file. Preprocessing involves removing stop

Fig. 3 Word cloud generation for breast cancer dataset

Fig. 4 Plotting word frequencies for breast cancer dataset

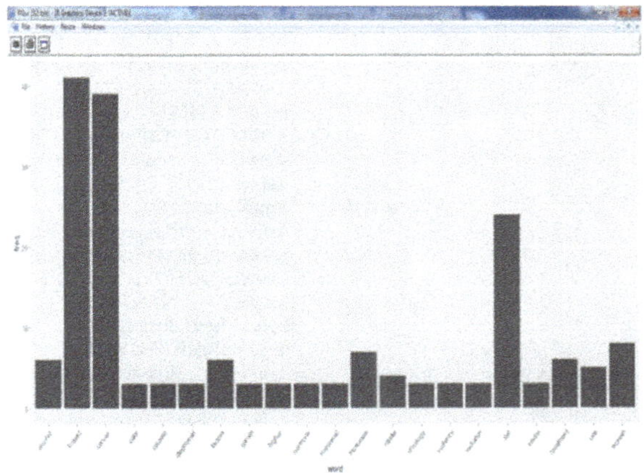

X- Axis: Word Y-Axis: Freq

Fig. 5 List of topics generated for breast cancer dataset

Topic0: Treatment(0.044103) Breast(0.044103) treatment(0.044103) radiation(0.044103) patient(0.044103) Women(0.044103) increased(0.044103) different(0.029451) surgery(0.029451) cells(0.029451) breast(0.029451) tumors(0.029451)
Topic1: factors(0.090679) cancer(0.05448) care(0.05448) oncology(0.05448) plan(0.03638) types(0.03638) grow(0.03638) causes(0.03638) women(0.018281) specializing(0.018281) medical(0.018281) treatments(0.018281) multidisciplinary(0.018281)
Topic2: status(0.069595) health(0.046474) therapy(0.046474) hormone(0.046474) Cancer(0.023353) variety(0.023353) professionals(0.023353) nurses(0.023353) social(0.023353) workers(0.023353) pharmacists(0.023353) postmenopausal hormone(0.023353)
Topic3: age(0.068718) breast cancer(0.068718) Chemotherapy(0.03453) radiationdestroy breast(0.03453) treatments(0.03453) healthy(0.03453) feel(0.03453) modified(0.03453) influenced(0.03453) capability(0.03453) spreading(0.03453)
Topic4: change(0.066446) nipple(0.066446) disease(0.066446) counselors(0.033388) sore(0.033388) Mouth sores(0.033388) size(0.033388) shape(0.033388) Weight gain(0.033388) Family(0.033388) history(0.033388) relatives(0.033388)
Topic5: breast(0.260042) cancer(0.176184) risk(0.151027) cancer(0.06717) increases(0.033627) diagnosed(0.025241) higher(0.025241) texture(0.00847) dimpling(0.00847) chances(0.00847) increase(0.00847) physical(0.00847)
Topic6: risk(0.124472) alcohol(0.124472) use(0.124472) women(0.099627) armpit(0.049938) cancer cell(0.025093) cause(0.025093) breast cancer(0.025093) factors (0.025093) develops breast(0.025093) lump(0.025093) affects(0.025093)

words from the documents. The topic of words with their probabilities value has been generated.

Figure 6 shows the list of topics generated for liver dataset. This dataset consists of liver cancer text file. Preprocessing involves removing stop words from the documents. The topic of words with their probabilities value has been generated.

Fig. 6 List of topics generated for liver dataset

Topic0: **liver(0.20934) body(0.069935) hepatic(0.046701) blood(0.046701) human(0.046701) organ(0.023467) section(0.023467) recover(0.023467) Diabetes(0.023467)**
Topic1: **pain(0.172521) discomfort(0.10365) clinical(0.069215) pain.(0.03478) swollen(0.03478) stage(0.03478) trials(0.03478) left(0.03478) Fatigue(0.03478)**
Topic2: **tumor(0.063988) right(0.063988) cancers(0.04273) lung(0.04273) Cancer(0.021471) Symptoms(0.021471) Lasers(0.021471) Family(0.021471)**
Topic3: **injected(0.051486) alcohol(0.051486) males(0.051486) females(0.051486) smoke(0.051486) Jaundice (0.025871) skin(0.025871) tongue(0.025871)**
Topic4: **cancer(0.147099) liver(0.147099) risk(0.107899) hepatitis(0.039298) people(0.039298) higher(0.039298) developing(0.039298) individuals(0.029498) Liver(0.019698)**
Topic5: **Side(0.062734) effects(0.062734) vomiting(0.062734) Liver(0.031523) Treatment(0.031523) Causes(0.031523) cancer(0.031523) treatment(0.031523)**
Topic6: **liver(0.212798) cancer(0.191539) cells(0.04273) L-carnitine(0.04273) deficiency(0.04273) acetaminophen(0.04273) medication(0.021471) damaged(0.021471)**

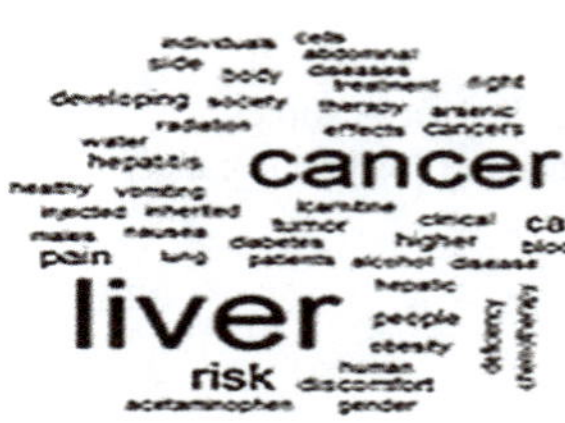

Fig. 7 Generation of word cloud for liver dataset

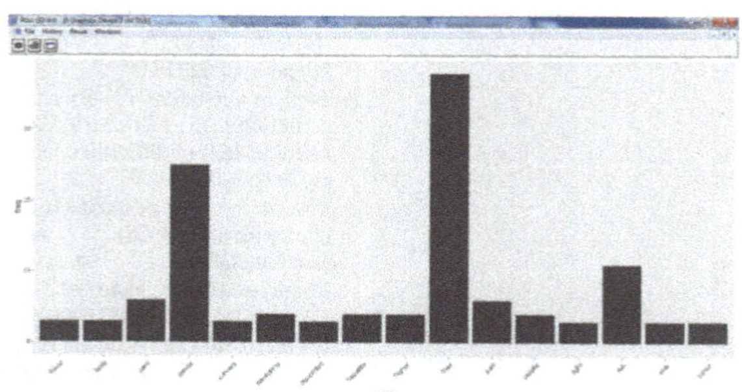

X-Axis: Word Y-Axis: Freq

Fig. 8 Plotting word frequencies for liver dataset

Figure 7 shows the word cloud which is generated by running the entire corpus of preprocessed liver corpus. Word cloud indicates the weights and frequently occurred words in the document. Liver is the word most frequently occurred.

Frequently occurring words have been plotted in Fig. 8 by using Word Plot in R tool. It was performed by setting the minimum frequency limit as 100 so the words which are repeated more than 100 times are listed. Liver is the medical terms occur frequently in the entire corpus.

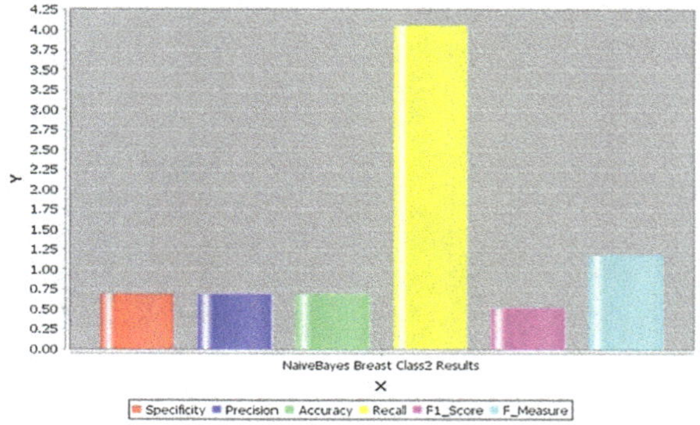

Fig. 9 Naive Bayes breast cancer class2 results

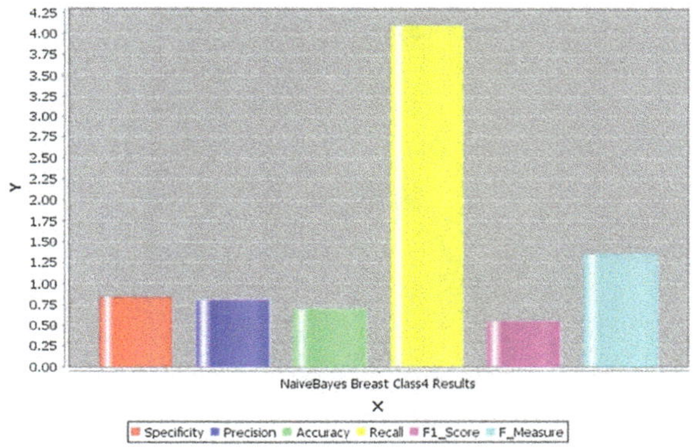

Fig. 10 Naive Bayes breast cancer class4 results

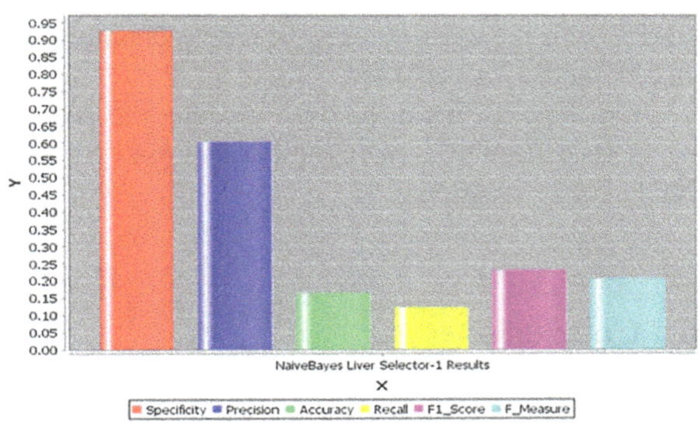

Fig. 11 Naive Bayes liver disorders selector 1 results

Evaluation Metrics like Precision, Recall, Accuracy, F Measure, Specificity, and F1 Score that are calculated for breast cancer class label 2 results are recorded in Fig. 9.

Evaluation Metrics like Precision, Recall, Accuracy, F Measure, Specificity, and F1 Score that are calculated for breast cancer class label 4 results are recorded in Fig. 10.

Evaluation Metrics like Precision, Recall, Accuracy, F Measure, Specificity, and F1 Score that are calculated for liver disorders class label selector 1 results are recorded in Fig. 11.

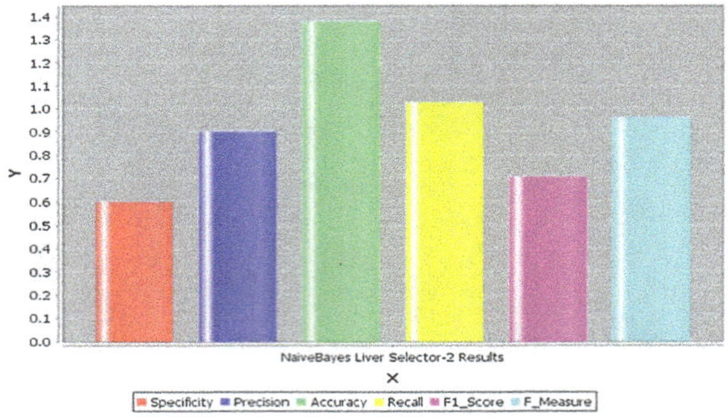

Fig. 12 Naive Bayes liver disorders selector 2 results

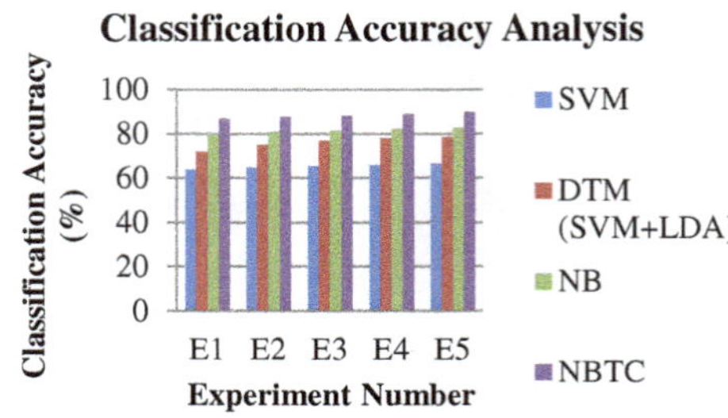

Fig. 13 Classification accuracy comparison analysis

Evaluation Metrics like Precision, Recall, Accuracy, F Measure, Specificity, and F1 Score that are calculated for liver disorders class label selector 2 results are recorded in Fig. 12.

Figure 13 shows the accuracy analysis based on the comparison of the proposed NBTC with the existing classification techniques namely SVM, SVM with LDA, and Naïve Bayes classifier.

From Fig. 13, it can be observed that the proposed NBTC algorithm provides better classification accuracy due to the use of LDA and temporal constraints with Naïve Bayes classifier.

6 Conclusion

In this paper, a machine learning approach to classify medical dataset has been proposed for effective analysis by combining Naïve Bayes classifier with LDA and temporal constraints. In addition, this model uses topic modeling techniques to display the list of topics from the textual documents stored in corpus. Moreover, patterns of words are measured by considering them as latent topics which are retrieved from large collection of documents that are designed to automatically organize using LDA technique. Word cloud has been generated for textual documents in order to provide effective storage. The data set has been analyzed using classification algorithms and proved that the proposed model performs better decision-making when it is compared with the existing systems.

References

1. Chen C, Buntine W, Ding N, Xie L, Du L (2015) Differential topic models. IEEE Trans Pattern Anal Mach Intell 37(2):230–242
2. Leema N, Khanna Nehemiah H, Arputharaj Kannan (2016) Neural network classifier optimization using differential evolution with global information and back propagation algorithm for clinical datasets. J Appl Soft Comput 49:834–844
3. Jane N, Nehemiah HK, Arputharaj K (2016) A Q-backpropagated time delay neural network for diagnosing severity of gait disturbances in Parkinson's disease. J Biomed Inform 60: 169–176
4. Ganapathy S, Sethukkarasi R, Yogesh P, Vijayakumar P, Kannan A (2014) An intelligent temporal pattern classification system using fuzzy temporal rules and particle swarm optimization. Sadhana 39(2):283–302
5. Saranya MS, Selvi M, Ganapathy S, Muthurajkumar S, Sai Ramesh L, Kannan A (2016) Intelligent medical data storage system using machine learning approach. In: 2016 Eighth international conference on advanced computing (ICoAC), IEEE, pp. 191–195
6. Blei D, Andrew Y, Jordan MI, Lafferty J (eds) (2003) Latent dirichlet allocation. J Mach Learn Res 3(4–5):993–1022
7. Blei D (2012) Probabilistic topic models. Commun ACM 55(4):77–84
8. Farid DM, Al-Mamun MA, Manderick B, Nowe A (2016) An adaptive rule-based classifier for mining big biological data. Expert Syst Appl 64:305–316
9. Nahato KB, Harichandran KN, Arputharaj K (2015) Knowledge mining from clinical datasets using rough sets and backpropagation neural network. Comput Math Methods Med 1–13
10. Farid DM, Rahman MZ, Rahman CM (2011) Adaptive intrusion detection based on boosting and naive Bayesian classifier. Int J Comput Appl 24(3):12–19
11. Lawrence R, Bunn A, Powell S, Zambon M (2004) Classification of remotely sensed imagery using stochastic gradient boosting as a refinement of classification tree analysis. Remote Sens Environ 90(3):331–336
12. Farid DM, Rahman CM (2013) Assigning weights to training instances increases classification accuracy. Int J Data Min Knowl Manage Process 3(1):13–25
13. Liu H, Yu L (2005) Toward integrating feature selection algorithms for classification and clustering. IEEE Trans Knowl Data Eng 17(4):491–502
14. Alghamdi R, Alfalqi K (2015) A survey of topic modeling in text mining. Int J Adv Comput Sci Appl (IJACSA) 6(1)
15. Zeng J, Cheung WK, Liu J (2013) Learning topic models by belief propagation. IEEE Trans Pattern Anal Mach Intell 35(5):1121–1134
16. Cheng X, Yan X, Lan Y, Guo J (2014) Btm: topic modeling over short texts. IEEE Trans Knowl Data Eng 26(12):2928–2941

Cloud-Based Intelligent System for Supply Chain Management: A Future Roadmap for SCM Technologies

Manish Mohan Baral and Amitabh Verma

Abstract Cloud computing will offer the adaptability of adaptably outsourcing programming framework for giving chain joint effort and foundation wants in an exceedingly higher way. The cloud computing is reshaping the information and data and correspondence innovation (ICT) store system and making a more interesting ICT condition and related advances like virtualization, Software as a Service (SaaS), Platform as a Service (PaaS) and Infrastructure as a Service (IaaS) have been touted as the "accompanying tremendous thing" and "unmistakable favorable position" for endeavors to improve their best line and essential concern. Proficient supply chains are an imperative need for some organizations. Inventory network administration follows up on operational procedures, different and merged data streams and cooperation forms with an assortment of business accomplices. This computing is transforming into an adoptable advancement in most of the relationship with its dynamic flexibility and utilization of virtualized resources as an organization through cloud. It will probably significantly affect the numerous ventures in future days to come. It is an amazing option for the businesses which are having lack of assets so as to work their data frameworks successfully without spending any more capital for the PCs and system gadgets. Endeavors of late years are typically confronting this decent variety by making and arranging focal data framework arrangements.

Keywords Cloud computing · Supply chain · Software as a Service (SaaS)
Platform as a Service (PaaS) · Infrastructure as a Service (IaaS)
Virtualized resources · Scalability

M. M. Baral (✉) · A. Verma
Department of Management, Birla Institute of Technology Mesra, Ranchi, India
e-mail: manishmohanbaral.31@gmail.com

A. Verma
e-mail: vermainfo123@gmail.com

V. Nath and J. K. Mandal (eds.), *Nanoelectronics, Circuits and Communication Systems*, Lecture Notes in Electrical Engineering 511,
https://doi.org/10.1007/978-981-13-0776-8_2

1 Introduction

Supply chain refers to all the players involved in the process and the exercises carried out to make a product item or a service accessible for purchase by end users. The advancement ventures related to supply chain management have been connected with performance measurement, engineering outline for supply chains and productivity improvements. Various types of frameworks have been presented. For instance, production network operations reference (SCOR) display goes for an institutionalized approach for creating and benchmarking diverse sorts of chains. Collaborative supply chain planning arranging is another endeavor to underline the collaboration between organize elements for accomplishing worldwide ideal or achieving global optimum [1]. The innovation advance in data innovation and the expanding utilization of Internet in regular business have made conceivable outcomes for programming-based supply chain management. There are a few designers starting from various schools, for example, ERP, application joining or integration and mathematical supply chain optimization. Still the essential destinations continued as before: to bring down stock levels and upgrade client benefit by means of enhanced readiness of assembling [1].

Cloud computing (CC) is a figuring worldview in which errands are allotted to a mix of associations, programming and administrations got to over a system and services can be utilized over a network. This system of servers and associations is by and large known as "the cloud". "Cloud" is a virtualized pool of processing assets. Clients can get to assets as they require them. CC is changing into an adoptable headway in most of the relationship with its dynamic flexibility and use of virtualized resources as a relationship through cloud. It will likely have a significant impact on many industries in future days to come. CC offers critical preferences especially for the decentralized and inexactly coupled nature of worldwide SCs, because of the way that IT forms are winding up increasingly steady and adaptable, e.g., through scalability and virtualization [2]. It can be a superb option for the businesses which has deficiency of assets keeping in mind the end goal to work their data frameworks adequately without spending any more capital for the PCs and system gadgets. Cooper et al. characterized SCM as "the combination of key business forms from end client through unique providers that gives items, administrations, and data that include an incentive for clients and different partners" [3]. In parallel to material streams and collaboration, a request arranged intra-and between hierarchical data supply is required. Along these lines, the test lies in choosing, adjusting, and utilizing appropriate IT arrangements. Here is the starting point of Cloud Computing. The National Institute of Standards and Technology (NIST) characterizes Cloud Computing as "a model for empowering omnipresent, advantageous, on-request arrange access to a mutual pool of configurable figuring assets (e.g., systems, servers, stockpiling, applications, and administrations) that can be quickly provisioned and discharged with negligible administration exertion or specialist organization collaboration" [4].

1.1 Background of Supply Chain Management

The literature survey which was done to consider the past investigates said that before 1950s, coordinations were into training in military [5]. It was expected to do with procurement, upkeep or support, and transportation of military offices, materials, and work compel. The investigation and routine with regards to physical dispersion and logistics rose in 1960s and 1970s [6]. Amid 1950s, there happened a few changes that may be delegated as an essential “Change”. The significance of logistics expanded extensively, once physical conveyance administration in collecting corporations was perceived as a unique authoritative capability [7]. The SCM concept was coined in the early 1980s by experts in logistics [8]. Amid 1950s, there happened a few changes that may be delegated an essential “Change” [9] (Fig. 1).

SCM has turned out to be such an “interesting issue” that it is hard to get a periodical on assembling, conveyance, promoting, client administration [11]. Few other authors had defined SCM as flow of goods or products, some in terms of management process, integrated system and some as management philosophy [12].

An example to explain a basic supply chain process is shown in Fig. 2.

The supply chain partners include suppliers, manufacturers, distributors, retailers, and customers. The main focus of the chain is customer satisfaction [13].

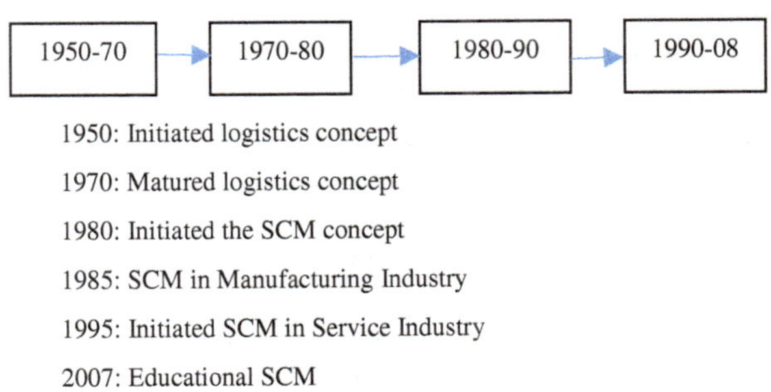

Fig. 1 Evolutionary timeline of SCM [10]

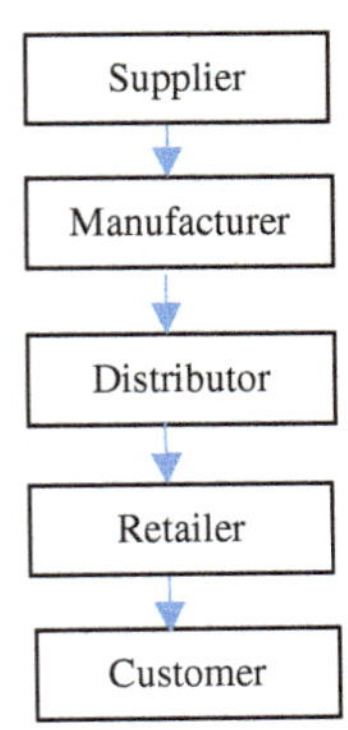

Fig. 2 The basic supply chain [13]

SCM was at first related to the stock administration inside a production network. This idea was later widened to incorporate administration of all capacities inside the network. As indicated by Chopra and Meindl [13], "SCM draws in the administration of streams between and among stages in a store network to limit add up to cost". This suggests SCM incorporates organization of streams of items, data, and fund upstream and downstream in the production network. Production network is a solitary procedure. The essential goal of the procedure is to incorporate and deal with the souring, stream and control of materials by utilizing all out framework point of view over numerous capacities and different levels of providers [14]. SCM consolidates logistics into the key choices of the business [15]. Various Total Quality Management (TQM) techniques can be used to improvise the whole process in the supply chain.

2 Literature Review

The eventual fate of distributed computing lies whose significant objective is diminishing its cost administrations while expanding handling throughput, unwavering quality, accessibility and adaptability and diminishing processing time [16]. Cloud computing is another change in outlook which includes computing resource services and information stockpiling. Thus, the world is rapidly changing toward an approach of get-together relative applications for millions to interface as a relationship as opposed to keep running on their PCs. Gartner defined a style of figuring in which tremendously adaptable IT-related limits are given as a help to outside customers using web advancements [17]. Erdogmus considered cloud computing a pool of through and through versatile, amazing structure is set up for enabling end-customer applications that are charged by utilize [18]. Sultan defined IT constraints that are asked for, provisioned, passed on, and ate up powerfully completed the web [19].

In the SaaS gathering, there is a system by that different programming applications square measure given by the applying expert relationship as a rental over the web, utilizing cloud structure and associations discharged by Salesforce.com client quality association (CRM), Google Apps, Oracle Siebel on Demand and Microsoft BPOS. The PaaS classification speaks to mists that entrance a scope of PC, database, and capacity works inside a virtualized stage gave over the web and administrations discharged by Salesforce.com, Microsoft Azure, and Google App Engine. At long last, the IaaS classification is the conveyance of PC framework as an administration. It is an arrangement show up in which an affiliation outsources the mechanical assembly used to help tasks, including limit, equipment, servers, and structure association parts. The specialist co-op claims the gear and is in charge of lodging, running, and looking after it. The customer generally pays on a for every utilization premise and administrations are given by AMAZON.COM AWS, SUN NETWORK.COM, IBM Blue Cloud, and Verizon CaaS. This concept of cloud computing can bring a revolution especially for the small and medium sized firms. Therefore, there is a need to identify the factors which will influence organization's production network operations and will help to improve the efficiency of the supply chain process.

3 Research Methodology

Research methodology is a systematic analysis of methods which is applied in the field of study. Methodology adopted in the present study incorporates both exploratory as well as descriptive research design. There are two approaches to conduct a research:

Qualitative approach: The researcher investigates the goal and studies it on the basis of previous grounded theory.

Quantitative approach: The researcher investigates the goal and studies it on the basis of experimental and computational methods.

There are two types of samplings: (i) Random sampling (ii) Nonrandom sampling

Random sampling is a technique in which all the items in the population have a probability of being selected from the sample. There are four types of random sampling: Simple random sampling, systematic sampling, stratified sampling, and cluster sampling.

Nonrandom sampling is a technique in sampling personal knowledge and opinions are used to identify the items from the population that are to be included in the sample. It depends on someone's expertise.

3.1 Method of Data Collection

In this analysis, primary data and secondary data were used. Primary data was gathered during the field visit in Ranchi area. After market research, exhaustive literature review was done for preparing the research paper. The field of study comprised of various SMEs located in and around Ranchi area.

The sampling technique we have used here is simple random sampling.

3.2 Survey Method

The method used for the analysis was market survey which helped in securing detailed information from a sample of respondents. Market survey is the most widely used market research. It is the best method to understand the need and demand of the customers by interacting directly with them in the market.

3.3 Sample Size for Survey

The sample size for the study was 50.

3.4 Statistical Method

In this research, linear regression and multiple regression techniques were used for the analysis of data to work out the strength of the link between dependent and independent variables. The statistical tool used for the analysis of the data is SPSS 16.0.

4 IT with SCM

Information Technology (IT) has become a great solution to many problems and also has many advantages. The technology does not require a huge capital investment and expensive infrastructure. It can also be operated from any part and anywhere. Due to the availability of cheap Internet, IT is becoming more popular because it helps to reduce the cost and lead times. It helps in linking the players which are involved in the supply chain process. As a result, these supply chain players have better control on the inventory and keep a track of it at various levels of supply chain and better utilization of manpower. The role of IT has helped many Start-up companies who have less funds. In the improvement and support of supply chain's data frameworks, both programming and equipment must be tended to. Equipment incorporates PC's info/yield gadgets and capacity media. Programming incorporates the whole framework and application program utilized for handling exchanges administration control, decision-making, and strategic planning. Due to the involvement of IT in supply chain, the e-commerce companies are selling goods at lower price because the players involved in the supply chain process are reduced and the delivery is also done faster which results in customer satisfaction.

5 SCMaaS

It stands for supply chain management as a service. Using this, we can create a platform through cloud to improve the SCM service and also increase the transparency. If we implement this in the supply chain process, it will help to manage the inventory at each level, reduce the cost and speed. Cloud computing applications for capacities, for example, stock administration, warehousing, renewal arranging, arrange handling, acquirement, transportation stack building, armada administration and transportation course arranging transportation will show up with expanding normality sooner rather than later. Cloud computing development will influence organization's production network operations in a few ways:

- Cost savings: Cloud computing is charged as either a utility or a membership benefit and depends on administrations utilized.
- Speed and timeliness: Cloud administrations are up and running in days rather than the weeks or months it may take to convey a customary in-house programming and equipment framework. This lessens the slack between understanding a business require and fulfilling the need.
- Flexibility is the capacity to change without disturbance to the business.
- Scalability is the capacity to shrink and extend as business needs dictate.
- Distance decreases the effect of geographic separation on a worldwide market.
- Company-wide integration: The integrative idea of the cloud stage implies that everybody will have the capacity to see a similar ongoing data at whatever point they require it.
- New competitors: Cloud computing can engage new organizations to develop themselves in a concise time allotment, without basic enthusiasm for establishment. This could incredibly influence the forceful scene of many set up exhibit pioneers.
- Speed to showcase for new items and administrations: the pace at which new income creating items and administrations are presented has put consistent weight on supply chains in recent years.
- Large-scale transformation: the danger from new rivalry and the expanded pace of new item and administration presentation could drive organizations with conventional, framework serious supply chains to rethink themselves. Supply chains will probably turn out to be more unique, more adaptable and fit for supporting the money-related destinations of boards and shareholders.
- Planning and forecasting: Cloud base tools are used planning and forecasting of a particular activity using basic analytics and statistical demand forecasts.

6 Research Question

From the literature survey as well as field survey, there arises an enquiry "Whether cloud-based data frameworks speak to a superior contrasting option to set up an IT bolster for supply chain management"

7 Conceptual Framework

Considering the factors, the following conceptual model is constructed to analyze organization's production network operations (Fig. 3).

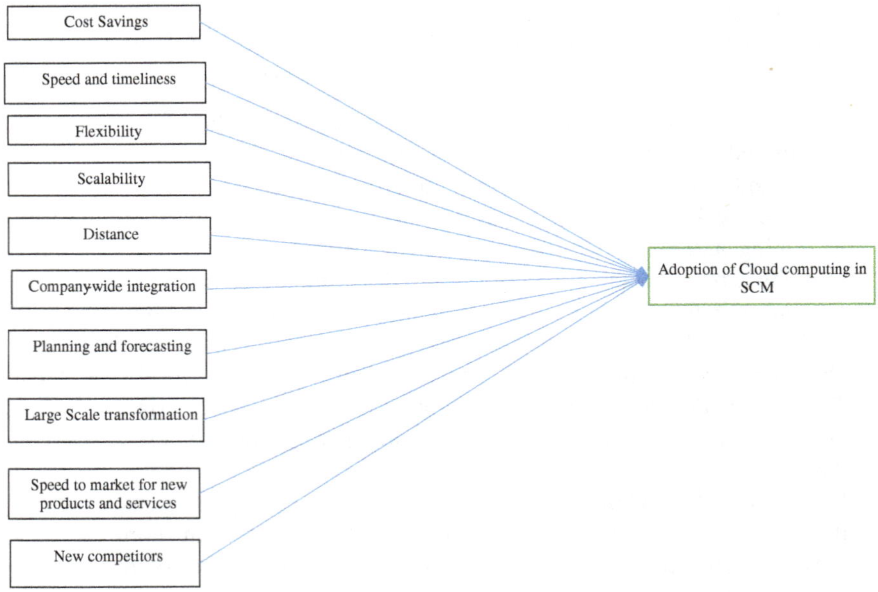

Fig. 3 Conceptual framework model

8 Analysis and Interpretation

A. Ho: Cost savings is independent on adoption of CC in SCM
   H1: Cost savings not independent on adoption of CC in SCM (Table 1)
   As the estimation of R^2 is more than 0.5, we can draw the inference that the model is a good fit. Hence, we accept the alternate hypothesis and reject the null hypothesis.
B. Ho: Speed and timeliness is independent on adoption of CC in SCM
   H1: Speed and timeliness not independent on adoption of CC in SCM (Table 2)
   As the estimation of R^2 is more than 0.5 so we can draw the inference that the model is a good fit. Hence, we accept the alternate hypothesis and reject the null hypothesis.
C. Ho: Flexibility is independent on adoption of CC in SCM
   H1: Flexibility is not independent on adoption of CC in SCM (Table 3)

Table 1 Relationship of cost saving with adoption of cloud computing in SCM

Model summary				
Model	R	R^2	Adjusted R^2	Std. error of the estimate
1	0.731[a]	0.535	0.515	0.34395

[a]Cloud computing in SCM: (Constant), Cost savings

Table 2 Relationship of speed and timeliness with adoption of CC in SCM

Model summary				
Model	R	R^2	Adjusted R^2	Std. error of the estimate
1	0.739[a]	0.542	0.531	0.34375

[a]Cloud computing in SCM: (Constant), speed and timeliness

Table 3 Relationship of flexibility with adoption of CC in SCM

Model summary				
Model	R	R^2	Adjusted R^2	Std. error of the estimate
1	0.736[a]	0.528	0.501	0.34312

[a]Cloud computing in SCM: (Constant), flexibility

Table 4 Relationship of scalability with adoption of CC in SCM

Model summary				
Model	R	R^2	Adjusted R^2	Std. error of the estimate
1	0.771[a]	0.571	0.551	0.34431

[a]Cloud computing in SCM: (Constant), scalability

Table 5 Relationship of scalability with adoption of cloud computing in SCM

Model summary				
Model	R	R^2	Adjusted R^2	Std. error of the estimate
1	0.767[a]	0.569	0.566	0.34418

[a]Cloud computing in SCM: (Constant), distance

As the estimation of R^2 is more than 0.5, we can draw the inference that the model is a good fit. Hence, we accept the alternate hypothesis and reject the null hypothesis.

D. Ho: Scalability is independent on adoption of CC in SCM
   H1: Scalability is not independent on adoption of CC in SCM (Table 4)
   As the estimation of R^2 is more than 0.5, we can draw the inference that the model is a good fit. Hence, we accept the alternate hypothesis and reject the null hypothesis.
E. Ho: Distance is independent on adoption of CC in SCM
   H1: Distance is not independent on adoption of CC in SCM (Table 5)
   As the estimation of R^2 is more than 0.5, we can draw the inference that the model is a good fit. Hence, we accept the alternate hypothesis and reject the null hypothesis.
F. Ho: Company-wide integration is independent on adoption of cloud computing in SCM

H1: Company-wide integration is not independent on adoption of cloud computing in SCM (Table 6)
As the estimation of R^2 is more than 0.5, we can draw the inference that the model is a good fit. Hence, we accept the alternate hypothesis and reject the null hypothesis.

G. Ho: Planning and forecasting is independent on adoption of Cloud computing in SCM
H1: Planning and forecasting is not independent on adoption of Cloud computing in SCM (Table 7)
As the value of R^2 is more than 0.5, we can draw the inference that the model is a good fit. Hence, we accept the alternate hypothesis and reject the null hypothesis.

H. Ho: Large-scale transformation is independent on adoption of CC in SCM
H1: Large-scale transformation is not independent on adoption of CC in SCM (Table 8)
As the estimation of R^2 is more than 0.5, we can draw the inference that the model is a good fit. Hence, we accept the alternate hypothesis and reject the null hypothesis.

I. Ho: Speed to showcase for new items and administrations is independent on adoption of CC in SCM
H1: Speed to showcase for new items and administrations is not independent on adoption of CC in SCM (Table 9)

Table 6 Relationship of company-wide integration with effect of CC in SCM

Model summary				
Model	R	R^2	Adjusted R^2	Std. error of the estimate
1	0.789[a]	0.611	0.591	0.34512

[a]Cloud computing in SCM: (Constant), company-wide integration

Table 7 Relationship of planning and forecasting with effect of CC in SCM process

Model summary				
Model	R	R^2	Adjusted R^2	Std. error of the estimate
1	0.7276[a]	0.523	0.510	0.34213

[a]Cloud computing in SCM: (Constant), planning and forecasting

Table 8 Relationship of large-scale transformation with adoption of CC in SCM

Model Summary				
Model	R	R^2	Adjusted R^2	Std. error of the estimate
1	0.757[a]	0.557	0.537	0.34438

[a]Cloud computing in SCM: (Constant), large-scale transformation

Table 9 Relationship of speed to showcase for new items and administrations is independent on adoption of cloud computing in SCM process with adoption of cloud computing in SCM

Model summary				
Model	R	R^2	Adjusted R^2	Std. error of the estimate
1	0.731[a]	0.535	0.515	0.34395

[a]Cloud computing in SCM: (Constant), speed to showcase for new items and administrations is independent on adoption of cloud computing in SCM

Table 10 Relationship of new competitors with adoption of cloud computing in SCM process

Model summary				
Model	R	R^2	Adjusted R^2	Std. error of the estimate
1	0.767[a]	0.569	0.566	0.34418

[a]Cloud computing in SCM: (Constant), competitors is independent on adoption of Cloud computing in SCM process

Table 11 Relationship of all the factors mentioned with adoption of CC in SCM

Model summary									
Model	R	R^2	Adjusted R^2	Std. error of the estimate	Change statistics				
					R^2 change	F change	df1	df2	Sig. F change
1	0.715[a]	0.711	0.699	0.17132	0.711	816.42	2	47	0.000[a]

[a]Cloud Computing in SCM: (Constant), cost savings, speed and timeliness, flexibility, scalability, distance, company-wide integration, new competitors, speed to market for new products and services, large-scale transformation, planning, and forecasting

As the estimation of R^2 is more than 0.5, we can draw the inference that the model is a good fit. Hence, we accept the alternate hypothesis and reject the null hypothesis.

J. Ho: New competitors is independent on adoption of CC in SCM
   H1: New Competitors is not independent on adoption of CC in SCM (Table 10)
   As the estimation of R^2 is more than 0.5, we can draw the inference that the model is a good fit. Hence, we accept the alternate hypothesis and reject the null hypothesis.
K. Conceptual Model Summary (Table 11).
   From the above table, we can see that the R^2 value is more than 0.5 which means that regression line has a good fit of data. So, we can draw the inference that the identified factors will definitely affect company's supply chain operation. Hence, if cloud computing is used as a tool in the whole supply chain process then it will increase the transparency in the process.

9 Conclusion

SCM and cloud computing has been widely researched during the last decade. But most of the research work which has been done is conceptual and theory based. The emphasis is mostly on conveying the correct items or administrations in the correct amount, to the ideal place and perfect time and furthermore with greatest advantages. Using cloud with SCM, inform the logistics to reduce or decrease the waiting time and increase the transparency and efficiency which will finally result in the customer satisfaction and also generate profit for the organization which has been shown in this research paper by using quantitative method.

References

1. Helo P, Szekely B (2005) Logistics information systems: an analysis of software solutions for supply chain co-ordination. Ind Manage Data Syst 105(1):5–18
2. Jede A, Teuteberg F (2016) Towards cloud-based supply chain processes: Designing a reference model and elements of a research agenda. Int J Logist Manage 27(2):438–462
3. Cooper MC, Lambert DM, Pagh JD (1997) Supply chain management: more than a new name for logistics. Int J Logist Manage 8(1):1–14
4. Botta A, De Donato W, Persico V, Pescapé A (2016) Integration of cloud computing and internet of things: a survey. Future Gener Comput Syst 56:684–700
5. Ballou R (1978) Basic business logistics. Prentice-Hall, Englewood Cliffs, NJ
6. Heskett JL, Glaskowsky NA Jr., Ivie RM (1973) Business logistics, 2nd edn. The Ronald Press, New York, NY, pp 14–21
7. Heskett J, Ivie R, Glaskowsky N (1964) Business logistics, management of physical supply and distribution. The Ronald Press Company, New York, NY Heskett
8. Oliver RK, Webber MD (1992) Supply-chain management: logistics catches up with strategy. In: Christopher M (ed) Logistics: the strategic issues, Chapman & Hall, London O'Brien
9. Robert BH, Vijay RK, Choon TK (1998) Supply chain management: supplier performance and firm performance. Int J Purchasing Mater Manage 2–9
10. Habib M, Jungthirapanich C (2008) An integrated framework for research and education supply chain for the universities. In: Proceedings of the 4th IEEE international conference on management of innovation and technology, IEEE Computer Society, Piscataway, USA, pp 1027–1032. ISBN: 978-1-4244-2329-3
11. Ross D Frederick (1998) Competing through supply chain management. Chapman & Hall, New York, NY
12. Tyndall G, Gene C, Gopal W Partsch, Kamauff J (1998) Supercharging supply chains: new ways to increase value through global operational excellence. Wiley, NY
13. Chopra S, Meindl P (2001) Supply Chain Management. Prentice Hall, NJ
14. Robert M, Trent R, Handfield R (1994) Purchasing and supply chain management, South-Western College Publishing, Cincinnati, OH, Chapter 8
15. Carter JR, Ferrin BG (1996) The impact of transportation costs on supply chain management. J Bus Logist 16(1):189–212
16. Hayes B (2008) Cloud computing. Commun ACM 51:9–11
17. Gartner (2010) Cloud computing inquiries at Gartner. Available at: http://blogs.gartner.com/thomas_bittman/2009/10/29/cloud-computing-inquiries-at-gartner
18. Erdogmus H (2009) Cloud computing: does Nirvana hide behind the nebula? IEEE Softw 26:4–6
19. Sultan N (2010) Cloud computing for education: a new dawn? Int J Inf Manage 30:109–116

Effect of Processing Combined MFCC and DSCC Features with QCN for Hindi Vowel Classification in Noisy Environments

Shipra and Mahesh Chandra

Abstract This paper comes up with "Hybridized Delta Spectral Cepstral Coefficients" or HDSCC features. The HDSCC feature is obtained by processing concatenated Mel Filter Cepstral Coefficients and Delta Spectral Cepstral Coefficients (DSCC) with Quantile based Dynamic Cepstral Normalization technique (QCN). The strength of proposed HDSCC feature set has been evaluated for Hindi Vowel classification in noisy environments. The results have been compared with that obtained with MFCC and QCN-MFCC features. HDSCC feature set outweighs classification accuracy compared to that obtained with MFCC features and QCN-MFCC features in various operating conditions. Quantitatively for clean database results revealed an improvement of 20.33 and 15.93% as compared to that obtained with MFCC and QCN-MFCC features respectively for context independent (CI) cases. For context dependent (CD) cases an improvement of 18.59 and 11.69% has been detected for MFCC features and QCN-MFCC features respectively. For noisy database also HDSCC features demonstrate better efficiency than baseline features. The experiments for the classification of Hindi Vowels, revealed as high as 97.87 and 94.85% average % Vowel Classification Accuracy (% VCA) in noisy environments for CD cases and CI cases respectively. The maximum comparative improvement in % VCA for CI cases is noted down as 29.45 and 25.88% over MFCC and QCN-MFCC features respectively while the same for CD cases has been detected as 27.09 and 24.12% over MFCC and QCN-MFCC features respectively. All results are obtained on Matlab R2009b with HMM classifier.

Keywords MFCC · DSCC · QCN · HDSCC · HMM

Shipra (✉) · M. Chandra
Department of Electronics and Communication Engineering,
Birla Institute of Technology, Mesra, Ranchi 835215, India
e-mail: mishra.shipra.bit@gmail.com

M. Chandra
e-mail: shrotriya@bitmesra.ac.in

V. Nath and J. K. Mandal (eds.), *Nanoelectronics, Circuits and Communication Systems*, Lecture Notes in Electrical Engineering 511,
https://doi.org/10.1007/978-981-13-0776-8_3

1 Introduction

With the accessibility of cellular phones to almost every section of people the research in the field of Automatic Speech Recognition has shown an inclination towards application specific ASR systems [1, 2].Thus, in the recent past several ASR applications have been commercially available with sufficient efficiency so that it becomes useful for a layman. But the ever-existing challenge of ASR is yet present i.e. the ASR systems work very smoothly in guarded environments but in complex environments having noise, reverberations etc. the success rate of system deteriorates significantly [3].

ASR systems follow a basic design prototype consisting of front end and back end processing. Front end includes feature extraction techniques while back end comprise of various kinds of acoustic, language and pronunciation models. From Phoneme classification point of view Feature extraction is the process of extracting minimum information from a phoneme which gives maximum discrimination between phoneme classes [4, 5]. Thus, selection of features plays deciding role in determining the reliability of an ASR system. Also, for commercial success the ASR system should be equally efficient in different environmental conditions; therefore, while selecting the features it should be ensured that the features are robust to various environmental conditions.

The most accepted acoustic representation in the territory of ASR is Cepstral-based parameters on account of their inherent advantages such as concise representation of features, orthogonality etc. [6]. However, the Cepstral based features based ASR system fail in complex environments for the reason that generally these features incorporate Logarithmic operation resulting in addition of noise in subsequent frames of speech signal thus changing the spectral properties of speech [7].Various researchers in an endeavor of making ASR systems more robust to adverse conditions; have proposed enhancement or modification of various forms of Cepstral processing [7, 8].

Here, a new means for compounding two Cepstral based feature set namely MFCC and DSCC is introduced. DSCC removes the noise associated with the speech sample in the spectral domain thus application of logarithmic operation in later stage of processing is in a way operates on clean speech sample. Further the repercussion of any residual noise is reduced by application of QCN technique [9] on the compound MFCC+DSCC feature set. The resulting QCN based MFCC +DSCC feature set is attributed as Hybridized DSCC features or HDSCC features.

The remaining part of the paper is arranged as: Sect. 2 gives an outline of our proposed feature set; Sect. 3 describes the experimental setup; Sects. 4 and 5 discuss the results attained and conclusions drawn respectively based on the experiments performed.

2 Proposed Feature Set

DSCC features are appended with MFCC features and resulting feature set is processed with Quantile based Dynamic Cepstral Normalization technique (QCN). The resulting feature set is termed as hybridized DSCC features or HDSCC. This Hybridization approach is motivated by the factors summarized below;

[I] Delta Cepstral Features was proposed in ICASSP 1986 by S. Furui. Adding Delta Cepstral Features to MFCC features improves the robustness of resulting feature set on account of casting a depiction of temporal dependencies to the Hidden Markov Model frames. Delta Cepstral Coefficients improve the recognition efficiency because they capture speech contextual information.

[II] Prominent researchers have pointed out that as the SNR level degrades; the advancement in recognition efficiency with Delta features becomes less significant.

[III] Recently proposed DSCC features has experimentally authenticated that introducing Delta spectral features to MFCC processing boosts up signal power at low SNR values.

[IV] QCN technique has been purposed for unrestricted equalization of Lombard Effect. QCN technique is used by previous researchers in combination with several feature extraction techniques in difficult auditory environments and it has emerged as an efficient noise compensation technique [10].

The computation of proposed HDSCC features is denoted by Fig. 1. The algorithm follows the classic steps of spectral investigation of speech signal. Initially a pre-emphasis filter; defined by following equation; is applied on the speech signal,

$$H(z) = 1 + az^{-1}, a = 0.97 \tag{1}$$

This pre-emphasis filter is required because due to the physiology of speech production system the voiced speech have a negative spectrum slope of 20 dB/decade. The pre-emphasis filter enhances the signal spectrum by approximately 20 dB/decade and hence spectrally flattens the speech signal. Pre-emphasis filter also improves the perceptually relevant portion of speech spectrum which is above 1 kHz.

After pre-emphasis stage the framing of speech is acquired so that Fourier Transform may be applied to the speech signal as speech is a quasi-periodic signal. Since this paper works with Hindi Vowel classification therefore the frames are of duration 16 ms with 10 ms overlapping are taken. Hamming window is applied on all frames to weight speech samples towards the center of the frame. Hamming Window is mathematically represented as,

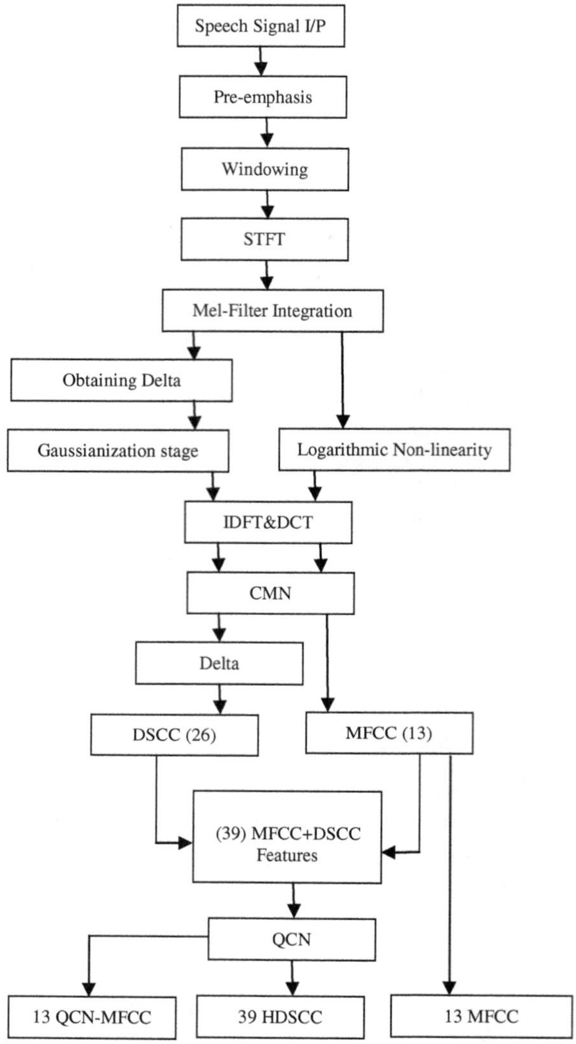

Fig. 1 Block diagram representation of combination of MFCC and DSCC with QCN technique

$$\begin{aligned} w(n) &= 0.54 - 0.46 \cos\frac{2\pi n}{N-1},\ 0 \le n \le N-1 \\ w(n) &= 0,\ \text{elsewhere.} \end{aligned} \tag{2}$$

The Fourier Transform of windowed signal is calculated using a DFT size of 1024. Mel Filter bank is used for mapping the spectrum of speech signal to a frequency scale emulating human auditory system [11]. The output of Mel-filterbank decomposes the signal into a set of discrete set of spectral samples which are vectors of power values $p(n)$ for individual frame of speech signal. This vector contains intelligence analogous to that present in advance levels of auditory system. From this stage of processing two different paths are followed for computing MFCC and DSCC. MFCC computation involves calculating Cepstral coefficients from $p(n)$ which is simply calculating the Inverse DFT of the logarithm of $p(n)$. Further Discrete Cosine Transform; DCT is applied to the Cepstral coefficient to de-correlate the feature vectors. Cepstral Mean Normalization [12] stage is added to the MFCC processing to curtail the effect of noise on feature vectors as we aim to work on noisy database. From the resulting feature vector 13 MFCC coefficients are selected excluding the energy term. DSCC processing involves applying Delta operation on $p(n)$ to obtain Delta Spectral Coefficients in agreement with the following equation,

$$D_p(n) = p[n+m] - p[n-m]. \tag{3}$$

Here, n is the index of the analysis frames and in practice m is approximately 2 or 3.

At this stage the distribution of Delta Spectral coefficients is highly non-gaussianized and hence unsuitable for further processing. To Gaussianize the Delta Spectral coefficients a Histogram Equalization stage [13] is added to the DSCC processing. The Cepstral coefficients are calculated from the culminating spectrum as in MFCC processing followed by CMN stage to yield DSCC features. Delta coefficients are added to the 13-dimensional DSCC resulting in 26 DSCC coefficients.

The 13 MFCC and 26 DSCC features are combined as a 39 dimensional feature vector referred to as MFCC+DSCC features. MFCC and MFCC+DSCC features are processed with Quantile based Dynamic Cepstral Normalization Technique resulting in 13 QCN-MFCC and 39 QCN based MFCC+DSCC features that we refer to as Hybridized DSCC features or HDSCC features.

3 Experimental Setup

Proposed HDSCC features are evaluated for Hindi Vowel Classification task. The performance of suggested feature set is compared with baseline MFCC feature set and QCN-MFCC feature set. The experimental setup as explained by Fig. 2 is based on the classic structure of Speech recognition system. It has three major sections; Database Preparation, Feature Extraction and Classification.

Fig. 2 Block diagram of experimental set up

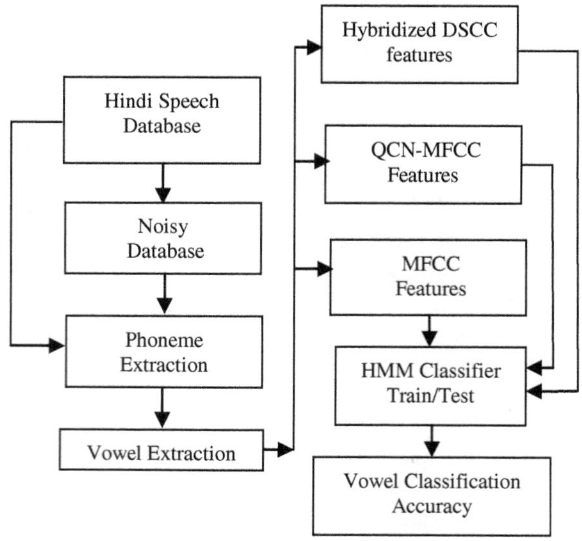

3.1 Database Preparation

For Hindi Vowel Classification task, a database of Hindi Phonemes is a pre-requisite. Hindi Phoneme database is prepared from TIFR Hindi Speech database [14]. TIFR database consists of 1000 sentences spoken by 100 speakers.

These sentences are designed at TIFR, Mumbai, India in such a manner that they capture maximum number of phonemes. The sentences are categorized in two sets. The first set has two sentences containing most of the Hindi phonemes. These two sentences are called dialect sentences. The second set is distinct for each speaker. The recordings are done at CEERI, New Delhi, India. Manually segmented phoneme boundaries are provided with TIFR database. Out of the 100 speakers 70 speakers; 47 male and 23 female are randomly chosen for creating Hindi Phoneme database. Hindi Phonemes are extracted from the TIFR database with the help of phoneme boundaries given with the database.

For evaluating the performance of proposed HDSCC features in noisy environment car noise and white noise at SNR levels of −5, 0, 5 and 10 dB are artificially added to TIFR database. Again Hindi Phonemes are extracted from the noisy database as per the phoneme boundaries provided with the database.

3.2 Feature Extraction

Firstly 13 MFCC, 13 QCN-MFCC, and 39 HDSCC features are computed for all of the Hindi Phoneme following the Feature Extraction algorithm explianed in Sect. 2.

3.3 Classification

The nature of our task is suited for Hidden Markov Model classifier. One HMM model is created having spherical covariance with 3 emitting states and 4 Gaussian Mixture components. The system calculates a model that has transitions containing the probabilities to one phoneme state from the remaining phonemes, as well as probabilities of each of the emissions from a distinct state. After the model is developed, an input of phoneme values is put into a vector which in turn sums the HMM transition probabilities from each state. In the end, whichever Vowel Class model has the highest probability will classify as the Vowel Class. Vowel Classification Accuracy (VCA) is computed by the formula:

$$\%\,\text{VCA} = \frac{\text{total test samples}}{\sum \text{test samles}} \times 100 \tag{4}$$

4 Results

4.1 Performance Evaluation for CI Cases

Average Vowel Classification accuracy of proposed HDSCC features for CI cases under various environments has been arranged in Table 1 and Fig. 3 along with that of MFCC and QCN-MFCC features. As it can be seen from Table 1 HDSCC features give much better recognition efficiency compared to MFCC features. The maximum improvement obtained is of 29.45 and 28.84% for car-noise and white-noise respectively for SNR level of −5 dB, while the least improvement of 7.0 and 6.42 has been detected for car noise and white noise respectively for SNR level of 0 dB. Similar results have been observed by comparing the average % VCA obtained with QCN-MFCC and HDSCC features. The maximum relative gain has been obtained for car-noise at SNR of −5 dB as 25.88% while for white noise at SNR −5 dB the relative gain is 24.68%. The minimum relative gain has been observed for SNR level 0 dB for car noise and white noise as 5.53 and 4.07% respectively. The best improvements are shown in bold and the least improvementsáare shown in bold italic.

4.2 Performance Evaluation for CD Cases

The results attained by performing same set of experiments for context dependent cases are arranged in Table 2 and graphically shown by Fig. 4. The maximum relative gain in this case also is observed for SNR level of 0 dB for MFCC and

Table 1 Comparative % VCA for MFCC, QCN-MFCC and HDSCC features for CI cases under clean and noisy conditions

Dataset	Features				
	MFCC	Relative gain w.r.t HDSCC	QCN-MFCC	Relative gain w.r.t HDSCC	HDSCC
Clean	76.12	20.33	80.50	15.95	96.45
Carnoise −5 dB	65.4	**29.45**	68.97	**25.88**	94.85
Carnoise 0 dB	65.12	***7.00***	66.59	***5.53***	72.12
Carnoise 5 dB	64.07	20.36	66.2	18.23	84.43
Carnoise 10 dB	63.6	25.27	65.8	23.07	88.87
Whitenoise −5 dB	62.27	**28.84**	66.43	**24.68**	91.11
Whitenoise 0 dB	66.77	***6.42***	69.02	***4.07***	73.19
Whitenoise 5 dB	70.4	18.81	71.98	17.23	89.21
Whitenoise 10 dB	64.6	23.79	66.11	22.19	88.39

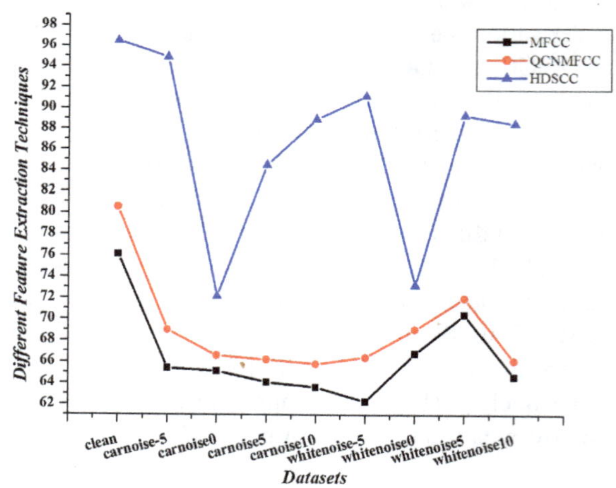

Fig. 3 Comparative % VCA for MFCC, QCN-MFCC and HDSCC features for CI cases under clean and noisy conditions

QCN-MFCC features in car noise and white noise environments. But the maximum relative gain of 23.91% for car noise has been observed at SNR 10 dB while for white noise a gain of 27.09% has been observed at SNR −5 dB for MFCC features. The results follow same pattern for QCN-MFCC features also.

Table 2 Comparative % VCA for MFCC, QCN-MFCC and HDSCC features for CD cases under clean and noisy conditions

Dataset	Features				
	MFCC	Relative gain of HDSCC	QCN-MFCC	Relative gain of HDSCC	HDSCC
Clean	78.08	18.59	84.98	11.69	96.67
Carnoise −5 dB	72.98	23.34	77.34	18.98	96.32
Carnoise 0 dB	72.27	4.18	73.98	2.47	76.45
Carnoise 5 dB	72.02	23.76	74.87	20.91	95.78
Carnoise 10 dB	72.11	**23.91**	73.67	**22.35**	96.02
Whitenoise −5 dB	69.92	**27.09**	72.89	**24.12**	97.01
Whitenoise 0 dB	72.65	6.62	77.11	***2.16***	79.27
Whitenoise 5 dB	76.21	21.66	82.01	14.86	97.87
Whitenoise 10 dB	74.01	20.76	75.23	19.54	94.77

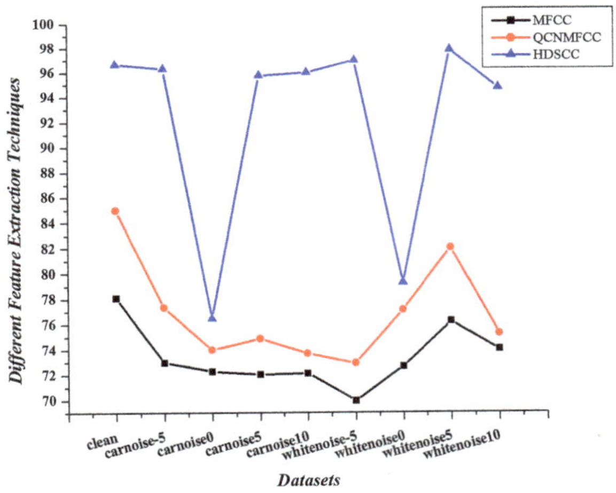

Fig. 4 Comparative % VCA for MFCC, QCN-MFCC and HDSCC features for CD cases under clean and noisy conditions

5 Conclusions

In this paper a new Hybridized Feature Extraction technique HDSCC features has been proposed. The proposed HDSCC feature set has exhibited very good Vowel Classification Accuracy (97.87% for CD case).) in comparison with the work carried out by Biswas and co-worker for WERBC feature set (91.24%). It has also been noted that for CD cases the performance of HDSCC features is relatively uniform showing a variation of 2.24% in varying operating conditions except for SNR level of 0 dB.

For CI cases the performance of HDSCC feature set show a variation in the range of 84.43 to 96.45%. In this case the best performance has been noted for clean database (96.45%) closely followed by that at SNR −5 dB for car noise (94.85%).

An analysis of the performance of HDSCC feature set compared to MFCC features for CI cases shows remarkable improvement in VCA in all the operating environments contemplated here. The relative gain in VCA is as good as 29.45% in car noise environment at SNR −5 dB. Similar analysis with respect to QCN-MFCC features shows superlative advancement of 25.88% for SNR −5 dB. Similar analysis for CD cases shows the largest advancement of 27.09% for MFCC features and 24.12% for QCN-MFCC features respectively.

References

1. Biswas A, Sahu P, Bhowmick A, Chandra M (2015) AAM based features for multiple camera visual speech recognition in car environment. Procedia Comput Sci 57:614–621
2. Kaynak MN, Zhi Q, Cheok AD, Sengupta K, Jian Z, Chung KC (2004) Lip geometric features for human computer interaction using bimodal speech recognition: comparison and analysis. Speech Commun 43(1):1–16
3. Potamianos G, Neti C, Luettin J, Matthews I (2004) Audio visual automatic speech recognition: an overview. Issues Visual Audio-Visual Speech Process 22:23
4. Biawas A, Sahu P, Chandra M (2014) Admissible wavelet packet features based on human inner ear frequency response for Hindi consonant recognition. Comput Electr Eng 40(4): 1111–1122
5. Farooq O, Datta S, Shrotriya MC (2010) Wavelet sub-band based temporal features for robust Hindi phoneme recognition. Int J Wavelets Multiresolut Inf Process 8:847–859
6. Wong E, Sridharan S (2001) Comparison of linear prediction cepstrum coefficients and mel-frequency cepstrum coefficients for language identification. In: Proceedings of IEEE internatinal symposium on intelligent multimedia, video and speech processing, pp 95–98
7. Kim C, Stern RM (2016) Power-normalized cepstral coefficients (PNCC) for robust speech recognition. IEEE/ACM Trans Audio Speech Lang Process 24(7):1315–1329
8. Kumar K, Kim C, Stern RM (2011) Delta-spectral cepstral coefficients for robust speech recognition. In: ICASSP 2011 IEEE international conference on acoustics, speech and signal processing, ICASSP-2011, pp 4784–4787
9. Bořil H, Hansen JHL (2010) Unsupervised equalization of Lombard effect for speech recognition in noisy adverse environments. IEEE Trans Audio Speech Lang Process 18(6): 1379–1393
10. Mishra S, Chandra M (2016) Hindi vowel classification using QCN-PNCC features. Indian J Sci Technol 9(38). https://doi.org/10.17485/ijst/2016/v9i38/102972
11. Davis SB, Mermelstein P (1980) Comparison of parametric representations for monosyllabic word recognition in continuously spoken sentences. IEEE Trans Acoust Speech Signal Process—ASSP 28:357–366
12. JainP, Hermansky H (2001) Improved mean and variance normalization for robust speech recognition. In: Proceesings of the IEEE international conference on acoustics, speech, and signal processing

13. Harvilla MJ, Stern RM (2012) Histogram-based subband power warping and spectral averaging for robust speech recognition under matched and multistyle training. In: Proceesings of the IEEE international conference on acoustics, speech, and signal processing, pp 4697–4700
14. Samudravijaya K, Rao PVS, Agrawal SS (2002) Hindi speech database. In: International conference on spoken language processing (ICSLP00), Beijing, pp 456–459

The Impact of Knowledge Management and Data Mining on CRM in the Service Industry

Sanjiv Kumar Srivastava, Bibhas Chandra and Praveen Srivastava

Abstract The marketing trend in the service sector is changing at a rapid pace due to fierce competition and ever-growing innovation in the field of information technology. The marketing approach has transformed from product-centric concept to customer-centric concept. Adopting customer-centric strategy has become the essential component of business philosophy especially in the service sector, where customers are treated like guests. The whole concept of business performance is based on the philosophy of hospitality. Efforts are on satisfying individual needs of the customer and thus creating personal relationship by offering customised service. Customer relationship management (CRM) has replaced the traditional concept of marketing. Successful implementation of CRM program requires effective database management system to facilitate acquiring and maintaining customer information and thus applying this information to extend customised service based on individual needs thereby, supporting customer satisfaction and retention. Hence, it is evident that technology-based CRM has become essential for the survival and growth of business organisation mainly, the hotel sector and maintaining effective CRM strategy has its origin in knowledge management and data mining technique.

Keywords Customer relationship management · Relationship marketing
Knowledge management · Data mining

1 Introduction

Creating and maintaining relationship with the customers has become the most essential tool of marketing and goal of business. A strong relationship with customer brings intimacy which results in customer retention which is cost-effective

S. K. Srivastava (✉) · P. Srivastava
Department of Hotel Management, BIT Mesra, Ranchi, Jharkhand, India
e-mail: sanjeev_srivastava72@yahoo.com

B. Chandra
Department of Management Studies, IIT(ISM) Dhanbad, Dhanbad, Jharkhand, India

V. Nath and J. K. Mandal (eds.), *Nanoelectronics, Circuits and Communication Systems*, Lecture Notes in Electrical Engineering 511,
https://doi.org/10.1007/978-981-13-0776-8_4

than attracting new customer. In case of service industry, customer relationship management (CRM) is a crucial factor to improve its performance. The economic environment over past few years is dynamically changed and it continues to be due to advent of innovative technology connecting people around the world making the world a global village. Customers are now a day more knowledgeable than ever before. The development has led the organisation to change their way of interaction with the customer. The firms are forced to understand their customers and their changing needs to survive in the market as loyalty of customer cannot be guaranteed. For a successful business organisation, one has to be proactive in anticipating the desire of their potential customer and respond suitably. A successful enterprise places their customer first and at centre while devising their marketing strategy as any change in customer behaviour can lead unpredictable profitability and result in marketing failure.

Since 1980, relationship marketing has taken the centre stage and has become the fundamental concept of business marketing. It is imperative for all business organisations especially the service sector to adopt the philosophy of CRM to be competitive and gain an edge over the competitors. CRM is technology-driven concept of identifying, attracting, managing and retaining potential customer. In many research studies, it has been concluded that CRM has positive impact on customer satisfaction and customer retention.

In recent years, information technology has played a great role and influenced change in the behaviour of consumer, organisation and industries worldwide. The market is transformed from labour intensive to technology intensive. Growth and productivity increase is no longer only dependent on human resource only, but technology is playing significant role for it.

Service industry is not untouched with recent development and change in technology, with bigger role of IT, it has enabled the system more sharing and interactive. This has facilitated the hospitality industry to improve performance and productivity through better CRM with the application of technology.

The contribution of knowledge management and data mining is exceptional for effectiveness of CRM in customer satisfaction and success of service sector. It is significant in many ways as it helps in cost reduction by customer retention, growth in revenue generation and improvement and management of better quality relationship.

Service sectors accumulate huge quantity of data in the process of business and information gathered can be processed quickly and accurately by data mining and knowledge acquired can be helpful in decision-making for customer satisfaction. Most of the service organisations have already applied data mining technology to create database. This database management system facilitates in knowing the taste and preference, likes and dislikes and other personal attributes of customer which support in rendering customised service on individual basis.

1.1 Customer Relationship Management

CRM is a concept that facilitates organisation to deliver service on individual basis, i.e. customised service. Customised service is possible when there is frequent interaction and in the process, valuable information is accumulated regarding the customer. Swift [1] defined CRM as *'Enterprise approach to understanding and influencing customer behaviour through meaningful communications in order to improve customer acquisition, customer retention, customer loyalty, and customer profitability'* [1].

CRM is a strategy facilitating an organisation to provide customised service to each individual customer leading to intimacy development that fosters personal relationship. The personalised service creates a sense of belongingness as the customer is cared for and hence creates a new business opportunity based on taste and preferences, likes and dislikes and profile of the customer through effective interaction [2]. CRM integrates human resource, business processes and technology to meet the expectation of the customer efficiently and improving business and customer relationship. It is more of a customer-centric business approach and all the activities of the organisations are aimed at achieving customer satisfaction. It is convenient and costs less to sell to retained customer than to a new customer who has encountered for the first time. Organisations are now devising strategies to achieve customer satisfaction and customer retention by adopting customer focused service oriented practices.

It is a concept of business strategy, philosophy or culture of managing customer for long-term customer value and organisational profitability. CRM is the integration of strategies and process backed by software in view of achieving competitive edge and customer loyalty. It is process to learn and understand more about the customer.

The goal of CRM is to achieve customer satisfaction by providing customised service, organisation profitability by customer retention, increase performance and productivity standard by collecting feedback and suggestion and hence improvement in service quality, cost reduction and savings and achieve competitive edge in the market.

Schierholz et al. [3] advocate CRM as an instrument used to understand customer relationships is an investment that improves competitive position in the market by increasing customer loyalty. Their study concluded positive and significant association between CRM with customer loyalty [3].

Altran [4] came with a conclusion that enterprise with effective CRM supported by effective technology maintains their customer database in a better way suffer less in economic crisis [4].

CRM can be classified into three components: operational, collaborative, and analytical CRM [5].

Operational CRM is a CRM component that improves the effectiveness of regular every day customers operations. The process is automated concerned with frequent interaction and communication with customer. This is the first stage of data generation integrating all the intercoordinated with all the functional areas. Here, the organisation on requirements of customer fulfils them.

Collaborative CRM is a communication point that occupies the central role of linking business establishment with their customers, stake holders and others who have share in management and operational functions. It allows all the stake holders to share and disseminate information. This component provides efficiency in operation and creates opportunities to improve service process leading to customer satisfaction. This component helps in creating relationship.

Analytical CRM focusses on the analysis of collected data by analytical tool which supports formulating strategy and making decisions related to identification, attraction and retention of customers. The data collected are processed to make it more meaningful and interpreted to prepare reports. This component enables to maintain customer relationship by analysing customer's behavioural pattern. Analysing data facilitates prediction of future trends of customer accurately.

According to Swift [1], Parvatiyar and Sheth [6] and Kracklauer et al. [7], the four dimensions of CRM are customer identification, customer attraction, customer retention and customer development. The four dimensions enable complete in-depth understanding of customers helping in optimising customer value to the organisation.

The following are some of the studies related to CRM on customer satisfaction and loyalty:

Haridasan and Venkatesh [8] investigated the impact of CRM on customer loyalty. The study was conducted on seven mobile service provider company Chennai, Tamil Nadu on 490 mobile users. It was concluded that the impact of CRM was positively correlated with customer loyalty [8].

Yao and Khong [9] made investigation about CRM effectiveness on customer satisfaction in the banking sector of Taiwan. The study was conducted on 31 general managers of 42 commercial banks. In their research findings, it was revealed that CRM implementation is positively associated with customer satisfaction; and there is significant linkage among IT capability, interaction rate management and recovery management with customer satisfaction [9].

Khaligh et al. [10] investigated the effect of CRM on customer retention and loyalty in the telecom industry of Iran. He surveyed 200 consumers of telecom services and found that the structure of the strategy should be flexible specially and the policies of price benefit the telecom firms and increase the customer loyalty [10].

The effectiveness of CRM on customer satisfaction and retention and ultimately performance of service sector can only be achieved if it is well integrated and supported by effective IT software system across the establishment covering various operational and functional departments and sub-departments. The integration of various sections will enable to gain real-time information to meet the expectation of customer through knowledge management and data mining technology.

2 Knowledge Management and Data Mining

2.1 Knowledge Management CRM

Knowledge management and data mining process technology is an essential significant component of CRM. CRM is a technology-driven strategy, and the soul of this concept lies in the process of knowledge management and data mining to achieve success in CRM effectiveness. In the present scenario, the service sector integrates knowledge management aspect with CRM as the knowledge which is gathered in the form of data from customer plays a significant role in the success of CRM. Knowledge is an internal dimension of CRM collected from the customer and is about the customer and applied for the satisfaction for the customer. The knowledge gained by service provider through various channels, touch points by means of interaction is applied for designing and developing strategies in the service delivery process. Service sector is making use of this information (Knowledge) to offer better quality services to its clients, understand the importance and advantage of knowledge management. Yichen and Chienshihen [11] in their his work titled 'having knowledge in creation of method for CRM' stated that customer knowledge and their knowledge management are key factors that have impact on organisations performance [11]. Knowledge management involves collecting and gathering information, storing the knowledge information, distribution of knowledge and extracting benefits from the knowledge by designing service based on the knowledge. The role of knowledge management has become important in the context of present customer-centric marketing environment [12]. Knowledge management is the management of organisations information and knowledge which supports consistent decision-making ability of the managers. Integrating knowledge with CRM data facilitates customer-centric decision-making. CRM can be effective if it is integrated with knowledge management system to enable management to evaluate their performance based on customer satisfaction and profitability [12, 13]. Therefore, it is evident that knowledge management contributes significantly in CRM, which the service sector can use it to build and maintain relationship with the customer that can enable to gain competitive edge in the market [14]. Successful and effective CRM can be identified if customer information can be realised into customer knowledge. Knowledge management supports business organisations in establishing better CRM and it contributes positively in its performance [14]. The importance building customer relationship is most essential for service sector and hence, knowledge management can help significantly to gain competitive advantage to such sector. Various studies have been conducted by scholars to establish positive relationship between knowledge management and customer satisfaction.

3 Data Mining

The recent development in information technology over the past few years has made unprecedented change in the style of functioning of the business organisation especially in the service sector where the core concept of business philosophy is to provide hospitality service to the customer which needs long and continued relationship. The IT has penetrated almost every organisation. The massive application of IT has enabled the provision of database management where huge amount of data is stored. The effective utilisation of database needs a systematic method of collecting, storing and analysing data.

In a general sense data mining is the process of searching valuable information from large volumes of data. It is the exploration and analysis of large quantity of data using computational technology to discover meaning pattern and rules.

The hospitality industry is customer service oriented organisation that collects and gathers large amount of data from property management system (PMS), central reservation systems (CRS), guest loyalty program databases and point of sale (POS) leads to data mining application which play significant role in the industry. The volume of data created through Internet and web site are difficult to be handled and therefore, the need for data warehousing and data mining system arises for hospitality industry. Data mining technology in service establishments converts customer information into valuable customised service process decisions to improve long-term relationship and profit and it is an effective and efficient technology for a service firm to transform customer information into customised and dynamic marketing decisions to improve long-term profit. A data mining technique also reveals new pattern of purchasing and customer behaviour.

Data mining helps in constructing models of customer behaviour by use of statistics and appropriate data mining technique. It helps to identify potential market segment for potential customer through analysis of various factors that predicts similar buying behaviour. Hence, it can be applied for attracting potential and profitable customer.

Data mining provides opportunity to database marketers to come closer to their customers by inputs related to customer's need, likes and dislikes, taste and preferences. The only condition is that the necessary data input should exist in the database system. It can be used to identify possible customers who have intention of switching over and necessary action may be taken to retain them. It can be applied for more revenue generation by upselling by identifying customers' needs and desires and offering service as per desire. It can be also used to meet the changing needs of customer by offering additional services.

Various studies have also been conducted by scholars to establish relationship between data mining, knowledge management and CRM, which is summarised in Table 1.

Table 1 Review of previous literatures

Year	Authors	Title	Journal	Findings
2016	Uma [15]	Social CRM: A Survey on Various Techniques for Text Analysis in Social Media	International Journal of Computer Science and Mobile Computing	Social media platforms are source of huge amount of valuable information expressed by opinions and thoughts that can be captured and analysed for attracting and retaining existing customer. Sentimental analysis and natural language processing along with other techniques can be very effective in data mining for texts available on the platform of social media
2016	Abdavi et al. [16]	The impact of knowledge management factors on improving the performance of success of Sports customer relationship management	Journal of Novel Applied Sciences	The research was conducted in six metropolitan cities of hotels of Iran which have sports service sector. Positive correlation was found between the components of KM and success of CRM
2016	Jeyashree [17]	An analysis of CRM through data analysis	International Journal of Advance Research in Computer Science and Management Studies	Analysing various customer retention techniques to enhance customer relationships through data mining. Helps to understand buying pattern and classify customers
2016	Nair et al. [18]	Improving Customer Relationship Management Using Data Mining	International Journal of Scientific & Engineering Research	The paper develops a model for Hotel Management System to maintain records of the requirement of the guests, suggests automated system for all the services provided by the hotel and email-based communication system to retain the customers

(continued)

Table 1 (continued)

Year	Authors	Title	Journal	Findings
2016	Sharma and Sharma [19]	A Study on Data Mining on Algorithms for Tourism Industry	International Journal of Latest Trends in Engineering and Technology	The paper presents an overview of different data mining algorithms used for tourism industry. It explains the application of algorithm for various aspects of business operations
2015	Femina and Sudheep [20]	An Efficient CRM Data Mining Framework for the Prediction of Customer Behaviour	Procedia Computer Science.	Proposed a data mining CRM framework using two models for predicting customer behaviour for the banking sector. Multilayer perception neural network (MLPNN) has better accuracy than Naive Bayes (NB). The data collected is for general client's behaviour, classification of gender is significant as their behaviour differs
2015	Tama [21]	Data Mining for Predicting Customer Satisfaction in Fast Food Restaurant	Journal of Theoretical and Applied Information Technology	Data mining concept based on decision tree and neutral network was applied to find out the determinants having impact on customer satisfaction. Staff behaviour, cleanliness of restaurant and food presentation were found to be the significant predictor of customer satisfaction
2015	Kumar [22]	Increasing the Efficiency of CRM Process Using Data Mining Practices	International Journal of Advance Research in Computer Science Management Studies	The paper investigates to improve the efficiency of CRM suitable software and technology related to data mining should be selected. Integration of CRM and data mining will respond promptly to customer's

(continued)

Table 1 (continued)

Year	Authors	Title	Journal	Findings
				requirement and improve service quality
2014	Maheswari et al. [23]	Role of Data Mining in CRM	International Journal of Engineering Research	The paper designs a mining model and explains the merits of Chi Square Automatic Interaction Detection (CHAID) method for classification as compared to Classification and Regression Trees (CART)
2013	Anand et al. [24]	Customer Relationship Management using Adaptive Resonance Theory	International Journal of Computer Application	Adaptive Resonance Theory (ART) can be used for the classification of customer and can handle large volume of customer data leading to understanding and needs of customer
2013	Yadav et al. [25]	Knowledge Management in CRM Using Data Mining Technique	International Journal of Scientific & Engineering Research	Application of various data mining techniques (algorithm) like clustering, classification, association, prediction and correlation to enterprise database to enhance CRM through knowledge management. Requirement of Knowledge management for different sector of business is different
2013	Sadath [26]	Data Mining in E-Commerce: A CRM Platform	International Journal of Computer Applications	The paper examines a three-tier architecture integrating data mining in Internet platform and model of value assessment. A close integration between e-commerce and data mining will

(continued)

Table 1 (continued)

Year	Authors	Title	Journal	Findings
				be the best method to improve strategy, prediction and decision-making for better relationship
2009	Ngai et al. [27]	Application of data mining techniques in customer relationship management: A literature review and classification	Expert Systems with Applications	The paper examines 87 articles concerned with CRM and Data mining. Majority of the paper described neutral network for classification, clustering and prediction followed by decision tree and association tools. The classification model is the most commonly used method for forecasting the future behavioural pattern of the customer. The research was conducted between 2000 and 2006 and only seven online databases were used

4 Data Mining and Analytical CRM

Data mining techniques of analysing customer data are getting more popularity in analytical CRM. Application of data mining to facilitate CRM analytical dimension is becoming an emerging trend [27]. This technique can be applied to achieve competitive advantage to acquire customer and maximise customer value by analysing and understanding customer characteristics and behaviour. Association, classification, clustering, forecasting, regression, sequence discovery and visualisation cover the main data mining techniques.

Association explains the relationships between attributes stored in a database. Classification is used to mapping a data item into a predefined class of category [28, 29]. Clustering is process of mapping a data item into a categorical class or clusters determined from the data, i.e. putting group of customers into one group with similar attributes of purchasing behaviour [28, 30].

Forecasting anticipates the future value of return from the pattern of customer attributes based on records. Predictor is the key element of forecasting analysis [30].

Regression facilitates real value prediction variable [28, 31]. Sequence discovery helps in explaining and identifying relationships between items for due course of time [28, 31, 32].

Visualisation is presenting data of complex nature in combination with other data mining model to provide better understanding of identified relationship [33, 34].

The identification of appropriate data mining techniques to interpret useful information from a large volume of customer databases is the important stage in data mining process.

The research work conducted by Swift [1], Parvatiyar and Sheth [35] and Kracklauer et al. [7]. Ngai et al. [27] proposed that for customer identification, classification and clustering techniques are the most preferred; to attract customers, classification technique is preferable; to retain customers, association and classification are frequently preferred; about Concerning customers' development, association technique is preferred. Usually, combination of more than one data mining techniques is used frequently to support CRM analytical dimension [27].

Following are some of the findings concluded by some of the research scholar on the subject CRM and data mining technology in service sector.

Danubianu et al. [36] in his research work titled 'improving CRM in hotel industry by data mining techniques' concluded that data mining technique can be useful in predicting behaviour of the guest and helpful in taking marketing decision in identifying and contacting customer, deciding to whom to offer incentive and kind of relationship to be maintained [36].

Camilovic [37] in his study on 'Data Mining and CRM in Telecommunications' concluded that telecommunication industry is highly competitive, and they receive massive amount of data. Customers are their assets and to retain customers competitive advantage must be achieved [37]. To serve the purpose, the telecom industry must apply data mining technology to understand the behaviour of customer and frequently interact with the customer to identify their needs and desire and satisfy them by advanced and flexible service. Data mining technology can support them to achieve these objectives by enabling customer segmentation and churn prediction. Data mining can be proved to be an effective tool for developing CRM strategy facilitating telecommunications enterprise to keep their customers happy.

Kim et al. [38] investigated on 'Data Mining Applications in Hospitality Industry' and concluded that using data mining technique was helpful in identify profitable customer, create more loyal customer and maximise profit in case of both restaurants and hotels [38].

Liao et al. [39] in his study on 'Mining customer knowledge for tourism new product development and CRM' on Phoenix Tours International, in Taiwan concluded that data mining helps in new product designing to meet the changing needs of customer in tourism sector and supports in attracting and retaining customer by improving CRM. Mining technique contributes significantly by developing innovative tour package by segmenting customers purchase behaviour and pattern.

It helps in identifying the factors and variables influencing the choice of customer demand [39].

Wang and Wu [40] in his study investigated about CRM of agricultural bank of China by applying data mining technique. The sample included 53,872 customers, classified into three types. The test sample was 51,749 customers. The study collected data about the basic information of the bank customer and details of transaction between customer and the bank. It was found that the implementation of data mining technology into the bank's CRM achieved CRM target with high quality, and service quality was improved effectively, assigned by the bank to target customers and minimise the cost of operation of the bank [40].

Lin [41] in his work in the research paper 'A Technique for Enhancing Customer Relationships in the Service Industry' 'application of data mining in CRM in Taiwan's hotel industry', 50 hotels were invited for study, the directors of 33 hotel agreed to participate and they confirmed that they are aware of data mining and using it. The data mining technique is aligned with needs of the hotel industry. According to them data mining was very helpful in managing strong and long-term relationship with the customers. The study concluded that this technique is useful in understanding the behaviour pattern of the customer through better and effective interaction with the customers [41]. To achieve competitive advantage in CRM, data mining is must for hospitality industry.

Jenabi and Mirroshandel et al. [42] study on 'Using Data Mining Techniques for Improving Customer Relationhip Management CRM', the data are collected from 2011 to 2014 in from automotive industry in Saipa Yadak, Iran. The data set consisted of 126 maintenance features. After pre-processing, data set was left with 2849 records and 23 features for data mining. The study concluded that data mining technique helps in obtaining accurate information about potential customer. It also suggests that data mining is useful in knowing the characteristics and the needs of customer so as satisfy customer and make them loyal and achieve more profit [42].

Tama [21] in his research study in fast food restaurants in Palembang, Indonesia applied data mining for predicting customer satisfaction in fast food restaurant. 400 respondents were surveyed by offering questionnaires with ten input variables such as taste, serving, dessert, restaurant environment, washing basin, waiting time, staff behaviour, menu pricing, discount and store location. Finally, 340 samples were selected for analysis. He concluded that the basic customer service attributes like staff behaviour, restaurant's cleanliness, and food presentation were significant predictor in assessing customer satisfaction with more accuracy.

Nair et al. [18] in her study on 'Improving CRM Using Data Mining' in a Hotel Management System concluded that by using data mining technique customised service is improved and it effectively supports decision activity, thereby, improving CRM.

Chen et al. [43] in his study titled 'Data Mining Application in CRM of Credit Card Business', data were collected from credit card customer about their transaction record from 2003 to 2004 from banks. The number of purchase data collected was 10,304,500 and 1,146,200 for the year 2003 and 2004 respectively and finally, 1,063,000 were used as dataset for each year in the study. The study

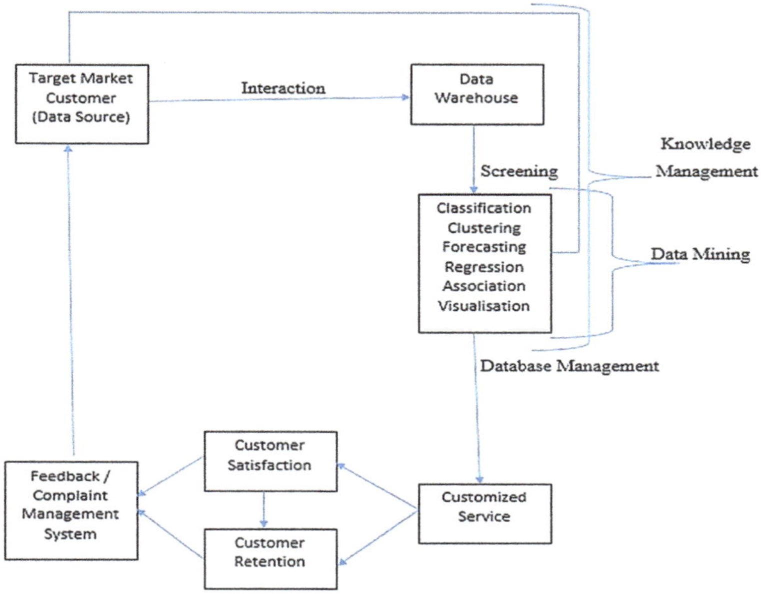

Fig. 1 Proposed framework for knowledge management—data mining based CRM cycle

concluded that data mining technique was useful in identifying the needs of customer and change in marketing strategy was required. 80% of customers have little contribution in bank's profit and only 3.3% customers were highly profitable and loyal to the bank. So, data mining technique is useful helping management to take corrective marketing actions to improve business performance [43].

XI and Chen [44] in their his study 'Application of Data Mining Technology in CRM System of Commercial Banks', they he concluded that data mining is used to analyse the huge data effectively in the CRM system of commercial banks. Data mining technique transforms information into knowledge, and hence benefits the bank for better decision-making [44].

Complaints or feedback plays a very important role in CRM for service industry. It creates opportunities for service provider to further improve their service. It ensures that a success or failure of data mining technique applied is communicated to the service provider. Immediate action based on feedback strengthens the trust of customer leading to customer satisfaction and hence ensuring customer retention.

Based on above review of literatures, a proposed framework for knowledge management and data mining leading to effective CRM is given in Fig. 1.

5 Suggestion and Conclusion

The world market has moved from product-centric approach to customer-centric approach. This has facilitated relationship marketing and hence, the growth in CRM has remarkably occupied the centre stage of marketing philosophy. The interaction and communication with the support of advanced IT technology enabled the organisation to have in-depth view of their customer like taste and preference, likes and dislikes, purchasing behaviour and pattern. Organisations can take advantage of this opportunity and offer customised service to the customer. The customised service creates strong and continued relationship with the customer for a long time that supports retention and loyalty.

The CRM concept is more desirable for service sector where the complete concept of business and marketing is based on hospitality. Customer satisfaction is treated as mission and vision of the service sector. CRM becomes more effective when it is technologically backed by knowledge management and data mining. The information collected through various touch points is transformed into knowledge through data mining. The knowledge creation by the process of data mining is valuable for the management to take marketing decisions aligned with customer needs. Data mining helps in identifying potential customer and various attributes associated with customer like behaviour and pattern of purchasing and their changing needs. It facilitates making appropriate marketing decision on one-to-one individual basis. From the above literature study, it is concluded that knowledge management and data mining play a significant role in improving quality of service, increasing effectiveness of CRM, improving customer satisfaction and retention of customer. Smaller and medium size service sector should also introduce data mining technology to compete and get competitive advantage in the market.

References

1. Swift RS (2001) Accelerating customer relationships: using CRM and relationship technologies. Prentice Hall PTR, Upper Saddle River, N.J
2. Peppers D, Rogers M, Dorf B (1999) Is your company ready for one-to-one marketing. Harvard Bus Rev 77(1):151–160
3. Schierholz R, Kolbe LM, Brenner W (2007) Mobilizing customer relationship management: a journey from strategy to system design
4. Altran U (2010) CRM pays off in times of economic crisis, CIS Services
5. Dych J (2001) The CRM handbook: a business guide to customer relationship management, 1st edn. Addison-Wesley Professional, USA
6. Parvatiyar A, Sheth JN (2001) Customer relationship management: emerging practice, process, and discipline. J Econ Soc Res 3:1–34
7. Kracklauer AH, Mills DQ, Seifert D (2004) Customer management as the origin of collaborative customer relationship management. In: Collaborative customer relationship management: taking CRM to the next level, Springer

8. Haridasan V, Venkatesh S (2011) CRM implementation in Indian telecom industry—evaluating the effectiveness of mobile service providers using data envelopment analysis. Int J Bus Res Manage 2(3):120–127
9. Yao HI, Khong KW (2011) Effectiveness of customer relationship management on customer satisfaction in the commercial banks of Taiwan. Contemp Manage Res 7(2):105–116
10. Khaligh A, Miremadi A, Aminilari M (2012) The impact of eCRM on loyalty and retention of customers in Iranian telecommunication sector. Int J Bus Manage 7(2):150–162
11. Yichen L, Chienshihen S (2006) A Knowledge-enabled procedure for customer relationship management. Ind Mark Manage 35:446–456
12. Winer RS (2001) A framework for customer relationship management. Calif Manag Rev 43(4):89–107
13. Fahey L (2001) Linking E-business and operating processes: the role of knowledge management. IBM Syst J 40(4):889–907
14. Sin LY, Alan CH, Yim FH (2005) CRM conceptualization and scale development. Eur J Mark 39(11/12):1264–1290
15. Uma N (2016) Social CRM: a survey on various techniques for text analysis in social media. Int J Comput Sci Mob Comput 5(1):313–319
16. Abdavi F, Pashaie S, Azarin YB (2016) The impact of knowledge management factors on improving the performance of success of sports customer relationship management. J Novel Appl Sci 5(3):86–91
17. Jeyashree S (2016) An analysis of CRM through data mining methods. Int J Adv Res Comput Sci Manage Stud 4(5):1–6
18. Nair A, Khasgiwala R, Mishra S (2016) Improving customer relationship management using data mining. Int J Sci Eng Res 7(2):1161–1163
19. Sharma GK, Sharma P (2016) A study on data mining algorithms for tourism industry. Int J Latest Trends Eng Technol 7(1):580–587
20. Femina BT, Sudheep EM (2015) An efficient CRM-data mining framework for the prediction of customer behaviour. Procedia Comput Sci 46:725–731
21. Tama BA (2015) Data mining for predicting customer Satisfaction in fast-food restaurant. J Theor Appl Inf Technol 75(1):18–22
22. Kumar CS (2015) Increasing the efficiency of CRM process using data mining practices. Int J Adv Res Comput Sci Manage Stud 3(7):106–111
23. Maheswari RU, Mahesan SS, Tamilarasan D, Subramani AK (2014) Role of data mining in CRM. International J Eng Res 3(2):75–78
24. Anand M, Khan Z, Shukla RS (2013) Customer relationship management using adaptive resonance theory. Int J Comput Appl 76(6):43–47
25. Yadav S, Desai A, Yadav V (2013) Knowledge management in CRM using data mining technique. Int J Sci Eng Res 4(7):465–472
26. Sadath L, Data mining in E-commerce: a CRM platform. Int J Comput Appl 68(24):32–37
27. Ngai E, Xiu L, Chau D (2009) Application of data mining techniques in customer relationship management: a literature review and classification. Expert Syst Appl 36(2, Part 2):2592–2602
28. Mitra S, Pal S, Mitra P (2002) Data mining in soft computing framework: a survey. IEEE Trans Neural Networks 13(1):3–14
29. Chen Y, Hsu C, Chou S (2003) Constructing a multi-valued and multi-labeled decision tree. Expert Syst Appl 25(2):199–209
30. Berry MJA, Linoff GS (2004) Data mining techniques: for marketing, sales, and customer relationship management, 2nd edn. Wiley Computer Publishing
31. Giraud-Carrier C, Povel O (2003) Characterising data mining software. Intell Data Anal 7(3): 181–192
32. Berson A, Smith S, Thearling K (1999) Building data mining applications for CRM. McGraw-Hill, New York
33. Shaw M (2001) Knowledge management and data mining for marketing. Decis Support Syst 31(1):127–137

34. Turban E, Sharda R, Delen D (2010) Decision support and business intelligence systems, 9th edn. Prentice Hall
35. Parvatiyar A, Sheth JN (2002) Customer relationship management: emerging practice, process and discipline. J Econ Soc Res 3:6–23
36. Danubianu M, Cristian V (2008) Improving customer relationship management in hotel industry by data mining techniques. http://www.s156.central.ucv.ro/annals/v7_2008/0036v7-015.pdf
37. Camilovic D (2008) Data mining and CRM in telecommunications. Serb J Manage 3(1): 61–72
38. Kim S, Lin L-C, Wang Y (2009) Data mining applications in the hospitality industry. http://www.igi-global.com/chapter/data-mining-applications-hospitality-industry/10852
39. Liao SH, Chen YJ, Deng MY (2010) Mining customer knowledge for tourism new product development and customer relationship management. Expert Syst Appl 37:4212–4223
40. Wang Y, Wu DS (2011) Research of the bank's CRM Based on data mining technology. Commun Inf Sci Manage Eng 1(4):30–35
41. Lin TC (2012) A technique for enhancing customer relationships in the service, IPCSIT, 30, IACSIT Press, Singapore
42. Jenabi G, Mirroshandel SA (2013) Using data mining techniques for improving customer relationship management. Eur Online J Nat Soc Sci 2(3)
43. Chen RS, Wu RC, Chen JY (2017) Data mining application in customer relationship management of credit card business. http://www.researchgate.net/publication/4174543, https://doi.org/10.1109/compsac.2005.67
44. XI YP, Chen M (2017) Application of data mining technology in CRM system of commercial banks. In: Proceedings of 2017, 2nd international conference on electrical and electronics: techniques and applications (EETA 2017)
45. Long CS (2013) Impact of CRM on customer satisfaction and loyalty. Asian Soc Sci 9(10)

Efficient Computer-Aided Diagnosis of Alzheimer's Disease and Parkinson's Disease—A Survey

R. S. Nancy Noella and J. Priyadarshini

Abstract Dementia is a broad category of brain-related diseases that continues for a long term and severely affects thinking and daily functioning of a human being. Among different types of dementia the fatal type of brain problems are Alzheimer's Disease (AD) and Parkinson's Disease (PD). More than 70% of cases are reported as dementia is in the Alzheimer's category. In AD, the patient's brain gets severely damaged, especially the outer part of the brain like cerebral cortex, hippocampus, ventricles, etc. The AD patients have enlarged ventricles, shrinkage in hippocampus and cortex. PD is also a common dementia after AD. In PD, the patient's mid-brain gets damaged, i.e., substantia nigra. The proposed work presents an efficient automation for the detection of the AD and PD with Machine Learning Techniques (MLT). To detect the presence of PD and AD, two different types of brain image databases have to be selected: Positron Emission Tomography (PET) and Single Photon Emission Computed Tomography (SPECT) database images, both of them contain data for AD and PD patients in comparison with the healthy brain images. From the input image, different features have to be extracted like statistical moments, geometrical moments, texture features, etc. Then Region of Interest (ROI) has to be selected to differentiate disease-affected areas. The results have to be generated automatically by comparing input image with the trained samples in the database. The proposed system concentrates on applying the MLT for segregating the outer part of brain with central part of brain for diagnosing the AD and PD in comparison with the healthy brain data.

Keywords Alzheimer's Disease (AD) · Parkinson's Disease (PD) Computed Tomography (CT) · Positron Emission Tomography (PET) Single Photon Emission Computed Tomography (SPECT) · Computer-Aided Diagnosis (CAD) · Region of Interest (ROI) · Machine Learning Techniques (MLT) · Fisher Discriminant Ratio (FDR)

R. S. Nancy Noella (✉) · J. Priyadarshini
School of Computing Science and Engineering, VIT University, Chennai 600127, India
e-mail: nancynoella@ymail.com

J. Priyadarshini
e-mail: priyadarshini.j@vit.ac.in

V. Nath and J. K. Mandal (eds.), *Nanoelectronics, Circuits and Communication Systems*, Lecture Notes in Electrical Engineering 511,
https://doi.org/10.1007/978-981-13-0776-8_5

1 Introduction

Alzheimer's Disease is one of the major types of dementia which affects more than 50 million aged people throughout the world. The drastic change in number of AD-affected patients per year and their growing dependence on other individuals provokes an important social repercussion. AD severely affects daily routines and cognitive functioning of the patients, by which patient's individual ability blurs [1]. The nerve cells in the brain die and it stops functioning. The death or partially functioning of these nerve cells causes changes in patient's memory, character and thinking ability. The AD-affected patient's ability to do the basic body functions like swallowing, walking, etc. also get affected. It is a chronic neurodegenerative disease usually starts slowly and continues over time. The primary symptoms of AD are difficulty in recalling recent incidents, names, etc. and later symptoms include impaired judgment, disorientation, confusion, behavioral changes and difficulty in speaking, swallowing and walking. The effects of Alzheimer's Disease affect not only in terms of familiar affliction and dependence but also economically too. In AD, the patient's brain gets severely damaged, especially the outer part of the brain like cerebral cortex, hippocampus, ventricles, etc. The AD patient will have enlarged ventricles, shrinkage in hippocampus and cortex. For the efficient diagnosis of different types of dementias, functional imaging techniques are mainly used. The early diagnosis of dementias still remains as a demanding task because it is carried out based on the data provided by a careful clinical examination carried out by clinical experts (Fig. 1).

Parkinson's Disease (PD) is a progressive disorder of the nervous system that severely affects movement [2]. More than 10 million people are affected by PD in worldwide. The PD disorder mostly restricts free movement. So, the patient feels difficulty in moving, walking, etc. Parkinson's Disease is the second most common neurodegenerative disorder surpassed only by Alzheimer's Disease. If a proper

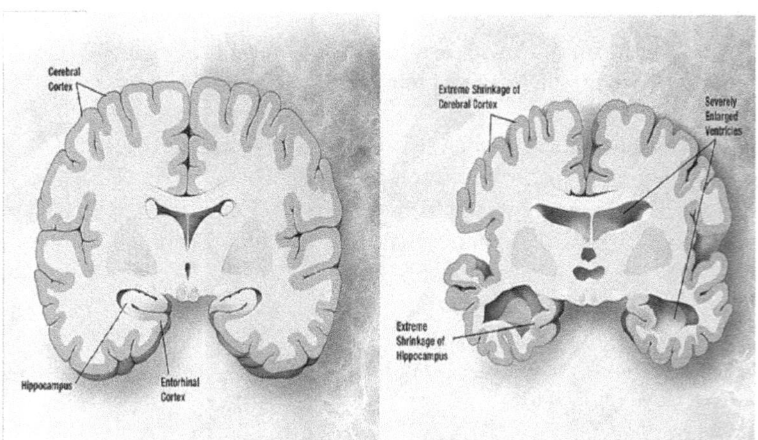

Fig. 1 Healthy brain versus Alzheimer's brain

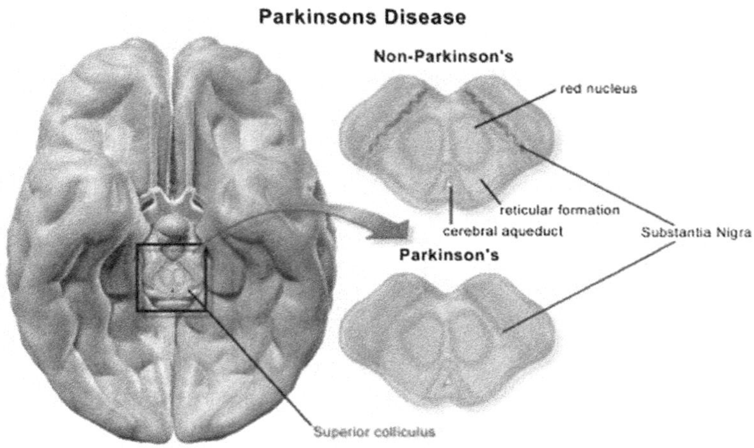

Fig. 2 Healthy brain versus Parkinson's brain

diagnosis carried out at the early stage, it can result in saving patient's life. In PD Alpha—synuclein starts to aggregate in the mid-part of the brain called as substantia nigra. This causes degeneration of the nerve cells that produce dopamine. In PD, the patient's mid-brain gets damaged, i.e., substantia nigra. The PD patient will have diminished substantia nigra (Fig. 2).

Automation is mainly focused on the proposed work so that researchers in the field of medical image processing can improve the accuracy in detection of any abnormalities in brain. Recently, engineers working on digital image processing are introducing software improvements by enabling computers to help the clinical experts by providing an amount of precise medical information which add more sense to the diagnosis work. A CAD system helps doctors in the analysis of medical images and it is an interdisciplinary technology combines the elements of computer vision and artificial intelligence with radiological image processing. CAD is mainly used by physicians in their diagnosis work that provides a blueprint of critical areas; moreover it helps in similar pattern identification of the disease which has been unnoticed by a physician.

The design and validation of Alzheimer's and Parkinson's Disease detection, with proper description of feature selection, extraction, classification and any other abnormalities in the brain are the main challenges of this work.

2 Literature Survey

The different works designed in the related area are studied and listed below:

Evanthia E. Tripoliti et al. developed a work on automated diagnosis of diseases based on classification: dynamic determination of the number of trees in random forests algorithm. The main advantage is that it runs efficiently on large databases. But that work failed to meet proper classification on datasets [3].

Carmen Paz Suarez Araujo et al. designed work on a new gating neural ensemble for automatic assessment of the severity level of dementia using neuropsychological tests. The main advantage of the work was easiness in training sample data sets. But they failed to meet the estimate level of confidence in the discrimination process [4].

Amira Ben et al. had worked on early diagnosis of Alzheimer's diseases using support vector machine. The main advantage of that method was it minimized the number of convex functions. The disadvantage in that work was complicated kernel selection [5].

Oriehi Edisemi Destiny Anyaiwe et al. had developed a work on weighted Manhattan distance classifier; SELDI data for Alzheimer's Disease diagnosis. In that work weighted Manhattan Distance Classifier is used. The highlight of the work was complex concepts can be learned by local approximation using simple procedures. But the work was computationally expensive [6].

Muhammad Aksam Iftikhar et al. proposed a work on automated detection of Alzheimer's Disease and mild cognitive impairment using ensemble classification. The benefits of that method were handling of high-dimensional spaces as well as large number of training samples. The main drawback was that in most cases it required much iteration to truly obtain high accuracy [7].

Alexander Luke Spedding et al. had developed a work on the detection of Alzheimer's Disease from structural MRI using an LDA and probability-based classifier. Naive Bayesian Classifier was used in that work. The main advantage of the work was that it required only small amount of training data to estimate the parameters. The main disadvantage was loss of accuracy because of class conditional independence [8].

A.Valli et al. had done work in Parkinson's Disease diagnosis using image processing techniques. Statistical Parametric Mapping (SPM) is the method they adopted in their work. The main advantage of that work was the control of false positive rate (errors). The main drawback of the algorithm was high time consumption [9].

Yue Peng et al. proposed work on Entropy Chain Multi-Label Classifiers for Traditional Medicine Diagnosing Parkinson's Disease. In that work, they used classifier chains algorithm. The advantage of that work was low memory consumption and low run time complexity. But the disadvantage of the method was low efficient compared to automated approaches [10] (Table 1).

Ali H. Al-Fatlawi et al. had worked on deep belief network for the efficient diagnosis system for Parkinson's Disease. The main advantage was simple sample distribution. The disadvantage was handling of time series data in the networks was a very complicated [11].

Made Satria Wibawa et al. had done work on performance evaluation of combined feature selection and classification methods based on voice feature in diagnosing Parkinson's Disease. In that work, they used Principal Component Analysis Method. The advantages of that work were lack of redundancy and reduction of

Table 1 Summary of literature survey

Author and year	Title	Proposal	Advantages	Disadvantages	Proposed work
Evanthia E. Tripoliti et al.	Automated diagnosis of diseases based on classification	Random forests algorithm	Suitable for large data sets	Improper classification	Works on large data set and proper classification
Carmen Paz Suarez Araujo et al.	A new gating neural ensemble for automatic assessment of the severity level of dementia using neuropsychological tests	Neural ensemble classifier	Ease in training data sets	Misclassification	Haar training is simple and proper classification
Amira Ben et al.	Diagnosis of Alzheimer diseases in early step using SVM	Support vector machine	Less number of functionality	Complicated Kernel selection	Training and classification can be done with minimum number of functionality
Oriehi Edisemi Destiny Anyaiwe et al.	Weighted Manhattan distance classifier; SELDI data for Alzheimer's Disease diagnosis	Weighted Manhattan distance classifier	Ease in training data sets	Complicated classification	Proper classification with fewer steps
Muhammad Aksam Iftikhar et al.	An ensemble classification approach for automated diagnosis of Alzheimer's Disease and mild cognitive impairment	Ensemble classifier	Suitable for high dimensional data	More number of iterations for classification	Suitable for any type of data and classify the diseases with few number of steps
Alexander Luke Spedding et al.	An LDA and probability-based classifier for the diagnosis of Alzheimer's Disease from structural MRI	Naive Bayesian classifier	Less number of training data is enough for classification	Less accurate	More accuracy compared to Naïve Bayesian Classifier
A. Valli et al.	Parkinson's Disease diagnosis using image processing techniques. Statistical parametric mapping	Statistical parameter mapping	Less error rate	Time consumption is more for classification	More accuracy

(continued)

Table 1 (continued)

Author and year	Title	Proposal	Advantages	Disadvantages	Proposed work
Yue Peng et al.	Entropy Chain Multi-Label Classifiers for traditional medicine diagnosing Parkinson's disease	Classifier chain	Low memory consumption	Low efficient	Highly efficient
Ali H. Al-Fatlawi et al.	Efficient diagnosis system for Parkinson's Disease using deep belief network	Deep belief network	Sample Distribution is easy	Time series of data in the network is complicated	Classification carried out with fewer steps
Made Satria Wibawa et.	Performance evaluation of combined feature selection and classification methods in diagnosing Parkinson's Disease based on voice feature	Principal component analysis method	Lack of redundancy and reduction of noise	Comparatively low accuracy	More accuracy

noise. But they failed to reach accurate result compared to other different algorithms [12].

The literature survey infers that the detection of AD and PD from same brain image data set is not processed. So, automation for detection of AD and PD is proposed in this work.

3 Proposed Work—Discussion

The proposed work aims in diagnosing the AD and PD among people. The existing works portrays various methods to identify the AD and PD. A work on diagnosis of AD using Decision Tree (DT) for classifying the stages of Alzheimer's Disease already implemented in [13] and have attained an accuracy of 80%.

Now this work is planning to extend the work with advancement in the techniques using Machine Learning Algorithm for diagnosing both AD and PD. By MLT the results of diagnosis will have more accuracy. The proposed systems plan to use Haar training and Ada boost classifier for training the data set. The overall architecture for the proposed system is in Fig. 3. The outer part of the brain is diagnosed for AD and central part for PD. The outer and central part segregation is done using MLT for more accuracy in which the existing system has not work. The proposed system concentrates on applying the MLT for segregating the outer

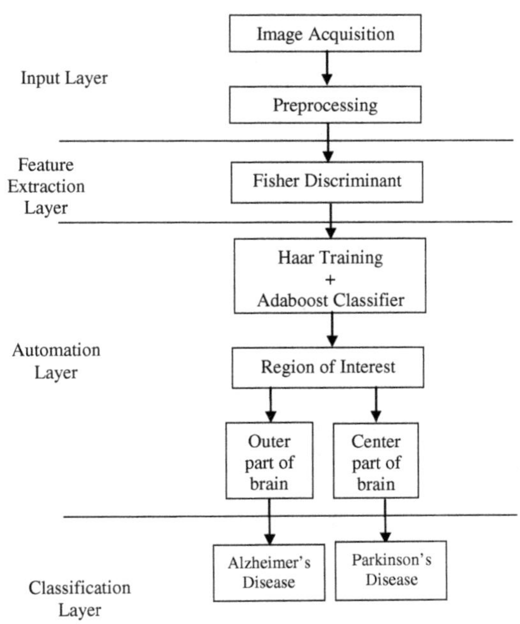

Fig. 3 Architectural diagram of diagnosis of AD and PD

Layer		Description
Input layer	Image Acquisition	PET & SPECT Images
	Preprocessing	Intensity Normalization
Feature Extraction layer		Fisher Discriminant Ratio
Automation Layer	Training	Haar Training + Ada boost Classifier
	Region of selection	Outer Region and Center Region
Classification Layer		Alzheimer's Disease
		Parkinson's Disease
		Healthy Brain

Fig. 4 Different layers included in the diagnosis of AD and PD

part of brain with central part of brain for diagnosing the AD and PD in comparison with the healthy brain data and AD- and PD-affected brain images.

Figure 4 shows the different layers included in the diagnosis procedure and the description of each step is as follows:

3.1 Image Acquisition

For the proposed system, the images selected are PET and SPECT. Computed Tomography is a diagnosing test use to build detailed images of different parts of human body like bones, internal organs, blood vessels, and soft tissues. The cross-sectional images created by the CT scan can be reformatted in multiple planes. This can also produce three-dimensional images that can be transferred to electronic media, seen on a computer monitor or printed on film. Emission Computed Tomography images are usually used in biomedical research field and medical field for the past few years. These emissions based functional images reproduce a blueprint of physiological functions along with anatomical structure images and this provides data about physiological phenomena and its position in the body. Emission computed tomography types include: Positron Emission Tomography (PET) [14] and Single Photon Emission Computed Tomography (SPECT) [15]. In this work, PET and SPECT are the two different brain image datasets used for brain image acquisition. PET is a nuclear medicine functional imaging practice used to view metabolic processes in the body. The diagnosis is done by emitting pairs of gamma rays indirectly introduced into the body of the patient on a biologically active molecule by a positron emitting radionuclide (tracer). Three-dimensional images of tracer concentration within the body are then constructed by computer analysis. SPECT is other gamma rays based nuclear medicine tomographic imaging technique. This is like other conventional nuclear medicine planar imaging using a gamma camera (scintigraphy). It provides the trusted three-dimensional information to the clinical experts. These data are typically showed as cross-sectional slices and it can be easily reformatted or manipulated according to the requirement. Both PET and SPECT systems are noninvasive nuclear medicine imaging techniques. PET and SPECT images create 3D medical image of functional processes in the body, such as blood pressure or metabolism of glucose by means of tracers. In both methods, the detected emissions are processed and a 3D image of the region is acquired by consequent computer-based investigation.

3.2 *Preprocessing*

The selected brain images have to be preprocessed. The preprocessing method used here is intensity normalization. In the further steps, input image has to be compared with the healthy brain and other brain data sets to detect the presence of AD or PD. For that the input data have to be normalized in terms of intensity to make comparison between the images according to their voxel normalized intensity levels. To avoid the problems that can be caused due to the presence of voxels with maximum intensity level, the intensity normalization is processed. The peak intensity values could not process properly due to noise. Sometimes there is a chance of images to get badly normalized if the normalization process carried out based on wrong noisy voxels. In the proposed work, intensity normalization is carried out based on the mean value of voxels with the highest intensity values in used to avoid the possible normalization errors.

3.3 *Feature Extraction*

The Fisher Discriminant Ratio (FDR) method for feature extraction is described by its separation capability and this helps for the suitable extraction of features from the given set of images. For the two set value, FDR can be defined as follows:

$$\mathrm{FDR} = \frac{(\mu_1 - \mu_2)^2}{(\sigma_1)^2 + (\sigma_2)^2} \tag{1}$$

where μ_1, μ_2 and σ_1, σ_2 in Eq. (1) denote the two class set values of mean value and variance for each input variable, respectively. The cumulative scattering in each class decreases or the ratio value increases as the difference of the mean values of each two classes increases for a given set of variables as seen in Eq. (1). The voxels that satisfy a specific FDR threshold level get nominated as the most discriminative variables in the functional images selected for analysis. Together with the selection of voxels, the feature extraction using FDR helps in dimensionality reduction of the process by choosing the voxels within the FDR threshold alone. This results in selection of minimum number of variables in each observation and so the complexity of process gets decreased.

3.4 Automation Layer

The training and classification are done by Haar Training and AdaBoost Classifier. The proposed system improves the output by enhancing the existing Haar training method using the AdaBoost classifier.

3.4.1 Haar Training

The Haar trainer's is a weak learner because its detection quality is slightly better than random guessing [16]. In spite of this, the training technique is still preferred. Wavelet coding is suitable for the applications where tolerable degradation and scalability are important. Haar wavelet transform decomposes the input signals into a set of the basis function are called wavelets. It identifies the relevant feature results which helps in easier, faster and better understanding of images. Relevant information of input data can be predicted by feature extraction. Algorithms are used to isolate and detect the shapes and desired portions significantly. The quality of the process of feature extracting affects the classification process. Thresholding helps to obtain the normal or abnormal images. Most effective techniques are used to isolate the object by converting in binary image from gray level and image with high contrast levels.

3.4.2 AdaBoost Algorithm

The AdaBoost algorithm [17] is a well-known method to build ensembles of classifiers with very good performance. AdaBoost algorithm along with decision trees gives excellent performance can be considered as a hybrid model for classification. The AdaBoost algorithm, proposed in [17], learns a combination of the output, $H_m(x)$ to produce the final decision of classification given by,

$$H(x) = \text{sign} \sum_{m=1}^{M} \alpha_m H_m(x) \tag{2}$$

where M is the maximum number of classifiers, α_m is the computed weight factor of each classification in Eq. (2). The weight distribution of the training set patterns is reorganized between iterations according to the accuracy of classification of the previous classifiers. The weight of the misclassified patterns is increased for the next iteration, whereas the weight of the correctly classified patterns is decreased. The next classifier is trained with a re-weighted distribution. The amount of change on the weight of each pattern is proportional to the classification error of the pattern. The training set is always the same in each iteration and each input pattern is given a weight according to its misclassification by the previous classifiers. On subsequent iterations, the weight of the misclassified patterns by the ensemble is increased, whereas the weight of the correctly classified patterns decreases. This allows for the current classifier to be focused on the most difficult patterns, that is, the ones that were not well classified by the previous classifiers.

3.4.3 Region of Interest and Classification

The Region of Interest (ROI) has to be selected select to differentiate disease-affected areas. The two diseases AD and PD can be differentiated by the difference in the voxels present in the input brain image by comparing the healthy brain. Here, the machine learning techniques have to be applied for segregating the outer part of the brain with central part of the brain. If in the classification, the varying pixels are seen most in the outer region, then the patient is having Alzheimer's Disease. If the variation is seen most in the center part of the input image while classifying, then the patient is having Parkinson's Disease. If no variation and the voxels quality matches with the healthy brain in the dataset, then the patient is not having any dementia.

4 Conclusion

The proposed work aims in detecting the AD and PD among people. The existing works portray various methods to identify the DA and PD. The efficient Computer-Aided Diagnosis (CAD) for the early diagnosis of the AD and PD with Machine Learning Techniques (MLT) proposed in the paper will surely achieve more accuracy than any other existing system. The proposed work has special specifications like automated system that will improve the quality of the previous work.

References

1. Petrella JR, Coleman R, Doraiswamy P (2003) Neuroimaging and early diagnosis of Alzheimer disease: a look to the future. Radiology 226:315–336
2. Smith SL, Lones MA, Bedder M, Alty JE, Cosgrove J, Maguire RJ, Pownall ME, Ivanoiu D, Lyle C, Cording A, Elliott CJH (2015) Computational approaches for understanding the diagnosis and treatment of Parkinson's disease. IET Syst Biol 9(6):226–233
3. Tripoliti EE, Fotiadis DI, Manis G (2012) Automated diagnosis of diseases based on classification: dynamic determination of the number of trees in random forests algorithm. IEEE Trans Inf Technol Biomed 16(4):615–622
4. Araujo CPS, Báez PG (2012) GaNEn: a new gating neural ensemble for automatic assessment of the severity level of dementia using neuropsychological tests. Lecture notes in computer science, vol 6928, pp 264
5. Rabeh AB, Benzarti F, Amiri H (2016) Diagnosis of alzheimer diseases in early step using SVM (Support Vector Machine). In: 13th international conference computer graphics, imaging and visualization, pp 364–367
6. Anyaiwe OED, Wilson GD, Singh GB, Geddes TJ (2017) Weighted manhattan distance classifier, SELDI data for Alzheimer's disease diagnosis, evolutionary computation, pp 257–262
7. Iftikhar MA, Idris A (2016) An ensemble classification approach for automated diagnosis of Alzheimer's disease and mild cognitive impairment. In: 2016 international conference on open source systems and technologies (ICOSST), pp 78–83
8. Spedding AL, Di Fatta G, Saddyt JD and the Alzheimer's Disease Neuroimaging Initiative (2015) An LDA and probability-based classifier for the diagnosis of Alzheimer's disease from structural MRI. In: IEEE international conference on bioinformatics and biomedicine (BTBM), pp 1405–1411
9. Valli A, Dr. Wiselin Jiji G (2014) Parkinson's disease diagnosis using image processing techniques a survey. In: IJCSA, vol 4, no 6, pp 57–67
10. Peng Y, Fang M, Wang C, Xie J (2015) Entropy chain multi-label classifiers for traditional medicine diagnosing Parkinson's disease. In: 2015 IEEE international conference on bioinforrnatics and biomedicine (BIBM), pp 1722–1724
11. Al-Fatlawi AH, Jabardi MH, Ling SH (2016) Efficient diagnosis system for Parkinson's disease using deep belief network. IEEE pp 1324–1330
12. Wibawa MS, Nugroho HA, Setiawan NA (2015) Performance evaluation of combined feature selection and classification methods in diagnosing Parkinson disease based on voice feature. In: 2015 international conference on science in information technology (ICSITech), pp 126–131
13. Nancy Noella RS, Sreeja SS (2013) Computer aided diagnosis of Alzheimer's disease based on NMF-decision tree classifier. In: International conference on innovative trends in computing and technology
14. Bailey DL (2005) Positron emission tomography. Springer Verlag, New York
15. Mignotte M, Meunier J (2000) Three-dimensional blind deconvolution of SPECT images. IEEE Trans Biomed Eng 47(2):274–280
16. Yu C, Liu K-L, Meng W, Wu Z, Rishe N, Member IEEE (2002) A methodology to retrieve text documents from multiple database. In: IEEE transactions on knowledge and data engineering, vol 14, no 6, pp. 13–47
17. Warneke D, Kao O (2011) Exploiting dyanmic resource allocation for efficient parallel data processing in the cloud. IEEE Trans Parallel Distrib Syst 22(6):985–997

Design and Analysis of Dedicated Power Converter for Hybrid Electric Vehicles

Brij Mohan Prasad and P. R. Thakura

Abstract The interest has grown in hybrid electric vehicles owing to the necessities of decreasing the environmental pollution and fuel saving. Hybrid electric vehicle (HEV) consists of electric motor, power electronic converters, batteries and microcontrollers and internal combustion engine (ICE). A Voltage Source Inverter (VSI) has been designed in the power electronics laboratory dedicated to HEV. The experimentation was done using three phase balanced star connected load and a three-phase induction motor drive. The various waveforms were obtained and comparison was made between two different loads. Total Harmonic Distortion (THD) has been calculated and found to be in the range of 9.44%.

Keywords Hybrid electric vehicle · Power MOSFETs · Voltage source inverter Microcontroller · Opto isolator · Three phase load · Induction motor

1 Introduction

The awareness has increased in Hybrid electric vehicles due to the problems of environmental pollution, fuel consumption, greenhouse effect and depleting oil and gas reserves [1–3]. Vehicles using internal combustion engine (ICE) were present since a long time back. The disadvantages of an internal combustion engine vehicle were its large fuel consumption and also it was largely air polluting [4].

Electric vehicle (EV) is a vehicle which is energy efficient and non-polluting [5]. The electric vehicles (EVs) are divided into three categories: Battery electric vehicles (BEVs), Hybrid Electric vehicles (HEVs) and Fuel cell electric vehicles (FCEVs) [5, 6]. Hybrid means a combination of two species. The HEVs use an ICE and an

B. M. Prasad (✉) · P. R. Thakura
Department of Electrical and Electronics Engineering,
Birla Institute of Technology, Mesra, Ranchi 835215, Jharkhand, India
e-mail: brijprasad72@yahoo.com

P. R. Thakura
e-mail: prthakura@bitmesra.ac.in

V. Nath and J. K. Mandal (eds.), *Nanoelectronics, Circuits and Communication Systems*, Lecture Notes in Electrical Engineering 511,
https://doi.org/10.1007/978-981-13-0776-8_6

electric motor and remove the drawbacks of ICEVs and the BEVs. There are two energy storage systems in an HEV of which one should be electrical. The energy sources can be gasoline, batteries, ultracapacitors, etc. The merits of HEV are its long driving range and rapid refueling time. HEVs produce very less emission and are more fuel economic than an ICEV having the same driving range [7]. Also in an HEV the kinetic energy can be recaptured during braking [1, 4, 7]. The BEVs and the HEVs are commercially available. The FCEVs are under development and are not cost effective. The BEVs have become less popular and FCVs are yet to become popular. The HEVs have become popular and are growing [8].

The electric drive used is the underlying technology for a Hybrid Electric Vehicle. The electric drives that are feasible for HEVs are induction, switched reluctance and Permanent Magnet Brushless drives [9]. The power electronic modules in an HEV consist of rectifiers, inverters and dc/dc converters [10]. The objective is to attain high efficiency, lesser dimensions and reduce the cost of the systems which include the power converter and the electric machines and the other electronic components [11].

An inverter in an HEV generates three phase ac voltage from a chemical battery which is fed to the electric motor [12]. Thyristors were used as switching devices for a three-phase inverter but the use of thyristors has now become obsolete because of large commutation circuit. In the present time transistors are mostly used in inverter circuits. IGBT and MOSFET inverter circuits are now gaining popularity [12, 13].

This paper describes about the development of hardware of three phase VSI for an HEV. In section II the hardware scheme of the VSI is discussed. In Sect. 3 the simulation of three phase VSI for 180° conduction mode is done and the simulation results are obtained. Section 4 gives the description of the hardware experimental set up. Section 5 discusses the hardware results. The output line voltage and phase voltage waveforms are obtained. Section 6 gives the conclusion of the work.

2 Hardware Scheme of VSI

The hardware scheme of VSI comprises of:

A. Power circuit of the three phase VSI
B. Firing circuit for the MOSFETs
C. Arduino Uno micro controller
D. Power supply.

A. Power circuit of VSI:

A three—phase VSI is used to obtain adjustable voltage adjustable frequency ac supply from a dc voltage source. It comprises of six MOSFETs. Figure 1 gives the power circuit of a three phase VSI. MOSFET used is IRFP 460 which has voltage and current ratings of 500 V and 20 A respectively. A capacitor is connected across the dc source which keeps the input dc voltage stable and also removes the harmonics fed back to the dc source. The capacitor used has got a rating of 470 μF, 250 V. The advantages of using power MOSFETs are that it can be used for high frequency switching and there is no second breakdown effect in PMOSFETs as compared to a BJT.

B. Firing circuit for the MOSFETs:

TLP 250 is used for gate drive circuit of power MOSFET. TLP 250 is given firing pulses from the micro controller. TLP 250 consists of an LED and a photo transistor and transfers electrical signals by the use of light. It is an optical isolator that isolates the MOSFETs and the microcontroller. It prevents high voltages from affecting the system receiving the signal. TLP pin diagram is shown in Fig. 2. The output of the TLP is applied to the gate (G)—source (S) terminals of the MOSFET as shown in Fig. 2.

C. Arduino Uno Microcontroller:

The Arduino Uno is a microcontroller board based on the Atmega 328. Six firing pulses are given from the microcontroller to the six TLPs and then the TLP outputs

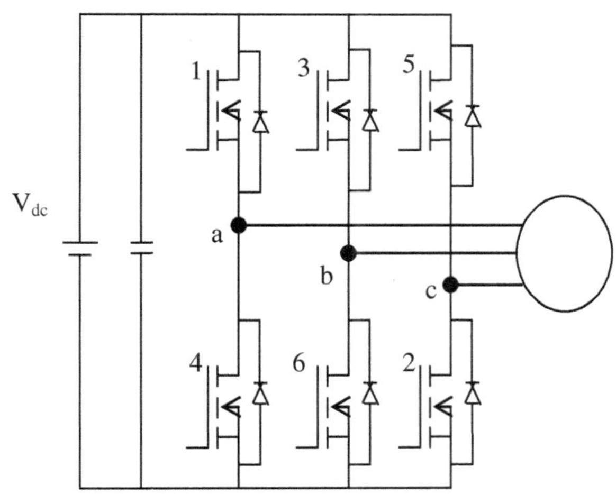

Fig. 1 Power circuit of three phase VSI

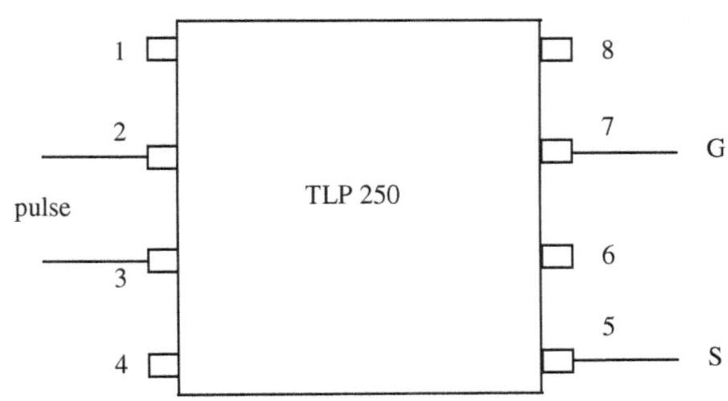

Fig. 2 Pin diagram of TLP 250 opto coupler

are fed to six IRFP 460 MOSFETs. For one cycle of 360°, each step is of 60° which means that the firing pulses from the microcontroller are such that the MOSFETs are fired at an interval of 60° according to their number sequence.

Programming of the microcontroller is done. A computer program is written for the microcontroller for the operation of the VSI in 180° conduction mode. Six pins are used as PWM outputs. The stepped output wave has a frequency of 50 Hz. Each firing pulse has a duration of 1/50 = 20 ms with 10 ms ON and 10 ms OFF period which corresponds to 180° conduction mode of each MOSFET and then an OFF period of 180° of this MOSFET. To prevent any two MOSFETs in the same leg from conducting simultaneously a time delay is introduced between one MOSFET turns OFF and the other MOSFET turns ON. The time delay used is 3.33 ms.

The working voltage is 5 V dc and suggested input voltage is 7–12 V. The amplitude of the pulses from the microcontroller is 5 V. When these pulses are given to the TLP 250 s their amplitude changes to 12 V. Then these TLP outputs are fed to six IRFP 460 MOSFETs.

D. Power supply:

VSI requires dc supply which is given from a 0–200 V dc power supply. When a lamp load is used the supply is raised up to 100 V. When an induction motor is used the supply is raised up to 200 V.

For each of the TLP 250 a dc source of 12 V is required. The circuit for this is as shown in Fig. 3. A 13.76 V ac voltage is given to each of the diode bridge rectifier D35SBA80 through a six winding 230/13.76 V transformer and the output of each of the six bridge rectifiers is 13.76 V dc. A capacitor is connected in parallel with the bridge rectifier and the voltage at its output side is 14.76 V. This voltage is given to the input terminals of the LM 7812 voltage regulator which generates an output voltage of 12 V at its output terminals.

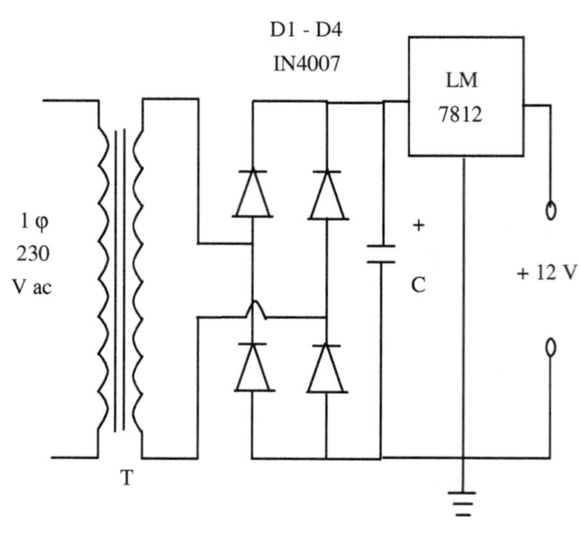

Fig. 3 Power supply of TLP 250

3 Psim Based Simulation and Simulation Results

Figure 4a gives the simulation diagram of three phase VSI for 180° conduction mode and Fig. 4b gives the results of simulation. The result of simulation going downwards is of input dc voltage V_s, gate pulses of MOSFET 1 & MOSFET 4 in the first leg of the three phase VSI, load line voltage, phase voltage of load and phase current. The simulation result shows that when MOSFET 1 is ON, MOSFET 4 is OFF and vice versa. Similar conduction pattern occurs in the second and third leg. The applied input dc voltage is $V_s = 400$ V and the line voltage has a peak value $V_L = V_s = 400$ V and is present for 120° duration in each half cycle. The load phase voltage has a peak value $V_{ph} = 2\ V_s/3 = 266.67$ V. Each load resistance has a value of 100 Ω and hence load phase current $I_{ph} = 266.67/100 = 2.67$ A for a resistive load.

4 Hardware Experimental Set up of Vsi

The components required for the development of the hardware were regulated dc power supply (0–200) V, capacitor of rating 470 μF, 250 V, six nos. of IRFP 460 power MOSFETs, Arduino Uno microcontroller, six nos. of TLP 250 opto couplers and digital storage oscilloscope. The components required for making 12 V dc supply for each TLP were six winding 230/13.76 V transformer, six nos. of bridge rectifier D35SBA80, six nos. of capacitor of rating 470 μF, 25 V and six nos. of LM 7812 voltage regulator. The output waveforms of microcontroller firing pulses, phase voltage of load and line voltage of load are obtained. The hardware set up is given in Fig. 5.

5 Experimental Results

A. Firing pulses of the Arduino Uno microcontroller

The firing pulses for MOSFETs 1, 2 and 3 are shown in Fig. 6a and for MOSFETs 4, 5 and 6 are shown in Fig. 6b. MOSFETs are fired at an interval of 60° according to their number sequence. Each MOSFET conducts for 180°. The adjacent MOSFETs in the upper group of MOSFETs conduct with a phase difference of 120°. Similarly, the adjacent MOSFETs in the lower group of MOSFETs conduct with a phase difference of 120°.

B. Experimental results with lamp load

Each phase of the star connected lamp load had three bulbs in parallel. Rating of each bulb was 200 W, 250 V. The voltage was increased and the values of line voltage and phase voltage were noted down for different values of input dc voltage.

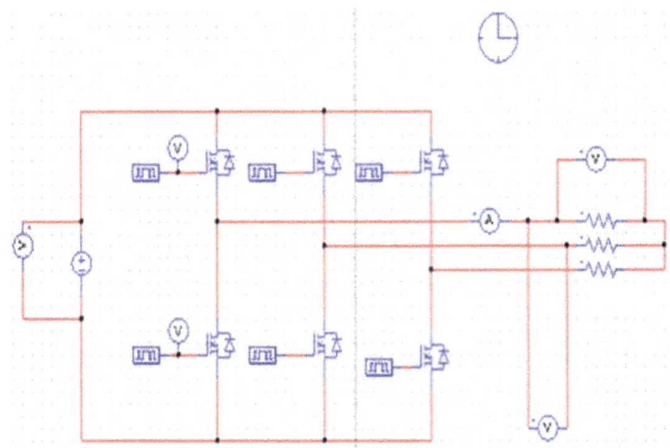

(a): Simulation circuit of 180° conduction mode of three phase VSI

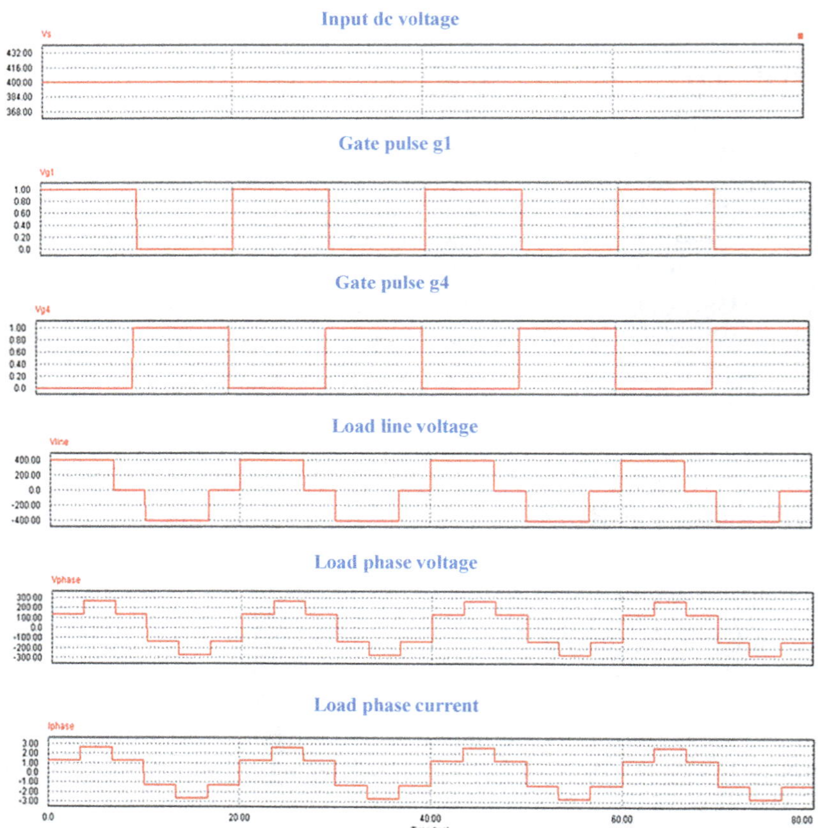

(b): Simulation results for 180° conduction mode of VSI

Fig. 4 **a** Simulation circuit of 180° conduction mode of three phase VSI, **b** simulation results for 180° conduction mode of VSI

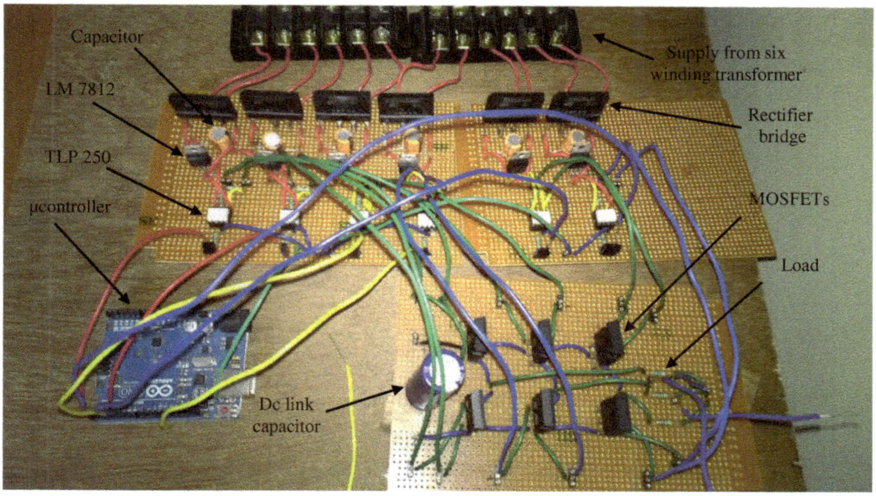

Fig. 5 Hardware for three phase VSI

The phase voltages are shown in Fig. 7a. Each cycle of output phase voltage has six steps and each step is of duration 60° which corresponds to a duration of 3.33 ms. The phase voltages of phase a, b and c are out of phase by 120°, with phase b lagging phase a and phase c lagging phase b. The load line voltage waveform is shown in Fig. 7b. For each half cycle line voltage is present for 120° (6.6 ms) duration and then there is a gap of 60° (3.33 ms) for which line voltage is zero. The line voltages differ from each other in phase by 120°, with phase b voltage lagging phase a and phase c voltage lagging phase b voltage by 120°.

The line to line voltages can be expressed by Fourier series as:

$$V_{\mathrm{ab}} = \sum_{n=1,3,5,}^{\infty} \frac{4V_{\mathrm{s}}}{n\pi} cos \frac{n\pi}{6} \sin n\left(\omega t + \frac{\pi}{6}\right)$$

$$V_{\mathrm{bc}} = \sum_{n=1,3,5,}^{\infty} \frac{4V_{\mathrm{s}}}{n\pi} cos \frac{n\pi}{6} \sin n\left(\omega t - \frac{\pi}{2}\right)$$

$$V_{\mathrm{ca}} = \sum_{n=1,3,5,}^{\infty} \frac{4V_{\mathrm{s}}}{n\pi} cos \frac{n\pi}{6} \sin n\left(\omega t + \frac{5\pi}{6}\right)$$

The rms value of the nth component of line voltage is

$$V_{\mathrm{L}n} = \frac{4V_{\mathrm{s}}}{\sqrt{2}n\pi} \cos \frac{n\pi}{6}$$

n = 1, gives fundamental line voltage
$V_{\mathrm{L1}} = \frac{4V_{\mathrm{s}} \cos 30^{\circ}}{\sqrt{2}\pi} = 0.7797\, V_{\mathrm{s}}$

Fig. 6 **a** Triggering pulse of MOSFET 1, 2 and 3. **b** Triggering pulse of MOSFET 4, 5 and 6

(a)

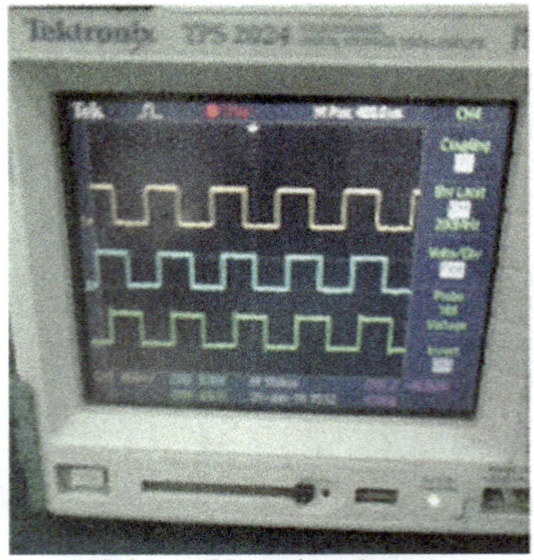

Voltage (V) vs time (msec)

(b)

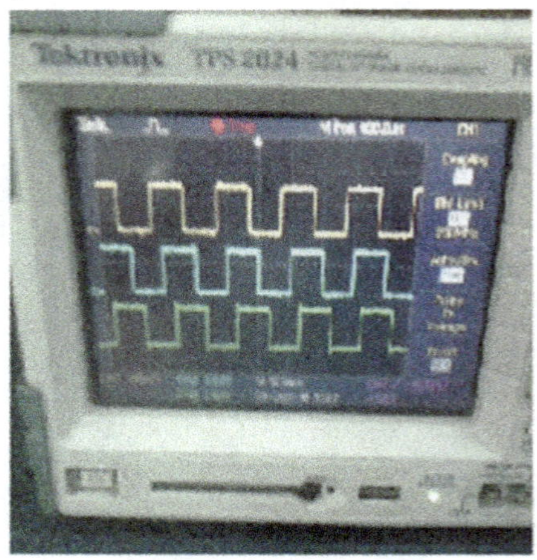

Voltage (V) vs time (msec)

Total harmonic distortion (THD) = $\frac{V_h}{V_{L1}} = \frac{\left(V_L^2 - V_{L1}^2\right)^{1/2}}{V_{L1}}$

V_h = sum of rms value of harmonics

The table given below shows the value of line voltage, phase voltage and THD for different values of input voltage and current (Table 1).

C. Experimental results with three phase induction motor

Rating of induction motor was 3 phase, Y connected, 415 V, 50 Hz, 1415 rpm, 1.8 A and 0.8 p.f. The input dc voltage was increased and the values of input voltage and input current were noted down. The induction motor started rotating at a voltage of 65.3 V and input current of 0.417 A. With the increase in voltage the

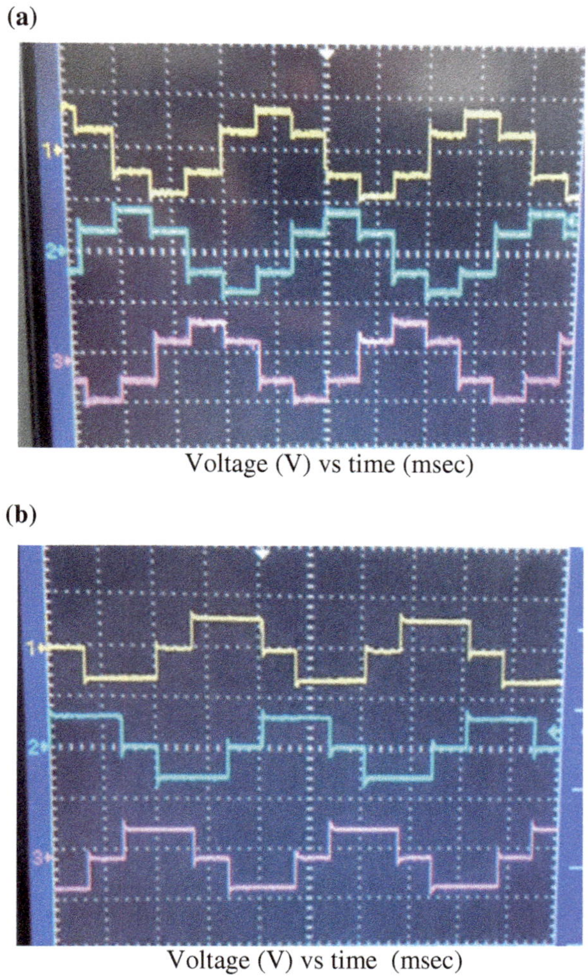

Fig. 7 **a** Phase voltages of phases a, b and c. **b** Load line voltages V_{ab}, V_{bc} and V_{ca}

Table 1 Experimental voltage and corresponding current

V_{dc}	I	V_{ph}	V_L	V_{L1}	THD
21.6	0.757	9.42	16.92	16.84	8.41
40	0.952	17.97	31.31	31.19	8.74
60	1.118	27.34	46.97	46.78	8.98
80.8	1.279	37.17	63.27	63	9.22
100.1	1.415	46.2	78.4	78.05	9.44

Table 2 Experimental voltage and current

V_{dc}	I
65.3	0.417
102	0.136
153	0.123
193	0.128

Table 3 Comparison between conventional converter and dedicated converter

	Conventional converter	Dedicated converter
a.	Thyristors are being used which have high voltage and high current ratings but the switching frequency is low	IGBTs and power MOSFETs are being used which have medium voltage and current ratings. Switching frequency is high
b.	Commutation circuit is needed which makes the converter bulky as well as it takes more space	Converter is less bulky and the space required is less
c.	Sinusoidal PWM is difficult with high power SCRs	Sinusoidal PWM can be done using modern high power semiconductor switches
d.	In previous work Total Harmonic Distortion is not taken care of	Total Harmonic Distortion (THD) is the figure of merit for a voltage source inverter. It must be below 10% and in this work it is of the range 9.44%

current was found to decrease. The table below gives the values of input current with input dc voltage (Tables 2 and 3).

6 Conclusion

Owing to the regulations on pollution and the consumption of fuel, hybrid electric vehicle will continue to gain interest from the people. This paper illustrated the hardware scheme for a three phase VSI. A three phase VSI is a significant part of a hybrid electric vehicle. The three phase VSI was developed using power MOSFET, microcontroller and opto isolator and was tested using three phase load and it is found that the phase voltage is stepped wave whose fundamental component will be sinusoidal. The analysis of waveform has been done and it is found that total harmonic distortion is in the range of 9.44% which is low and optimum value for a VSI. The output voltage waveform at the load terminals was found to be as expected theoretically. The voltage source inverter which is developed will be suitable for functioning of a hybrid electric vehicle.

References

1. Chan CC (2007) The state of the art of electric, hybrid and fuel cell vehicles. Proc IEEE 95 (4):704–718
2. Emadi A et al (2005) Topological overview of hybrid electric and fuel cell vehicular power system architectures and configurations. IEEE Trans Veh Technol 54(3), May 2005
3. Agarwal V, Dev M (2013) Introduction to hybrid electric vehicles: state of art. In: IEEE student's conference on engineering and systems, April 12–14
4. Ehsani M, Gao Y, Gay SE, Emadi A (2005) Modern electric, hybrid electric, and fuel cell vehicles: fundamental, theory and design. CRC Press, Florida
5. Chan CC (2002) The state of the art of electric and hybrid vehicles. Proc IEEE 90(2):247–275
6. Ehsani M, Rahman KM, Toliyat HA (1997) Propulsion system design of electric and hybrid vehicles. IEEE Trans Ind Electron 44(1):19–27
7. Chan CC, Chau KT (2008) Modern electric vehicle technology. Oxford University Press, New York
8. Umamaheshwari S, Thakura PR, Keshri RK (2011) Hardware development of voltage source inverter for hybrid electric vehicle. In: 2011, 1st International conference on electrical energy systems
9. Chau KT, Chan CC, Liu Chunhua (2008) Overview of permanent—magnet brushless drives for electric and hybrid electric vehicles. IEEE Trans Ind Electron 55(6):2246–2257
10. Bellur DM, Kazimierczuk MK (2007) Dc–dc converters for electric vehicle applications. In: Procedings of IEEE. ISBN 978-1-4244-0446-9
11. Emadi A et al (2008) Power electronics and motor drives in electric, hybrid electric and plug —in hybrid electric vehicles. IEEE Trans Ind Electron 55(6):2237–2245
12. Bimbhra PS (2015) Power electronics, 5th edn. Khanna Publishers, Delhi
13. Bose BK (2008) Modern power electronics and A. C. drives. Pearson Education, New Delhi

An Underground Mine Safety of Personnel's Using IoT

Keshav Sinha and Partha Paul

Abstract According to a survey of the year 2011, in India there are 600 coal mines are running and over one million peoples are working in those mines. In underground mines, there are various cases where people die due to the leakage of toxic gasses. Nowadays, the accident rate in underground coal mines will decrease with the help of sensor technology and quick gas testing results. There are various traditional methods for gas detection and testing such as methane lamp, carrying live canaries etc. The most common sensors for gas detection are electrochemical gas sensors, semiconductor sensors etc. But the reliability of this method is not accurate and impressive. In this survey paper, our prime focus on the safety of personnel by using IoT based on ZigBee techniques for monitoring the peoples in the underground mines and provide the immediate assistance and repositioning of personnel in an emergency situation.

Keywords ZigBee · Wireless network · Sensor · Gases · Pythagoras

1 Introduction

Many accidents occur in underground mines every year. The accident may cause by leakage of toxic gases [1], in the underground mines such as carbon monoxide (CO), methane (CH_4), carbon dioxide (CO_2) and inadequate quantity of oxygen. In India, the death rate of per 1000 people employed in CIL (Coal India Limited) is 0.27% in the year 1990 and it will decrease to 0.16% in the year 2014. So, for

K. Sinha (✉)
CSIR-CIMFR, Dhanbad, Jharkhand, India
e-mail: keshav.sinha@yandex.com

P. Paul
Department of Computer Science Engineering, Birla Institute of Technology, Mesra, India
e-mail: p_india@rediffmail.com

V. Nath and J. K. Mandal (eds.), *Nanoelectronics, Circuits and Communication Systems*, Lecture Notes in Electrical Engineering 511,
https://doi.org/10.1007/978-981-13-0776-8_7

balanced growth of the coal industry, it is necessary to develop the technologies to make mine accident free. There is certain condition according to which we monitor the level of toxic gases.

(i) Regular monitoring of gases.
(ii) Trending information about the atmosphere.
(iii) Give warning against the explosive and toxic atmosphere.

Under this criteria, no personnel's will enter into any place where there are poor ventilation and circulation of air. It always ensures the air inside the mines are breathable and free from toxins.

A. Hazardous Area in Underground Mines

In underground mines, the area is said to be hazardous [2] when the air of that area is a mixture of fire, toxic air, dust, and gas explosions. There are two types of gasses are present in mines (i) Combustible Gases and (ii) Toxic Gases.

1. Combustible Gases

According to the combustion, the gases present in underground mines are defined as (i) Type of gas, (ii) Temperature of gas, and (iii) Likelihood of gas. These three properties are present in gas for the flammability. At room temperature, there is a limit set for the explosion of the gases. They are of two types (i) LEL, and (ii) UEL.

(i) Lower Explosive Limit (LEL): At room temperature least amount of gas mixed with the air. And when it comes under the fire it will be exploded, but the range of explosion in very lesser amount.
(ii) Upper Explosive Limit (UEL): In this, a maximum amount of gas is mixed with the air that causes a rich explosion when it contacts with the fire.

2. Toxic Gases

There are several gases present in underground mines which cause harm to the personnel's long period of time or higher concentration will harm in a shorter period of time. Many countries provide the Threshold Limit Values (TLV) for this type poisonous gases. There are three types of threshold values (i) TWA, (ii) STEL, and (iii) CL. For better safety and health of the personnel's, we set this three threshold values.

(i) Time Weighted Average (TWA): In this, all the workers are uncovered for hours without knowing the effect of air.
(ii) Short-term Exposure Limit (STEL): In this, the personnel's is exposed for a few minutes in the air. And it should be done for every time.
(iii) Ceiling Limit (CL): In this, the workers are not uncovered for single seconds. Because due to the toxic behavior of the substance, its effects immediately.

B. Gas Detection Sensors

There is a various technique for the detection of the gases [3–5] in the mines. Here we discuss a few of them such as for Oxygen they use the flame lamp, and paramagnetic. For detection of Methane, they use catalytic oxidation, thermal conductivity, optical, acoustic, and flame lamp. For Carbon dioxide detection they use the optical and infrared sensor. For detection of Carbon monoxide, they use the Electrochemical, catalytic oxidation, semiconductor, and infrared. For detection of Sulfur dioxide, Nitric oxide, Nitrous oxide, Nitrogen dioxide and Hydrogen sulfide they use the electrochemical sensor. For hydrogen detection, they use the catalytic oxidation. This is some of the traditional approaches for gas detection in underground mines. But these techniques are done by manually and it will degrade the efficiency. In another hand IoT, based sensor technique is very fast and reliable.

C. ZigBee Technology

The combination of a wireless network with gas detection sensor will achieve a better result in future. For a selection of IoT [6, 7] technique we commonly used the short distance wireless networks. There are various technologies are present in the short distance, such as Bluetooth, RFID, ZigBee, Wi-Fi, IrDA, etc. The general difference between this techniques is while using Wi-Fi he cost is higher than other short distance technique. The distance of ZigBee is up to 10–75 m which is much higher the Bluetooth and IrDA but less than Wi-Fi. Due to this same reason ZigBee is a better technique for underground mining than other technique. The characteristics of ZigBee technique are:

(i) Low Cost,
(ii) Less power consumption,
(iii) Huge data capacity.

1. ZigBee Protocols

The technology consists of a layer of protocols was data is transferred from one layer to the other. ZigBee protocol consists of four layers.

(i). Physical Layer,
(ii). MAC Layer,
(iii). Network Layer,
(iv). Application Layer.

In Fig. 1, illustrate the ZigBee protocol infrastructure. According to the infrastructure the physical layer present bottom of the stack. The work of the physical layer is to collect the spread spectrum communication data. According to the IEEE802.15.4 physical layer obeying the rules of protocol for the selection of the different channels, change of state of channels, transmitting and receiving, and collection of information of channels.

One step up of physical layer there is a MAC layer the responsibility of MAC layer is to access the channel, generate the coordinates, support PAN network,

Fig. 1 ZigBee protocol stack

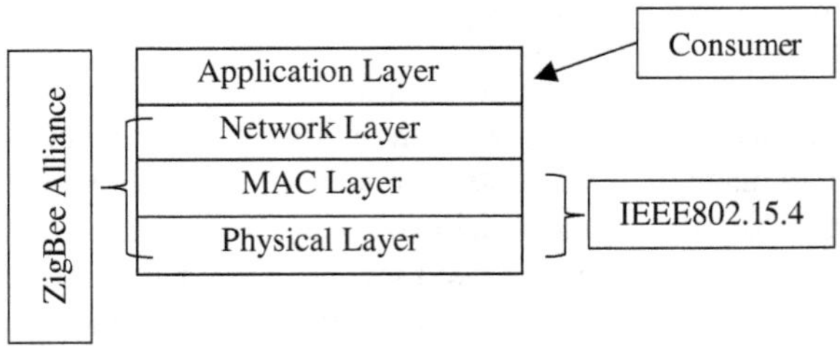

provide security in the communication, and provide support to connect different devices MAC. The data of MAC Layer are sent to the Network Layer. According to the MAC layer data network layer is established. The responsibility of the network layer is to create a new network, maintain the details of leaving and connected nodes, assign an address to the node, etc. these are the functionality of the network layer. The working of the application layer is to provide the services to the personnel. The application layer is a collection of APS, ZDO, and object of the application.

2. ZigBee Topology

ZigBee topology consists of shared type structure. The most suitable structure for ZigBee network is a star, Mesh, and cluster structure. The ZigBee device hardware consists of two types: (i) Full-Function Device (FFD) and (ii) Reduced Function Device (RFD). The working of FFD is in communication devices of the ZigBee whereas RFD is worked only on FFD network. Construction of ZigBee consists of both FFD and RFD. Where FFD acts as a coordinator, router and the nodes of the network. While RFD is acting as a node of the network. Figure 2, represents the different ZigBee network topology.

This is the some of the overviews of the underground mines techniques and the network topologies. While using this method will give the functionality to perform work in more precision and efficient way without losing lives of the personnel's. The organization of this paper is as follows: Sect. 2 describes the literature review followed by an Introduction in Sect. 1. Section 3 deals with the case study. In Sects. 4 and 5 performs the Research methodology and Discussion. Finally, in Sect. 6 conclusion is drawn.

Fig. 2 ZigBee network topology

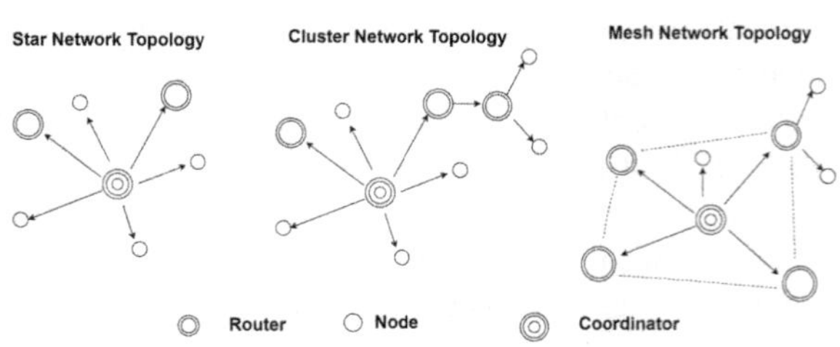

2 Literature Reviews

Different researchers proposed a various technique for the wireless sensor network. Some of them are very reliable for the underground mines. According to those papers, a survey literature is drawn for the better understanding of the problems faced in the underground mines. In Table 1, represent the various techniques presented by different researchers for solving the problem in underground mines.

3 Case Study

Under CSIR-CIMFR, I get the privilege to come in Kajora Area, ECL. According to the gas testing results of KKC (Khas Kajora Colliery), a survey is prepared for the safety of personnel's. In mines there are numerous toxic gases are present in the underground mines. And for the safety and precautionary of these gasses, it has to be tested on the daily basis routine as per the DGMS rules. Here in Kajora Area,

Table 1 Literature survey of different research papers

Research	Technique used	Methodology	Advantage	Disadvantage
Nie et al. [8]	ZigBee technique	Repositioning of personnel using the wireless network in an emergency situation	Very effective and less time-consuming	Speed of signal transmission is very slow
Wameke et al. [9]	Smart dust	Cubic-millimeter mote integrated to massively distributed sensor networks	Less intrusively, low cost, low energy consumption	Works for small computing device
Gupta et al. [10]	IoT and big data	Using IoT technique for optimization and big data for mining of data for better results	Boost the profit of production	No use for underground mining
Zhang et al. [11]	IoT	Monitoring and tracking the health of the workers	complete mechanization level was raised up	Not very efficient manner.
Wang et al. [12]	IoT and RFID	To get the coal mines staff attendance and real-time location	low cost, stable, economical and reliable	But works only for 3 meters.
Vermesan and Friess [13]	IoT and M2M applications	Using microelectronics and micromechanical sensing device for monitoring of temperature	Reliable and provide change in production industry	Works for automobiles and not tested for the mines
Lan [14]	Real time monitoring using IoT	IoT provides the sharing of information through terminal	This provides the accurate and greatly improved working skills	More time consuming

two chemists are available for the gasses testing **Mr. Sangram Panda**, Chemist and **Mr. Asim Koner**, Asst. Chemist which was examined and tested the gas samples. In Kajora lab, the testing is done by using Von Helden process and Graham & Lawrence methods for five major toxic gasses which will harm the personnel. The gases are CO_2, O_2, CH_4, CO, and N_2.

Test 1: Testing of CO_2, and O_2

Helden process is used for gas examination. In which we first use the KOH as a base which will absorb all CO_2 from the sample and this process is done by three times. At the second time, we use Potassium pyrogallate (C6H5KO3) which absorb all O_2 from the gas sample this also done by three times. Finally, we take a reading and subtract with the original sample.

Test 2: Testing of Methane (CH_4)

The presence of methane is tested by Helden process where we burn the gas sample in the presence of oxygen and it releases $CO_2 + H_2O$ and CO_2 is absorbed by KOH. This process consists of Hg, Platinum Wire, and KOH, which is used for the testing.

Test 3: Testing of Carbon Monoxide (CO)

Testing of CO is done by Graham & Lawrence apparatus. It consists of several phases. The first gas sample is passed through KOH and at 130°C it passes through Iodine Pentoxide (I_2O_5). The CO reacts with I_2O_5 and it liberates I_2. And I_2 is passed through KI (Potassium Iodide). And it forms KI_3 (Potassium Tri-iodide). The color of Potassium Tri-iodide + starch turns to blue. Then (Potassium Tri-iodide + starch) is titrated with Sodium thiosulfate $Na_2S_2O_3$ to determine the percentage of CO.

Normally, the gas testing is done on a daily basis by which we can monitor the level gases in the mines. In Fig. 3, shows that the normal condition, results of KKC colliery means by which we find that the percentage of different gases.

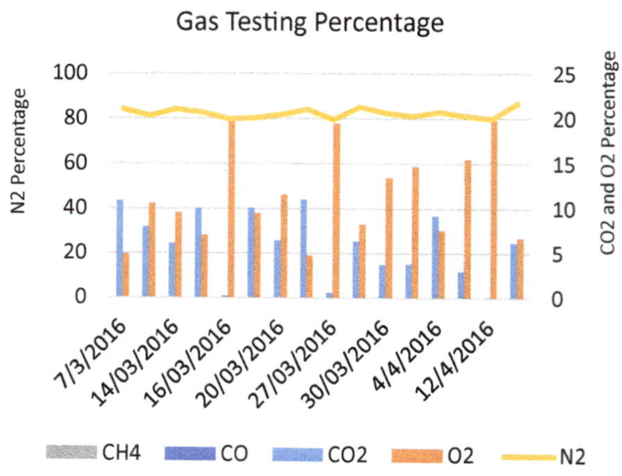

Fig. 3 Gas testing results for normal days

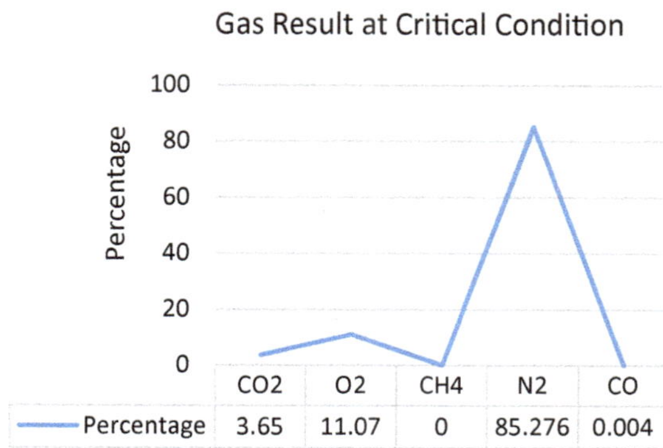

Fig. 4 Critical condition at KKC colliery

On 11th April 2016, firing occurs in the colliery. In Fig. 4, shows that the level of different gases in KKC colliery. At the fire, the level of O_2 is increased. The high level of oxygen cause blasting in the mines.

Here, in the graph shows that the 0.004% of carbon monoxide which will take the personnel's life in few seconds.

4 Research Methodology

After an exhaustive study of different techniques and case study, we would like to propose an IoT based framework for the safety of the personnel's which is shown in Fig. 5.

The step by step procedure for the proposed safety of personnel's [15] strategy is discussed as follows.

Step 1: In the initialization phase, the network is created in mines by using the routers, terminals, and a coordinator. The responsibility of the coordinator is to start the network and send the pulse signal to the router. All the router is joined to the wireless network and process the received signal. The functionality of a router is to find the network, the selection of the router, data transferring, sub-device joining.
Step 2: In ZigBee, the data are collected, starting node sets up multi-hop network construction begins. Every time the router send the stress signal request it will be granted and the message is sent to the parent node. The terminal node is used for extraction of physiological information. The information is converted into data and it is sent to the parent node.
Step 3: All the affected channels of physical layer are scanned. The Array list is prepared of all, the complaint channels. Select the idle one and assign the new ID.
Step 4: The Address node is assigned and then send the message to the rescue center.
Step 5: All the message is collected in the local processing unit where the decision is made. In processing unit, the positioning algorithm is applied to the personnel.

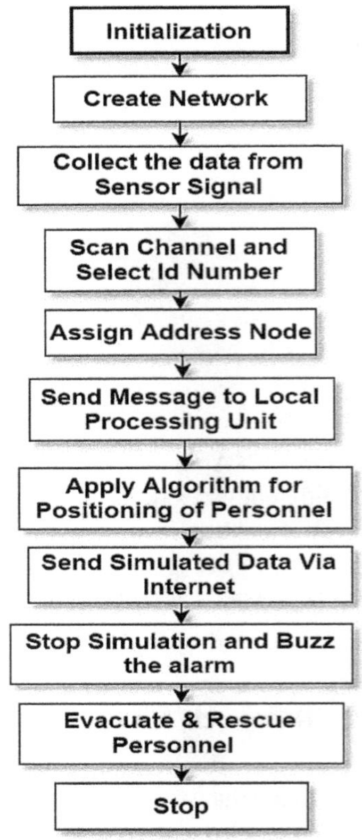

Fig. 5 Solution strategy for the safety of personnel's

Here, we proposed the RSSI (Received signal strength indication) algorithm for measuring the power of the wireless signal and provide the immediate support to the far end personnel. The idea behind for measuring the distance of the incoming signal is given by Eq. (1).

$$P_R = P_T \times G_T \times G_R \times \left(\frac{\lambda}{4\pi}\right)^2 \times \left(\frac{1}{d}\right)^n \tag{1}$$

Whereas,

P_R Receiving Device
P_T Transmitting Device
G_T Gains of Transmitter Antennas
G_R Gains of Receiver Antennas
λ Wavelength of the signal
d Distance between transmitter and receiver.

In a wireless network, the most popular technique for positioning [16, 17] is Multilateration [5] which serves as a primitive block. For a closer look at the concept behind it. Let us consider a point 'A' having the known position coordinate (X_i, Y_i) where (i = 1, 2, 3). And there is a node having an unknown position (X_u, Y_u). And the distance between two nodes is d_i (where i = 1, 2, 3) is calculated by using Pythagoras theorem for the set of three equations is given by Eq. (2).

$$(X_i - X_u)^2 + (Y_i - Y_u)^2 = d_i^2 \tag{2}$$

Whereas, i = 1, 2, 3

For solving the Eq. 2 in linear equation of X_u and Y_u form is shown as Eq. (3).

$$2 \times \begin{bmatrix} X_3- & X_1 & Y_3- & Y_1 \\ X_3- & X_2 & Y_3- & Y_2 \end{bmatrix} \times \begin{bmatrix} X_u \\ \hline Y_u \end{bmatrix} = \begin{bmatrix} (d_1^2 - d_3^2)- & (X_1^2 - X_3^2)- & (Y_1^2 - Y_3^2) \\ (d_2^2 - d_3^3)- & (X_2^2 - X_3^2)- & (Y_2^2 - Y_3^2) \end{bmatrix} \tag{3}$$

The matrix consists for known constraints. If the problem is redundant and the measurement is distant than it is determined by the Eq. (4).

$$\begin{aligned} & 2 \times \begin{pmatrix} X_n- & X_1 & \cdots & Y_n- & Y_1 \\ & \vdots & \ddots & \vdots & \\ X_n- & X_{n-1} & \cdots & Y_n- & Y_{n-1} \end{pmatrix} \times \begin{bmatrix} X_u \\ \hline Y_u \end{bmatrix} \\ & = \begin{bmatrix} (d_1^2 - d_n^2)- & (X_1^2 - X_n^2)- & (Y_1^2 - Y_n^2) \\ & \vdots & \\ (d_{n-1}^2 - d_n^3)- & (X_{n-1}^2 - X_n^2)- & (Y_{n-1}^2 - Y_n^2) \end{bmatrix} \end{aligned} \tag{4}$$

In underground mines, there is always a chance for the signal interference. So for measuring the distance, we used the four-side RSSI based positioning methods is shown in Fig. 6. To improve the reliability of the algorithm we adopt the averaging

Fig. 6 Four sided measurement

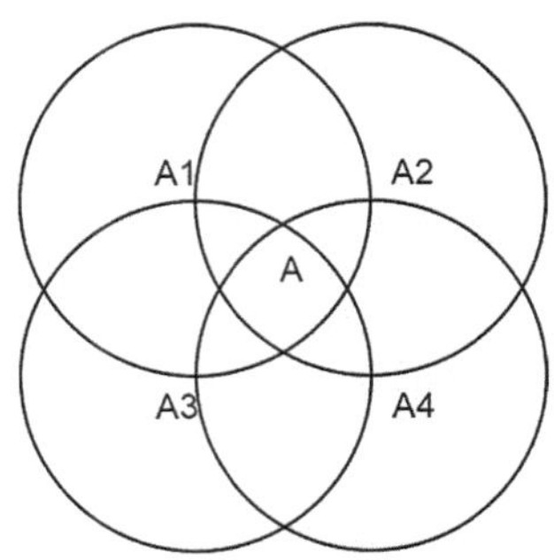

concept. Where we first take the ten measurement data and then the average of the result is taken. For moving forward, we always choose the immediate next of the secured node.

5 Discussion

In order to create a ZigBee network, there is a sensor kit belonging to the Silicon Laboratories, which is consist of 2.4 GHz, 802.15.4 development kit, IC C8051F121 microcontroller, and Chipcon of CC2420 2.4 GHz 802.15.4 for the transceiver. Which supports the USB interface, and JTAG programming. There are push buttons present for the LED and voltage regulator. In underground mines, the piezoelectric pulse sensor is used to monitor the blood pressure and the body temperature of the personnel. The working of the pulse sensor is to convert the pressure signal into the electrical signals. This signal is sent to the router node and by using coordinator node, it is sent to the rescue center. When the accident occurs in mine EMF will change. As a result [8], the noise interference occurred in the processing of the signal. If the pulse suddenly exceeds 200 then it cannot be measured, but after some time when the pulse is stabilized then we take the average value. In Fig. 7, represent the statistical pulse, the date for the personnel's where every 3 s pulse is recorded and the average is multiplied by 12, show that it converted in the 60 s.

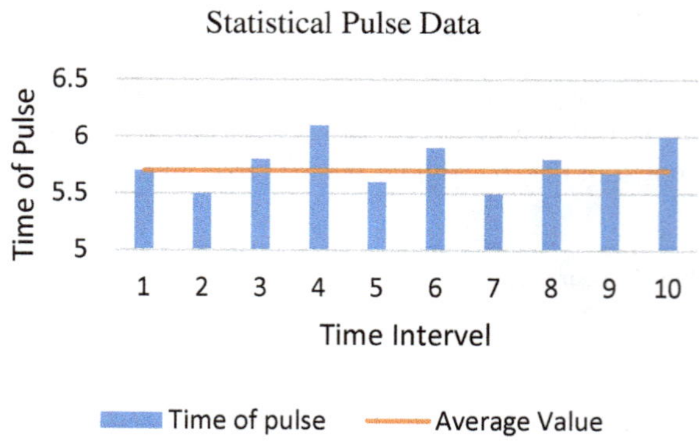

Fig. 7 Personnel's pulse data for every 3 s

6 Conclusion

Sensors help us to maintain the environment safe and healthy in underground mines. By taking continuous monitoring of toxic air and taking appropriate control measures. There are several sensors are present for gas monitoring, but none of them are reliable. But with the combination of the sensor with the wireless technology, we can achieve more stability and reliability in the data transmissions. Using four sides positioning algorithm we can cover almost all the areas of the underground mines and it enhances the positioning accuracy of the personnel. By using portable testing devices we send the data of body pulse, body temperature, and blood pressure to the emergency rescue center via a wireless network. By using IoT based wireless network increases the efficiency and increase the time to live rate during an emergency situation.

References

1. Raheem SR (2011) Remote monitoring of safe and risky regions of toxic gases in underground mines: a preventive safety measure. In: Postgraduate diploma thesis report, African Institute for Mathematical Sciences (AIMS), South Africa
2. Kumar A, Kingson TMG, Verma RP, Kumar A, Mandal R, Dutta S, Chaulya SK, Prasad GM (2013) Application of gas monitoring sensors in underground coal mines and hazardous areas. Int J Comput Technol Electron Eng (IJCTEE) 3(3)
3. Warburton PR (2000) Methods of Identifying the gas. US patent no. 6165347
4. Gupta SD, Kundu S, Mallik A (2012) A MEMS based carbon nanotube–field effect transistor as a gas sensor. Int J Recent Technol Eng 1:38–42
5. Huang J, Wan Q (2009) Gas sensors based on semiconducting metal oxide one-dimensional nanostructures. Sensors 9:9903–9924
6. Zhang F, Gu W (2010) Application of internet of thing in logistics information in coal mine. China Min Mag 101–104
7. Hang S, Ding EJ, Xu Z, Hua G (2010) Part two of lecture of internet of things and sensor mine-sensor mine. Digit Mine Integr Autom Mine Indus Mines Autom 19:129–132
8. Nie BS, Chen WX, Wang LK, Zhang RM, Wang C (2011) Internet of things-based positioning of coalmine personnel and monitoring of emergency state. In: International conference on digital manufacturing and automation, IEEE
9. Wameke B, Liebowitz B, Pister KSJ (2001) Smart dust: communicating with a cubic-millimeter computer. IEEE Comput Mag 34(1):44–51
10. Sharma A, Sharma A, Gupta S (2017) Int J Electron Electr Comput Syst 6(2), Feb 2017
11. Zhang Y, Fu G, Zhao Z, Huang Z, Li H, Yang J (2012) Discussion on application of IOT technology in coal mine safety supervision. Procedia Eng 43:233–237
12. Wang L, Nie B, Zhang R, Zhai S, Li H (2011) Zigbee-Based positioning system for coal miners. Procedia Eng, 26 2406–2414
13. Vermesan O, Friess P (2013) Internet of things: converging technologies for smart environments and integrated ecosystems
14. Lan RB (2014) Research on real time acquisition and control technology of mine excavation surface temperature based on the technology of the internet of things. Appl Mech Mater 686:272–276

15. Sun JP (2009) Networking of underground mine personnel position supervising and measuring system. Coal Sci Technol 37:77–79
16. Mitzner KD, Sternhagen J, Galipeau DW (2003) Development of a micromachined hazardous gas sensor array. Sens Actuators B 93:92–99
17. Sinha K, Priya A, Paul P (2016) Energy efficient scalable and reliable wireless sensor routing. Imp J Interdiscip Res 2:2454 1362

Implementation of Trapdoor Functionality to Two-Layer Encryption and Decryption by Using RSA-AES Cryptography Algorithms

Ravi Prakash, Premkumar Chithaluru, Deepak Sharma and P. Srikanth

Abstract This paper is going to deal with the implementation of two-layer encryption; RSA along with AES Cryptography Algorithm. It is a cryptography application coupled with a secret user directory where certain files are encrypted for better security. The user has access to his particular folder and can store any files inside it and also has an option of encrypting certain files. This paper implements trapdoor functionality to encrypt and decrypt files through a terminal using Cryptography.

Keywords RSA · AES · Cryptography · Trapdoor

1 Introduction

We live during a time and society that encompasses us with data, and progressively our everyday lives rely on this data and our capacity to control it. Tragically, at whatever point there are administrations for controlling data that has esteem, there will be deceitful components in the public eye that will look to subvert these administrations for their own particular advantage. This has prompted the advancement of research into data security.

Data security is the field of research that expects to shield data from pernicious assailants while yet enabling honest to goodness clients to control information uninhibitedly.

R. Prakash (✉) · P. Chithaluru · D. Sharma · P. Srikanth
SoCSE, UPES, Dehradun, Uttarakhand, India
e-mail: rprakash@ddn.upes.ac.in

P. Chithaluru
e-mail: chpremkumar@ddn.upes.ac.in

D. Sharma
e-mail: dksharma@ddn.upes.ac.in

P. Srikanth
e-mail: pshrikant@ddn.upes.ac.in

V. Nath and J. K. Mandal (eds.), *Nanoelectronics, Circuits and Communication Systems*, Lecture Notes in Electrical Engineering 511,
https://doi.org/10.1007/978-981-13-0776-8_8

Cryptography is the branch of data security, which covers the investigation of calculations and conventions that safe information. It is an assorted field, taking in components of electronic building, software engineering and arithmetic, including computational and mathematical number hypothesis, combinatory, gather hypothesis, and intricacy hypothesis [1].

Trapdoor capacity broadly used to encode information. In numerical values, like if *fn* is a trapdoor function, at that point there exists some unknown data a, b. By the end of goal that given $fn(a)$ and b, but the difficulty with register a. An instance of straight logical trapdoor number is "6895601 is after effect of two prime numbers. What these final prime numbers?" A default game plan is try-isolating 6895601 by a couple of trapdoor-generated prime numbers until getting the suitable reaction.

Regardless, if anyone is motivated that 1931 is one of the number, then we can find the suitable reaction by entering "6895601 ÷ 1931" going through any scaled down PC. Therefore, the delineation is not an instance of trapdoor—exhibit regular PCs are escalated most of relevant answers will get in few seconds—moreover, this example problem stated that, to upgrade by using the after effect of any two generously greater prime numbers [2].

Here, the client will approach a mystery registry in which he can store records, can without much of a stretch scramble, and unscramble the document through GUI [3]. Every client has singular index structure for putting away document and can bolt and cover up individual catalogue structure to give assist security, the client access to his specific organizer is segregated with a client ID and a password which is connected to his/her email ID. The client has the benefit to accumulate his overlooked client ID through the email and client can ask for an email to his email ID to recall the overlooked secret word.

2 Literature Review

Advance Cryptography report deals with the Cryptographic process explained through generalised functions are talking about which encryption and unscrambling is finished by the different calculations to like RSA calculation, Hash Functions and Security dangers to people and associations [1].

Trapdoor capacities came to unmistakable quality in cryptography in the mid-1970s with the distribution of distorted (or public-key) encryption procedures by Diffie, Hellman, and Merkle. In reality, Diffie and Hellman (1976) authored the term. A few capacity classes have been proposed, and it soon ended up plainly clear that trapdoor capacities are harder to discover than was at first idea. For instance, an early recommendation was to utilize plans in view of the subset entirety issue. This turned out—rather rapidly—to be unsatisfactory.

Starting at 2004, the best known trapdoor work (family) competitors are the RSA and Rabin groups of capacities. Both are composed as exponentiation modulo a composite number, and both are identified with the issue of prime factorization.

Capacities identified with the hardness of the discrete logarithm issue (either modulo a prime or in a gathering characterized over an elliptic bend) are not known to be trapdoor capacities, on the grounds that there is no known "trapdoor" data about the gathering that empowers the productive calculation of discrete logarithms.

A trapdoor in cryptography has the specific beforehand said meaning and isn't to be mixed up for a backhanded access (these are routinely used on the other hand, which isn't right). An auxiliary entry treated as an instrument that will be merged to a cryptographic evaluation (e.g., A particular key competition period estimation, mechanized checking figuring, et cetera.) or working system, for example, that stipends no less than one unapproved get-togethers to evade or subvert the security of the structure in some outline. In this all theory and application of Trapdoor functions deals with the usage of the trapdoor functions and working such functions in cryptography and pseudorandom number generation and abstract complexity theory [2].

Applied Cryptography by Bruice Scheniere describes many cryptography calculations, gives reasonable guidance on the most proficient method to actualize them in cryptographic programming, and shows how they utilized to take care of security issues [4].

RSA is one of the well-organized public-key system, RSA algorithm mechanism for deploying encryption and digital signatures. This algorithm is very simple to create for encryption and coding. RSA is named after three inventors—Rivest, Shamir, and Adelman.

RSA is a cryptographic system for public key encryption, and calculation comprehensively used for securing sensitive data, particularly while being sent over dubious framework, for instance, the Internet [5].

3 System Model

Trapdoor function encryption and decryption algorithm implemented using RSA algorithm. As RSA is most widely implemented algorithm for cryptography. The flowchart Fig. 1 defines the process illustrated in the diagram. If using the application for the first time, user needs to create an account, after successful authentication into the application the user will be asked to navigate through respective directory to browse the files.

The user can now perform following operations such as selecting encrypting and decrypting the files, after successful encryption/decryption the user locks the directory and logs out (Fig. 2).

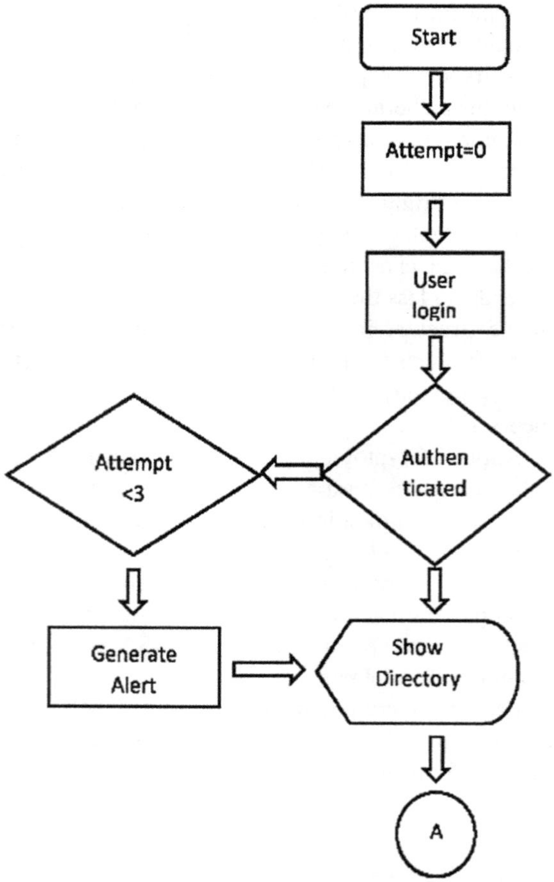

Fig. 1 Authentication phase

4 Problem Definition

Data being stored is both confidential and sensitive. Security in storing data has become a major problem for the users as the data stored is subjected to Backdoors, system hacking and data manipulation through viruses. Since, an operating system does not provide directory level locks for different users with encryption. There is a need for an application to have a directory level access and encrypt files inside the directory.

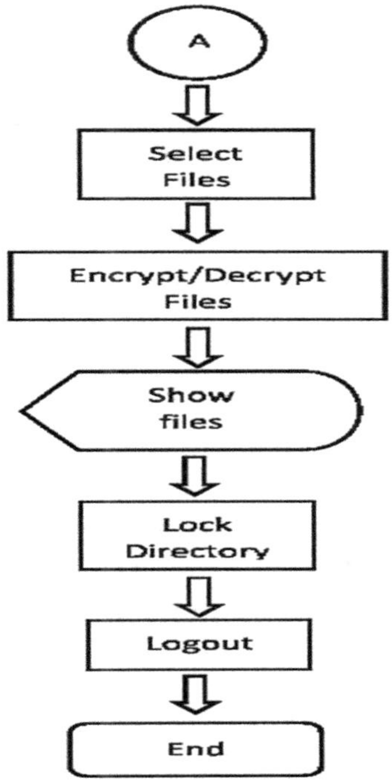

Fig. 2 Encryption/Decryption and storage phase

5 Algorithm

5.1 RSA Algorithm

Encryption

- Choose two particular substantial prime numbers say a and b. For security reasons, the both a and b numbers ought to be selected indiscriminately, and the both numbers are comparing with in the timeline. However, contrast takes much time, every digit to make figuring harder. All Prime numbers should productively to finding a prime numbers test.
- Calculate $m = ab$, is exploited for both universal society and isolated keys and the length generally lead into bits. Therefore, this is the procedure for getting key length.
- Calculate $\lambda(m) = lcm(\lambda(a), \lambda(b)) = lcm(a-1, b-1)$, where notation λ is the Carmichael's totient work. This esteem is kept private.
- Choose a number e with the end goal that

$$1 < c < \lambda(m) \text{ and } gcd\,(c, \lambda(m)) = 1;$$

i.e., c and $\lambda(m)$ are coprime. Where c having a short piece length and little Hamming weight brings about more productive encryption.

- Hence acquire people in general key i.e., (m, c).
- Now get the characters of the content record to be scrambled and by utilizing the above got key, ascertain the figure of the individual character where,

$$Cipher\_char = character^{\wedge}c \mod m$$

Decryption

Decryption methodology was simple and straightforward

- Hence by using the values of a, b and c chosen in the encryption algorithm, find the private key i.e.,(k, c).
- Where, $k \equiv c - 1(\text{mod}\lambda(m))$; i.e., k is the modular multiplicative inverse of $c(\text{modulo}\lambda(m))$.
- For solving k is given below,

$$k * c \equiv 1(\text{mod}\lambda(m)).$$

- By fetching the values of the text file to be decrypted and by using the above obtained key, find the character of the obtained cipher where,

$$character = Cipher\_char^{\wedge}k \mod m.$$

6 Simulation Tools

Java Swings used to making Graphical User Interface (GUI) for designing web applications and applets. Knowledge Swing with the NetBeans IDE is much faster and easier way to begin working with Swing. NetBeans IDE's GUI builder, a powerful feature that to construct GUI.

7 Conclusion

An effort is made to implement the Cryptography algorithm with a two layer of encryption, and to provide a user friendly application to create a distinct hidden directory for different users. The application is able to create multiple directories for different users where they can store any files and folders and lock as they wish. Further this application encrypts the text files as per the user's wish. The application is also able to recover forgotten password securely with the help of the email confirmation and access. The basic aim of the application is to provide a simple file vault for different users and implement encryption algorithm to encrypt their data which has been fulfilled by the application well.

8 Future Scope

The technique would be replaced with a better encrypting technique for high encryption and to provide high security by including real time notification of unauthorised access using SMS facility.

References

1. Advance Cryptography Report [Internet] c2012. Available from cryptography-full-report
2. Trapdoor Function [Internet] c2014 from https://en.wikipedia.org/wiki/Trapdoor_functions
3. Java Swings [Internet] c.2008. Available from http://docs.oracle.com/javase/tutorial/swing
4. Scheniere B (2002) Applied cryptography. Wiley, New York, pp 461–475
5. RSA (cryptosystem) [Internet] c2010. Available from https://en.wikipedia.org/wiki/RSA (cryptosystem)

Experimental Validation of Spectrum Sensing Techniques Using Software-Defined Radio

Bathula Siva Kumar Reddy

Abstract Software-defined radio (SDR) and cognitive radio (CR) system technologies provide added flexibility and offer improved efficiency to complete spectrum use. CR is obstreperous radio communication and networking technology. It is presently undergoing drastic development due to its potential to resolve many of the problems affecting existing systems. Sensing a spectrum is the crucial part of CR to sense the existence of primary users (PUs) or licensed user signals and exploit the spectrum holes. Hence, in this work, the testing of spectrum sensing technique for 16QAM and 64QAM transceivers on SDR testbed is carried out. Furthermore, this paper also addresses the practical signal detection and the impact of various filtering windowing methods on detected signals. Experimental tests are performed on SDR testbed, which employs universal software radio peripheral (USRP) N210 and GNU Radio as hardware and software platforms, respectively.

Keywords Cognitive radio · GNU radio · QAM · Software-defined radio
Spectrum sensing · USRP N210

1 Introduction

Development of microelectronics for wireless transceivers is flexible, dominant and manageable. It empowered the advancement of software-defined radio (SDR) to expertise [1], where the baseband processing is exclusively accomplished by the radio transceivers performs in software such as modulation/demodulation, error correction coding, and compression. The modern SDR is entitled as the heart of a cognitive radio (CR) [2]. CR can be segregated from a SDR by the applications implemented on the radio. As there is increasing insufficiency of spectrum resources, CR networks are current emerging part of wireless networking land-

B. Siva Kumar Reddy (✉)
Department of Electrical Engineering, Institute of Infrastructure
Technology Research and Management (IITRAM), Ahmadabad 380026, India
e-mail: bsivakumar100@gmail.com

V. Nath and J. K. Mandal (eds.), *Nanoelectronics, Circuits and Communication Systems*, Lecture Notes in Electrical Engineering 511,
https://doi.org/10.1007/978-981-13-0776-8_9

scape. The enabling key of CR is spectrum sensing because it identifies and exploits unoccupied frequency bands [3–5]. In our contribution, 64QAM transceiver is designed by energy detection spectrum sensing technique using SDR testbed.

Remaining paper is structured as follows: Sect. 2 presents the experimental testbed that employs a hardware, USRP N210, and a software, GNU Radio. Section 3 presents the system model, experimental results, and discussions. Section 4 concludes the results drawn from Sect. 3.

2 Experimental Testbed

The development and advancement of wireless communication field have been increasing very rapidly in the past few decades. All the present day electronic devices implement wireless communications in some way or the other, such as Bluetooth, Wi-Fi, WiMAX, or LTE. Each of these different radio systems has its own specific protocols. Consequently, these different radio systems had to be implemented using hardware configurations. The physical components used by the hardware radios cannot be easily modified. Therefore, a software is needed to implement some stages of hardware radio system to be more easily modified, configured, and developed for multiple systems. This new form of radio implementation is termed as software radio or software-defined radio (SDR). In this paper, universal software radio peripheral (USRP) N210 and GNU Radio are employed as hardware and software platforms, respectively as shown in Fig. 1.

There are three sections in an SDR architecture such as radio frequency (RF), intermediate frequency (IF, and baseband section. From Fig. 2, it can be observed that the smart antenna receives the RF signal and sends it to the hardware, that is inbuilt by various components such as daughterboard, ADC/DAC, FPGAs, DSPs, and ASICs. Where RF signal is changed over to baseband signal (digitized) by the hardware and afterward, baseband signal processing is done using GNU Radio software installed personal computer (PC), where all the signal processing blocks

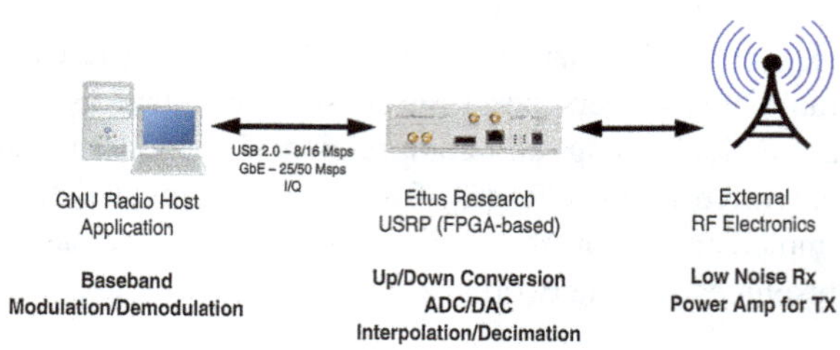

Fig. 1 Experimental testbed software-defined radio

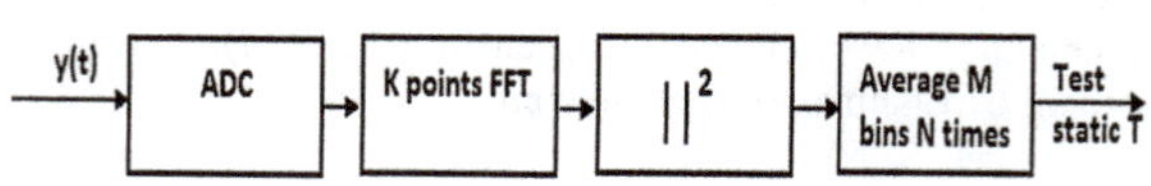

Fig. 2 Functional block diagram of Periodogram

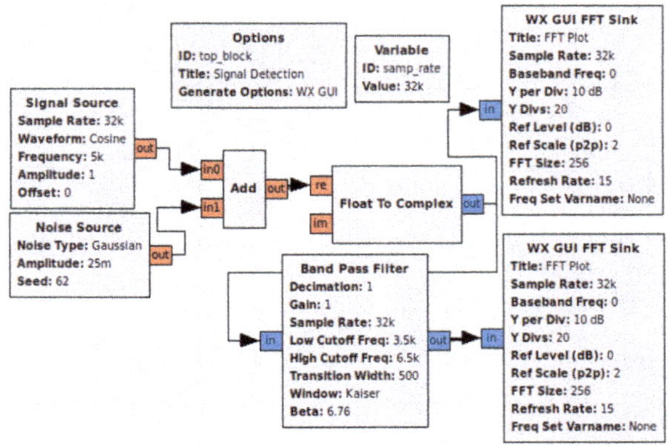

Fig. 3 GRC schematic for detecting primary users on spectrum

are designed. In addition to that, RF signal is converted into baseband signal in the receiver (Rx) path. The implementation of energy detection spectrum sensing is shown in Fig. 2 [6, 7] (Fig. 3).

3 Experimental Results and Discussion

For our experimentation, a noisy channel is created utilizing a Gaussian noise source and a cosine signal source and [8] as shown in Fig. 4. The parameters on each block can be reconfigurable at runtime also, which is a major advantage of using GNU radio. The original signal detection without filtering is shown in Fig. 4a. The impact of various windowing methods such as Hann, Hamming, Blackman, Rectangular and Kaiser windowing methods can also be observed in Fig. 4b, c, d, e, and f, respectively.

Figure 5 shows a GNU Radio Companion (GRC) Schematic drawn to implement energy detection spectrum sensing for 16QAM and 64QAM Transceivers on with multipath interference signals. Where different multipliers, adders, and signal sources are connected to create multipath interference for simulation purpose. Furthermore, experimental validation is carried out in Fig. 5, where the transmitted signal is passed through USRP sink and this signal is received by USRP Source for further processing. 16QAM and 64QAM modulated signals are captured on FFT sink and are shown in Fig. 6a, b, respectively. Multipath interfered transmitted signals for 16QAM and 64QAM modulation are shown in Fig. 7a, b, respectively. It can be observed that the transmitted signals have more distortion compared to modulated signals and this interference is included due to the impact of added noise (Table 1).

MPSK SNR Estimator block is employed to monitor and retrieve estimations of the signal SNR as shown in Fig. 5. A number sink block is added to see the

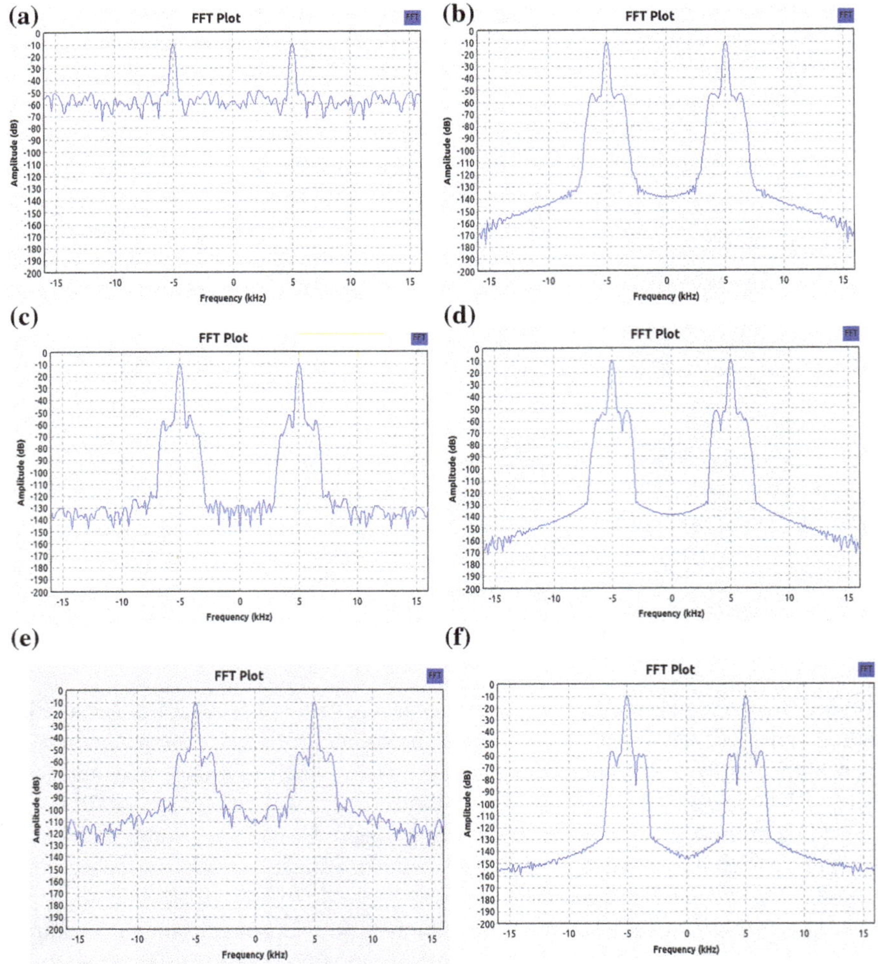

Fig. 4 **a** Original signal detection on FFT sink without employing BPF. **b** Detected signals by employing BPF for Hann windowing. **c** Hamming. **d** Blackman. **e** Rectangular. **f** Kaiser windowing techniques

numerical values of SNR. It can include multiply constant block to improve the SNR values of transmitted signals. The estimated SNR values are noted in Table 2. It is observed that the estimated SNR increases with the increase in the value of multiply constant block.

Energy detection spectrum sensing is performed on the received signal, as shown in Fig. 6, and the complex samples coming out from the USRP source are converted into their magnitude form. The output detected data is shown in Figs. 8 and 9. It can be observed the difference between spectrum sensing detection for original input signal and received signals in Figs. 8 and 9, respectively. High peaks

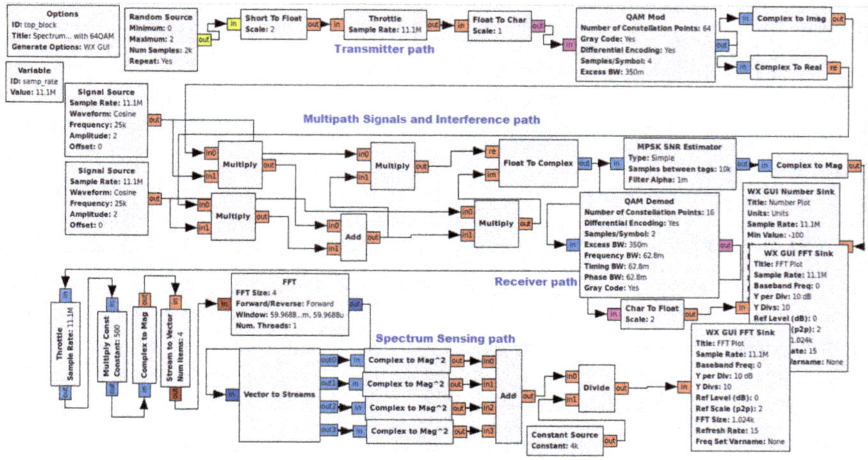

Fig. 5 GRC schematic for 16QAM and 64QAM transceivers with energy detection spectrum sensing with multipath interference signals

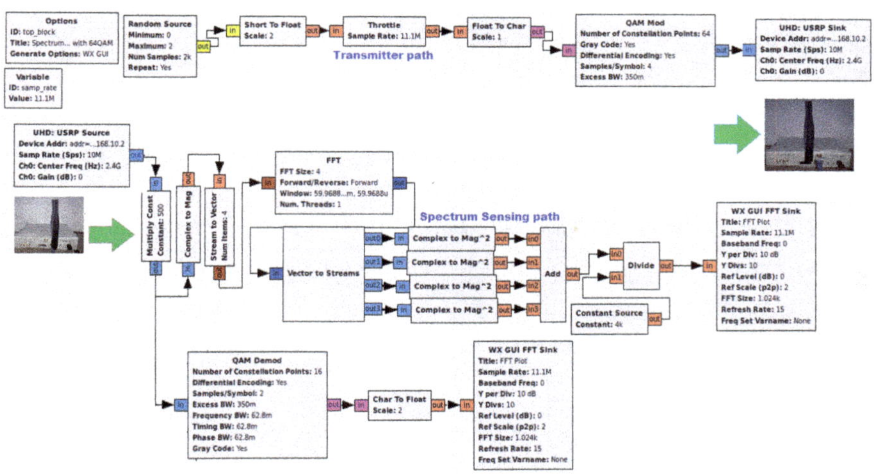

Fig. 6 GRC schematic for 16QAM and 64QAM transceivers with energy detection spectrum sensing over USRP hardware platforms

represent the presence of primary users (PUs) and low peaks represent the presence of PUs or presence of the secondary users. Received signals are passed through the demodulator and the output waveforms for 16QAM and 64QAM are shown in Fig. 9a, b.

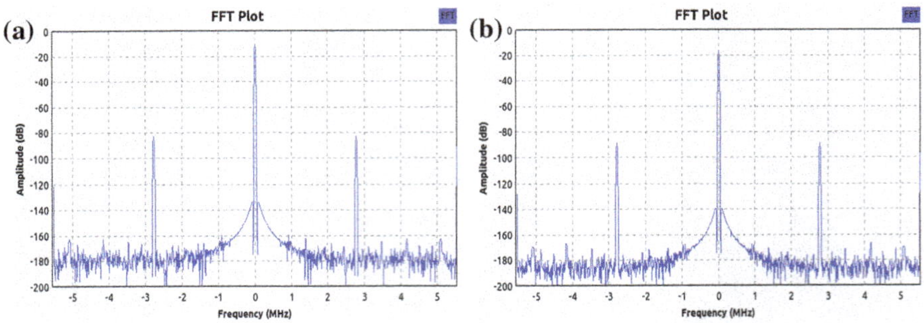

Fig. 7 Modulated signal on FFT sink for **a** 16QAM, **b** 64QAM modulation

Table 1 Experimental parameters

Parameter type	Value
Signal sources	Cosine signal source, random source
Noise	Gaussian
Filter	Bandpass filter (BPF)
Modulation type	16QAM and 64QAM
Spectrum sensing	Energy detection
Transmitter antenna	Tx
Receiver antenna	Tx/Rx
Sinks	FFT sink, scope sink and number sink

Table 2 Obtained transmitted signal SNR values for different multiply constant value

Multiply constant value	Estimated SNR value
1	1.00234
10	10.6085
50	52.809883
100	108.647

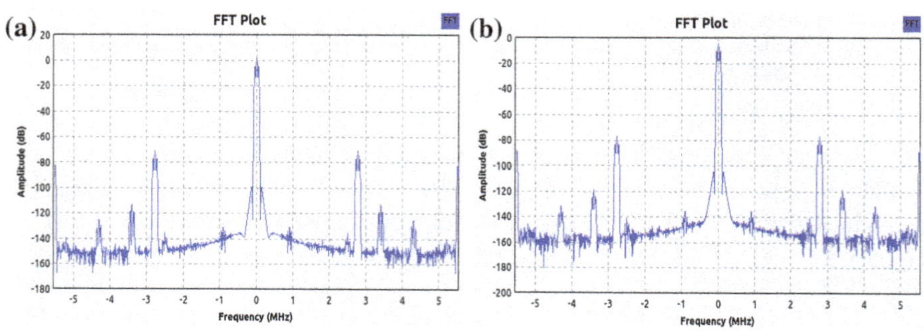

Fig. 8 Transmitted signals for **a** 16QAM, **b** 64QAM modulation

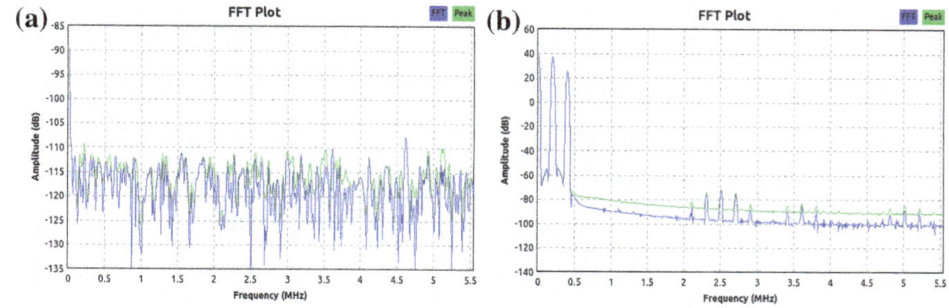

Fig. 9 Energy detection spectrum sensing for **a** original input signal **b** received signal

4 Conclusion

This paper addressed the design of spectrum sensing technique on QAM transceiver using SDR testbed that employed USRP N210 and GNU Radio. Hence, it reduced the cost of implementation by employing USRP N210 and GNU Radio. The impact of various filtering windowing methods is verified on detected signals using GNU Radio.

References

1. Mitola J (1992) Software radios-survey, critical evaluation and future directions. In: IEEE national telesystems conference, 19–20 May 1992, pp 13/15–13/23
2. Available: http://www.itu.int/net/newsroom/wrc/2012/features/software.aspx
3. Lee W, Cho D-H (2013) Channel selection and spectrum availability check scheme for cognitive radio systems considering user mobility. IEEE Commun Lett 17(3):463–466
4. Lee J, Andrews JG, Hong D (2013) Spectrum-sharing transmission capacity with interference cancellation. IEEE Trans Commun 61(1):76–86
5. Zhao Q, Swami A (2007) A decision-theoretic framework for opportunistic spectrum access. IEEE Wirel Commun Mag (Special Issue on Cognitive Wireless Networks) 14:14–20
6. Siva B, Lakshmi B (2014) BER analysis of energy detection spectrum sensing in cognitive radio using GNU radio. World Acad Sci Eng Technol Int J Electr Comput Energ Electron Commun Eng 8(11):1692–1698
7. Arslan H (2007) Cognitive radio, software defined radio, and adaptive wireless systems. Springer, London
8. Available: www.gnuradio.org

Smart Switch for Power Saving

Shaik Qadeer, Ajaz Fatima, Asfia Aleem and Amreen Begum

Abstract The rapid depletion of natural resources has increased the need for energy conservation. This project has been developed with the intention to enhance the power saving in public buildings such as educational institutes and hospitals. The basic principle of this project is to switch the power based on some conditions. It consists of two smart switches. The first detects the presence of an animate object and allows power supply as long as the animate object stays in the monitored area. The second switch takes a time input and provides supply for the requested period. A case study was performed in Muffakham Jah College of Engineering and Technology and it was observed that a total power of 65,780 KW per year could be saved.

Keywords PIR sensor · Microcontroller · Solid-state relay · Switch

1 Introduction

The growth index of any country indicates the generation and usage of electricity. The economic growth of the country is evaluated based on the energy conserved. Conservation of energy is a concern due to fast depleting natural resources. Therefore, it was decided to develop an electronic aid, which is helpful for the above purpose named as *Smart Switch*.

S. Qadeer (✉) · A. Aleem · A. Begum
EED, MJCET, Hyderabad, India
e-mail: haqbei@gmail.com

A. Aleem
e-mail: asfia.aleem@outlook.com

A. Fatima
EEE, MJCET, Hyderabad, India
e-mail: fatimaeee@mjcollege.ac.in

V. Nath and J. K. Mandal (eds.), *Nanoelectronics, Circuits and Communication Systems*, Lecture Notes in Electrical Engineering 511,
https://doi.org/10.1007/978-981-13-0776-8_10

A large amount of electricity is wasted due to human ignorance. We have developed this switching system to ensure no wastage of electricity due to negligence. In order to calculate the amount of electricity, which can be saved, we performed a case study in the Electrical Department of our Institute. It is observed that this system provides inexpensive and easy conservation. The rest of the work is organized as follows: In Sects. 2.1 and 2.2, the principle of operation of the proposed system is discussed, whereas Sect. 2.3 covers the components used and Sect. 2.4 discusses the firmware details. Finally, Sect. 2.5 gives the results of a case study. The conclusion is covered in Sect. 3.

2 Proposed System

2.1 Principle of the Proposed System 1

The basic operation of this device is that when a person enters the monitored area, the Passive Infrared Sensor (PIR sensor) absorbs infrared radiation emitted by that person and is activated. The PIR sensor senses the motion of a human body hence recognizing the presence of a person and it turns on the power supply to the panel. The load remains ON until it senses the presence. Any kind of device, i.e., lamp, fan, alarm, etc. can be used as a load. Once the motion ceases, the switch goes into OFF mode [1].

2.2 Principe of the Proposed System 2

The rooms such as the classrooms, conference hall, and auditoriums use the electrical power for only a given duration of time. In this proposed system, a push to ON switch is implemented which gives the power supply to the control for a given session of time. Soon after the completion of the time duration specified, it will automatically disconnect the power supply to the room. Therefore, in case the person forgets to turn OFF the load in the room before leaving, it will turn off when it reaches the specified time. In this way, we can save the power up to some extent.

In the next two section the hardware and software details of the proposed work are covered. Section 2.3 covers the hardware whereas Sect. 2.4 gives the firmware / software details.

2.3 Components of the Proposed System

The circuit diagram of the proposed power saving kit shown in Figs. 1 and 2 reveals that the following are the significant components of the given proposal:

- PIR Sensor (Automatic Human Detection)
- Solid-state relay
- Manual request switch
- Controller unit (8051 Microcontroller)

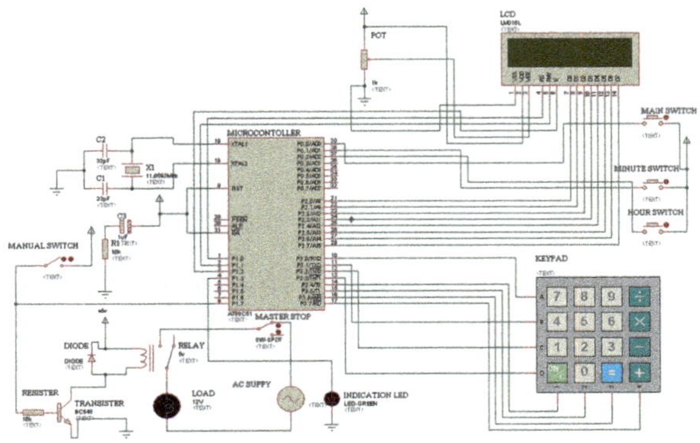

Fig. 1 Schematic diagram of the Proposed System 1

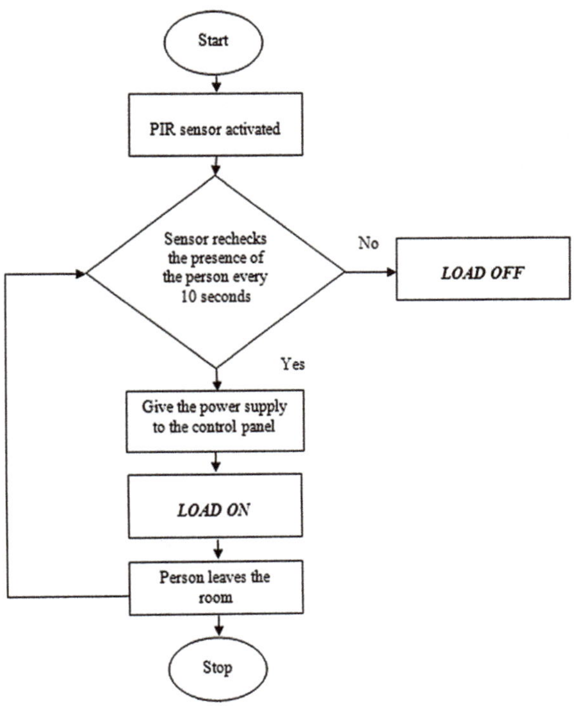

Fig. 2 Flow chart of the Proposed System 1

(1) *Automatic Human Detection*: This feature can be obtained by using Passive Infrared Sensors (PIR-based motion detector is used to sense movement of warm-blooded objects). These are extensively used for both indoor and outdoor applications such as automatically activated lighting systems or security systems. These are called "PIR" or "PID", for "passive infrared detector" [2].
(2) *Solid-State Relay*: A solid-state relay (SSR) is an electronic switching device which is used here for switching the appliances ON or OFF, when a small external voltage is applied across its control terminals. A current or a voltage signal is used as a control signal to trigger the sensor incorporated in the relay. The relay can be designed to switch either AC or DC to the load [3].
(3) *Manual Request Switch*: This component uses a push button to request the electrical power to the control unit (i.e., 8051 Microcontroller) for a given duration of time.
(4) *Controller Unit*: The 8051 is an 8-bit microcontroller used for simple embedded system application [4]. The language used to program this controller is embedded C on the simulation environment [KEIL] on Proteus [5, 6].

2.4 Firmware Detail of the Proposed System

The device shown in Fig. 1 senses the presence of a person and controls the power supply to the panel accordingly. The working is explained with the help of a flowchart as shown in Fig. 2 and the simplified circuit diagram Fig. 6 is developed to clarify the operations. The system is activated when the PIR sensing starts. If the presence is sensed, the control panel is given power supply and the load is turned ON. When the presence is not detected for 10 s the load is turned OFF, i.e., the sampling time of sensing is 10 s.

The working of the proposed system 1 is depicted by the flowchart shown in Fig. 2 and the schematic diagram in Fig. 1 is simplified in Fig. 5.

The device shown in Fig. 3 is based on the timer operation. The working is explained with the help of a flowchart as shown in Fig. 4 and the simplified circuit

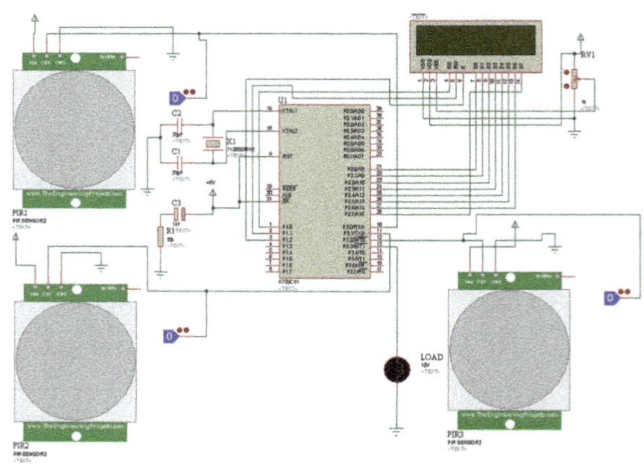

Fig. 3 Schematic diagram of the Proposed System 2

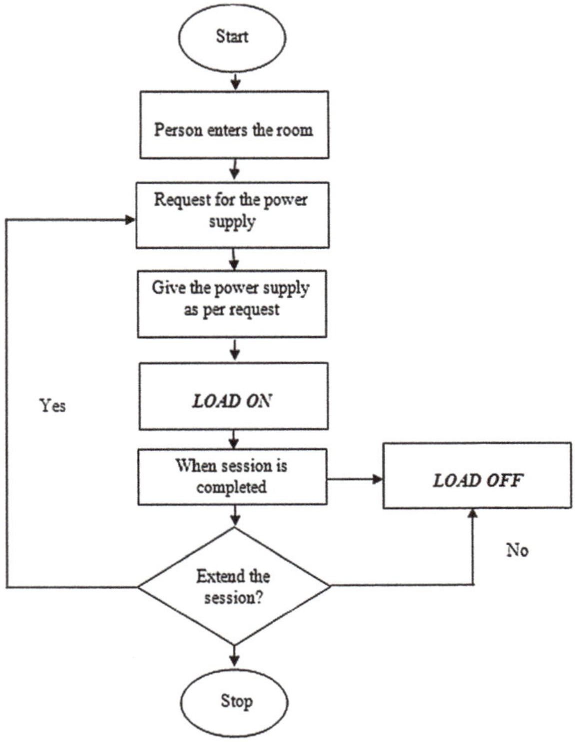

Fig. 4 Flowchart of the Proposed System 2

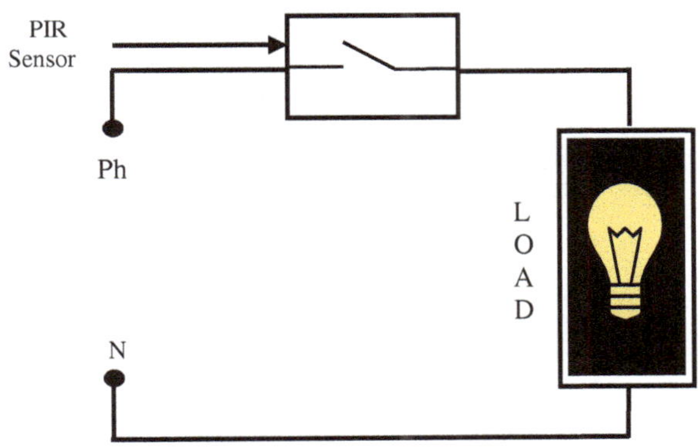

Fig. 5 Simple circuit diagram for the proposed system

diagram Fig. 5 is developed to clarify the operations. The system is activated when the set time starts. The power is supplied till the end of set time limit, and then it turns OFF.

The working of the proposed system 2 is depicted by the flowchart shown in Fig. 4 and the schematic diagram in Fig. 3 whose simplified representation is shown in Fig. 5.

2.5 The Code of the Proposed System

The code [7] for the proposed system 2 is as shown below: The code is self explanatory, as every line is added with explanatory comment.

```
void main () // main function for operation
{
    unsigned char a1, a2, time; // initialization
  m= 1; // main switch is high
  s1= 1; // minute switch is high
  s2= 1; // hour switch is high
  led1= 0; // load is low
  led2= 0; // indication for delay
  m=0; // main switch is low
  s1=0; // minute switch is low
  s2=0; // hour switch is low
  led1=0; // load is low
  lcd_init(); // lcd initialization
  lcd_cmd(0x80); // lcd command
  delay(50); //software delay
  send_data("welcome"); // lcd data
  delay(500); // delay
  while(1)
  {
    if(m==1)
    {
      lcd_cmd(0x80); // lcd command
              delay(50); // delay
      send_data("Enter session"); // lcd data
          delay(500);  // delay
          lcd_cmd(0x01); // lcd command
        delay(50);  //delay
        send_data("s1 for min"); // lcd data
        delay(500); // delay

        lcd_cmd(0xc0); // lcd command
        delay(50); // delay
        send_data("s2 for hour"); // lcd data
      delay(500); // delay
      delay(500); // delay
```

```
lcd_cmd(0x01); // command
     delay(50); // delay
while(1)
        {
if(s1==1)
   {
   lcd_cmd(0x01); // command
   delay(50); // delay
   send_data("Req min"); // lcd data
   delay(50);
   a1 = get_key_pad(); //get keypad values
   a2= get_key_pad();
   time=(a1*10)+a2;  // Timer calculation
   led1=1; // load on
   delaymin(time); // load delay
   led1=0; // load off
   lcd_cmd(0x80); //command
   delay(50);
   send_data("Thank you");
   break;
}
if(s2==1)
{
   lcd_cmd(0x01);
   delay(50);
   send_data("Req hours");
   delay(50);
   a1 = get_key_pad(); // get keypad values
   a2= get_key_pad();
   time=(a1*10) + a2; // Timer calculation for hours
   led1=1;
   delayhour(time); //load delay
   led1=0;
   lcd_cmd(0x80);
   delay(50);
   send_data("Thank you");
   break;
 }
}
}
}
}
```

Fig. 6 Simple circuit diagram for the proposed system

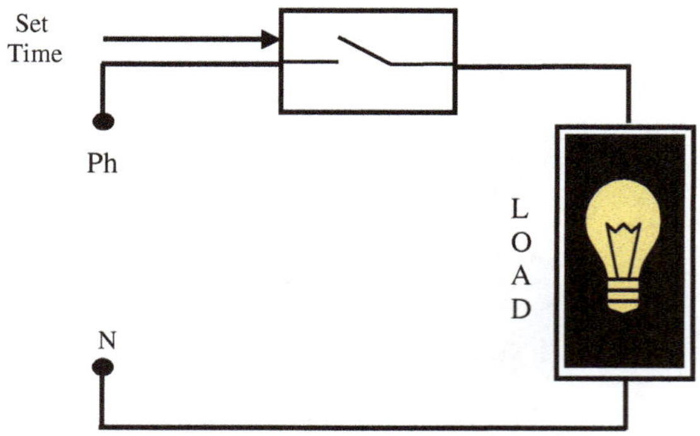

2.6 The Results of the Proposed System

The system was first simulated using Proteus software. Later, the prototype made was checked. Finally, it is installed. The flow of execution of proposed research work is as shown in Fig. 7.

The device in Fig. 2 works on the principle of the timer. When the request is sent through a push to ON button, the signal is sent to the microcontroller and then to the relay switch, which turns ON the supply to the panel and then, it is energized for a given duration of time and then turns OFF automatically.

To predict the possible saving achieved from the proposed system in Fig. 1, the approximate audit is applied on the staffrooms of MJCET. It can be seen that in each staffroom, the nearby load wattage is observed to be 540 KW for an hour. The

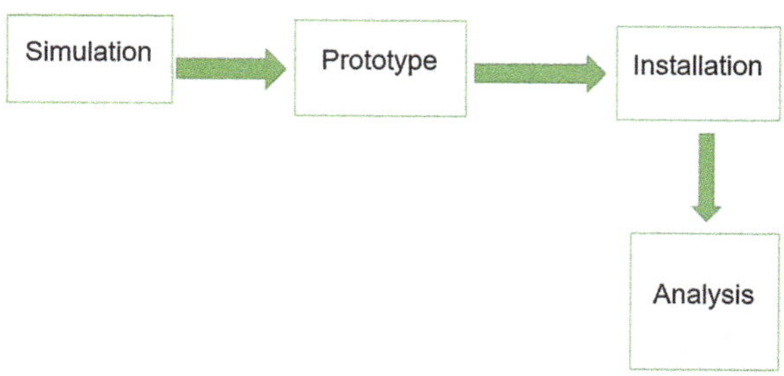

Fig. 7 Execution flow for the proposed work

Table 1 Power consumed by 25 staffrooms

	Power consumed (KW)		
Load in each staffroom	Day	Month	Year
540	27	702	7,722

Table 2 Power consumed by 50 classrooms

	Power consumed (KW)		
Load in each classroom	Day	Month	Year
880	88	2,288	26,168

audit of six departments of MJCET considering the number of staffrooms to be 25 is as shown in Table 1.

It can be observed from the analysis in Table 1 that the load in each staffroom is considered to be 540 W. Then the power consumed by 25 staffrooms in a day, taking into account two wastage hours, is 27 KW. Further, for 26 working days per month (excluding four Sundays per month), the power consumed is 720 KW. Hence, in a year 7,722 KW power is consumed.

To predict the possible saving achieved from this proposed system in Fig. 2, the approximate audit is applied on the classrooms loads of MJCET. It can be seen that in each classroom, the nearby load wattage is observed to be 880 KW for an hour. The audit of six departments of MJCET considering the number of classrooms to be 50 is as shown in Table 2.

It can also be observed from the analysis in Table 2, the load in each classroom is considered to be 880 W. Then the power consumed by 50 classrooms in a day, taking into account two wastage hours, is 88 KW. Further, for 26 working days per month (excluding four Sundays per month), the power consumed is 2,288 KW. Hence, in a year, 25,168 KW power is consumed.

Thus, the saving achieved after using the proposed system is the sum of Tables 1 and 2, i.e., 7,722 + 25,168 = 32,890 KW (per year).

3 Conclusion

In this chapter, a soft method of energy conservation is presented and a smart switch is designed. Two implementations are presented for staff rooms and (classrooms) auditorium.

Basically, the human ignorance is avoided and thus energy is conserved. For the case study shown, the power saved is 32,890 KW/year. With a tariff of 5 INR, the amount saved is 1.65 Lakhs per year. The above can be achieved using a simple hassle free economical proposed device which approximately costs 500 INR.

Acknowledgements The authors would like to thank the MJCET management for sponsoring this research work. The first author would also like to thank Mr. M. Abrar Ahmed and Ms. Samiya Iftekhar (both student of EED, MJCET) for their support to complete this work.

References

1. Bai Y-W, Ku Y-T (2008) Automatic room light intensity detection and control using a microprocessor and light sensors. IEEE Trans Consum Electron (Aug 2008)
2. Keller HJ (2009) 30 years of passive infrared motion detectors- a technology review, May 11, 2009
3. Akiyama S (1989) US Patent 4,804,866, 1989
4. Atmel 8051 Hardware Manual
5. Kanethkar Y (2016) Let us C++, 2nd edn. BPB Publications
6. Willams E (2014) Make: AVR programming, 2014 edn. Maker Media
7. Mazidi MA, Naimi S, Naimi S (2012) The AVR microcontroller and Embedded Systems: using assembly and C, 2012 edn. Prentice Hall Publication

SDN-Based Programmable RSU Selection Method in Heterogeneous Networks

Debdatta Kandar and Babu Sena Paul

Abstract In recent years, vehicular communication industries focus on the live information transfer with high-speed mobility of vehicles. Though vehicles are normally equipped with a single communication system, they may find different networks for communication during travel. In order to access, seamless connectivity vehicles needed to be connected in different networks and vertical handover comes into the picture. Even future generation wireless communication systems require support for multimode communication of Heterogeneous Networks (HetNets). This paper introduces a novel Road Side Unit (RSU) selection scheme for vehicular communication. The objective is to provide travelers with seamless connectivity and increase the network utilization. This scheme sense of balances between the interests of RSUs and On-Board Unit (OBU) via a suitably Software-Defined Network (SDN) control. The SDN assesses the HetNet performance based on user preferences and finds the proper RSU for the OBUs. This scheme maximizes the HetNet utilization and minimizes handover by properly balancing the load among the RSUs in the HetNet. The scheme enhances the performance by rejecting unnecessary requests. The simulation model claims that the proposed RSU selection scheme improves the HetNet functionality for vehicular communication in terms of Handover Served Ratio (HSR), OBU Served Ratio (OSR), and total throughput.

Keywords Road side unit · Software-Defined Network · OBUs served ratio
Handover Served Ratio · Intelligent transportation system

D. Kandar (✉) · B. S. Paul
Department of Electrical and Electronics Engineering, University of Johannesburg, Johannesburg, South Africa
e-mail: kdebdatta@gmail.com

B. S. Paul
e-mail: babusenapaul@gmail.com

V. Nath and J. K. Mandal (eds.), *Nanoelectronics, Circuits and Communication Systems*, Lecture Notes in Electrical Engineering 511,
https://doi.org/10.1007/978-981-13-0776-8_11

1 Introduction

The advance of future vehicular communication will plead for the collaboration of Heterogeneous Networks (HetNets), with the purpose of providing travelers with a variety of mobile applications in dynamic channel condition through a single device in pervasive computing. The present vehicular communication network lags in widespread coverage, whereas the existing wireless mobile communication systems, together with Worldwide Interoperability for Microwave Access (WiMAX), Long Term Evolution (LTE), the Global System for Mobile communication (GSM), General Packet Radio Service (GPRS), and Universal Mobile Telecommunications System (UMTS), can offer several mobile applications [1].

Though vehicular technology supports high data rates, coverage area is limited. On the other hand, few cellular technologies support longer distances for communication. Again, there subsist few networks, for example, WPAN and WLAN which provide high throughput surrounded by smaller coverage areas, like hotel, airport, campus, and other hot spots. Therefore a combination of cellular networks and vehicular network may result in high data rate, longer coverage with high mobility, and ultimately improved quality of service (QoS).

The key challenges of the HetNet for vehicular communications are to maintain flawless vertical handover of car OBUs without compromising QoS. The proper handover decision controls load balancing, supporting users' demand, and improving network performance. Therefore, the requirement for the development of a competent vertical handover method turns out to be crucial.

We talk about the typical similar kind of research studies in this section. The Network Selection Problem (NSP) considering QoS found in [2–10]. Researchers in [3, 4, 6, 8, 10] usually considered user's satisfaction. In [3, 6, 8], researchers discussed bit impractical single-user scenario. In [5, 7, 9, 10], more comprehensive multiuser scenarios was conferred. In [5], the authors proposed two different algorithms for user's and operator's satisfaction without combining them and with the goal of improvement of objective function. The objective function evaluates sum utility by combining of the user's and operator's utility function [7]. A bid and price criteria based NSP is discussed in [9]. NSP for multicast transmissions is explored in [10]. The Multi-RAT, Multimode Terminals (MMT) technique was explored in [11–14] for HetNet scenario.

In [15], Wang et al. proposed policy enabled NSP. In [16], J. Hou et al. discussed fuzzy logic based NSP. In [17], Lee et al. suggested that network can be selected based on the residual bandwidth. In [18], Ciubotaru et al. proposed smooth adaptive soft handoff algorithm (SASHA).

The remaining of this manuscript is arranged as follows. The architecture of HetNet for vehicular communication is mentioned in Sect. 2. Section 3 presents the MATLAB simulation results. Section 4 discusses the conclusion of this work.

2 SDN-Based HetNet Architecture for Vehicular Communication

In usual vehicular communication infrastructure, On-Board Units (OBU) installed on vehicles communicate among themselves through V2V communication, and communicate to the Road Side Units (RSU) and establish V2I communication. A typical HetNet consists of different technologies, namely LTE, WiMax, DSRC can be used for vehicular communication as shown in Fig. 1.

In this design, by combining SDN and OpenFlow will facilitate the control of RSUs or base stations' data plane and improve the HetNet performance. Therefore, efficient load balancing can be done with proper supervision of OBUs and RSUs. The main idea of SDN [19–21] is the create division between the control plane and the data plane as shown in Fig. 2. The data plane is used for the data forwarding while the control plane is used for the RSU selection.

2.1 SDN-Based RSU-OBU Model

SDN wireless node or OBUs include entire operability of an OpenFlow-enabled OBU in conventional OpenFlow networks. The SDN unit is empowered with the packet processing and the interface which receives the input from a detached control plane. An OpenFlow network consists of two building blocks, the OpenFlow controller and several OpenFlow-enabled switches or RSUs which can communicate themselves via a communication channel as shown in Fig. 2.

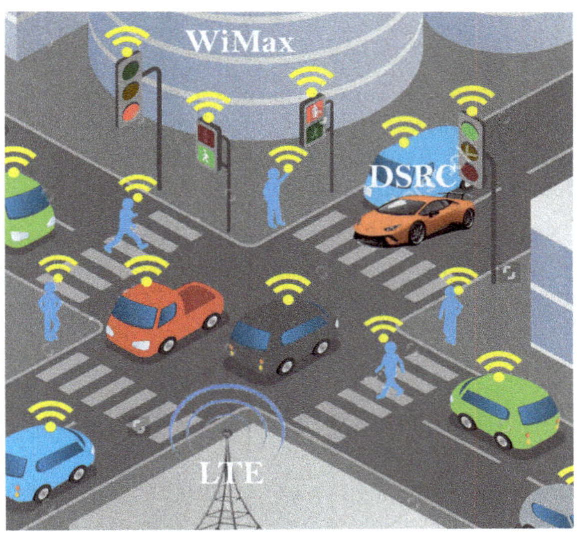

Fig. 1 HetNet vehicular communication scenario

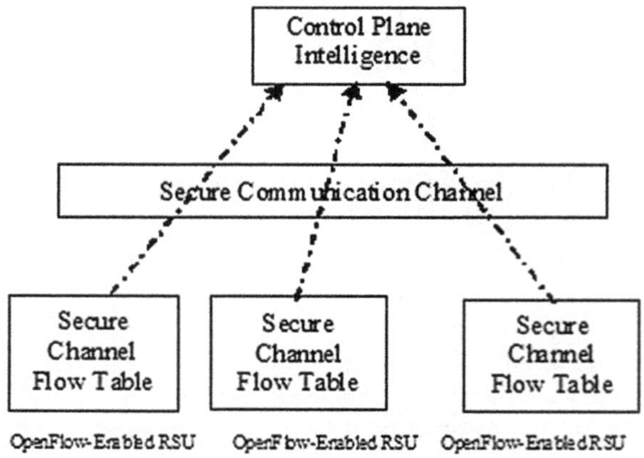

Fig. 2 OpenFlow components of SDN-based HetNet

RSUs contain a software program component or OpenFlow controller which manages the activities of a network by sending flowMods packets for the updation of the content of flow tables. Flow table holds a list of flow entries, which were composed of Match Fields, Priority, counters, instructions, timeouts, cookie and the actions associated. During packet arrival at the RSU, there is a lookup into the matching field of each entry. If matched, the packet will be processed according to the actions of the matching flow entry. Conversely, if not matched, a table-miss occurs as specified in the table-miss flow entry. For example, the OBUs can encapsulate the packet and send it to the controller through the secure channel or directly drop the packet [22].

The goal of SDN-based vehicular communication system is to take the benefit of SDN models to maximize the vertical handover scenarios and the system capacity. To facilitate an SDN-based HetNet for vehicular communication, this architecture consists of mainly three components, namely SDN Controller, SDN RSU, and SDN OBU.

SDN Controller: The logical central unit which controls the behavior of the HetNet.
SDN RSU: Centrally SDN-controlled static data plane units installed along roads.
SDN OBU: Centrally SDN-controlled data plane units that receive the control message to perform specific activities.

In this proposed RSU selection scheme, the Doppler effect, OBUs' Direction of Arrival (DOA) plays significant roles, as we considered a large coverage area of a vehicular HetNet scenario.

2.2 Mathematical Model for RSU Selection Method

We consider the high-speed car OBU for the formulation of mathematical model. The car OBU might access different base stations or RSUs which are centrally organized by SDN controller. Using Media Independent Handover (MIH) standard [23], the controller may allow the vertical handover [24, 25] of OBUs.

At any instant, let R be the set of m number of RSUs, $R = \{r_1, r_2, \ldots, r_m\}$, available to the OBUs. The RSU has f number of channels each having bandwidth β_i (in MHz) and the RSUs r_i ($r_i \in R$, $i = 1, 2, \ldots, m$) equally distribute its frequency [17].

Let B be the set of n number of OBUs, $B = \{b_1, b_2, \ldots, b_n\}$, request for services on move. OBU b_j ($b_j \in B$, $j = 1, 2, \ldots, n$) may communicate with one of its accessible RSUs which is called the current RSU of OBU b_j.

At time t, an adjacency matrix $\mathbf{\Phi}_{ij}(t)$ replicates the relationship between RSUs and OBUs which is defined as

$$\Phi_{ij}(t) = \begin{matrix} \\ r1 \\ r2 \\ . \\ rm \end{matrix} \begin{bmatrix} b_1 & b_2 & . & . & . & b_n \\ \Phi_{11} & \Phi_{12} & . & . & . & \Phi_{1n} \\ \Phi_{21} & \Phi_{22} & . & . & . & \Phi_{2n} \\ . & . & . & . & . & . \\ \Phi_{m1} & \Phi_{32} & . & . & . & \Phi_{mn} \end{bmatrix} \tag{1}$$

$$\Phi_{ij}(t) = \begin{matrix} 1, & \text{when } r_i \text{ and } b_j \text{ are connected} \\ 0, & \text{otherwise} \end{matrix}$$

In this HetNet, we consider only downlink scenarios. The OBU b_j is separated from RSU r_i by Euclidean distance $d_{ij}(t)$. At any instant t, the received signal power S_{ij} (dBm) at OBU b_j is defined by [26], where S_i (dBm) is the transmission power

$$S_{ij}(t) = S_i.h_{ij}.d_{ij}(t)^{--\alpha} \tag{2}$$

where h_{ij} the channel fading gain and the path loss exponent which differ due to channel situations ($\alpha > 2$).

At any instant t, while OBU b_j communicates with RSU r_i, the co-channel interference $Sx_j(t)$ (in dB) at OBU b_j due to the other RSUs r_x ($r_x \in R$, $x \neq i$) is defined as

$$Sx_j(t) = s_x.hx_j.\mathrm{d}x_j(t)^{--\alpha} \tag{3}$$

where $\mathrm{d}x_j(t)$ refer to the Euclidean distance between RSU r_x and OBU b_j at instant t, where the transmit power of RSU r_x is s_x in dBm and channel fading gain is hx_j. Assuming short distance, low power transmission of OBUs, and fast change in OBUs' position, the interference among OBUs is considered negligible [27].

The channel capacity $g_{ij}(t)$ in Mbps is defined as [28]

$$g_{ij}(t) = \beta_i.\log 2\left[1 + \frac{S_{ij}(t)}{\Sigma rx \in X_j Sx_j(t) + N_{ij}(t)}\right] \tag{4}$$

$N_{ij}(t)$ is the additive white Gaussian noise [29], $S_{ij}(t)$ is the received signal power at b_j at instant t in dBm, and X_j ($X_j \subseteq R$) be the set of RSUs that transmitting co-channel signal.

Let us assume OBU's data receiving rate is equal to the channel capacity and least amount bandwidth requirement of OBU b_j is γ_j (Mbps). If the current RSU (rc) and OBU b_j connection channel capacity does not satisfy $uc_j(t) < \gamma_j$ constraint, OBU b_j will execute vertical handover, otherwise $uc_j(t) \geq \gamma_j$, OBU b_j will continue with its current RSU(rc) connection.

The $U(t) = [u_1(t), u_2(t), \ldots, u_n(t)]$ vector is used to identify the handover or non-handover kind of OBUs at instant t.

$$u_j(t) \text{ is defined below, where } j = 1, 2, \ldots, n.$$

$$u_j(t) = \begin{matrix} 0, \text{difines handover OBU } b_j \text{ at instant } t, \\ 1, \text{difines non-handover OBU } b_j \text{ at instant } t, \end{matrix} \tag{5}$$

Let us consider the channel capacity $v_{ij}(t)$ between the RSU r_i and OBU b_j at time t, if r_i is selected as the RSU. According to Eq. (6), for handover OBU b_j ($u_j(t) = 0$), $v_{ij}(t)$ is $g_{ij}(t)$. For non-handover OBU b_j ($u_j(t) = 1$), $v_{ij}(t)$ is 0.

$$v_{ij}(t) = \begin{matrix} g_{ij}(t), & u_j(t) = 0, \\ 0, & u_j(t) = 1 \end{matrix} \tag{6}$$

Let $V_j(t) = (v_{1j}(t), v_{2j}(t), \ldots, v_{mj}(t)$. The SDN controller selects a new RSU for b_j according to the values of $V_j(t)$. The result of RSU selection method at time t is defined by $W_j(t) = (w_{1j}(t), w_{2j}(t), \ldots, w_{mj}(t))$. The $w_{ij}(t)$ is defined below, where $i =$ 1, 2, … m.

$$w_{ij}(t) = \begin{matrix} 1, & r_i \text{ is the new RSU of } b_j \text{ at time } t, \\ 0 & \text{otherwise} \end{matrix} \tag{7}$$

During the selection of new RSU r_i, the remaining bandwidth of r_i has to satisfy the minimum bandwidth requirement of b_j.

3 Simulation Model and Results

Here in this work, we study the RSU selection problem in HetNet scenario considering RSU selection using centrally controlled SDN. The mathematical model of the RSU selection method is mentioned earlier. In this design, one OBU is allowed to connect to one RSU only at a time. At first, OBUs assess the Signal to Interference and Noise Ratio (SINR) to the current RSU and permit to send a handover request if it is lower than a certain level. At first, handover request frames are sent by the OBU to the accessible RSUs with values either zero or the channel capacities [30]. Non-handover OBUs also send request frames with zero value also contain "Hello" messages to its available RSU.

Then RSU sends EchoRequest frames to the SDN controller to initialize the RSU selection. In reply, the SDN controller sends back the EchoResponse Frames to those RSUs. RSUs send the Packet-In messages which contain the data fields to the SDN controller. The data field contains channel capacity, the bandwidth requirement, and direction of arrival information of the OBU. Then, the RSU evaluates the elements of V_j(t) based on the data field received from RSUs. The SDN controller staying at the central part of the HetNet forms the request matrix $V_j(t)$ after evaluating the data field received from different RSUs. The $V_j(t)$ elements are equal to zero for non-handover OBUs, and the value of the corresponding column in $V_j(t)$ will be set to be zero which indicates that no new RSU will be selected for the OBUs. The unassigned elements in $V_j(t)$ are set to zero corresponding to their unaccessible RSUs for handover OBUs. Throughout RSU selection phase, the SDN controller selects the RSU for OBU after the formulation of handover request matrix $V_j(t)$. The SDN controller calculates the RSU selection result of $W_j(t)$ using the RSU 0–1 programming problem. In this study, we give less handover priority, to the moving away vehicles and give more priority to the moving in vehicles towards the RSUs. We consider the moving OBUs towards RSU as positive handover OBU and moving OBUs away from RSU as negative handover OBU. The positive OBUs get priority for the RSU selection process. We develop RSU selection method in Matlab software and the results are discussed next.

3.1 Study of Handover Served Ratio (HSR)

In this work, we give emphasis on the study of Handover Served Ratio (HSR) instead of vertical handover number. The vertical handover increases with the increase in vehicles in case of vehicular communication scenario. The vertical handover mostly varies with the vehicle speed and vehicles' direction of motion. We studied HSR with the OBUs number as shown in Fig. 3. The HSR is defined as the ratio of handover OBUs with a number of handover requested OBUs. Initially, HSR observed is low but increased very fast as the number of OBUs increased and finally close to one which indicates almost all the handover requested OBUs are connected to some RSU. If any RSU is congested with OBU, the load automatically balances and is transferred to other RSU. The initial HSR low value is because of the simulation begin with zero value and after some time it will fluctuate around one. This initial phase can be ignored. The situation is true for Fig. 4 too.

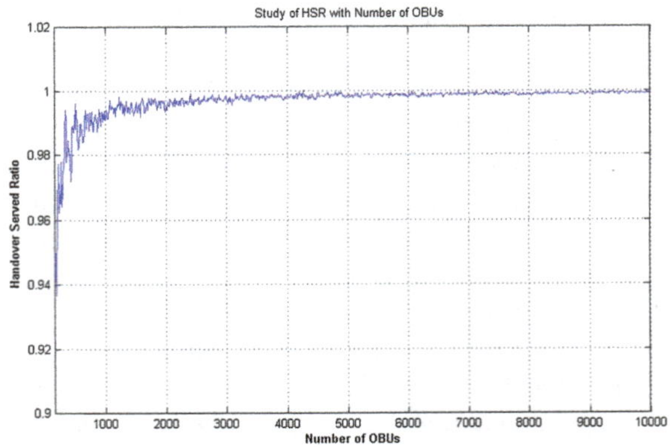

Fig. 3 HSR performance with OBUs' number

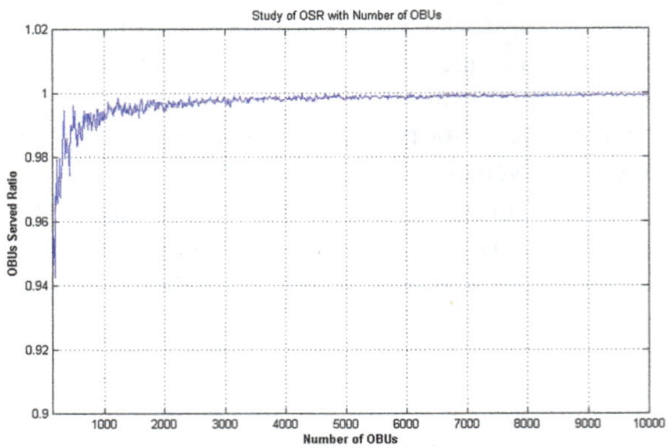

Fig. 4 OSR performance with OBUs' number

3.2 *Study of OBUs Served Ratio (OSR)*

We study here OBU Served Ratio (OSR) with the number of OBUs. The OSR is the ratio of OBUs which have the connections (including handover and non-handover OBUs) with a number of OBUs as mentioned in Eq. (8) [30]

$$\text{OSR} = \left[\sum_{j=1}^{n} u_j(t) + \sum_{i=1}^{m} 0 \sum_{j=1}^{n} w_{ij}(t)\right] / n \quad (8)$$

where $\sum_{j=1}^{n} u_j(t)$ is the number of non-handover OBUs which got new RSU connections and $\sum_{i=1}^{m} 0 \sum_{j=1}^{n} w_{ij}(t)$ is the number of handover OBUs which has old RSU connections and n is the number of OBUs which indicates that few OBUs still have no connection).

We observed the variation of OSR with a number of OBU as shown in Fig. 4. At first, OSR observed is low but increased as the number of OBUs increased and finally close to one, which indicates almost all the OBUs are connected to the

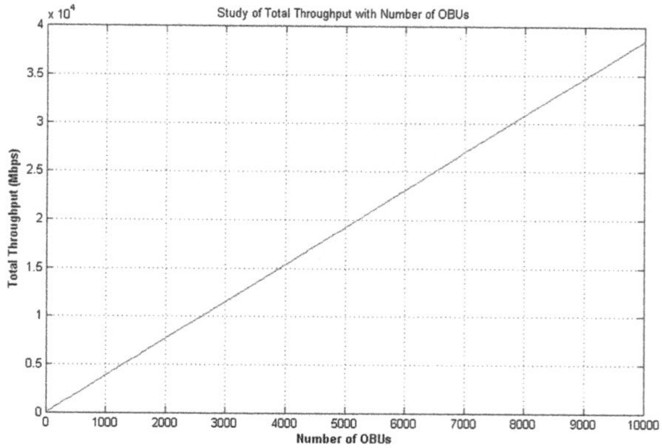

Fig. 5 Total throughput performance with OBUs' number

RSUs. Still, there are few OBUs which are not having any connection with the RSU. These non-connected OBUs are not having connection due to the multipath channel condition. The SDN-based system performance is far better as the system is controlled centrally. In case of without SDN-controlled network scenario, these performance parameters (Figs. 3 and 4) decrease with the increase in the number of OBUs and are very low ($\ll 1$).

3.3 Total Throughput

In this section, we studied the served throughput provided by the SDN-based HetNET for vehicular scenario. The served throughput is defined as the sum of channel capacities of connected OBUs (handover and few non-handover OBUs) under HetNet scenarios. At any time, one RSU can provide one channel to one OBU only. In this HetNet design model, channel capacities between 1 and 5 Mbps were considered. Here in this simulation, we considered ten RSU can support 10,000 channels (Fig. 5).

4 Conclusion

The work highlights vertical handover issues for vehicular communication in HetNet scenario. As per our knowledge, we are the first researchers to apply SDN for RSU selection in vehicular communication. Instead of studying of variation of number of handover with a number of OBUs, we studied HSR and OSR variation. The ratios indicate the amount of service obtained out of the amount of handover requested by OBUs. As still there may be few OBUs which are not having connection due to channel condition. As the vehicles move very fast and their directions of motion change rapidly, the handover increases as the number of OBUs increases. Handover requirements also vary with the dynamic multipath channel condition. So, the performance study should be compared with the same kind of infrastructure with the same kind channel condition. In this design, we restricted handover for the moving away vehicles rather give priority to moving towards vehicles. In the future work, we are planning to prepare a handover priority matrix of OBUs' whose ultimate goal is to provide at least one channel to all the OBUs.

References

1. Song Q, Jamalipour A (2008) A quality of service negotiation-based vertical handoff decision scheme in heterogeneous wireless systems. Eur J Oper Res 191:1059–1074. https://doi.org/10.1016/j.ejor.2007.07.011
2. Falowo OE, Chan HA (2010) Heuristic RAT selection policy to minimize call blocking probability in next generation wireless networks. Wirel Commun Mob Comput 10(2): 214–229
3. Lopez-Benitez M, Gozalvez J (2011) Common radio resource management algorithms for multimedia heterogeneous wireless networks. IEEE Trans Mob Comput 10(9):1201–1213
4. Falowo OE, Chan HA (2012) Dynamic RAT selection for multiple calls in heterogeneous wireless networks using group decision-making technique. Comput Netw 56(4):1390–1401
5. Nguyen-Vuong Q-T, Agoulmine N, Cherkaoui EH, Toni L (2013) Multicriteria optimization of access selection to improve the quality of experience in heterogeneous wireless access networks. IEEE Trans Veh Technol 62(4):1785–1800
6. Liou Y-S, Gau R-H, Chang C-J (2014) A bargaining game based access network selection scheme for HetNet. In: Proceedings of the 1st IEEE international conference on communications (ICC '14), pp 4888–4893. IEEE, Sydney, Australia, June 2014
7. Kosmides P, Rouskas A, Anagnostou M (2014) Utility-based RAT selection optimization in heterogeneous wireless networks. Pervasive Mob Comput 12:92–111
8. Kumar A, Mallik RK, Schober R (2014) A probabilistic approach to modeling users' network selection in the presence of heterogeneous wireless networks. IEEE Trans Veh Technol 63(7):3331–3341
9. Konka J, Andonovic I, Michie C, Atkinson R (2014) Auction based network selection in a market-based framework for trading wireless communication services. IEEE Trans Veh Technol 63(3):1365–1377
10. Sui N, Zhang D, Zhong W, Wang C (2016) Network selection for heterogeneous wireless networks based on multiple attribute decision making and Evolutionary Game Theory. In: Proceedings of the 25th wireless and optical communication conference (WOCC '16), pp. 1–5,May 2016
11. Lim G, Xiong C, Cimini LJ, Li GY (2014) Energy-efficient resource allocation for OFDMA-based multi-RAT networks. IEEE Trans Wirel Commun 13(5):2696–2705
12. Yu G, Jiang Y, Xu L, Li GY (2015) Multi-objective energy efficient resource allocation for multi-RAT heterogeneous networks. IEEE J Sel Areas Commun 33(10):2118–2127
13. Miao J, Hu Z, Yang K, Wang C, Tian H (2012) Joint power and bandwidth allocation algorithm with QoS support in heterogeneous wireless networks. IEEE Commun Lett 16(4): 479–481
14. Fu B, Wei Z, Yan X, Zhang K, Feng Z, Zhang Q (2015) A game-theoretic approach for bandwidth allocation and pricing in heterogeneous wireless networks. In: Proceedings of the IEEE wireless communications and networking conference (WCNC '15), pp 1684–1689, IEEE, New Orleans, La, USA (Mar 2015)

15. Wang HJ, Katz RH, Giese J (1999) Policy-enabled handoffs across heterogeneous wireless networks. In: Proceedings of the 2nd IEEE workshop on mobile computing systems and applications, WMCSA '99, Feb 25–26, 1999, pp 51–60
16. Hou J, O'Brien DC (2006) Vertical handover-decision making algorithm using fuzzy logic for the integrated radio-and-OW system. IEEE Trans Wirel Commun 5(1):176–185
17. Lee W, Cho D-H (2011) Enhanced group handover scheme in multiaccess networks. IEEE Trans Veh Technol 60(5):2389–2395
18. Ciubotaru B, Muntean G-M (2009) SASHA—a quality oriented handover algorithm for multimedia content delivery to mobile users. IEEE Trans Broadcast 55(2):437–450
19. Mckeown N (2009) Software-defined networking. INFOCOM keynote talk, Apr 2009
20. McKeown N, Anderson T, Balakrishnan H, Parulkar G, Peterson L, Rexford J, Shenker S, Turner J (2008) Openflow: enabling innovation in campus networks. SIGCOMM Comput Commun Rev 38(2):69–74
21. OpenFlow Switch Specification, Version 1.4.0 (Wire Protocol 0x05). Available: https://www.opennetworking.org/images/stories/downloads/sdn-resources/onf-specifications/openflow/openflow-specv1.4.0.pdf
22. Ku I, Lu Y, Gerla M, Ongaro F, Gomes RL, Cerqueira E (2014) Towards software-defined VANET: architecture and services. In: Proceedings of 13th annual mediterranean ad hoc networking workshop (MED-HOC-NET), (IEEE Xplore), pp 103–110
23. Taniuchi K, Ohba Y, Fajardo V, Das S, Tauil M, Cheng Y-H, Dutta A, Baker D, Yajnik M, Famolari D (2009) IEEE 802.21: media independent handover: features, applicability, and realization. IEEE Commun Mag 47(1):112–120
24. Akhunzada A, Ahmed E, Gani A, Khan M, Imran M, Guizani S (2015) Securing software defined networks: taxonomy, requirements, and open issues. IEEE Commun Mag 53(4): 36–44
25. Andrews JG, Baccelli F, Ganti RK (2011) A tractable approach to coverage and rate in cellular networks. IEEE Trans Commun 59(11):3122–3134
26. Lee W, Cho D-H (2011) Enhanced group handover scheme in multiaccess networks. IEEE Trans Veh Technol 60(5):2389–2395
27. Yang K, Gondal I, Qiu B, Dooley LS (2007) Combined SINR based vertical handoff algorithm for next generation heterogeneous wireless networks. IEEE Golbecom, 4483–4487
28. Shannon CE (2001) A mathematical theory of communication. ACM SIGMOBILE Mob Comput Commun Rev 5(1):3–55
29. Song Q, Jamalipour A (2005) A network selection mechanism for next generation networks. In: IEEE international conference on communications (ICC), vol 2, pp 1418–1422
30. Qiang L, Li J, Ji Y, Huang C (2016) A novel software-defined networking approach for vertical handoff in heterogeneous wireless networks. In: Wireless communications mobile computing online in Wiley Online Library (wileyonlinelibrary.com). https://doi.org/10.1002/wcm.2690

Design and Implementation of an Energy-Efficient, Low-Cost Robotic Solar Tracker

K. Damayanti, T. Sunil Reddy, B. M. Reddy, Avireni Srinivasulu and SM-IEEE

Abstract In the recent past, solar panel (Konar and Mandal in IEE Proc A Sci Meas Technol 138:237–241, 1991, [1]) has been under use to convert solar energy into electrical energy. The solar systems can be connected to electrical grids and also be used as a stationary system. The maximum solar power received by the earth is around 84 TW, of this, 12 TW of power is consumed every day by the world. A concept has been proposed (Gruber in Proceedings of IEEE power electronics specialists conference, pp 128–138, 1970, [2], Snyman and Enslin in Proceedings of 20th annual IEEE power electronics specialists conference, vol 1, pp 67–74, 1989, [3], Yang et al. in IEEE Trans Circuits Syst I Regul Pap 59:1546–1556, 2012, [4]) that by using Arduino extract maximum energy from the solar panel (Ram and Rajasekar in IEEE Trans Power Electron 32:8486–8499, 2017, [5]). If the solar panels are placed perpendicular to the Sun maximum amount of solar energy can be extracted (Kim et al. in Proceedings of the 31st intersociety energy conversion engineering conference, vol 3, pp 1673–1678, 1996, [6], Luque-Heredia et al. in Proceedings of 3rd world conference on photovoltaic energy conversion, vol 1, pp 857–860, 2003, [7]). So, it is important to track the Sun's location and position of the solar panel. In the case of fixed solar panels, energy conversion at morning and evening is less, but it is more at noon (Liu et al. in Proceedings of IEEE 3rd international future energy electronics conference and ECCE Asia (IFEEC 2017—ECCE Asia), pp 932–935, 2017, [8]). As at noon, the solar panel is placed perpendicular to the sun. The proposed work is to design a system to track the sun's position (Kobayashi et al. in IEEE Trans Industr Electron

K. Damayanti · T. Sunil Reddy · B. M. Reddy
Department of Electronics & Communication Engineering,
V.F.S.T.R University (Vignan's University), Vadlamudi 522213, Andhra Pradesh, India
e-mail: damayanthi5088@gmail.com

T. Sunil Reddy
e-mail: tetala.sunilreddy96@gmail.com

A. Srinivasulu (✉)
Department of Electronics & Communication Engineering,
JECRC University, Jaipur 303905, Rajasthan, India
e-mail: avireni_s@yahoo.com

V. Nath and J. K. Mandal (eds.), *Nanoelectronics, Circuits and Communication Systems*, Lecture Notes in Electrical Engineering 511,
https://doi.org/10.1007/978-981-13-0776-8_12

53:495–499, 2006, [9]). Photoresistors will be used as sensors in this system. The system consists of light-dependent resistors, Arduino Uno, RC servomotors, and a solar panel. The proposed method will give an output up to 20% more energy than solar panels without entering tracking systems.

Keywords Maximum energy · Solar pannel · LDR · Solar tracker

1 Introduction

For varied applications throughout the world, conversion of sunlight into electrical energy is the most resourceful method through photovoltaic technology. It is used for electrification, water pumping, satellite communication, etc. This energy efficiency depends on climatic conditions such as temperature and solar radiation [8]. The majority of the solar systems are fixed solar modules. Solar photovoltaic system is used to convert solar energy into DC electricity which is one among the simple and clean energy conversions. In Malaysia, primary sources of electrical energy are coal and oil for a long period of time and they are now extinct. So that an alternative renewable energy [4, 11, 12, 13, 14, 15, 16, 17, 24] system is created to produce electricity through solar energy panels.

1.1 Solar Tracker [18, 1]

There are many solar tracking systems available in the world. They can be classified as

(a) Single-axis tracking system [13]
(b) Dual-axis tracking system [16, 20]

In single-axis tracking system, only one servomotor is used which gives an output of 0–180° as the Sun moves from east to west. But in dual-axis tracking system, two servomotors are used in *x*, *y*-axis position in a 2 D plane. In this tracking system, each servomotor has an output of 0–180° and with the combination of these two servomotors gives an output of 0–360°.

The most advantage of our project is tracking the sun's position using light-dependent resistors (LDRs) which can be seen in Fig. 1. Here we have used solar panel, LDRs, resistors, Arduino Uno board, and RC servomotors for a perfect solar tracking system [11, 21]. Using this solar tracking system, we can extract more power when compared to the fixed solar tracking system. We can extract 100% more energy from the Sun, when the solar panel is placed perpendicular to the Sun. Maximum amount of solar energy can be extracted [7, 10].

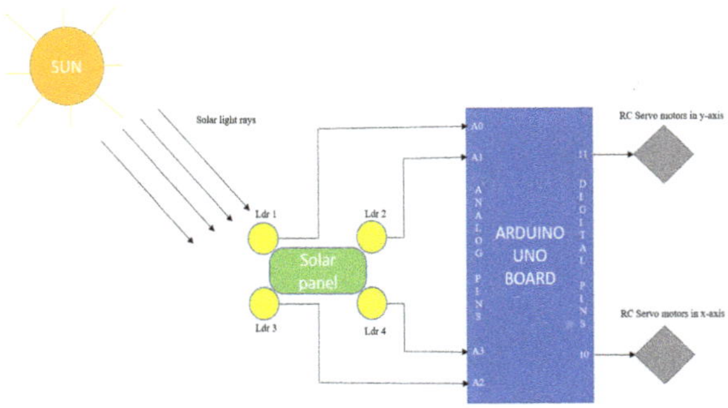

Fig. 1 Block diagram

1.2 Methodology

This project is summed up in two parts, i.e., hardware and software. Hardware consists of LDRs, solar panel, and RC servomotors. Software consists of Arduino Uno board which includes programming. A novel algorithm has been proposed for comparing the four LDR's and for rotation of the RC servomotors.

2 Hardware Implementation

2.1 Hardware [16]

First, LDRs have high resistance. When a high-intensity light falls on the LDR's then its resistance drops and allows current to flow through it. LDR's are light sensitive devices. In this solar tracking system, LDR sensors measure sunlight intensity and are taken as input signal.

When the solar rays fall on the four LDRs then the output of the LDRs is compared and based on this comparison, we get output as rotation of RC servomotors.

2.2 Solar Panel

Solar panel consists of photovoltaic cells [17, 22, 23] which convert solar energy into electricity as shown in Fig. 2. It consists of a junction which divides protons and electrons into equal halves. When the solar rays hit the solar panel, the protons and electrons move away from this junction and generate current flow between two oppositely charges of layers. The position of the Sun changes time to time so it is required to position the solar panel perpendicular to Sun for an increase in the efficiency of photovoltaic cell [5, 13, 1, 24, 25].

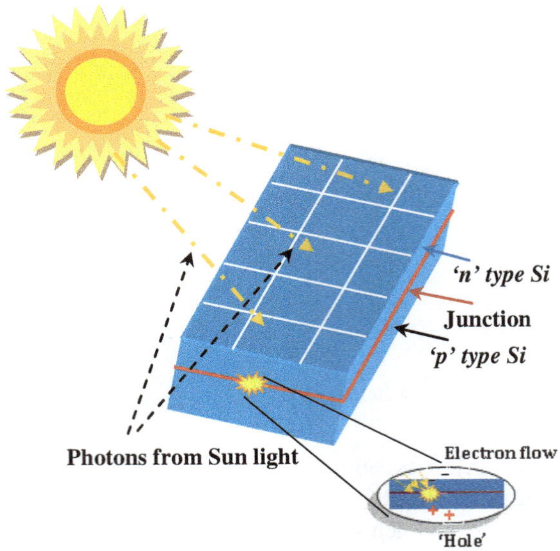

Fig. 2 Solar panel

2.3 RC (Radio Control) Servomotors

Servomotor contains a DC motor [26] and servo arm (output shaft) which is connected to a motor with gear system and a circuit which controls the shaft's position and the main objective of this is to control the angular position of the RC servo. PWM signal is given as input to voltage converter (IC LM 358) and output to amplifier [18].

In Fig. 3, output shaft is given as input to potentiometer. Output of the potentiometer is given as input to amplifier which compares the current position of the shaft with the desired position of the shaft, i.e., PWM signal and if any difference is traced out, it is known as error signal and it is amplified and is used to drive the dc motor in the required direction by eliminating the error. The output shaft then gets the desired position when the error is zero.

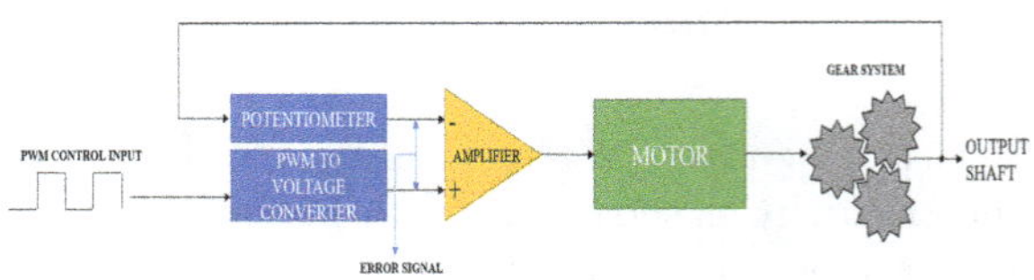

Fig. 3 RC servomotor

2.4 Arduino Uno Board and Code

Arduino (IDE) is the open source software which makes it easy to write the code and upload this to the board. This runs on Linux, Mac OS X, and Windows. The code can be written in Java, which is based on processing and other softwares. Any version of the Arduino can be used in this software which can be seen in the control flowchart of Fig. 4.

Here, Arduino board is used to interface LDRs with servomotors.

Code

```
#include <Servo.h>
Servo bottomservoxaxis;
Servo topservoyaxis;
int bottomservox = 0;
int topservoy = 0;
int topservoymaxLimit = 160;
int topservoyminLimit = 0;
```

Fig. 4 Control logic flowchart [11]

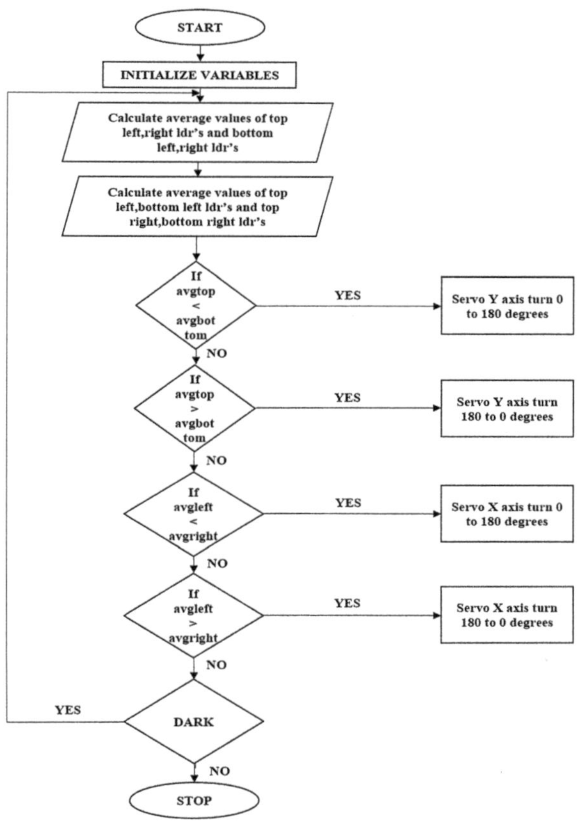

```
int ldrattopleft = A0;
int ldrattopright = A1;
int ldratbottomleft = A2;
int ldratbottomright = A3;
int topleft=0;
int topright=0;
int bottomleft=0;
int bottomright=0;
int averageoftop=0;
int averageofbottom=0;
int averageofleft=0;
int averageofright=0;
  void setup ()
  {
topservoyaxis.attach(11);
topservoyaxis.write(0);
bottomservoxaxis.attach(10);
bottomservoxaxis.write(0);
Serial.begin(9600);
delay(2000);
  }
 void loop()
 {
bottomservox = bottomservoxaxis.read();
topservoy = topservoyaxis.read();
topleft = analogRead(ldrattopleft);
topright = analogRead(ldrattopright);
bottomleft = analogRead(ldratbottomleft);
bottomright = analogRead(ldratbottomright);
averageoftop = (topleft + topright) / 2;
averageofbottom = (bottomleft + bottomright) / 2;
averageofleft = (topleft + bottomleft) / 2;
averageofright = (topright + bottomright) / 2;
Serial.print("average of top:");
Serial.println(averageoftop);
Serial.print("average of bottom:");
Serial.println(averageofbottom);
Serial.print("average of left:");
Serial.println(averageofleft);
Serial.print("average of right:");
Serial.println(averageofright);
delay(1000);
  if (averageoftop<averageofbottom)
  {
Serial.println("Servo y-axis turn 0 to 180 degrees");
```

```
topservoyaxis.write(topservoy +5);
    if (topservoy>topservoymaxLimit)
     {
topservoy = topservoymaxLimit;
     }
delay(10);
   }
   else if (averageoftop>averageofbottom)
   {
Serial.println("Servo y-axis turn 180 to 0 degrees");
topservoyaxis.write(topservoy -5);
    if (topservoy<topservoyminLimit)
   {
topservoy = topservoyminLimit;
   }
delay(10);
   }
   else
   {
topservoyaxis.write(topservoy);
   }
   if (averageofleft<averageofright)
   {
Serial.println("Servo x-axis turn 0 to 180 degrees");
bottomservoxaxis.write(bottomservox +5);
delay(10);
   }
   else if (averageofleft>averageofright)
   {
Serial.println("Servo x-axis turn 180 to 0 degrees");
bottomservoxaxis.write(bottomservox -5);
delay(10);
   }
   else
   {
bottomservoxaxis.write(bottomservox);
   }
delay(10);
  }
```

3 Experimental Results

The Arduino board, solar panel, servomotors, and connecting wires have been assembled and developed as an energy-efficient solar tracker. This solar tracker system can track the Sun moving from east to west or from northeast to southwest changes. As this solar tracker consists of two servomotors in *X*-axis and *Y*-axis, it covers 360° rotation and track in all directions (east, west, north, and south).

When the light intensity is more on B and B^{I} LDRs, the bottom servomotor remains constant and top servomotor rotates in clockwise direction. The snapshot of the same is shown in Fig. 5.

Similarly, when the light intensity is more on A and B-LDRs, the bottom servomotor rotates in clockwise direction and top servo remains constant. The snapshot of the same is shown in Fig. 6.

Fig. 5 More light intensity on B and B^{I} LDRs

Fig. 6 More light intensity on A and B LDRs

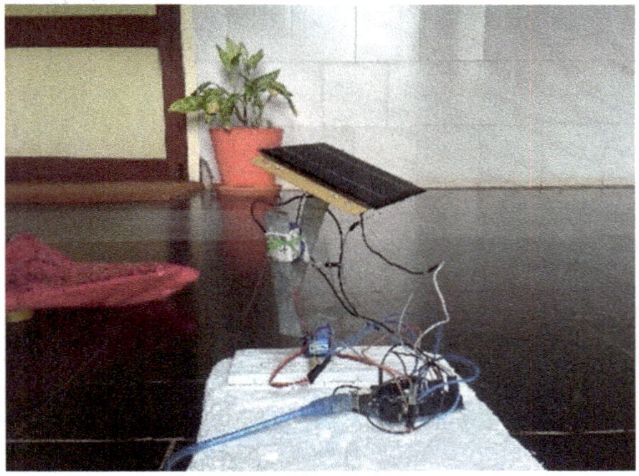

Fig. 7 More light intensity on A^I and B^I LDRs

Fig. 8 More light intensity on A and A^I LDRs

Likewise, when the light intensity is more on A^I and B^I LDRs, the bottom servo rotates anticlockwise direction and top servo remains constant. The snapshot of the same is shown in Fig. 7.

Similarly, when the light intensity is more on A and A^I LDRs, the bottom servomotor remains constant and top servo rotates in anti-clockwise direction. The snapshot of the same is shown in Fig. 8 (Table 1).

Table 1 The results of proposed model with other models

S.No.	Our project	Others project
1	Less expensive	More expensive
2	Easily portable	Not easily portable
3	360° tracking	180° tracking
4	20% more efficient	10% more efficient
5	Less complex to understand	More complex to understand

4 Conclusion

An easy and new code in C language has been generated in Arduino software for solar tracking system and worked out practically with an excellent result. This concept and controls can be used for big modules for tracking the Sun location. Hence, the proposed system shall be effectively used for the renewable energy. This can also be used with a compact model to generate power for individual houses, remote areas weather monitoring and tracking system, etc.

References

1. Konar A, Mandal AK (1991) Microprocessor based automatic Sun tracker. IEE Proc A Sci Meas Technol 138(4):237–241. https://doi.org/10.1049/ip-a-3.1991.0032
2. Gruber R (1970) High efficiency solar cell array peak power tracker and battery charger. In: Proceedings of IEEE power electronics specialists conference, pp 128–138. https://doi.org/10.1109/pecs.1970.7066248
3. Snyman DB, Enslin JHR (1989) Combined low-cost, high-efficient inverter, peak power tracker and regulator for PV applications. In: Proceedings of 20th annual IEEE power electronics specialists conference, vol 1, pp 67–74. https://doi.org/10.1109/pesc.1989.48474
4. Cristaldi L, Faifer M, Rossi M, Toscani S (2014) An improved model-based maximum power point tracker for photovoltaic panels. IEEE Trans Instrum Meas 63(1):63–71. https://doi.org/10.1109/TIM.2013.2277579
5. Kislovski AS (1990) Dynamic behaviour of a constant-frequency buck power cell in a photovoltaic battery charger with a maximum power tracker. In: Fifth annual proceedings on applied power electronics conference and exposition, pp 212–220. https://doi.org/10.1109/apec.1990.66414
6. Kim Y, Jo H, Kim D (1996) A new peak power tracker for cost-effective photovoltaic power system. In: Proceedings of the 31st intersociety energy conversion engineering conference, Washington, DC, USA, vol 3, pp 1673–1678. https://doi.org/10.1109/iecec.1996.553353
7. Luque-Heredia I, Martin C, Mananes MT, Moreno JM, Auger JL, Bodin V, Alonso J, Diaz V, Sala G (2003) A subdegree precision sun tracker for 1000X microconcentrator modules. In: Proceedings of 3rd world conference on photovoltaic energy conversion, Osaka, Japan, vol 1, pp 857–860
8. Liu H-D, Hua Lin C, Cheng W-X, Shih W-C, Chen L-R (2017) A MPPT control strategy of solar power systems for low irradiance conditions. In: Proceedings of IEEE 3rd international future energy electronics conference and ECCE Asia (IFEEC 2017—ECCE Asia), Kaohsiung, Taiwan, 3–7 June 2017, pp 932–935. https://doi.org/10.1109/ifeec.2017.7992165

9. Kobayashi K, Matsuo H, Sekine Y (2006) An excellent operating point tracker of the solar-cell power supply system. IEEE Trans Industr Electron 53(2):495–499. https://doi.org/10.1109/TIE.2006.870669
10. Abdelwahab H, Nawari M, Abdalla H (2017) Maximizing solar energy input for Cubesat using sun tracking system and a maximum power point tracking. In: Proceedings of international conference on communication, control, computing and electronics engineering (ICCCCEE), Khartoum, Sudan, 16–18 Jan 2017, pp 1–7. https://doi.org/10.1109/iccccee.2017.7866695
11. Jaen C, Pou J, Capella G, Arias A, Lamich M (2009) On the use of sun trackers to improve maximum power point tracking controllers applied to photovoltaic systems. In: Proceedings of compatibility and power electronics (CPE'09), Badajoz, Spain, 20–22 May 2009, pp 67–72. https://doi.org/10.1109/cpe.2009.5156015
12. Otieno CA, Nyakoe GN, Wekesa CW (2009) A neural fuzzy based maximum power point tracker for a photovoltaic system. In: AFRICON 2009, Nairobi, Kenya, 23–25 Sept 2009, pp 1–6. https://doi.org/10.1109/afrcon.2009.5308552
13. Thulasiyammal C, Sutha S (2011) An efficient method of MPPT tracking system of a solar powered Uninterruptible Power Supply application. In: Proceedings of 1st international conference on electrical energy systems, Newport Beach, CA, USA, 3–5 Jan 2011, pp 233–236. https://doi.org/10.1109/icees.2011.5725334
14. Yang C-Y, Hsieh C-Y, Feng F-K, Chen K-H (2012) Highly efficient analog maximum power point tracking (AMPPT) in a photovoltaic system. IEEE Trans Circuits Syst I Regul Pap 59 (7):1546–1556. https://doi.org/10.1109/TCSI.2011.2177008
15. Keyrouz F, Hamad M, Georges S (2013) A novel unified maximum power point tracker for controlling a hybrid wind-solar and fuel-cell system. In: Proceedings of 8th international conference and exhibition on ecological vehicles and renewable energies (EVER), Monte Carlo, Monaco, 27–30 Mar 2013, pp 1–6. https://doi.org/10.1109/ever.2013.6521526
16. Munna MS, Islam Bhuyan MA, Rahman KM, Hoque MA (2015) Design, implementation and performance analysis of a dual-axis autonomous solar tracker. In: Proceedings of 3rd international conference on green energy and technology (ICGET), Dhaka, Bangladesh, 11–11 Sept 2015, pp 1–5. https://doi.org/10.1109/icget.2015.7315104
17. Devika Menon MK (2016) A novel method to extract maximum power from solar panel of a grid connected phototvoltaic system using phase angle control and hysteresis current control. In: Proceedings of international conference on electrical, electronics, and optimization techniques (ICEEOT), Chennai, India, 3–5 Mar 2016, pp 3510–3516. https://doi.org/10.1109/iceeot.2016.7755357
18. Cantor C (1963) Fine sun tracker for advanced orbiting solar observatory. IEEE Trans Aerosp Navig Electron, Technical_Paper, pp 1.5.4-1–1.5.4-10. https://doi.org/10.1109/tane.1963.4502242
19. Ram JP, Rajasekar N (2017) A novel flower pollination based global maximum power point method for solar maximum power point tracking. IEEE Trans Power Electron 32(11):8486–8499. https://doi.org/10.1109/TPEL.2016.2645449
20. Fathabadi H (2017) Novel online sensorless dual-axis Sun tracker. IEEE/ASME Trans Mechatron 22(1):321–328. https://doi.org/10.1109/TMECH.2016.2611564
21. Nanda L, Dasgupta A, Rout UK (2017) Smart solar tracking system for optimal power generation. In: Proceedings of 3rd international conference on computational intelligence & communication technology (CICT), Ghaziabad, India, 9–10 Feb 2017, pp 1–5. https://doi.org/10.1109/ciact.2017.7977340
22. Tse KK, Ho BMT, Chung HS-H, Hui SYR (2004) A comparative study of maximum-power-point trackers for photovoltaic panels using switching-frequency modulation scheme. IEEE Trans Industr Electron 51(2):410–418. https://doi.org/10.1109/TIE.2004.825226
23. Kim I-S, Youn M-J (2005) Variable-structure observer for solar-array current estimation in a photovoltaic power-generation system. IEE Proc Electr Power Appl 152(4):953–959. https://doi.org/10.1049/ip-epa:20045245

24. Barsoum N (2009) Implementation of a prototype for a traditional solar tracking system. In: Proceedings of Third UKSim European symposium on computer modeling and simulation, Athens, Greece, 25–27 Nov 2009, pp 23–30. https://doi.org/10.1109/ems.2009.59
25. Kaur T, Mahajan S, Verma S, Priyanka, Gambhir J (2016) Arduino based low cost active dual axis solar tracker. In: Proceeding of IEEE 1st international conference on power electronics, intelligent control and energy systems (ICPEICES), Delhi, India, 4–6 July 2016, pp 1–5. https://doi.org/10.1109/icpeices.2016.7853398
26. Marian-Ştefan N, Radu-Florin M, Petre-Marian N, Ionuţ-Daniel S (2017) A Hardware-in-the-Loop control strategy of a permanent magnet synchronous motor used in positioning a solar tracker. In: Proceedings of international conference on optimization of electrical and electronic equipment (OPTIM) & 2017 International Aegean conference on electrical machines and power electronics (ACEMP), Brasov, Romania, 25–27 May 2017, pp 809–814. https://doi.org/10.1109/optim.2017.7975069

Fast Squaring Technique for Radix Vicinity Numbers for Radix $2^n \pm M$ with Reduced Computational Complexity

Arindam Banerjee, Arpan Deyasi, Swapan Bhattacharyya and Angsuman Sarkar

Abstract A fast squaring technique for the operands nearer to a particular radix is reported in this paper. The technique offers significantly less computational complexity which reduces the processing time to a large extent. Considering the applicability of the technique for a particular set of numbers which are in the vicinity of the corresponding radix, it is shown that the technique is equally acceptable for very large radix number. The proposed approach has been verified using Xilinx ISE Vertex-7 FPGA device. Simulated findings showed that with a significant increment in radix value (99.87%), the change of propagation delay is extremely small (2.33%), and corresponding increase in power dissipation is within tolerable range (29.84%). The results show the improvement in speed and computational complexity.

Keywords Squaring · Vicinity · Radix · FPGA

A. Banerjee · S. Bhattacharyya
Department of Electronics and Communication Engineering,
JIS College of Engineering, Kalyani, India
e-mail: banerjee.arindam1@gmail.com

S. Bhattacharyya
e-mail: swapanbhattacharyya@ieee.org

A. Deyasi (✉)
Department of Electronics and Communication Engineering,
RCC Institute of Information Technology, Kolkata, India
e-mail: deyasi_arpan@yahoo.co.in

A. Sarkar
Department of Electronics and Communication Engineering,
Kalyani Government Engineering College, Kolkata, India
e-mail: angsumansarkar@ieee.org

V. Nath and J. K. Mandal (eds.), *Nanoelectronics, Circuits and Communication Systems*, Lecture Notes in Electrical Engineering 511,
https://doi.org/10.1007/978-981-13-0776-8_13

1 Introduction

Multipliers and squarers play crucial role in designing ALU architecture for different types of processors used in digital signal processing, image processing, artificial neural network, multimedia application, etc. [1–9]. Squarer design using multiplier is already available in various literatures but to compute the square of large operands, significantly large computations are required which result in degradation of the performance of the processors. Patent has been made by invoking a novel method for graphics blending operation where squaring operation is made by multiplier with truncation circuit [10]. Digital squaring circuit is already incorporated for frame synchronization in digital radio [11]. Transcendental function designs [12] using hardware module have been realized by approximate squaring circuits.

If large computations are required, then the CPU usage time increases significantly. Moreover, the increase in computational complexity may cause severe interconnection problem as edge connection increases linearly with edge length, whereas circuit space requirement increases with the square of the factor [13]. This enhances the idea of functional modularization [14], which represents the necessity of designing radix system with higher order.

From last few decades, researchers are working on the performance improvement of the processors by modifying the structure of ALU by incorporating faster architectures of multipliers and squarer. Many researchers have designed multipliers in ternary and quaternary radix system [4–9]. But all the designs are dedicated to a particular radix system (radix 3 and 4). Authors have already proposed [15] a generalized ALU design scheme for $(2^n \pm m)$ radix system, and as per the knowledge of the authors, rare literatures are available on the proposed generalized radix system. The present work is dedicated to faster squarer design which is computationally effective for larger operands which are in close vicinity to the radix. For any radix system, the technique provides very fast square computation of comparatively large operands which are nearer to the corresponding radix. As per the literature survey made so far, the authors can claim that the proposed technique is new and there is a rare scope of comparison with the other literatures reported earlier. The authors have only focused on the proposal of squaring arithmetic for fast computation in the generalized radix system. Results show the importance of the present work in terms of reduced computational complexity. The results have been verified by using Xilinx ISE 14.1 simulator using Vertex-7 XC7VX330t device.

2 Mathematical Background

Let us consider an operand A whose squaring is to be determined. Here, A is in the range from 0 to (Radix-1), where Radix $= 2^n \pm m$, n and m are any positive integers. To achieve fast squaring, A must be in vicinity to the radix. Now let the residue be res = Radix − A.

$$A^2 = (\text{Radix} - \text{res})^2 = \text{Radix}(\text{Radix} - 2 \times \text{res}) + \text{res}^2 \tag{1}$$

$$= \text{Radix}(A - \text{res}) + \text{res}^2 \tag{2}$$

$$= (A - \text{res})\text{Radix}^1 + \text{res}^2\text{Radix}^0 \tag{3}$$

Here "res" should be such a number that its square does not exceed the radix, i.e., $\text{res}^2 < \text{Radix}$. Thus, the range of the numbers, for which the squaring can be performed using the proposed technique, is restricted. As for example, for Radix = 15, if 11^2 is performed then $\text{res} = \text{Radix} - A = 15 - 11 = 4$. So, $\text{res}^2 = 4^2 = 16 > \text{Radix}$. Therefore, such kind of squaring should be avoided. Thus, we need to set the range of operands for this proposed technique. Here, we have set a variable indicated as "maxvalue" above which the numbers are selected as operand. The examples given below are using the proposed squaring technique in different radices.

Example 1 **For Radix-4**: Consider the squaring of a number "3" which belongs to radix-4. It can be written as $3 = 4 - 1$. Here, $A = 3$, Radix = 4 and res = Radix − A = 1. So, $3^2 = 4(3 - 1) + 1 = 2 \times 4^1 + 1 \times 4^0 = (21)_4$.

Example 2 **For Radix-6**: Consider the squaring of a number "4" which belongs to radix-6. It can be written as $4 = 6 - 2$. Here, $A = 4$, Radix = 6, and res = Radix − A = 2. So, $4^2 = 6(4 - 2) + 2^2 = 2 \times 6^1 + 4 \times 6^0 = (24)_6$.

Example 3 **For Radix-15**: Consider the squaring of a number "12" which belongs to radix-15. It can be written as $12 = 15 - 3$. Here, $A = 12$, Radix = 15, and res = Radix − A = 3. So, $12^2 = 15(12 - 3) + 3^2 = 9 \times 15^1 + 9 \times 15^0 = (99)_{15}$.

In the above three examples, all the operands are nearer to the radix and belong to the range assigned above the "maxvalue". As per the rule, for obtaining the "maxvalue", set earlier, maxvalue = Radix − res_{min} for which $\text{res}^2_{\text{min}} \geq \text{Radix}$. For radix-4, maxvalue = 2; for radix-6, maxvalue = 3; and for radix-15, maxvalue = 11. Based on the rule, the proposed technique has been implemented.

If "res" is quite large to perform square for large operands, then it can be also treated as an operand in a convenient radix system and its squarer can also be calculated in a different technique proposed in this paper. As for example, if we choose the radix as 755, then $\text{res}^2_{\text{min}} \geq 755$ which implies that $\text{res}_{\text{min}} = 28$ because $\text{res}^2_{\text{min}} = 784$. Therefore, maxvalue = Radix − res_{min} = 755 − 28 = 727. If we choose the number 728 as operand "A", then res = Radix − A = 755 − 728 = 27 whose squaring needs few complex calculations to perform.

For large residue, to perform the squaring operation, the proposed approach is very effective. Let, $\text{res} = m \pm n$, where "m" is an integer ending with 5 and "n" is another integer. Then, "res" can be written as

$$\text{res} = p \times 10^1 + 5 \pm n \tag{4}$$

where $m = p \times 10^1 + 5$. Then, it can be written as

$$\text{res}^2 = (m+n)^2 = m^2 \pm 2mn + n^2 \tag{5}$$

$$= (p \times 10^1 + 5)^2 \pm 2n(p \times 10^1 + 5) + n^2 \tag{6}$$

$$= p^2 \times 10^2 + p \times 10^2 + 25 \pm 2n(p \times 10^1 + 5) + n^2 \tag{7}$$

$$= p \times (p+1) \times 10^2 + 25 \pm (2p+1)n \times 10^1 + n^2 \tag{8}$$

Example 4 In the above case, $\text{res} = 27 = 25 + 2$. Here, $p = 2$ and $n = 2$. Therefore, $\text{res}^2 = 2(2+1)10^2 + 25 + (2 \times 2 + 1) \times 2 \times 10^1 + 4 = 6 \times 10^2 + 25 + 10^2 + 4 = 729$.

3 Proposed Architecture

To realize the proposed technique, three sub-modules are required: (a) Maxvalue generator, (b) subtractor, and (c) residue squarer. The modules are described below.

Maxvalue generator: This module has the role to generate the "maxvalue" above which the operands are selected for squaring. The following algorithm generates the "maxvalue".

Algorithm 1: Maxvalue Generation Algorithm

```
Input: integer radix;
Output: integer maxvalue;
Variable: integer y, z;
y = 1;
do
{
z = y * y;
y = y + 1;
}
while (radix - z ≥ 0);
maxvalue = y - 1;
```

Subtractor: This module has the role to perform subtraction between two numbers. In the algorithm, there are two subtraction operations: (i) subtraction of operand from the radix to generate residue and (ii) subtraction of residue from the operand.

Residue squarer: This module generates the squaring operation of the residue. The algorithm for residue squaring is shown below. The algorithm is developed using Eq. (8).

Algorithm 2: Residue Squaring Algorithm (Function name ressqr)

```
Function integer ressqr (integer residue)
{
Input: integer residue;
Output: integer sqr;
Variable: boolean sign;
Variable: integer p, n, y, z, q, mat1[3];
y = mod (residue, 10);
z = residue - y + 5;
p = z / 10;
if(z >= residue)
{
n = z - residue;
sign = true;
} else
{
n = residue - z;
sign = false;
}
mat1(0) = (n * n) + 25;
mat1(1) = ((2 * p) + 1) * n;
mat1(2) = (p + 1) * p;
q = (mat1(2) * 100) + mat1(0);
if(sign = = false)
{
sqr = q + (mat1(1) * 10);
}
else
{
sqr = q - (mat1(1) * 10);
}
}
```

Here, the input of the function is the residue and the output is assigned by a variable "sqr" which is the square of the residue. The residue squaring operation has been explained elaborately in the following table. Here, the residue is divided into two digits as per Eq. (8). After using the squaring algorithm, the values are put in the corresponding positions as $p(p + 1)$ is placed in the Digit2 position, $(2p + 1)n$ is

Table 1 Digit-wise residue squaring operation

Digit	Digit3 (10^3)	Digit2 (10^2)	Digit1 (10^1)	Digit0 (10^0)
Residue operand			p	$5 \pm n$
Squaring algorithm		$p(p + 1)$	$(2p + 1)n$	$25 + n^2$
Final result	$p(p + 1)10^2 \pm p(p + 1)n10^1 + 25 + n^2 10^0$			

placed in the Digit1 position, and $(25 \pm n^2)$ is placed in the Digit1 position. Finally, we can get the result by combining the three position values (Table 1).

Combining the three modules described above, the overall algorithm shown below has been developed.

Algorithm 3: Overall Squaring Algorithm

```
Input: integer operand, radix;
Output: integer digit0, digit1;
Variable: integer maxvalue, y, z, k;
y = 1;
do
{
z = y * y;
y = y + 1;
}
while (radix - z ≥ 0);
maxvalue = y - 1;
if (operand > maxvalue) then
{
k = radix - operand;
digit0 = ressqr (k);
digit1 = operand - k;
}
else
{
digit0 = 0;
digit1 = 0;
}
```

4 Complexity Analysis

For a single-digit squaring in any radix system, we require large multiplication (multiplication of an operand by its copy). For $(2^n \pm m)$ radix system, if a large operand is chosen to perform its squaring operation, at first the operand is decomposed into several digits. If we consider the number of digits as "x", then we need x^2 product operation followed by $x(x-1)$ addition to generate the partial product matrix. After that, the matrix elements are added. There are "x" rows in the matrix each of which has "x" elements. Thus, it takes another $x(x-1)$ addition operations. So finally x^2 product operations and $2x(x-1)$ addition operations are required in the overall process. Therefore, the computational complexity is $O(x^2)$.

In the proposed technique, the "Maxvalue Generator" module requires exhaustive searching operation. The "Residue Squarer" takes six product operations and five addition (or subtraction) operations. Finally, two subtractions are required, one for generation of residue and the other for generation of square term as shown in Eq. (3). Except the "Maxvalue" generation, only six product operations and seven addition (or subtraction) operations are required. This is true for any radix system and any operand within that radix system. Thus, the complexity should be quite less and it is NP-complete.

5 Result Analysis

Table 2 shows the simulation results of the proposed design implemented in FPGA in different radix systems. The most noticeable thing is that the delay, power, and area are comparatively large when the radix is far from a perfect square number. When the radix is nearer to a perfect square number, then the computational complexity is less which results in less delay, power, and area. As for example, for radix- 67 and 79, the radices are nearer to 64 and 81, respectively. So the computational complexity is quite less. But the radix-4724 is far from the square number 4761. Therefore, the computational complexity is comparatively large which results in large delay, area, and power which can be observed from the table shown above. This is the beauty of the proposed technique.

6 Conclusion

In this paper, a very fast squaring algorithm in any radix system has been described for a particular range of operands which are in close vicinity to the corresponding radix. This is state-of-the-art architecture and can be used to perform the squaring operation of higher order operands in a very fast approach. The main focus of the work is to show the computational complexity of the architecture for larger operands which reflects from the abovementioned table. The area, delay, and power consumption have been estimated by simulation using Xilinx ISE 14.1 simulator, and the designs have been implemented using FPGA Vertex-7 device. The results shown above prove that the technique provides fast operation and also less computational complexity which play a pivotal role in multiple-valued VLSI circuit design.

Table 2 Simulation results for different radix systems

Radix	Area (in Slice no.)	Delay (in ns)	Power (in mW)
7	111	6.28	0.87
15	112	6.29	0.89
33	125	6.35	1.11
45	127	6.38	1.15
57	129	6.41	1.18
67	127	6.39	1.16
79	128	6.40	1.17
83	127	6.39	1.16
93	135	6.43	1.23
123	128	6.38	1.18
157	134	6.44	1.24
193	127	6.38	1.18
227	128	6.38	1.17
287	129	6.38	1.18
357	131	6.40	1.20
392	134	6.42	1.22
473	135	6.43	1.23
567	134	6.42	1.22
653	135	6.43	1.23
727	132	6.38	1.19
831	134	6.42	1.22
973	135	6.43	1.23
1029	127	6.38	1.18
2039	134	6.41	1.19
3139	134	6.41	1.19
4724	135	6.43	1.23
5376	135	6.43	1.24

References

1. C. Lazzari, P. Flores, J. Monteiro, L. Carro (2010) A new quaternary FPGA based on a voltage-mode multi-valued circuit. In: Proceedings of the conference on design, automation and test in Europe, Mar 2010
2. Eaton M (2012) Design and construction of a balanced ternary ALU with potential future cybernetic intelligent systems applications. International conference on cybernetic intelligent systems (CIS), pp 30–35
3. Raad N, Mansour MM (2011) A low power 32-bit quaternary tree adder. In: International conference on energy aware computing, pp 1–2
4. Ishizuka O, Ohta A, Tannno K, Tang Z, Handoko D (1997) VLSI design of a quaternary multiplier with direct generation of partial products. In: 27th international symposium on multiple-valued logic

5. Chu W-S, Current W (1994) Quaternary multiplier circuit. In: 24th international symposium on multiple valued logic
6. Datla SRPR, Thornton MA (2010) Quaternary voltage-mode logic cells and fixed-point multiplication circuits. In: International symposium on multiple valued logic
7. Hareesh P, Chakravathi K, Rao DT (2015) Fast multiplication with partial products using quaternary signed digit number system. Comput Intell Data Min 2:437–446
8. Vijeyakumar KN, Sumathy V, Devi MG, Tamilselvan S, Nair RR (2012) Design of hardware efficient high speed multiplier using modified ternary logic. Proc Eng 38:2186–2195
9. Vranesic ZG, Hamacher VC (1971) Ternary logic in parallel multipliers. Comput J 15 (3):254–257
10. Augustine SL, Buhler DC, Prouty BG (1999) Computer graphics system with improved blending. United States Patent, No. 5,896,136, Apr 20, 1999
11. Junell J, Mikko K (1995) Frame synchronization in a device receiving digital radio transmissions, European Patent application. 95108198.3, May 30, 1995
12. Farmwald PM (1981) High bandwidth evaluation of elementary functions. In: Proceedings of IEEE 5th symposium on. computer arithmetic, pp 139–142
13. Keyes R (1979) the evolution of digital electronics towards VLSI. IEEE J Solid State Circuits SC-14:193–201
14. Smith KC (1981) The prospects for multi valued logic: a technology and applications view. IEEE Trans Comp C-30(9):619–634
15. Banerjee A, Bhattacharyya S, Deyasi A (2016) High speed reconfigurable ALU design for radix ($2^n \pm m$). Adv Ind Eng Manag 5(2)

A Survey on Multiprocessor Scheduling Using Evolutionary Technique

Annu Priya and Sudip Kumar Sahana

Abstract In this paper, various conventional approaches are studied for the task scheduling, precedence–resource constrained, load balancing, and multiprocessor scheduling problems. In parallel machines the sequence of dependent execution setup time for the minimization of makespan in scheduling problems and prepared a concise review. Multiprocessor scheduling is an NP-hard problem, whereas scheduling algorithm schedules the tasks which may or may not be dependent on each other. There are several traditional approaches existing for processor scheduling such as modified critical path (MCP), dominant sequence clustering (DSC), and priority-based multichromosome (PMC). While using these approaches, we achieve partial solutions in less than the minimum computing time. In this paper, an innovative multiprocessor scheduling technique that is inspired by evolutionary techniques has been embodied.

Keywords DAG · Genetic algorithm · Ant colony optimization

1 Introduction

Scheduling algorithm allows us to manage the sequencing of tasks, confine the scheduled task, time management, and optimizing effort and assignments in any production or manufacturing process. Scheduling is used to allocate machinery resources to tasks. Processor scheduling is required for the high performance of computing systems. There are different heuristic methods existing for processor scheduling. Multiprocessor scheduling is considered as NP-hard problems, which are very difficult to solve with conventional techniques. Processor scheduling is

A. Priya (✉) · S. K. Sahana
Department of Computer Science Engineering, Birla Institute of Technology, Mesra, India
e-mail: annu.priya12@yahoo.com

S. K. Sahana
e-mail: sudipsahana@bitmesra.ac.in

V. Nath and J. K. Mandal (eds.), *Nanoelectronics, Circuits and Communication Systems*, Lecture Notes in Electrical Engineering 511,
https://doi.org/10.1007/978-981-13-0776-8_14

classified into two categories: local processor scheduling and global processor scheduling. Conventional techniques mostly use local processor scheduling. Global processor scheduling is grouped into several families depending on scheduling processor techniques. These heuristic scheduling methods are biologically inspired and can be further grouped into several classes of heuristic approaches like an ant colony, bee colony, genetic algorithm, particle swarm optimization, etc. Local processor scheduling is scheduled upon the single-processor platform and also uses list scheduling to construct a schedule, one cycle at a time, whereas in global scheduling it uses the multiprocessor environment. In real-time task scheduling problem on multiprocessor, it has goal of meeting the deadline, and infrequent arrival of task has implicit deadline and such type of scheduling problems is handled by global static processor scheduling techniques. Static processor scheduling is deterministic scheduling policy. It is classified into two subclasses: optimal processor scheduling and suboptimal processor scheduling. Optimal processor scheduling is used when the job executed at that scheduler should know information about the requirement of processor and state of processors. Optimal scheduling reached NP-complete problem, so researchers focused on suboptimal processor scheduling that is classified as approximation and heuristic algorithm. Approximation scheduling uses the computational techniques for searching of entire solution space for an optimal solution. Heuristic approaches such as ant colony optimization, genetic algorithm, etc. provide feasible solutions to a scheduling problem which cannot give the optimal solution. Global dynamic scheduling had better to physically be inherent in a single processor. In dynamic processor scheduling techniques, a number of processors vary during execution. Dynamic processor scheduling leads to degradation of performance because of high run-time overhead. It has two subclasses: physically non-distributed processor scheduling and physically distributed processor scheduling. Physically non-distributed system includes the responsibility for the task scheduling, whereas in physically distributed system the work involved in decision-making should be physically spread among the various processors. Physically distributed processor scheduling is divided into two classes: (i) cooperative processor scheduling and (ii) non-cooperative processor scheduling. In non-cooperative processor scheduling, each separate processor acts as autonomous entities and it works as a decision-making for various resources running independently which affects the decision of rest system. In cooperative system, the goal of each processor is to carry its own part of scheduling and its work toward a common system. In the case of static system, the decision is taken for the taxonomy tree which reaches to the optimal, and for the consideration of the system the suboptimal heuristic solution is to be taken. Non-cooperative processor scheduling is classified into two categories: (i) approximation processor scheduling and (ii) heuristic processor scheduling. In the case of approximate processor scheduling, it is presented for the independent tasks and runs on different processors with different speeds in a multiprocessor environment. Approximation scheduling provides the low polynomial time complexity to the problems. These types of scheduling processes are secure by obtaining the solutions which are closest to the optimal solution. Heuristic processor scheduling approaches are used to find the suboptimal solution in considerable computation time. Heuristic-based method provide nearest optimal solution, and as a result of its appearance, computation time in GA is very less. The rest of this paper is organized as follows: Sect. 2 included processor scheduling techniques. Section 3 contains the literature survey of existing methods for processor scheduling. Section 4 contains problem statement. Section 5 concludes the discussion of comparison of different algorithms. Section 6 concludes about the future implementation and outlines the future scope of this work.

2 Processor Scheduling

Processor scheduling is classified into two categories: local processor scheduling and global processor scheduling. Conventional techniques mostly use local processor scheduling. Figure 1 shows the classification of process scheduling.

2.1 *Global Processor Scheduling*

Global processor scheduling is grouped into several families depending on scheduling processor techniques. These heuristic scheduling methods are

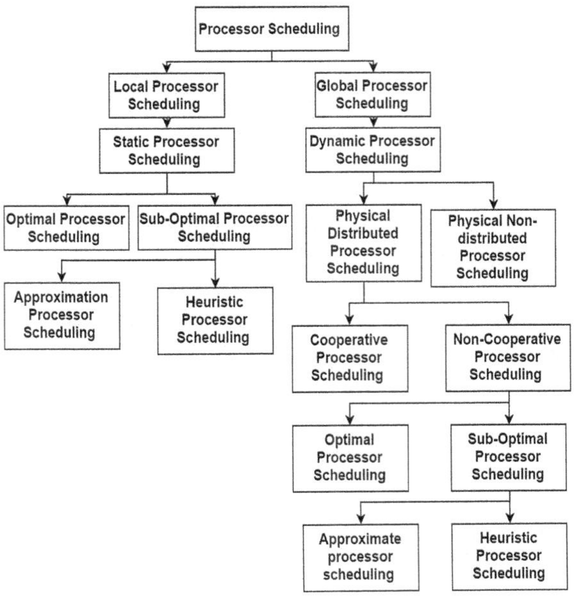

Fig. 1 Classification of processor scheduling

biologically inspired and can be further grouped into several classes of heuristic approaches like an ant colony, bee colony, genetic algorithm, particle swarm optimization, etc.

2.2 Local Processor Scheduling

Local processor scheduling is scheduled upon the single-processor platform and also uses list scheduling to construct a schedule, one cycle at a time, whereas global scheduling processor is scheduled upon multiprocessor platform. In real-time task scheduling problem, the goal of multiprocessor is to meet deadlines, whereas the arrival of tasks is infrequent which has implicit deadlines. To deal with this type of scheduling problem, a global static processor scheduling technique is used.

2.3 Static Processor Scheduling

Static processor scheduling [1, 2] is deterministic scheduling policy. Optimal processor scheduling is used when the job executed at that scheduler should know information about the requirement of the processor and state of processors. The static processor scheduling is divided into two categories: (i) optimal scheduling and (ii) suboptimal processor scheduling. Optimal scheduling reached NP-complete problem, and researchers focused on suboptimal processor scheduling that is classified as approximation and heuristic algorithm. Approximation scheduling technique uses the computational methods for searching of entire solution space for an optimal solution. Heuristic approaches such as ant colony optimization, genetic algorithm, etc. provide feasible solutions to a scheduling problem which cannot give the optimal solution. Suboptimal processor scheduling has two sub-branches: (i) approximate processor scheduling and (ii) heuristic processor scheduling. In multiprocessor environment, the approximate processor scheduling is used to plan the independent tasks where the speed of the processors varies. Approximation scheduling provides the low polynomial complexity to the problems. By using approximation scheduling, we are assured to achieve the optimal solutions. Heuristic processor scheduling approaches find the suboptimal solution in significant time. In heuristic-based approaches, the results of computational time are very optimal and the initial population size is considerably nearby to the solution.

2.4 Dynamic Processor Scheduling

In dynamic processor scheduling techniques [3, 4], the allocated number of processors varies during execution. Dynamic processor scheduling leads to the degradation of performance because of high run-time overhead. Dynamic processor scheduling is divided into two categories: (i) physical distributed processor scheduling and (ii) physical non-distributed processor scheduling. For responsible task scheduling, the physically non-distributed processor scheduling is required, whereas, while using the global dynamic scheduling which is inherent in a single processor and also known as a physically non-distributed, the work involved is to make a decision which should be physically distributed among the processor. The physical distributed processor scheduling is divided into two subcategories: (i) non-cooperative processor scheduling (ii) cooperative processor scheduling. In non-cooperative processor scheduling the individual processor acts as autonomous entities, arrives at decision-making, and makes a choice to select the resources which are independent in nature which affects the decision of the rest system, and in cooperative processor scheduling the goal of each processor is to carry out its own portion of task, whereas all processors are working toward the common system. The same discussion has to be presented in the static case and applies for the dynamic case also. When the taxonomy tree reaches the bottom, we have to consider an optimal [3], suboptimal, and heuristic [3] solutions for the multiprocessor.

3 Literature Survey

There are various research has been done in this field such as C. Jianer and L. Chung Yee proposed a model for processor scheduling in which several alternatives are used a process job and each alternative several machine process the assigned job to them. They invented pseudo-polynomial algorithm [3] to solve optimally two-machine processor problem and also provide a heuristic scheduling algorithm to solve the three-machine processor problem to minimize the completion time for all jobs. X. Yuming and L. Kenli address the task scheduling problems and suggested multiple priorities queuing genetic scheduling [5] for distributed system as well as parallel heterogeneous computing system. They used HEFT approach to search an optimal solution for mapping task of the processor. L. Shih-Tang et al. presented a modified ACO approach DDACS [6] to minimize precedence and resource constraint in multiprocessor scheduling problem. Here, to represent the scheduling problem, a matrix graph is adopted. This matrix graph is used to minimize makespan schedule. A. Hadi Lotfi et al. proposed new coarse-grain genetic algorithm [7] to schedule the tasks and reduce the solution search space and to prevent the speedy convergence between the subpopulation. The initial population is divided into multiple subpopulations and the experimental results of this paper show that the proposed technique will reduce the makespan and it also achieves a better scheduling method in comparison with the other existing methods such as MCP and genetic algorithm. K. Yan and Z. Zhenchao proposed an activity-based genetic scheduling algorithm [8] in which the scheduled tasks run on the heterogeneous grid system network that is signified by directed acyclic graphs (DAGs). First, this approach list all the nodes according to the scheduling algorithm to generate the initial population of GA and it also represents the possible operation sequences so that it reduces the coding space when compared to permutation representation. This approach assigns tasks as activity on the processor to improve quality of the random probability solution and the activity is added to crossover and mutation operator. In Table 1, represent the various techniques proposed by different researchers for solving the problem of processor scheduling.

There are various methods like DSC, MCP, LC, PGA, etc. used for scheduling algorithm. The linear clustering (LC) algorithm uses the recursive grouping of all the nodes in the critical path while zeroing all the edges on the path in one step. In the second step, it schedules the same processor for all partitions that do not execute concurrently. Dominant sequence clustering (DSC) uses the two major ideas: (i) to directly reduce the dominant sequence of the graph and (ii) to create an algorithm with low computational complexity. DSC keeps track of the dominant sequence to reduce parallel time. The complexity order of the algorithm is also reduced. We need to specify certain constraints and definitions to describe DSC such as (i) scheduled: if a processor has been assigned to a node, it should be scheduled; (ii) free: if the processor is unscheduled and all its predecessors are arranged, then the node is free; and (iii) partial free: if it is unscheduled and at least one of its predecessors is unscheduled, then the node is partially free. In this algorithm, the complexity is reduced by confining the range of edges to be zero. Priority-based genetic algorithm (PGA) is based on genetic algorithm, whereas the gene location is used to signify the task node and construct the scheduler among candidate, and the priority of the task node value is used. The proposed PGA method [9] first generates the initial random chromosome, whereas each chromosome is called a gene and each gene is used in the priority node in the DAG structure. According to this technique, it easily validates any encoding permutation to the corresponding scheduler, so that the most recent operator is used for the encoding. Modified critical path algorithm (MCP) explains that the precedence list is prepared on the basis of the "highest bot-level first" ordering. The reason behind is that if the same task has same priority, then to overcome with the critical path the algorithm breaks the task and ties by using the highest priority to the successor tasks and the second highest priority of its successors, and so on. The major objective of this algorithm is to minimize the implementation time so that it becomes more cost-effective. Calculate the cost function of the algorithm in the parallel time environment which is equal to

$$\mathrm{PT} = \max(\mathrm{ST}(nj) + Tj) \leq \max \mathrm{ST}(n_j) + \max Ti.$$

This above equation shows that if we minimize the initial time of the last task then we can get the result in the decline of the overall parallel time.

Table 1 Summary of research development in the field of processor scheduling

Research	Methodology	Technique used	Advantages	Disadvantages
Hadi lotfii et al. [7]	(i) Divided the population into multi-sub-population (ii) Reduce the searching speed (iii) Prevent the convergence and migration between the subpopulations	Coarse-grained genetic algorithm	Minimizing makespan Minimizing communication cost Maximizing CPU utilization	For further solution multi-population combinations of genetic algorithm required
Yan and Zhenchao [8]	(i) Here, GA represents the feasible operation and sequences (ii) Reduce the coding space	Activity-based genetic task scheduling algorithm	The quality of the solution will enhance	The iterative algorithm does not perform well
Ravreet and Gurvinder [11]	Leads to the suboptimal solution	Genetic algorithm (GA) is proposed for static	Minimizing the job completion time, completion time	Not feasible for the multiprocessor systems
Rami and Khalid [14]	(1) One way to prepare the chromosome (2) It uses a new methodical process for the crossover operator	Generic algorithm to resolve the task scheduling problem	(i) Minimize the task completion time (ii) Make the most of the throughput of the system (iii) Decrease the length of schedule	Scheduling tasks in homogeneous parallel multiprocessor systems.
Ricardo et al. [15]	Improved genetic algorithm with the introduction of some knowledge about the scheduling problem	Genetic algorithm-based scheduling problem heuristic in the crossover and mutation genetic	(i) Program's execution time is minimized (ii) Produce good quality solutions in shorter time	Works on homogeneous multiprocessor system
Rashid and Deniz [16]	Genetic algorithm used for optimization of multiprocessor in	Multi-population-based parallel genetic algorithm	(i) Reduced the implementation time	(ii) Improved the quality of scheduling solutions

(continued)

Table 1 (continued)

Research	Methodology	Technique used	Advantages	Disadvantages
	the presence of communication costs.			
Jianer and Chung Yee [3]	(1) Schedule jobs to minimize the completion time of all jobs (2) Machines process the job assigned	Pseudo-polynomial algorithm	Combine fully polynomial scheme and a heuristic to solve the three-machine problem	Integer programming cannot be solved
Yuming and Kenli [5]	Advantages of both evolutionary and heuristic-based algorithms	Multiple priorities queuing genetic algorithm (MPQGA)	(i) Minimizing makespan	Not good for larger task graphs and more processors
Ghafarian et al. [17]	Combined the use of ant colony and evolutionary metaheuristics to search	Cellular automata (CA)	Execution time is minimized.	Get an optimal solution before generation 20
Kumar et al. [18]	Comparative study of the performance of two algorithms GFTS and AFTS	Primary backup (PB)-based fault-tolerant scheduling (PBFTS) technique	Minimize the makespan	Getting the immediate optimal possible results for the non-preemptive scheduling on task sets
Shih et al. [10]	(1) Two-dimensional matrix is used for assigning the jobs on the processors (2) The dynamic rule applied to adjust the earliest starting time of jobs	Modified ant colony optimization (ACO)	(1) Solve the multiprocessor system scheduling problems with resource constraints (2) Minimize makespan	Works for homogeneous processors
Savas [19]	Applying a GA to adapt non-identical parallel machine scheduling problem	Genetic algorithm	Solve hard combinatorial optimization problems	(1) Get better result for small-scale problems (2) Dispatching the rule does not guarantee good result in various problems

4 Problem Statement

From the available literature, it is observed that performance of the different techniques on a particular scenario exhibits as shown in Fig. 2. In multiprocessor scheduling, the critical problem is how to provide the task precedence relation between tasks and processor so that program's execution time is minimized. In real time, the most scheduling problems are NP-hard in nature and it is a very complex problem. For that, performance from the start of the algorithm is very poor. This problem is extremely hard to solve. Most scheduling problems are NP-hard [10] in nature. Scheduling algorithms for processors in hard real time are also complex problem. For such large-scale scheduling problem, the performance of state-of-the-art algorithms is very poor. It is observed that evolutionary and swarm-based algorithms exhibit better performance for large-scale combinatorial problems. Our objective is to analyze, study, apply, and implement possible improvements in the said application using evolutionary and swarm-based

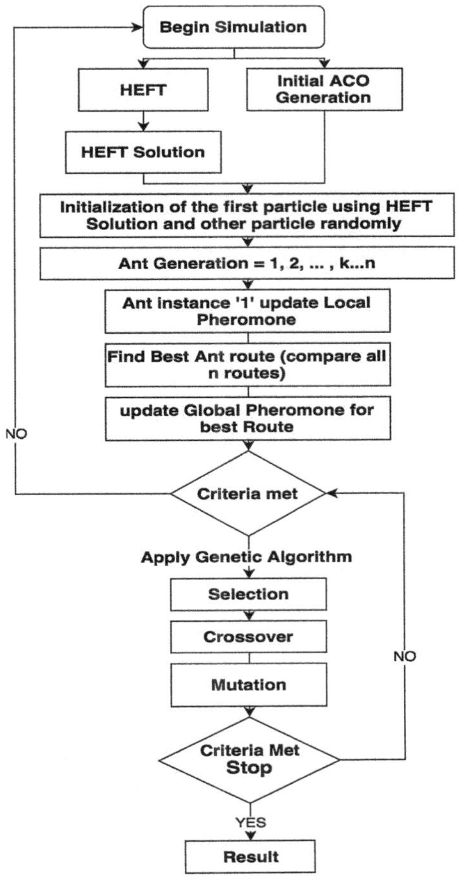

Fig. 2 Flowchart of hybrid system based on ACO and GA with HEFT for task scheduling on heterogeneous nodes and processors

algorithms. After an exhaustive study of the different research papers and case studies, we like to propose a hybrid system based on ACO and GA with HEFT for task scheduling on heterogeneous nodes and processors. The framework consists of two parts: in the first part, ACO and HEFT which are combined and used for finding the local pheromone and global pheromone, whereas in the second part the result of the ACO and HEFT is taken as input for the GA and applied to the selection, crossover, and mutation process.

5 Discussion

Evolutionary techniques effectively solve the scheduling problems. These methods apply the skills of evolutionary and it is observed that performance of the different techniques on particular scenario exhibits superiority of GA over traditional heuristic approaches as shown in Fig. 2. Generally, genetic algorithm is more efficient to solve NP-complete problem of multiprocessor scheduling problems. Here, in this paper, we have discussed some of the processor scheduling algorithms inspired by evolutionary techniques. A comparative study was carried out, and merit and demerit areas are highlighted. Previously, it is observed that increasing the number of task increases the value of makespan as compared to GA [11] and LA [12] methods. After that, coarse-grain genetic algorithm [13] shows that it has the lower makespan value of proposed algorithm than other algorithms. To make this comparison, various algorithms and different parameters have been taken for the execution.

Figure 3 shows that the makespan of MCP is higher than the other five algorithms, whereas DSC, DCP, and MD have the similar makespan, but the number of processors is higher for DSC to achieve the results of MD and DCP algorithm.

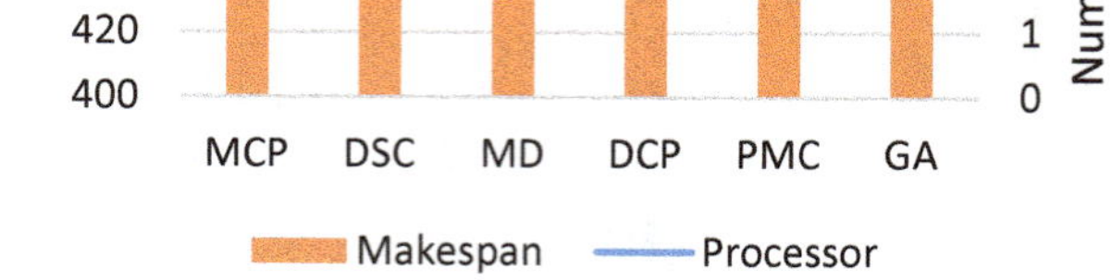

Fig. 3 Comparison between different algorithms

The GA and PMC have used the same number of processor to achieve the same makespan. The comparison of this technique gave the overall scenario for the processor scheduling environment.

6 Conclusion

In this paper, a survey is drawn between different traditional scheduling algorithms used in the parallel multiprocessor system. And it is proven that by using the evolutionary technique such as GA we get the better results. Here, in this paper, we proposed a framework for the multiprocessor task scheduling model using ACO and HEFT. The hybrid structure will generate the best possible result in conventional CPU time. And it also forms the suboptimal solution for allocating the tasks to the homogeneous parallel multiprocessor system. Performance of the hybrid ACO and HEFT scheduling algorithm will produce the better result than GA and other traditional approaches for job scheduling in a multiprocessor environment.

References

1. AI Na'mneh RA, Darabkh KA (2013) A new genetic-based algorithm for scheduling static tasks in homogeneous parallel systems. In: International conference on robotics, biomimetics, intelligent computational systems (ROBIONETICS) Yogyakarta, Indonesia, Nov 2013
2. Wu, Yu H, Jin S, Lin K-C, Schiavone G (2004) An incremental genetic algorithm approach to multiprocessors scheduling. IEEE Trans Parallel Distrib Syst 15(9):824–834
3. Jianer C, Chung-Yee L (1999) General multiprocessor task scheduling. Wiley, Hoboken
4. Apostolos G, Tao Y (1992) A comparison of clustering heuristics for scheduling directed acyclic graphs on multiprocessors. J Parallel Distrib Comput
5. Yuming X, Kenli L, Tung Truong K, Meikang Q (2012) A multiple priority queueing genetic algorithm for task scheduling on heterogeneous computing systems. In: IEEE 14th international conference on high-performance computing and communications 2012
6. Shih-Tang L, Ruey-Maw C, Yueh-Min H, Chung-Lun W (2007) Multiprocessor system scheduling with precedence and resource constraints using an enhanced ant colony system. Elsevier Ltd
7. Hadi lotfii A, Broumandnia A, Shahriar A (2010) Task graph scheduling in multiprocessor systems using a coarse grained genetic algorithm. In: IEEE 2nd international conference on computer technology and development (ICCTD 2010)
8. Yan K, Zhenchao Z, Pengwu C (2011) An activity-based genetic algorithm approach to multiprocessor scheduling. In: IEEE seventh international conference on natural computation
9. ReaKook H, Mitsuo G, Hiroshi K (2006) A performance evaluation of multiprocessor scheduling with genetic algorithm. ReaKook Hwang et al./Asia Pac Manag Rev 11(2):67–72
10. Shih T, Ruey MC, Yueh-Min H, Chung-Lun W (2007) Multiprocessor system scheduling with precedence and resource constraints using an enhanced ant colony system. Elsevier Ltd
11. Ravreet K, Gurvinder S (2012) Genetic algorithm solution for scheduling jobs in multiprocessor environment. IEEE

12. Jahanshahi M, Meybodi MR, Dehghan M (2009) A new approach for task scheduling in distributed systems using learning automata. In: Proceedings of the IEEE international conference on automation and logistics Shenyang, China, Aug 2009
13. Hadi L, Ali B, Shahriar L (2010) Task graph scheduling in multiprocessor systems using a coarse grained genetic algorithm. In: IEEE 2nd international conference on computer technology and development (ICCTD 2010)
14. Rami A, Khalid A (2013) A new genetic-based algorithm for scheduling static tasks in homogeneous parallel systems. In: IEEE international conference on robotics, biomimetics, intelligent computational systems (ROBIONETICS) Yogyakarta, Indonesia, Nov 25–27, 2013
15. Ricardo C, Afonso F, Pascal R (1999) Scheduling multiprocessor tasks with genetic algorithms. IEEE Trans Parallel Distrib Syst 10(8) (Aug 1999)
16. Rashid M, Deniz D (2016) A multi-population based parallel genetic algorithm for multiprocessor task scheduling with communication costs. In: IEEE symposium on computers and communication (ISCC)
17. Ghafarian T, Deldari H, Mohammad R (2009) Multiprocessor scheduling with evolving cellular automata based on ant colony optimization. In: IEEE proceedings of the 14th international CSI computer conference (CSICC'09), 2009
18. Kumar A et al (2014) Aco and Ga based fault-tolerant scheduling of real-time tasks on multiprocessor systems—a comparative study. IEEE
19. Savas_ Balin (2010) Non-identical parallel machine scheduling using genetic algorithm. Elsevier Ltd
20. Kwok Y, Ahmad I (1999) Static scheduling algorithm for allocating directed task graph to multiprocessors. ACM Comput Surv 31(4) (Dec 1999)

Fractional Sobel Filter Based Brain Tumor Detection and Segmentation Using Statistical Features and SVM

Minal Padlia and Jankiballabh Sharma

Abstract In this paper, a scheme for detection and segmentation of brain tumor from T1-weighted and fluid-attenuated inversion recovery (FLAIR) brain images is presented. To prevent the effect of noise and enhance texture of the brain image, fractional Sobel filter is used. Fractional order (α) of fractional Sobel filter provides additional flexibility in improving the segmentation results. Detection of asymmetry between hemispheres is achieved using Bhattacharya coefficients and mutual information. In order to detect the hemisphere containing tumor, histogram asymmetry method is applied. To segment the tumor region from the tumor hemisphere, the statistical features of a defined window are calculated and classified using support vector machine (SVM). Simulations are performed on the images, taken from the BRATS-2013 dataset, and performance parameters such as accuracy, sensitivity, and specificity for different values of α are computed. The simulation results show that the performance of proposed scheme is comparable to the nearest schemes compared.

Keywords Brain tumor · Segmentation · Fractional Sobel filter
Statistical features · Fluid-attenuated inversion recovery (FLAIR)
Support vector machine (SVM)

1 Introduction

Brain tumor is made up of tissues which grow abnormally in the brain. Because of the complex structure of brain, tumor detection and segmentation from brain are not too easy [1]. In the field of medical imaging, magnetic resonance imaging

M. Padlia (✉) · J. Sharma
Department of Electronics Engineering, Rajasthan Technical University, Kota, India
e-mail: minalpadlia@gmail.com

J. Sharma
e-mail: jbsharma@rtu.ac.in

V. Nath and J. K. Mandal (eds.), *Nanoelectronics, Circuits and Communication Systems*, Lecture Notes in Electrical Engineering 511,
https://doi.org/10.1007/978-981-13-0776-8_15

(MRI) gives good results because MRI provides good contrast of different soft tissues of brain [2, 3].

Several segmentation schemes such as feature-based [2–5], atlas-based [6], intensity-based [1, 7, 8], and cluster-based techniques [9] have already been proposed. Intensity-based techniques such as contouring [8], watershed transform [7], and region growing [1] are inflexible to apply for brain tumor detection and segmentation. Because of the intensity similarities between some normal tissues, i.e., cerebrospinal fluid (CSF) and tumor tissues, it can cause confusion within the scheme [10]. To overcome this problem, feature-based tumor segmentation techniques have been proposed. Among feature-based techniques [2], brain tumor segmentation by discriminative random field (DRF) technique is suggested. In this scheme, a set of alignment-based and multiscale image-based features has been extracted for segmentation. Multi-fractal features with modified AdaBoost algorithm-based technique for tumor texture characterization and segmentation have been proposed in [4]. An automatic scheme, detecting tumor hemisphere using histogram asymmetry and segmenting the tumor area through different classification methods like KNN, SVM, SRC, NSC, and the k-means clustering have been presented in [5]. An algorithm-based parallel axis bounding boxes between the hemispheres of a brain MR slice has been proposed in [11] that uses Bhattacharya coefficient to calculate histogram similarity between the test and reference images. A fractional-order edge detection operator used for the medical images of structure feature extraction has been suggested in [12].

None of the segmentation schemes uses fractional Sobel mask, Bhattacharya coefficient, and statistical features together for detection and segmentation of brain tumor tissues from MRI. Thus, this paper presents a novel scheme for the segmentation of brain tumor from T1-weighted and fluid-attenuated inversion recovery (FLAIR) MRI image using fractional Sobel mask, Bhattacharya coefficients, and statistical features with SVM. Fractional-order Sobel filter generates sharper gradient magnitude image with enhanced medium frequency texture details. Therefore, in this paper, fractional Sobel filter is used to enhance the texture features. Fractional-order parameter (α) of fractional Sobel filter provides additional degree of freedom for improvement of segmentation results. Asymmetry between hemispheres is detected using Bhattacharya coefficient, and mutual information is also applied to verify the results of Bhattacharya method. Histogram asymmetry analysis is used for the detection of hemisphere containing tumor. A sliding window is used here to extract statistical features, and classification of tumor tissues is performed through SVM.

The rest of the paper is arranged as follows. Section 2 describes the materials and methods. The proposed algorithm is described in Sect. 3. The experimental results are presented in Sect. 4, and the conclusion is presented in Sect. 5.

2 Materials and Methods

2.1 *Fractional-Order Sobel Filter*

In gradient-based detection scheme, the fractional-order derivatives are used to find the maximal point with the help of first-order derivative [12]. The gradient of an image $A(s,t)$ at location (s,t) based on first-order derivative can be defined as

$$\nabla A(s,t) = \sqrt{G_s^2 + G_t^2} \tag{1}$$

Here, G_s and G_t are gradient magnitude in s and t directions, respectively.

Here, Grunwald–Letnikov (G-L) definition of fractional-order Sobel filter is used. A fractional-order Sobel filter is obtained through the first-order Sobel filter as [12]

$$G_s^\alpha = \frac{1}{2}\left(\frac{\partial^\alpha A(s+1,t-1)}{\partial s^\alpha} + 2\frac{\partial^\alpha A(s+1,t)}{\partial s^\alpha} + \frac{\partial^\alpha A(s+1,t+1)}{\partial s^\alpha}\right) \tag{2}$$

$$G_t^\alpha = \frac{1}{2}\left(\frac{\partial^\alpha A(s-1,t+1)}{\partial t^\alpha} + 2\frac{\partial^\alpha A(s,t+1)}{\partial t^\alpha} + \frac{\partial^\alpha A(s+1,t+1)}{\partial t^\alpha}\right) \tag{3}$$

Here, α is the order of fractional-order Sobel filter. With the help of fractional gradient derivatives, fractional-order Sobel convolution filter is described as follows.

Here,

$$C_k^\alpha = \frac{\Gamma(\alpha+1)}{\Gamma(k+1)(\alpha-k+1)} \tag{4}$$

where gamma function is used and $k = 0, 1, 2, 3 \ldots$

2.2 *Statistical Feature Extraction Schemes*

2.2.1 First-Order Statistical Features

Mean, entropy, kurtosis, RMS, standard deviation, skewness, and variance are valuable first-order statistical features. Mean is the average value of the intensity of the image. Entropy exposes randomness of the intensity value. Kurtosis shows the uniformity of the histogram. RMS is a square root of mean square. Standard deviation is a measure of a set of data that differs from its mean. Skewness measures asymmetry of the histogram around mean. Variance shows the variation in intensity around mean. Formulas are as follows [5]:

$$\text{Mean:}\quad \mu = \sum_{j=0}^{V-1} jP(j) \tag{5}$$

$$\text{Standard deviation:}\quad \sigma = \sqrt{V^{-1}\sum_{j=1}^{V} (j-\mu)^2} \tag{6}$$

$$\text{RMS:}\quad \text{rms} = \sqrt{V^{-1}\sum_{j=1}^{V} (j)^2} \tag{7}$$

$$\text{Variance:}\quad \sigma^2 = V^{-1}\sum_{j=1}^{V} (j-\mu)^2 \tag{8}$$

$$\text{Entropy:}\quad H = -\sum_{j=0}^{V-1} P(j)\log_2[p(j)] \tag{9}$$

$$\text{Skewness:}\quad \mu_3 = \sigma^{-3}\sum_{j=0}^{V-1} (j-\mu)^3 P(j) \tag{10}$$

$$\text{Kurtosis}:\mu_4 = \sigma^{-4}\sum_{j=0}^{V-1} (j-\mu)^4 P(j) \tag{11}$$

where V is the maximum gray value of the brain image and $P(j)$ is the probability density function of the intensity levels which is taken from [5]:

$$P(j) = g(j)/N \tag{12}$$

where $g(j)$ is the total number of pixels that have intensity value j, and N is the total number of pixels in image.

2.2.2 Second-Order Statistical Features

- *Gray-Level Co-occurrence Matrices (GLCMs)*

A second-order statistical method of examining texture which considers the spatial relationship of pixels is known as GLCM and also called as gray-level spatial dependence matrix [5]. In this scheme, $d = 1, 2$ and $\theta = 0°, 45°, 90°, 135°$

are used for the computation. Four GLCM texture features are demonstrated as follows:

$$\text{Contrast}: \sum_{j=0}^{V-1}\sum_{k=0}^{V-1} |j-k| C_{jk} \tag{13}$$

$$\text{Correlation}: \sum_{j=0}^{V-1}\sum_{k=0}^{V-1} \frac{jkC_{jk}-\mu_s\mu_t}{\sigma_s\sigma_t} \tag{14}$$

$$\text{Energy}: \sum_{j=0}^{V-1}\sum_{k=0}^{V-1} (C_{jk})^2 \tag{15}$$

$$\text{Homogeneity}: \sum_{j=0}^{V-1}\sum_{k=0}^{V-1} \frac{C_{jk}}{1+|j-k|} \tag{16}$$

Here, C_{jk} is the element of the co-occurrence matrix.

2.3 Support Vector Machine (SVM)

Support vector machine is a supervised machine learning algorithm which is used for classification. Classification is used to differentiate the two classes by finding the hyperplane. Objective of SVM classifier is to group feature vectors into classes that have similar values of feature [13].

Let the training data *c* consists of *p* samples each in *m* dimensions and belong to class *d*, which is communicated as

$$(c_1, d_1)\ldots(c_i, d_i)\ldots(c_n, d_n), c \in R^m, d \in \{+1, -1\} \tag{17}$$

In this scheme, SVM plans data (c_i, d) into an infinite dimensional hyperplane $\varphi(c_i, d)$ using Gaussian kernel function, and the decision rule is defined by $sign(f(c))$. With the help of weight vector *w* and bias *b*, the discriminate function $f(c)$ generates the optimum hyperplane decision boundary.

$$f(c) = w\Phi(c) + b \tag{18}$$

The optimum values of *w* and *b* are predicted by solving the following problem:

$$\min_{w,\xi}\left\{\frac{1}{2}||w||^2 + C\sum_{j=1}^{n}\xi_j\right\}; y_j(w \times \Phi(x_j) + b) \geq 1 - \xi_j, \xi_k \geq 0 \tag{19}$$

where C is the parameter of regularization and ξ_j is the slack variable. This problem is solved by Lagrange optimization through dual formation, which produces optimum value for weight vector w and bias b. Details of this algorithm are given in [13].

3 Proposed Algorithm

In this section, an algorithm of tumor detection and segmentation from T1-weighted and FLAIR MRI brain images is presented, which includes filtering of brain image, intensity normalization, separate brain image into hemisphere, detection of asymmetry between hemispheres, detection of tumor hemisphere, feature extraction, classification, and performance evaluation.

Brain tumor is first processed through filter process in which fractional-order Sobel filter is used. The frequency characteristics of fractional-order Sobel filter maintain low-frequency contour feature from the smooth areas. This filter also retains high-frequency marginal feature from frequently altered gray levels and enhances medium frequency texture features [12]. Fractional-order filter also has a good noise immunity and high signal-to-noise ratio. Due to low computation complexity, 3 × 3 fractional Sobel mask for $k = 2$ (shown in Fig. 1) is used in this scheme.

Due to the intensity variation in filtered brain image, intensity normalization is required for quantitative texture analysis. In this scheme, histogram normalization method is used which tends to shift and stretch the filtered image histogram to include all gray-level values in the image [13]. Intensity normalization is defined as

Fig. 1 a, b Fractional Sobel filter

(a)

$(-1)^k C_k^\alpha / 2$	$(-1)^k C_k^\alpha$	$(-1)^k C_k^\alpha / 2$
$\vdots$	$\vdots$	$\vdots$
$(\alpha^2 - \alpha/4)$	$(\alpha^2 - \alpha/2)$	$(\alpha^2 - \alpha/4)$
$-\alpha/2$	$-\alpha$	$-\alpha/2$
$1/2$	1	$1/2$

(b)

$(-1)^k C_k^\alpha / 2$	......	$(\alpha^2 - \alpha/4)$	$-\alpha/2$	$1/2$
$(-1)^k C_k^\alpha$	......	$(\alpha^2 - \alpha/2)$	$-\alpha$	1
$(-1)^k C_k^\alpha / 2$	.....	$(\alpha^2 - \alpha/4)$	$-\alpha/2$	$1/2$

$$f(s,t) = \frac{f_{\max} - f_{\min}}{h_{\max} - h_{\min}}(h(s,t) - h_{\min}) + f_{\min} \tag{20}$$

where $h(s,t)$ is the histogram of the filtered image, $h_{\max}$ and $h_{\min}$ are maximum and minimum gray scale level, $f(s,t)$ is the new histogram, and $f_{\max}$ and $f_{\min}$ are the new maximum and new minimum intensity values.

After intensity normalization, separation of brain image into two hemispheres is processed. Brain image is first divided into two hemispheres with the help of centroid and longest axis of the brain. Initially, brain image is separated from the background by converting gray-level brain image into binary image. Thereafter, center mass algorithm is applied to find the center of the brain and longest axis of brain is determined. Thus, the brain is separated into two parts from the centroid.

Now, asymmetry between two hemispheres is detected through Bhattacharya coefficient. Asymmetry of hemispheres is also compared with their neighborhood in order to ensure the presence of tumor in any of the hemispheres. Bhattacharyya coefficient is defined as [11]

$$BC(s,t) = \sum_{j=1}^{n} \sqrt{s_j t_j} \tag{21}$$

where n indicates number of bins in histogram and s and t are samples. Here s_j and t_j denote histograms of block in upper and lower halves. Next, computation for asymmetry will be computed using the following formula"

$$As = 1 - BC \tag{22}$$

where As denotes the asymmetry caused by tumor.

The borderline of the brain is determined just to calculate Bhattacharya coefficient of brain tissue, and it will not consider the background. Here, the borderline is determined through erosion of the binary image by taking disk-shaped structural element (SE) of size one and subtracting eroded image from the binary image. Bhattacharya coefficient between the hemispheres is calculated by taking blocks of size ¼ (for convenience) of brain image. For each block, Bhattacharya coefficient is calculated with its symmetric block in the other hemisphere. If asymmetry is present in between these blocks, the value of Bhattacharya coefficient for that block differs from the other blocks. In order to verify the efficacy of Bhattacharya coefficient, mutual information is also used to detect the asymmetry. Mutual information is defined as [12]

$$I(U;L) = H(U) + H(L) - H(U,L) \tag{23}$$

where

$$H(U) = -\sum_{u} P_U(u) \log_2 P_U(u) \tag{24}$$

$$H(L) = -\sum_{l} P_L(l) \log_2 P_L(l) \tag{25}$$

$$H(U, L) = -\sum_{ul} P_{UL}(u, l) \log_2 P_{UL}(u, l) \tag{26}$$

Here, $H(U)$ = entropy of upper hemisphere, $H(L)$ = entropy of lower hemisphere, $H(U, L)$ = joint entropy, and P is the probability distribution function.

The process of detection of asymmetry by Bhattacharya coefficient for each hemisphere block is same as for the mutual information.

Thereafter, the location of tumor is obtained by computing histogram asymmetry analysis of each block. Block which has less similar histogram than the nearby blocks is treated as block containing tumor. Thus, tumor is detected.

Further, statistical features are extracted to train SVM for classification of tumor and non-tumor region. In this process, some windows of size 45 × 45 are taken from the selected hemisphere. The size of window is selected empirically by experiment for improved results. The borderline and longest axis line are used to limit the windows only in hemisphere's tissue. Statistical features are extracted from these windows for training purpose. For test set of the classifier, a same sized sliding window is used, which pixel by pixel sweeps through the selected hemisphere and first- and second-order statistical features are extracted from selected hemisphere through sliding windows. Here, seven first-order statistical features entropy, kurtosis, mean, RMS, standard deviation, skewness, and variance are extracted and four second-order gray-level co-occurrence matrix (GLCM) statistical features at distance d = 1, 2 and angle θ = 0°, 45°, 90° and 135° are extracted. So the offsets, which are obtained by the distance and angle of GLCM features, are (0 1), (−1 1), (−1 0), (−1 −1), (0 2), (−2 2), (−2 0), (−2 −2). The four second-order GLCM features are contrast, correlation, energy, and homogeneity. Thus, for each GLCM feature, these eight offsets are used. Therefore, total 7 first-order and 32 second-order statistical features make 39 statistical feature vectors (Fig. 2).

Now, SVM is used as a classifier for the classification of tumor and healthy tissues. In SVM, original input space is planned into a higher dimensional space with the help of kernel functions [13]. For training, an equal number of tumor windows are taken as that of healthy windows to balance the training set. Here, linear kernel function is used to classify tumor and healthy tissues. The block diagram of this proposed scheme is as follows (Fig. 3).

The summarized steps of the proposed scheme are as follows:

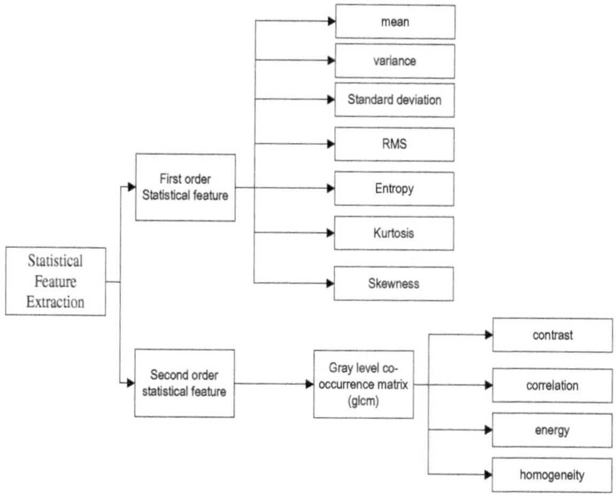

Fig. 2 Statistical feature extractions

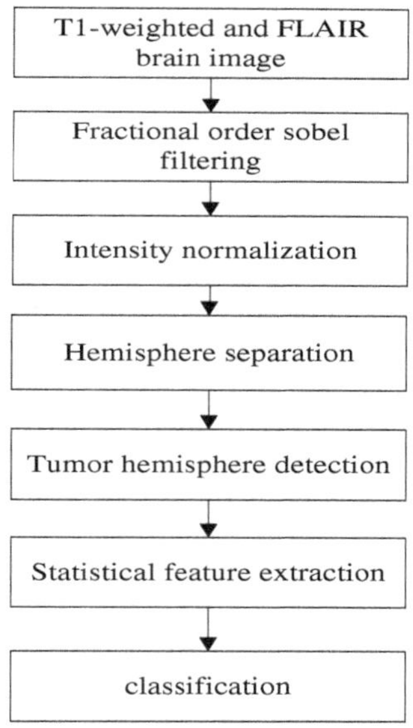

Fig. 3 Flowchart of the proposed system

Step 1: First brain image is taken for filtration through fractional Sobel filter.

Step 2: Filtered image is normalized through histogram normalization.

Step 3: With the help of centroid and longest diameter of brain, normalized image is divided into two brain symmetric hemispheres. In order to detect the asymmetry between hemispheres, Bhattacharya coefficient and mutual information between these hemispheres are determined.

Step 4: The hemisphere containing tumor is determined by computing histogram of each 45×45 window to determine the location of tumor in the hemisphere.

Step 5: Now, first- and second-order statistical features are extracted from the tumor hemisphere.

Step 6: The tumor classification methodology is then applied to each window, and if the window contains tumor, then central pixel of the window is labeled as 1 and if it does not contain tumor, then the central pixel is labeled as 0.

4 Result and Discussion

Simulation results of the proposed scheme implemented using MATLAB software are presented in this section. T1-weighted and FLAIR brain MRI images of brain patients 0001, 0004, 0013, and 0017 with high grade glioma are taken from BRATS-2013 dataset. The proposed scheme is compared with [5], and accuracy, sensitivity, and specificity performance parameters are evaluated for different values of α. For the detection of asymmetry between two hemispheres, a threshold of value 40 is taken for Bhattacharya coefficient and for the mutual information, threshold of value 3.5 is taken. These thresholds are taken by comparing neighborhood windows. From results (Tables 2, 3, 4, and 5), it is observed that different values of α give better performance parameters for different patients (Table 1).

Original T1-weighted and FLAIR MRI brain images for patient 0001, 0004, 0013, and 0017 are shown in Figs. 4a and 5a, respectively. Fractional Sobel mask of size 3 $\times$ 3 and different fractional orders α are applied on the original image shown in Figs. 4b and 5b. Normalized image obtained by intensity normalization is shown in Figs. 4c and 5c. Figures 4d and 5d show the borderline of the brain image, which is obtained by erosion and subtraction. Figures 4e and 5e and 4f and 5f show the upper and lower hemispheres of the brain image. Figures 4g and 5g show the segmented tumor region. The final computation results of performance parameters, i.e., accuracy, sensitivity, and specificity for all four patients and for values of α from 0 to 1, are listed in Tables (2, 3, 4, and 5). Performance parameters sensitivity, specificity, and accuracy are used to quantify the reliability of test [14, 15]. Parameters are shown below:

$$\text{sensitivity} = \frac{T_P}{F_N + T_P} 100\% \tag{27}$$

Table 1 Comparison of performance parameters for best value of α with Nabizadeh et al. [5]

Patient	α	Accuracy	Sensitivity	Specificity
0001	0.1	98.01	86.59	98.26
0004	0.1	99.19	60.90	99.92
0013	0	98.03	63.08	99.64
0017	0.2	98.97	90.61	99.11
[5]	–	96 $\pm$ 0.7	94 $\pm$ 0.3	95 $\pm$ 0.3

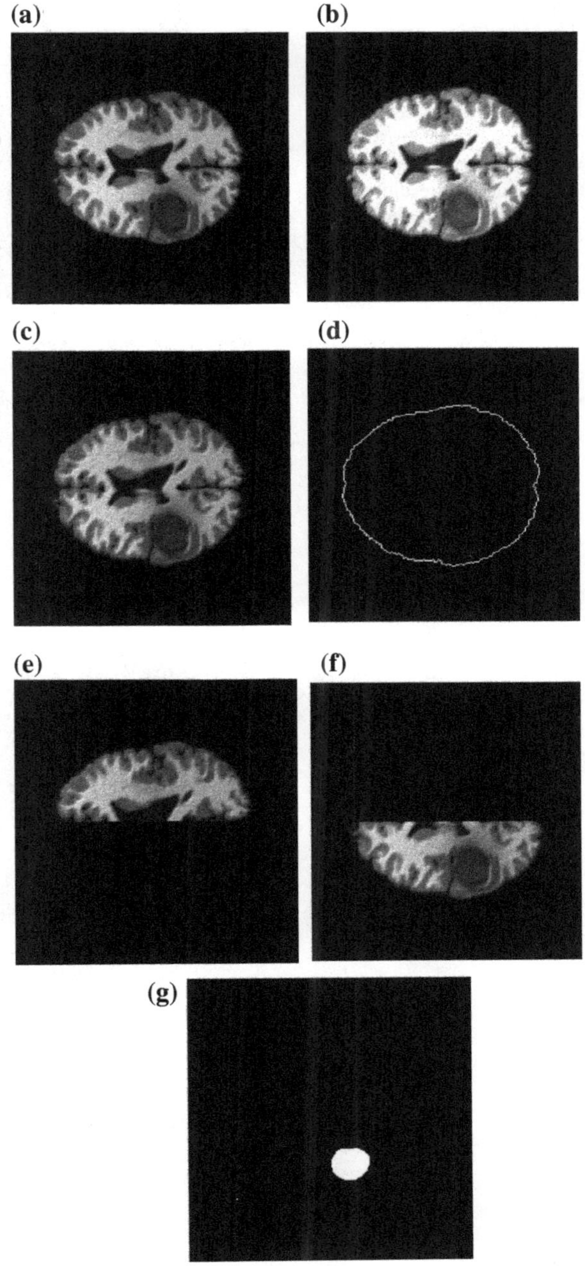

Fig. 4 Patient 0004 T1-weighted image. **a** Original image, **b** result of applying fractional Sobel mask of original image (gradient magnitude image) for $\alpha = 0.1$, **c** intensity normalization, **d** finding the borderline, **e**, **f** resulting brain hemispheres, **g** segmented tumor through SVM classifier

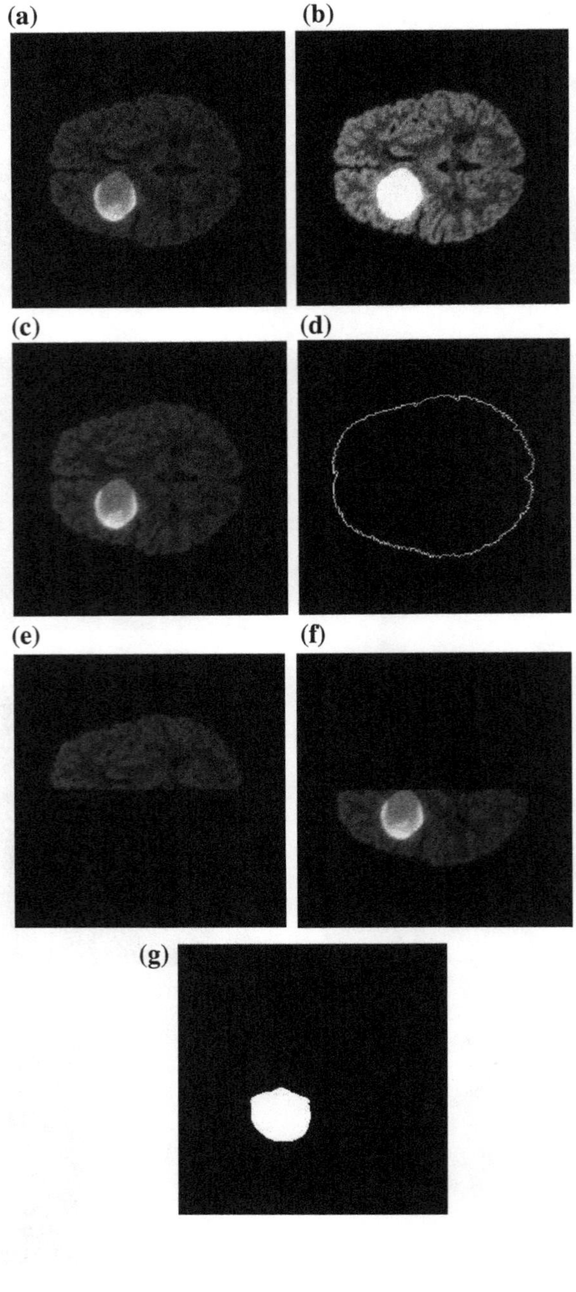

Fig. 5 Patient 0001 FLAIR Image. **a** Original image, **b** result of applying fractional Sobel mask of original image (gradient magnitude image) for $\alpha = 0.1$, **c** intensity normalization, **d** finding the borderline, **e**, **f** resulting brain hemispheres, **g** segmented tumor through SVM classifier

Table 2 For patient 0001 flair image

α	Accuracy	Sensitivity	Specificity
0	98.24	85.60	98.52
0.1	**98.01**	**86.59**	**98.26**
0.2	98.07	71.56	98.65
0.3	98.22	70.07	98.84
0.4	98.45	70.70	99.06
0.5	98.67	70.56	99.29
0.6	98.73	71.34	99.33
0.7	98.78	72.26	99.36
0.8	98.90	70	99.53
0.9	98.91	59.64	99.77
1	98.34	29.12	100

Table 3 For patient 0004 T1-weighted image

α	Accuracy	Sensitivity	Specificity
0	99.18	60.41	99.91
0.1	**99.19**	**60.90**	**99.92**
0.2	99.20	60.74	99.93
0.3	99.14	56.21	99.95
0.4	95.48	73.41	95.89
0.5	98.95	44.03	99.99
0.6	98.85	45.76	99.86
0.7	98.39	19.50	99.88
0.8	87.00	80.74	87.12
0.9	86.59	99.75	86.35
1	85.89	99.67	85.63

Table 4 For patient 0013 T1-weighted image

α	Accuracy	Sensitivity	Specificity
0	**98.03**	**63.08**	**99.64**
0.1	97.97	60.47	99.70
0.2	97.94	60.02	99.69
0.3	97.87	56.93	99.75
0.4	97.93	60.06	99.67
0.5	97.43	54.29	99.41
0.6	95.68	50.43	97.76
0.7	95.33	45.95	97.61
0.8	95.69	53.52	97.63
0.9	96.31	61.87	97.89
1	96.01	62.08	96.67

Table 5 For patient 0017 flair image

α	Accuracy	Sensitivity	Specificity
0	99.30	64.71	99.90
0.1	99.02	79.69	99.35
0.2	**98.97**	**90.61**	**99.11**
0.3	98.85	68.77	99.37
0.4	98.97	70.75	99.46
0.5	99.26	86.10	99.48
0.6	99.32	82.76	99.60
0.7	99	61.64	99.64
0.8	99.11	65.16	99.69
0.9	98.93	43.05	99.89
1	98.85	43.77	99.80

$$\text{specificity} = \frac{T_N}{F_P + T_N} 100\% \tag{28}$$

$$\text{Accuracy} = \frac{T_N + T_P}{F_P + T_P + F_N + T_N} 100\% \tag{29}$$

Here, T_P is tumor truly identified as tumor, T_N is normal brain truly identified as normal, F_P is normal brain falsely identified as tumor, and F_N is tumor falsely identified as normal. The highlighted sections in Tables 2, 3, 4, and 5 gives approximately good values of accuracy, sensitivty and specifity.

Variation of performance parameters for different values of α is shown as follows:

5 Conclusion

In this paper, a scheme is implemented that is able to detect and segment the tumor from the T1-weighted and FLAIR MRI sequences. Fractional-order Sobel filter is used here to provide improved statistical features. Fractional Sobel filter gives multiple choices to choose the order to get the better result for tumor segmentation. Bhattacharya coefficient and histogram asymmetry are used in this scheme to detect the tumor hemisphere. First- and second-order statistical features are extracted to train the SVM. The results show that at different fractional-order α values, this scheme gives comparable results of sensitivity, specificity, and accuracy parameters when compared with [5].

References

1. Wu J, Poehlman S, Noseworthy MD, Kamath MV (2008) Texture feature based automated seeded region growing in abdominal MRI segmentation. In: International conference on biomedical engineering and informatics, vol 2, pp 263–267, May 2008
2. Lee CH, Schmidt M, Murtha A, Bistritz A, Sander J, Greiner R (2005) Segmenting brain tumor with conditional random fields and support vector machines. In: International conference on computer vision, pp 469–478
3. Corso JJ, Yuille A, Sicotte NL, Toga AW (2007) Detection and segmentation of pathological structures by the extended graphshifts algorithm. Med Image Comput Comput Aided Interv 1:985–994
4. Islam A, Reza SMS, Iftekharuddin KM (2013) Multifractal texture estimation for detection and segmentation of brain tumors. IEEE Trans Biomed Eng 60(11):3204–3215
5. Nabizadeh N, Kubat M (2015) Brain tumors detection and segmentation in MR images: Gabor wavelet vs. statistical features. Comput Electr Eng 45:286–30 (July 2015)
6. Bauer S, Seiler C, Bardyn T, Buechler P, Reyes M (2010) Atlas-based segmentation of brain tumor images using a Markov random field-based tumor growth model and non-rigid registration. In: Engineering in Medicine and Biology Society (EMBC), pp 4080–4083, Aug 2010
7. Hill PR, Canagarajah CN, David RB (2003) Image segmentation using a texture gradient based watershed transform. IEEE Trans Image Process 12(12):1618–1633
8. Somasundaram K, Kalavathi P (2011) Skull stripping of MRI head scans based on Chan-Vese active contour model. Int J Knowl Manag e-Learn 3(1):7–14
9. Qurat-Ul-Ain GL, Kazmi SB, Jaffar MA, Mirza AM (2010) Classification and segmentation of brain tumor using texture analysis. In: Recent advances in artificial intelligence, knowledge engineering and data bases, pp 147–155, Feb 2010
10. Kharrat A, Benamrane N, Messaoud M, Abid M (2009) Detection of brain tumor in medical images. In: 3rd IEEE international conference on signals, circuits and systems, pp 1–6
11. Saha BN, Ray N, Greiner R, Murtha A, Zhang H (2012) Quick detection of brain tumors and edemas: a bounding box method using symmetry. Comput Med Imaging Graph 2:95–107
12. Tian D, Wu J, Yang Y (2014) A fractional-order edge detection operator for medical image structure feature extraction. In: Control and decision conference, pp 5173–5176, May 2014
13. Cortes C, Vapnik V (1995) Support vector networks. Mach Learn 20:273–297
14. Zhu W, Zeng N, Wang N (2010) Sensitivity, specificity, accuracy, associated confidence interval and ROC analysis with practical SAS implementations. In: NESUG proceedings: health care and life sciences, Baltimore, Maryland, pp 19, Nov 2010
15. Haghighat MBA, Aghagolzadeh A, Seyedarabi H (2011) A non-reference image fusion metric based on mutual information of image features. Comput Electr Eng 37(5):744–756

Cladding Mode Analysis of Photonic Crystal Fiber Using Scalar Effective Index Model

Anupam Kumar and Manoj Kumar

Abstract Cladding modes of the Photonic crystal fiber (PCF) is investigated under weakly guiding approximation. The propagation constants of the modes are calculated and their respective field patterns are simulated. The modes are nearly confined in the outer silica ring. This analysis also reveals the penetration of evanescent fields in outer medium which would be helpful in the design of the devices for sensing application.

Keywords Photonic crystal fiber · Cladding mode · Modal field
Evanescent fields · Effective index model

1 Introduction

In the past decade, Photonic crystal fiber (PCF) has emerged as alternative to conventional optical fiber in communication as well as sensing application. PCFs are the micro structured single index optical fiber consisting of defected core enclosed by hexagonal array of cylindrical air holes running along the whole length of fiber [1]. The electromagnetic wave is guided in these fibers either by total internal reflection [2] or by the photonic bandgap effect [3]. In contrast to conventional optical fibers, PCFs have tailor made properties like endlessly single mode, variable mode area, high birefringence and anomalous dispersion.

There are already numerous algorithms used to study PCFs, such as the localized basis function method [4], Block-iterative frequency-domain methods [5], finite element method (FEM) [6–8], finite difference method [9] and the plane-wave expansion method (PWM) [10]. Although these methods provide us with rigorous solution however they require a large computation time and large memory.

A. Kumar (✉) · M. Kumar
Department of Physics, Ranchi College, Ranchi, India
e-mail: anupamiitd@gmail.com

M. Kumar
e-mail: profmanoj@rediffmail.com

V. Nath and J. K. Mandal (eds.), *Nanoelectronics, Circuits and Communication Systems*, Lecture Notes in Electrical Engineering 511,
https://doi.org/10.1007/978-981-13-0776-8_16

Several models have also been proposed such as scalar effective index method (SEIM) [11] and fully vectorial effective index method (FVEIM) [12] which allow us replace PCF with equivalent convention step index fiber. The cladding index of such equivalent step index fiber is equal to effective index of fundamental space filling mode (FSM). Most of these studies are intended to study the core mode of the fiber. However PCFs are widely used for sensor application utilizes core as well as cladding modes of the fiber. Thus it becomes extremely important to study the behaviour of cladding modes sustained in PCFs.

In this paper Cladding mode analysis of PCF using scalar effective index model is demonstrated. One of the most important parameter in studying the above models is FSM [1, 2] which is the modes supported in the array of air holes with highest propagation constant. The modal index of such modes is termed as n_{fsm}. Throughout the study weakly guiding is assumed to be valid.

2 The Model

The PCF essentially consist of core made of silica surrounded by array of air holes followed by another layer of solid silica. The typical structure of LMA-8 is as shown in Fig. 1. The light is guided in the core region which is surrounded by holey region whose effective index is lower than the solid silica core. The holey region is replaced by the effective index of FSM. FSM of a PCF is well thought-out to be the mode with the largest modal index of that propagates in the infinite photonic crystal lattice structure which surrounds the PCF's core. The characteristic transcendental equation for the determination of effective index [13] is given by Eq. (1).

$$W\frac{I_1(aW)}{I_0(aW)}\left(J_0(aU) - Y_0(aU)\frac{J_1(UR)}{J_1(UR)}\right) = -U\left(J_1(aU) - Y_1(aU)\frac{J_1(UR)}{J_1(UR)}\right) \quad (1)$$

where J_1, Y_1 and I_1 are the Bessel's function of first kind, Bessel's function of second kind and modified Bessel's function of first kind respectively and a is radii of hole while R is radii of equivalent circular lattice. Several value of R has been proposed however $R = \Lambda/2$ yields the results which agree with numerical results [12] reported so far, where Λ is the distance between two holes commonly referred as pitch of the fiber. Also U and W are given as

$$U = \sqrt{\left(k^2 n_{\text{silica}}^2 - \beta^2\right)}$$

and

$$W = \sqrt{\left(\beta^2 - k^2 n_{\text{air}}^2\right)}$$

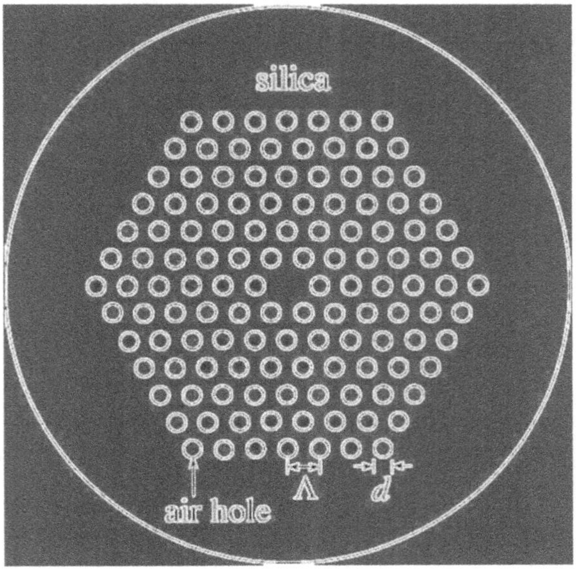

Fig. 1 Typical structure of LMA-8

The light guided by the core remains isolated from the surroundings and field of the core mode penetrates to only first layer of holes. Thus the light guided core has no chance of reaching the outer surface of the fiber. However if the light is somehow launched in outer silica ring it is guided in between air (outer medium) and holey region of effective index n_{fsm}. Such modes propagating in the surface of PCF can interact with the surroundings which are used for sensing application. The modes in these fibers are although see a large index change at outer surface however at the inner interface the effective change in index is small. Thus weakly guiding approximation can be applied.

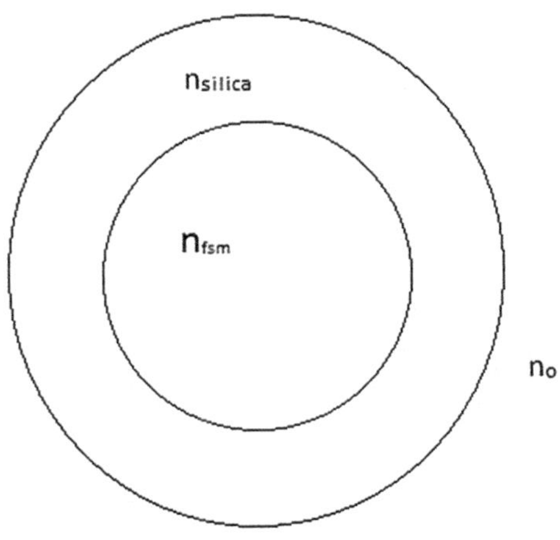

Fig. 2 Three layer model

The light is guided by the outer silica which is bounded by two lower refractive index media by phenomena of total internal reflection at two interface namely air-silica and air-n_{fsm}.

In order to study the linearly polarized (LP) modes of PCF supported in the outer silica ring we model it with three layer structure. Thus the modelled structure consists of four layers, however the core is isolated from the cladding mode and there is hardly a penetration of cladding mode field in the core. The equivalent three layers model is represented in Fig. 2.

3 Scalar Model: LP-Modes

The electromagnetic fields, propagating in a cylindrically symmetric fiber structure along its axis must satisfy the homogeneous vector wave equation [14].

$$\frac{\partial^2\psi}{\partial r^2}+\frac{1}{r}\frac{\partial\psi}{\partial r}+\frac{1}{r^2}\frac{\partial^2\psi}{\partial\varphi^2}+\left[k_0^2 n(r)^2-\beta^2\right]\psi=0 \tag{2}$$

where k_0 is free space propagation constant, $n(r)$ is the refractive index, β is modal propagation constant and ψ represents the electric or magnetic field. Under the limit of weakly guiding approximation, azimuthal dependency of the field in an axially symmetric fiber is given by

$$\phi(\varphi)=\begin{cases}\sin(l\phi+\alpha)\\ \cos(l\phi+\alpha)\end{cases} \tag{3}$$

And radial part of field dependency is expressed in terms of Bessel's function

$$R(r)=\begin{cases}A_l I_l\left(\frac{ur}{R}\right) & r<a\\ B_l J_l\left(\frac{vr}{R_o}\right)+C_l Y_l\left(\frac{vr}{R_o}\right) & a<r<b\\ D_l K_l\left(\frac{wr}{R_o}\right) & r>b\end{cases} \tag{4}$$

where

$$u=R\sqrt{\beta^2-k_0^2 n_{\text{eff}}^2}$$

$$v=R_o\sqrt{k_0^2 n_{\text{silica}}^2-\beta^2}$$

and

$$w = R_o\sqrt{\beta^2 - k_0^2 n_a^2}$$

When propagation constant β is in between n_{fsm} and n_{silica} the modes are guided along the fiber.

Here, ψ represents E as well as H field and β is the modal propagation constant. Under the weakly guiding approximation, the longitudinal field components, E_z and H_z are assumed negligible as compared to the transverse components (E_r, $E\phi$, H_r, H_ϕ). The LP modes are the sum of similar modes which are nearly degenerate. Thus ψ in Eq. (1) then represents the total field polarized normal to the fiber axis rather than one of the separate components of E or H field.

The fields and their derivatives must be continues at the interface of the two regions which yields the transcendental equation.

$$\left(u\frac{I_l'(u)}{I_l(u)} - v'\frac{J_l'\left(\frac{vR}{R_o}\right)}{J_l\left(\frac{vR}{R_o}\right)}\right)\left(w\frac{K_l'(w)}{K_l(w)} - v\frac{Y_l'(v)}{Y_l(v)}\right)J_l\left(\frac{vR}{R_o}\right)Y_l(v)$$
$$= \left(u\frac{I_l'(u)}{I_l(u)} - v'\frac{Y_l'\left(\frac{vR}{R_o}\right)}{Y_l\left(\frac{vR}{R_o}\right)}\right)\left(w\frac{K_l'(w)}{K_l(w)} - v\frac{J_l'(v)}{J_l(v)}\right)Y_l\left(\frac{vR}{R_o}\right)J_l(v) \qquad (5)$$

Equation (5) determines the eigenvalue of propagation constant β.

The value of the effective index of cladding is one of the important parameter as it determines the wavelength over which a holey fiber is single-mode [2].

Till now, studies on cladding mode have been done using computational methods which require large computation time. Another limitation in these computations is the large memory uses. In this scalar analytical approach, we have shown in this paper that at least some of effective index values resembles significantly from those computed using more precise computational method. We will also present information on the modal field pattern and their depth of penetration. As far as we know, very few computational results have been published in the field of photonic crystals and holey fibers for modes supported by cladding. Our results appear to provide a physical insight on these modes.

4 Results and Discussions

The cladding modes of LMA-8 is simulated on MATLAB which are shown in Fig. 3. It is clear from the field pattern that modes are confined in the annular region. The LP_{01} mode in shown in the top corner clearly shows the larger penetration in the inner medium in comparison to outer medium. This is due to the fact of different refractive index on either side. The various modes simulated are LP_{11}, LP_{21}, LP_{32}, LP_{53} and LP_{75} are shown in the figure.

Fig. 3 Cladding mode fields

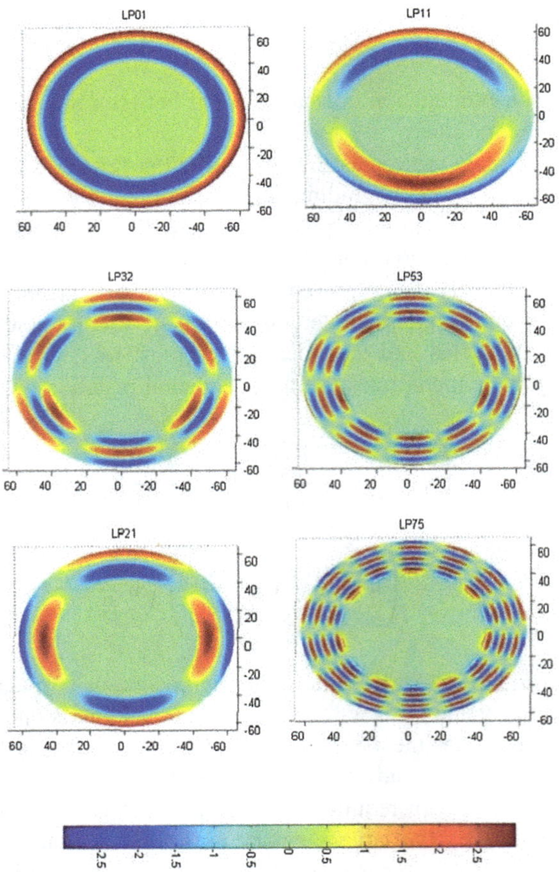

Some of the Modal propagation constants obtained by solving the eigenvalue equation are tabulated in Table 1. In the mode designation LP_{lm}, l is the order of mode and m is the m_{th} root of the eigenvalue equation.

For a given value of l, different values of propagation constants are determined with different radial field distribution. With similar radial field distribution there is also variation in propagation constant.

5 Conclusions

We have investigated the characteristics of guided cladding modes in a PCF. Using the weakly guiding approximation, the field distribution and modal propagation constants of the LP modes of hexagonal PCF LMA-8 were calculated. The analysis of scalar modes reveals that the field remains confined in the outer silica ring.

Table 1 Different modal propagation constants

1	m = 1	2	3	4	5	6
0	5.9508 35018	5.9464 95471	5.9392 83096	5.9292 41119	5.9164 72466	5.9013 13717
1	5.9508 01684	5.9464 61202	5.9392 48525	5.9292 0627	5.9164 37178	5.9012 77124
2	5.9507 01773	5.9463 58371	5.9391 44797	5.9291 01713	5.9163 31308	5.9011 67367
3	5.9505 35555	5.9461 86907	5.9389 71862	5.9289 27418	5.9161 54844	5.9009 84508
4	5.9503 03481	5.9459 46696	5.9387 2964	5.9286 83336	5.9159 07765	5.9007 2864
5	5.9500 06169	5.9456 37586	5.9384 18016	5.9283 69397	5.9155 90039	5.9003 99886
6	5.9496 44402	5.9452 59394	5.9380 36847	5.9279 85513	5.9152 01627	5.8999 98385
7	5.9492 1911	5.9448 11923	5.9375 85956	5.9275 31575	5.9147 42477	5.8995 24285

However decaying evanescent fields are present both at inner and outer surface. The interaction of these fields with the gases, liquids and nano-materials is possible. The investigation gives a general understanding of the modal properties of PCFs, which should be useful in design of optical sensors based on PCFs utilizing cladding modes and evanescent waves. The study may further explore new possibilities such as testing the optical properties of nano-materials by interacting with evanescent wave.

References

1. Knight JC, Birks TA, Russell PSJ, Atkin DM (1996) All-silica single-mode optical fiber with photonic crystal cladding. Opt Lett 21(19):1547–1549
2. Birks TA, Knight JC, Russell PSJ (1997) Endlessy single-mode photonic crystal fiber. Opt Lett 22:961–963
3. Knight JC, Broeng J, Birks TA, Russell PSJ (1998) Science 282:1476
4. Knudsen E, Bjarklev A (2003) Modelling photonic crystal fibres with Hermite–Gaussian functions. Opt Commun 222(1–6):155–160
5. Johnson SG, Joannopoulos JD (2001) Block-iterative frequency-domain methods for Maxwell's equations in a planewave basis. Opt Express 8:173–190
6. Saitoh K, Koshiba M (2002) Full-vectorial imaginary-distance beam propagation method based on a finite element scheme: application to photonic crystal fibers. IEEE J Quantum Electron 38:927–933
7. Cucinotta A, Selleri S, Vincetti L, Zoboli M (2002) Holey fiber analysis through the finite element method. IEEE Photon Technol Lett 14:1530–1532
8. Brechet F, Marcou J, Pagnoux D, Roy P (2000) Complete analysis of the characteristics of propagation into photonic crystal fibers, by the finite element method. Opt Fiber Technol 6:181–191

9. Zhu Z, Brown T (2002) Full-vectorial finite-difference analysis of microstructured optical fibers. Opt Express 10:853–864
10. Arriaga J, Knight JC, Russell PS (2003) Modeling photonic crystal fibers. Physica E 17:440–442; Koshiba M, Saitoh K (2004) Applicability of classical optical fiber theories to holey fibers. Opt Lett 29:1739–1741
11. Li Y-F, Wang C-Y, Hu M-L (2004) A fully vectorial effective index method for photonic crystal fibers: application to dispersion calculation. Opt Commun 238:29–33
12. Li Y, Wang C, Chen Y, Hu M, Liu B, Chai L (2006) Solution of the fundamental space-filling mode of photonic crystal fibers: numerical method versus analytical approaches. Appl Phys B 85:597–601
13. Li Y, Wang C, Hu M (2004) A fully vectorial effective index method for photonic crystal fibers: application to dispersion calculation. Opt Commun 238:29
14. Snyder AW, Love JD (1983) Optical waveguide theory. Chapman & Hall, New York

A Hybrid Forward Selection Based LASSO Technique for Liver Cancer Classification

M. J. Abinash and V. Vasudevan

Abstract Big data deals with the terrific volume of data. The rapid development of data increases in many fields such as business, biological, stock market, scientific, etc., and our day to day activities. So while handle the growing data can causes many issues in managing, analyzing, storing, capturing and some others. The term microarray data comes under the biological data, consisting of large amount of gene instances. Bioinformatics deals with the interdisciplinary of computer science or information technology with the biological data. Using this the knowledge discovery can be done through analyzing these biological data with the software tools. In this, the knowledge discovery can done through machine learning techniques. For classification using these biological data the feature selection is an essential one. The selecting features work with the classifier model to improve their performance. In this paper a wrapper based sequential forward selection (SFS) with LASSO (least absolution shrinkage and selection operator) hybrid method are used for selecting the variables for cancer classification. It will improve the classifier model for better classification.

Keywords Big data · Gene expression · Sequential forward selection (SFS)
LASSO (least absolution shrinkage and selection operator)
SVM (Support vector machine)

M. J. Abinash (✉) · V. Vasudevan
Department of Information Technology, Kalasalingam University, Krishnankoil, Srivilliputtur, Tamil Nadu, India
e-mail: mj.abinash@gmail.com

V. Vasudevan
e-mail: vasudevan_klu@yahoo.co.in

V. Nath and J. K. Mandal (eds.), *Nanoelectronics, Circuits and Communication Systems*, Lecture Notes in Electrical Engineering 511,
https://doi.org/10.1007/978-981-13-0776-8_17

1 Introduction

Actually big data is the huge amount of data it can grown higher recent years because of the data usage by our mankind. It may be in any format such as tables, voice, video, web, XML sheets, etc. While managing these data can led to many problems such as storing, capturing, analyzing, handling, retrieving and so on. This can occur in many fields such as medical, business, biological, scientific, statistics and some others. The main issues such as analyzation, classification, clustering, etc. The term molecular biology focuses on structure and behavior of the microarray technology that regulate the protein synthesis and their task [1]. The bioinformatics is the combination of biological and the information technology field, the biological problems are resolved using the different domains in the bioinformatics. It mainly focus on the analyzing, knowledge discovery and the diagnosis. One example is the cancer classification of analyzing the micro array data with the help of genes [2]. Measuring the expression levels of microarray can made through the chip consisting of thousands of genes. The chip is in 2 cm and it consists of thousands of micro spots. High dimensionality is the big challenge of gene expression data in large classification problems. And these having the redundancy which led to poor classification accuracy [3]. One of the supervised computer learning technique called SVM (Support vector machine) is well in data classification. The microarray data can be evaluated well with the performance of using SVM [4]. The optimal hyperplane which maximizes the margin which performs classification using data [5]. The cancer classification are done with fewer gene expressions, the *T*-test and class separability are used for gene selection then, SVM and KNN (K nearest neighbor) is used for classification [6]. In past the techniques such as are correlation, information gain, genetic algorithm (GA), etc., are the feature selection approaches are used for increase in accuracy and then several classifiers are used in predicting the results such as bayes network, artificial intelligence, SVM, KNN.

Here we propose a novel feature selection technique such as a wrapper based sequential forward selection (SFS) with least absolution shrinkage and selection operator (LASSO) is a hybrid one is used for variable selection and regularization in order and SVM is used for classification. This gives the better feature selection among the existing approaches.

The excess of this paper is formulated with related works, some existing feature selection methods, proposed work, results and discussions and final the conclusion with future work.

2 Related Works

Leng et al. [7] proposed the feature selection process is carried over the dataset using the GA in conjunction with the K-nearest neighbor algorithm. The wrapper based GA with KNN can provide the better result. Here the implementation is carried through the c# and the .net. The accuracy got in the classification is up-to 80%.

Medjahed et al. [8] proposed first the micro-array dataset is given then the SVM-RFE algorithm for selecting the k genes and then the Binary Dragonfly (BDF) a new benchmark for the feature selection process. In this step the optimal gene subset is selected and then the SVM classification is carried and then the evaluation on classification error rate arrived then again the step is continued until reach the higher accuracy.

Samb et al. [9] presented the RFE-SVM is being used to feature select approach for classification. The actual process carried here is first the given set is converted into the subsets. And then the evaluation is carried in the subset and the stopping criterion is available if the goodness occur then passed to result validation if not then rerouted to the subset generation again to regenerate and to re-validate. Thus the RFE-SVM provides the high efficient results.

Vinod and Vasudevan [10] suggested a method to improve the accuracy of classification by providing good features to classification algorithm. They remove insignificant features and quality features are identified by HCFS (Highly Correlated Feature Set Selection) method and for classification they use hierarchical learning approach.

Zhuo et al. [11] proposed the hyper spectral image get classified after the feature selection process carried over using the GA based wrapper method using the SVM. The test results were compared and finally the result after the feature selection provides 92% of accuracy and where else the non feature selected one has the accuracy at the rate of 88% only. Thus the wrapper based SVM provides the enhanced result.

Wang et al. [6] proposed the cancer classification done in two steps with few number of genes expressions. First is the ranking scheme for genes and next is the accuracy for classification, divide and conquer method is used to got accuracy then, *t*-test and class separability is used for scoring.

Ghorai et al. [12] offered the feature selection is done with the help of minimum maximum redundancy with relevance (MRMR) ranking method and the wrapper method and proximal classifier with nonparallel plane (NPPC) is used for computational classifier on the gene data sets.

Osareh and Shadgar [13] proposed the signal-to-noise ratio feature ranking by SFS and with principal component analysis (PCA) for feature extraction and KNN, SVM and the probabilistic techniques based on neural networks for computational classifiers used to diagnose the breast cancer.

Zheng and Liu [14] proposed the LASSO gene selection with Dantzig selector is the automatic gene selection techniques helpful in selecting the informative genes with linear regression analysis for cancer diagnosis.

Zhongxin et al. [15] they proposed for high dimensional microarray data based feature selection, it involves two steps calculate the genes using mutual information from that select the candidate genes and the informative genes are selected using the LASSO method this makes the feature selection better to yield classification accuracy.

Yuchun et al. [16] they proposed SVM-RFE based feature selection. It is a two stage process, in the first stage it remove the redundant, irrelevant, noisy genes by pre-filtering process it avoid redundancy and it involves multiple iterations and in the second stage all the subsets are selected and remove the irrelevant gene at each iteration. Here classification done with the linear SVM and later it compared with correlation for better accuracy under ROC.

3 Gene Feature Selection Methods

3.1 Sequential Forward Selection (SFS)

In SFS [17] it works with the form of greedy search algorithm, it add the essential feature start on the empty set for every step. The feature with the highest rank has been selected and the features with the lowest rank has been removed. The search performs the optimal subset for the limited number of features. By the evaluation of the large number of states near the empty set be potentially evaluated. It search towards the full or empty sets and the search space is in the form of an ellipse. If once feature is added it cannot be removed is the major disadvantage of SFS.

Algorithm:

1. Start with the empty set Y0 = {□}
2. The best feature is select next X+ = argmax[J(Yk + X)]; x¢Yk
3. Update Yk + 1=Yk + X+; k = k + 1
4. Goto 2

3.2 Least Absolute Shrinkage and Selection Operator (LASSO)

In [18] LASSO is a regression method it performs both variable selection and regularization for linear regression in order proposed by Tibshirani et al. [19]. It is related to Ordinary Least Squares (OLS) regression and the Residual Sum of Squares (RSS) with absolute values that minimizes the sum is less than a constant. It is similar in ridge regression as an additional constraint, it sum the squared values to the constraint as a coefficients. This allows the variable selection in LASSO makes a simple modification to perform. So the shrinkage has some coefficients shrunk to zero exactly.

$$\text{Lasso} = \min\left(\sum_{i=1}^{n}\left(Y_i - \beta_0 - \sum_{j=1}^{p}\beta_j x_{ij}\right)^2 + \lambda\sum_{j=1}^{p}|\beta_j|\right)$$

4 Proposed Methodology

4.1 Hybrid Forward Selection Based LASSO (HFSBL)

The proposed work in Fig. 1 focuses on the effective feature selection technique such as HFSBL to form the new feature selection technique it helpful in good performance by the combination of both SFS and LASSO that force to improve the classifier performance.

The gene selection are done with HFSBL it has the capability to improve in the performance. SFS forces in adding the variables and LASSO avoids poor discretization in redundant labels. Both works with the comparable. The feature selection performed with the SFS and LASSO. After the feature selection the extracted features is set to the classifier for training.

4.2 Feature Selection Method

The HFSBL based procedure as follows:

Input: Given Ranked Set J = [ ], Given Feature Subset S = [1, … e]

Until all subsets are ranked Redo it
Input variables are added from set S

Output: Reduce Set

For each attribute add the set
And to regularize with subset
Identify the feature with small rank and irrelevant
Eliminate it
Repeat Steps
Update

Reduce Set

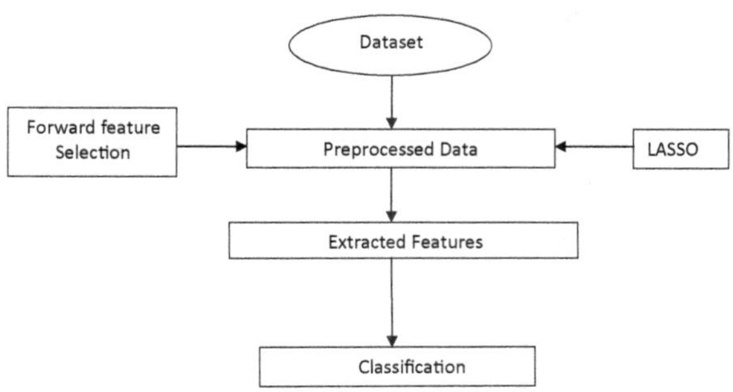

Fig. 1 Proposed methodology for HFSBL

4.3 SVM for Classification

SVM is a supervised machine learning technique [5] and is a powerful technique in classification and regression tasks. Vapnik et al. [20] outlined SVM first based on the statistical learning methods for classifying the data. Here in SVM kernel mapping is possible than the transformations of high dimensional space from the sample space. And the linear classification is possible in SVM. The classifier is separated by constructing the hyperplane. The main aim of the SVM is to maximize the margin between the support vector and the optimal hyperplane. There four kernel functions are used in the SVM they are linear, polynomial, sigmoid and RBF. The original data can be called as the support vectors by separating the surface only on the subset. The one or more planes are placed on the data by achieving the good results for classification. Based on the results the functional margin is achieved by the largest distance of the plane with any class to the nearest data points. Then the classifier with generalization error will be small because of the functional margin is large.

4.4 Dataset Description

This dataset description, we apply SFS with LASSO for variable gene selection and for classification we using SVM applied on the liver cancer dataset.

The provided dataset obtained from http://genome-www.stanford.edu/hcc/ [21] comprises the two classes having non tumor and HCC. The dataset contains the expression levels of 1648 important genes for 156 samples. Among them 74 are non tumor livers and 82 are HCCs. For training and testing the samples in the dataset is divided into 78 samples each for both purpose randomly. This dataset consists of some missing values. This missing values are imputing by mice method using R language. Nowadays R programming are mostly used in the machine learning techniques and for statistical analysis.

5 Results and Discussion

From the above described dataset is used for classification done with SVM based on hybrid feature selection. After proper ranking by SFS will reduces the complexity of computation process, this result given to LASSO that terminates the redundancy. This approach shows the small amount of classification error because the LASSO method rest not on the classifier but on the gene space. The following results show the classification accuracy. The classification accuracy for conventional SVM using SVM-RFE and BBF is 97.22% is compared with our feature selection approach SFS and LASSO is 98.21% (Fig. 2 and Table 1).

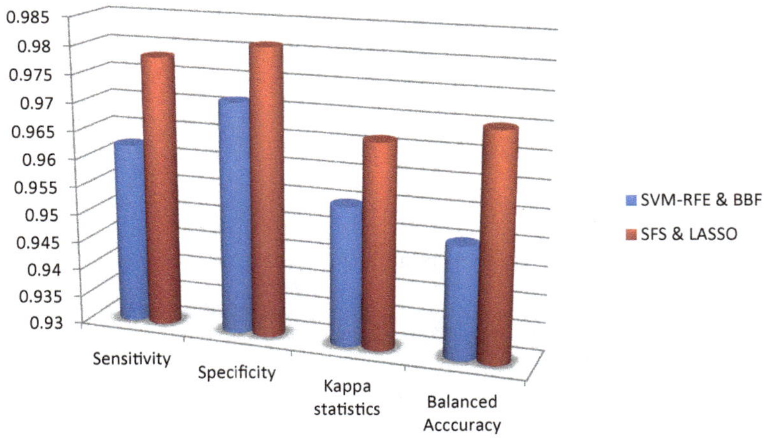

Fig. 2 Graphical representation of classification accuracy

Table 1 Comparative measures and classification accuracy of two hybrid algorithms

Algorithm	Sensitivity	Specificity	Kappa statistics	Balanced accuracy	Classification accuracy	Class
SVM-RFE and BBF	0.9622	0.9712	0.9548	0.95	97.22	Positive/ Negative
SFS and LASSO	0.9782	0.9811	0.9664	0.97	98.21	Positive/ Negative

Finally the performance measure and classification accuracy can be mentioned as below.

Here TP is the true positives it belongs to the malignant it point to the correct classifications, benign is the true negatives as TN, FP is false positives it is benign as malignant is the negative samples and the false negatives FN is malignant as benign.

$$\text{Sensitivity} = (\text{TP}/\text{TP} + \text{FN})$$

$$\text{Specificity} = (\text{TN}/\text{TN} + \text{FP})$$

$$\text{Accuracy} = (\text{TP} + \text{TN}|\text{TN} + \text{TP} + \text{FN} + \text{FP})$$

6 Conclusion

From the above work, encompasses the hybrid approach for gene selection it combines the SFS with LASSO was proposed for gene selection. Compared to previous related works, this proposed work combined with the SFS and LASSO shows the better accuracy and performance on the experimental results done on the liver cancer datasets. For the future work the new classifier model will going to propose for the huge datasets using supervised learning technique.

References

1. Pham TD, Crane DI (2006) Analysis of microarray gene expression data. Bentham Science Ltd
2. Golub T, Slonim D, Tamayo P, Huard C, Gaasenbeek M, Mesirov J, Coller H, Loh M, Downing J, Caligiuri M, Bloomfield C, Lander E (1999) Molecular classification of cancer: class discovery and class prediction by gene expression monitoring. Science 286:531–537
3. Yu L, Liu H (2004) Redundancy based feature selection for microarray data. In: Proceedings of the 10th ACM SIGKDD international conference on knowledge discovery and data mining, USA, pp 737–742
4. Chu F, Wang L (2005) Applications of support vector machines to cancer classification with microarray data. Int J Neural Syst 15(6):475–484
5. Guyon J, Weston S, Barnhill S, Vapnik V (2002) Gene selection for cancer classification using support vector machines. Available at: http://axon.cs.byu.edu/Dan/778/papers/Feature%20Selection/guyon*.pdf, last visited 2002
6. Wang L, Chu F, Xie W (2007) Accurate cancer classification using expressions of very few genes. IEEE Trans Comput Biol Bioinform 4(1):40–53
7. Leng J, Valli C, Armstrong L (2010) A wrapper-based feature selection for analysis of large data sets. In: International conference on computer and electrical engineering (ICCEE 2010)
8. Medjahed SA, Saadi TA, Benyettou A, Ouali M (2017) Kernel-based learning and feature selection analysis for cancer diagnosis. Appl Soft Comput 51:39–48
9. Samb ML, Camara F, Ndiaye S, Slimani Y, Esseghir MA (2012) A novel RFE-SVM-based feature selection approach for classification. Int J Adv Sci Technol 43
10. Vinod DF, Vasudevan V (2016) A filter based feature set selection approach for big data classification of patient records. In: IEEE Xplore Digital Library
11. Zhuo L, Zheng J, Wang F, Li X, Ai B, Qian J (2008) A genetic algorithm based wrapper feature selection method for classification of hyperspectral images using support vector machine. In: Geoinformatics 2008 and joint conference on GIS and built environment
12. Ghorai Z et al (2011) Cancer classification from gene expression data by NPPC ensemble. IEEE Trans Comput Biol Bioinform 8(3):659–671
13. Osareh A, Shadgar B (2010) Machine learning techniques to diagnose breast cancer. In: 5th international symposium on health informatics and bioinformatics (HIBIT), pp 114–120
14. Zheng S, Liu W (2011) An experimental comparison of gene selection by Lasso and Dantzig selector for cancer classification. Comput Biol Med 41:1033–1040
15. Zhongxin W, Gang S, Jing Z, Jia Z (2016) Feature selection algorithm based on mutual information and Lasso for microarray data. Open Biotechnol J 10(1):278–286
16. Yuchun T, Yan-Qing Z, Zhen H (2007) Development of two-stage SVM-RFE gene selection strategy for microarray expression data analysis. IEEE/ACM Trans Comput Biol Bioinform 4 (3):365–381
17. Ladha L et al (2011) Feature selection methods and algorithms. Int J Comput Sci Eng (IJCSE) 3(5)

18. Matteo Cassotti and Francesca Grisoni. Milano Chemometrics and QSAR Research Group—Department of Environmental Sciences, University of Milano-Bicocca, P.za della Scienza 1 – 20126 Milano (Italy)
19. Tibshirani R (1996) Regression shrinkage and selection via the Lasso. J R Stat Soc B 58: 267–288
20. Vapnik V (1998) Statistical learning theory. Wiley, New York
21. Gene expression patterns in liver liver cancers. http://genome-www.stanford.edu/hcc/

Feature Subset Selection Using IULDA Model for Prediction

Smita Pallavi, Akshay Kumar and Utkarsh Mohan

Abstract With the decrease in tariff rates and growing popularity of telecom options, competition in the field for abstracting customers and expanding market is becoming fiercer. It is evident by research that the cost levied by losing a customer from the telecom affiliation is sixfold more drastic than the profit that of adding a new one. The proposed Indexed Uncorrelated Linear Discriminant Analysis (IULDA) classification model for customer churn prediction effectively handles increased amount and dimensionality of data and has been tested on L-class problems of UC Irvine Machine Learning Repository and real dataset of the train sample—5,200 customers, the calibration sample—3,680, and the test sample—4,500 observations. The objective evaluation of the investigated methods was measured by precision, specificity, sensitivity, and accuracy by implementing the MATLAB tool. The accuracy of the IULDA model was 95% for UCI churn datasets and 72.4% for real customer datasets, respectively.

Keywords Feature subset selection · Churn prediction · Dimensionality reduction Reliability index

1 Introduction

Dimensionality reduction is a technique employed to eliminate irrelevant, redundant attributes from a dataset. The two forms are feature extraction and feature selection. In *feature extraction* [1], features are reduced by projecting data into a lower dimensional space. Thus, created set of new attributes varies from the older feature set. *Feature selection* implies selecting the most relevant attributes whereby smaller subsets of the original features are selected. No additional space is required. For reducing amplification, removing unwanted data, increasing accuracy, and

S. Pallavi (✉) · A. Kumar · U. Mohan
Department of CSE, Birla Institute of Technology Mesra,
Patna Campus, Ranchi, Bihar, India
e-mail: smita.pallavi@bitmesra.ac.in

V. Nath and J. K. Mandal (eds.), *Nanoelectronics, Circuits and Communication Systems*, Lecture Notes in Electrical Engineering 511,
https://doi.org/10.1007/978-981-13-0776-8_18

improving result understandability, feature subset selection can be used effectively [2]. Finding an optimal feature subset is usually intractable and NP-hard [3] as shown in many applications.

Data making and mining have predominantly become massive with the progress of information technology since nineteenth century. Extensive research has established that to retain customers is a lesser expensive task than acquiring new customers. Retention managers have to devise strategies with prior knowledge to gain acceptable returns. So, customer churn determinates the customers who might leave or analyze the possibility of retention of existing customers. The important terminologies to be noted are as follows:

Customer churn is described as the phenomenon exhibited by customers to cease the service contract with the current company and deviate from the services offered by other companies.

Churn analysis is the method of calculation of the customer attrition and is depicted by the difference in the number of customers at the end and beginning of a time period. It denotes the customers preferring to discontinue company's service or product.

Churn rate is the metric measure of customers who prefer not to renew their subscription of the company's product. It is computed as proportion of the customers discontinuing or leaving a service and the total number of customers (Figs. 1 and 2).

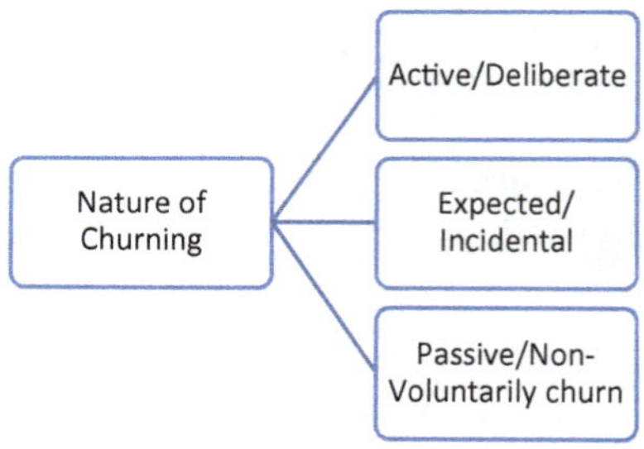

Fig. 1 Categorization of customer churners

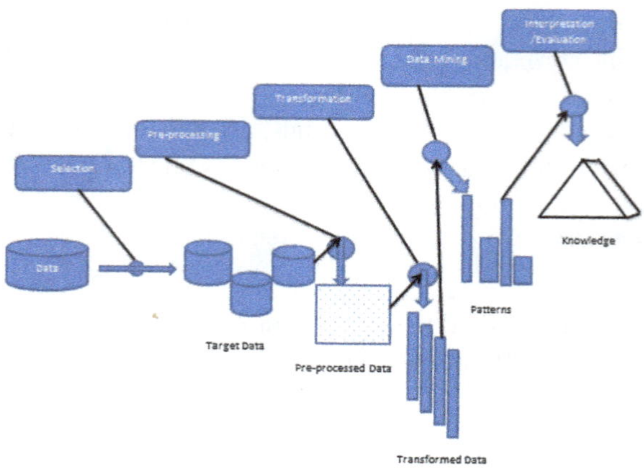

Fig. 2 The data–knowledge interpretation model source: data mining—recommender systems 0.7.5 documentation

Table 1 Hypothesis addressed in the research work

Index	Problem statement
P1	To gather and preprocess customer demographic predictors on a big data platform
P2	To predict customer churn through discriminant analysis and identify the optimum key attributes of churn using simulated customer datasets
P3	How do comparatively perform the class separability using with and without mean and standard deviation and computed within- and between-class scatter from the proportional and the balanced training data

This paper attempts to find optimum predictive indicators which tend to affect most of the environment among many available data values. The results of churn classification may be integrated with other experimental observations and improvise to generate greater significant indicators as part of knowledge interpretation (Table 1).

The problem statement can hence be addressed as follows:

The research paper is organized as follows: Sect. 2 contains the related literature and the tools' description. Section 3 outlines the experimental model and algorithm devised, while Sect. 5 analyzes the assessment criteria.

2 Related Literature

[Gavril Toderean, Ionut Brandusoiu, 2016] [4] presented an advanced DM methodology using support vector machines algorithm with four kernel functions which were used to predict the churning rate of existing 3333 customers. The authors classified 21 churn attributes of people using mobile communications, resulting in binary-valued classes Yes/No. Based on the result, churn prediction was claimed to be performed with accuracy of 88.56% by SVM polynomial kernel function, while the other three kernel functions had the performance accuracy nearing 80%.

[Muhammad Raza Khan, J. Manoj, A. Singh, and J. Blumenstock, 2015] [3] devised a unified analytic framework by assigning "Churn Score" for detecting the likelihood of churn, within a predefined amount of time. Brute force approach was used to feature reduction in a metric set of large number of overlapping features from customer transaction logs. The features are then run through classification-supervised learning algorithms that can accurately predict subscriber churn. For a dataset of roughly 1,00,000 subscribers from a South Asian mobile operator observed for 6 months, an approximate of 90% accuracy was achieved.

[X. Zhang, Z. Liu, X. Yang, W. Shi, and Q. Wang, 2010] [5] investigated the effects of interaction among customers and their social network connections on the accuracy of churn prediction. The authors did a comparative study of traditional attribute-based, network attribute, and combined attributes models and deduced that inclusion of social network attributes into predicting models can greatly improve

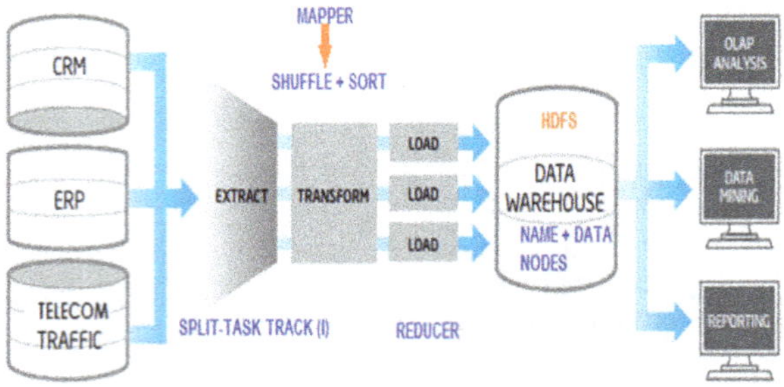

Fig. 3 The big data preprocessing task (authors compilation)

the prediction accuracy. The network attributes may also turn out to be useful complements to the traditional attributes.

[Al-Jazzar M, 2012] [6] in his dissertation studied and compared two discrete classification methods of multinomial logistic regression and linear discriminant analysis and evaluated the behavior of models. The performance was evaluated by overall classification accuracy, and the quality of prediction was investigated in terms of sensitivity and specificity, AUC with the k-fold partitioning technique, the "leave-one out" and "hold out" procedures.

[Pohar M., Blas M., and Turk S, 2004] [7] in their research work exhibited that for normal distribution of the explanatory variables and categorical variables, linear discriminant analysis is a more appropriate method and the results are comparable with logistic regression only when the number of categories is small. The results of LDA and MLR are close whenever the normality assumptions are not too badly violated.

Our attempt is to modify and empower LDA for churn prediction using reputation index for devising churn likeness (Fig. 3).

ETL tools move enormous data like telecom customers information by performing three operations:

- Extraction of data from sources such as ERP or CRM applications. Data from several sources, multiple file formats (CSV, XML), or legacy systems mark ETL solution start off.
- Transformation of data into a common format by data manipulations like splitting, translating, merging, sorting, and pivoting abiding by standard ISO format.
- Loading of the data into the data warehouse for analysis. This step can be done in batch processes or row by row in a real-time scenario [8].

Hadoop Distributed File System (HDFS) gives scalability and fault tolerance by dividing large files into blocks (usually 64 or 128 MB) and by providing APIs for MapReduce applications to write the data and read data in parallel [8].

3 Linear Discriminant Analysis

The mathematical model is represented as follows:

Let x be a vector of the input set X with M elements. X_l designs subsets of X, thus

$$X^{\mathrm{T}} = \bigcup_{l=1}^{N} X_l \tag{1}$$

where

N number of the classes and
X^{T} represents the transpose of the vector x.

The cardinality of the subsets X_l is denoted by $\sum_{l=1}^{N} n_l = M$.

Consider the linear combination of equations predicting Y_i from $X_1 \ldots X_k$ as given by

$$\begin{aligned} Y_1 &= \Phi_{11}X_1 + \Phi_{12}X_2 + \ldots \Phi_{1K}X_K \\ Y_2 &= \Phi_{21}X_1 + \Phi_{22}X_2 + \ldots \Phi_{2K}X_K \\ &\ldots\ldots \\ Y_K &= \Phi_{K1} + X_1 + \Phi_{K2}X_2 + \ldots \Phi_{KK}X_K \end{aligned} \tag{2}$$

Thus, Y_i has a population variance given by

$$\mathrm{Cov}(Y_1, Y_2) = VC = \frac{1}{M}\sum_{l=1}^{K}\sum_{j=1}^{ni} \phi(X_{lj})\phi'(X_{lj}) \tag{3}$$

The assumptions of discriminant analysis as taken are as follows:

- Sample data size exceeds# of predictor variables.
- Sampled data is multivariately normalized distributed.
- Grouped data has homogeneity of variance–covariance matrix.
- Extreme outliers impacting increase in variability are opted out.
- Non-multicollinearity of data leads to independent sampling of subject.

The transformation of data vector and test vector can be handled by any of the two approaches:

Class-dependent transformation: It involves the maximization of the ratio of between-class variance to within-class variance for class separability. It uses two optimizing criteria to transform [9].

Class-independent transformation: This approach involves maximizing the ratio of overall variance to within-class variance. This uses only one optimizing criterion to transform.

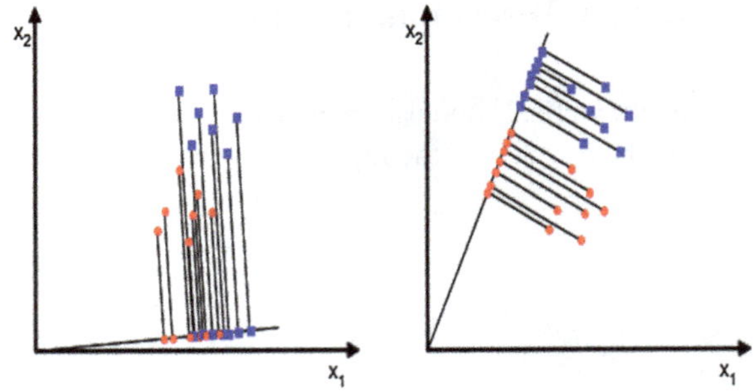

Fig. 4 Projection vector that maximizes the separability of the scalars

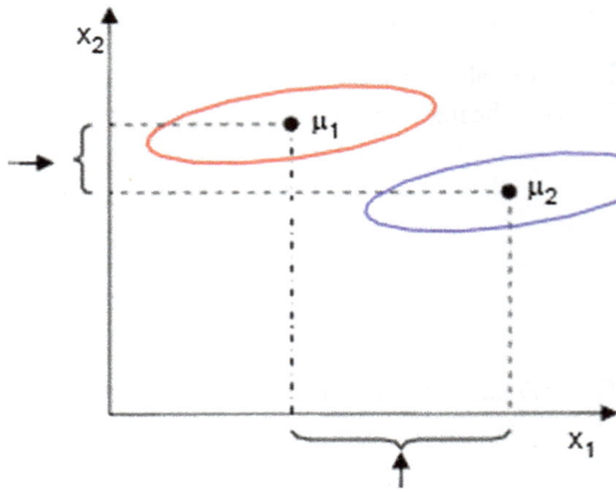

Fig. 5 Projection analysis without taking into account the standard deviation

The feature extraction is illustrated as follows: Given a feature space $Xi \in R_N$, map $Y = F(X)$: $R_N \rightarrow R_M$ with $M < N$ such that the transformed feature vector $Yi \in R_M$ preserving the information or structure in R_N (Fig. 4).

The mean vector of each class in X and Y feature space is

$$\mu_i = \frac{1}{N}\sum X \text{ and } \mu = \sum_{Y \subseteq \omega}^{n} y = \frac{1}{N}\sum_{X \subseteq \omega}^{n} w^{\mathrm{T}} X = w^{\mathrm{T}}\, \mu_i \tag{4}$$

The linear discriminant (LDA) is evaluated as the linear function of $\boldsymbol{w}^{\mathbf{T}}\boldsymbol{X}$ maximizing the criterion function (Figs. 5 and 6):

$$J(w) = |\widetilde{\mu}1 - \widetilde{\mu}2|^2 / \left(\widetilde{S}_{12} + \widetilde{S}_{22}\right) \tag{5}$$

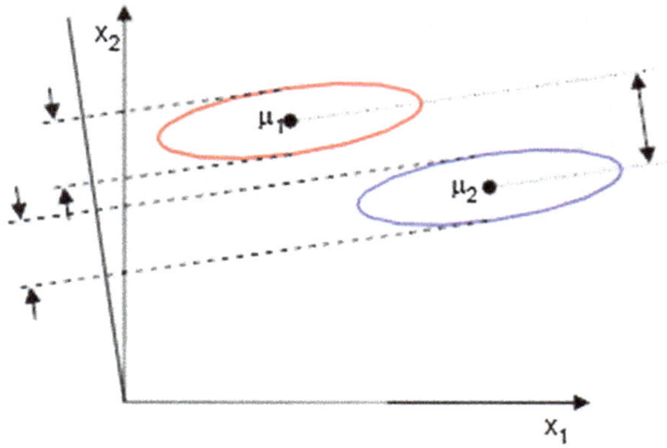

Fig. 6 Projection analysis considering mean and SD

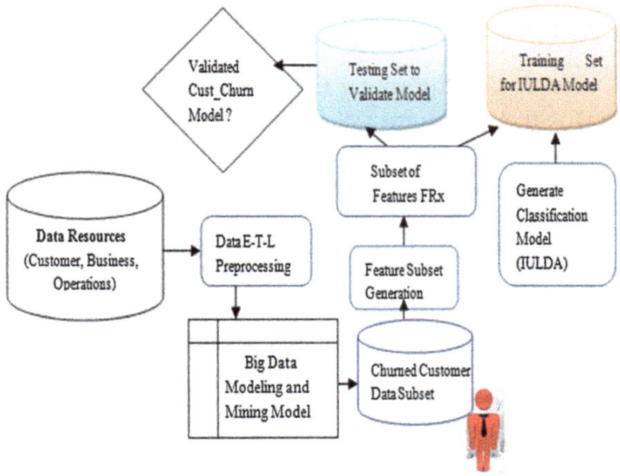

Fig. 7 Model framework of IULDA prediction

4 Experimental Model and Algorithm

The objective of this model algorithm is to enable dimensionality reduction at the same time preserving the class discriminatory information to the most optimal.

For a set of M-dimensional samples $\{x(1), x(2),.., x(N)\}$,

$N1$ of which belong to class $\omega1$(Non-Churners) and $N2$ to class $\omega2$(churners). Here, M is the number of attributes that we have obtained after refining the whole information and N is the number of samples taken for training dataset (Fig. 7).

5 Mathematics Used in the Model

Fisher's interclass separability criterion It is a grouped multivariate feature subset evaluation measure that evaluates feature subsets based on the proportion of within-class to between-class distances and the original feature set is used, instead of creating new features. The criterion is expressed as

$$F_{\text{SC}} = K\left([X_{\text{intra}} + X_{\text{inter}}]^{-1} \cdot X_{\text{inter}}\right) \in [0, 1] \quad (6)$$

The average of intra-class distance is found using the within-class scatter matrix, defined as

$$X_{\text{intra}} = \sum_{i=1}^{\text{C}} \sum_{j=1}^{P} X(i,j) - \mu\text{C} \cdot (X(i,j) - \mu\text{C})' \quad (7)$$

where X_{intra} is the scatter matrix within-class, P is the number of points in class C, $X(i,j)$ is the ith point in class C, and μC is the mean of points in class C.

The matrix is designated as

$$S_W = \sum_{i=1}^{k} \sum_{j=1}^{Ni} (\mu_{ij} - \bar{\mu}_i)(\mu_{ij} - \bar{\mu}_i)^{\text{T}} \quad (8)$$

The interclass distance is measured as

$$X_{\text{inter}} = \sum_{i=1}^{\text{C}} \frac{\text{P}}{\text{N}} \cdot (\mu\text{C} - \mu) \cdot (\mu\text{C} - \mu)' \quad (9)$$

where μ is the mean of all the points.

The matrix is designated as

$$S_B = \sum (\mu j - \mu_3)^* (\mu j - \mu_3)^{\text{T}} \quad (10)$$

Fisher criterion in terms of Sw and S_B is as

$$\text{F}_{\text{SC}}(w) = w^{\text{T}} S_B W / w^{\text{T}} S_W W \quad (11)$$

Reputation Index Trust Model

Calculation of RI is done for each customer network by the following:

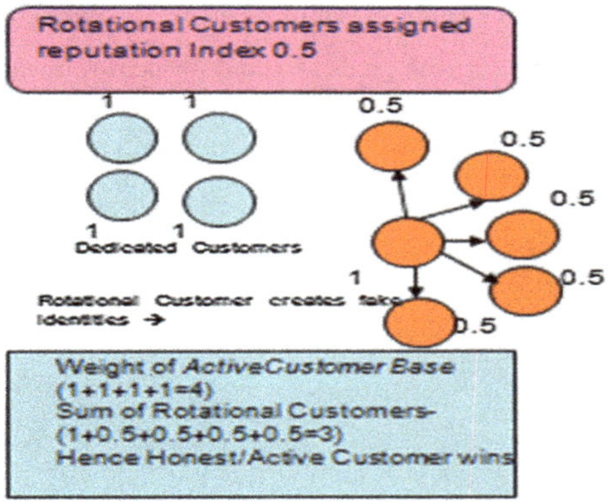

Fig. 8 Index factor for reputation count

1. Number of customer retained in same network after proactive marketing incentives.
2. Length and duration of calls at different periods of the day.
3. Count of plans registered by the customer.

Now we can calculate the reputation index based on the average of all above factors.

In the figure below, we show an example of how trust model can help in preventing the churners (Fig. 8).

Algorithm: *Indexed Uncorrelated LDA*

Input: **Case Class** (Ch, Non-Ch)

Step I: **Initially, scan and load** the dataset from the parallel (storage servers).

Step II: Perform **ETL** to the big data cube

Step III: Generate two new sub-training datasets T1 and T2 as training and testing data in ratio 70:30

Step IV: Calculate the mean value of the attributes for churned and non-churned customers

Step V: Calculate the Fischer's criterion F_{SC}

Class-Dependent Optimization → $\text{Criterion}_j = \text{inv}(\text{cov}_j) * S_B$

Class-Independent Optimization → $\text{Criterion} = \text{inv}(Sw) * S_B$

Step VI: Calculate index factor for each uncorrelated customer attribute

Step VII: Comparison of the values and attribute classification to one of the predefined sets of class.

Step VIII: Validate the model on subset T2 based on the accuracies of percentage of correctly classified example in sub-training dataset T1

Table 2 Feature subset of predictor variables comparison

Sample set with predictors, Independent variables				
Customer demographics				
Area	Customer's location, indicated by a two-letter abbreviation			
Name	Age	Gender	Marital status	
Cust_ID	Address	Contact#		
Call statistics				
Type of call		Local, National, International by 0,1,2, respectively		
Type	Day minutes	Day calls	Evening Minutes	
Night_mins		Night calls	Evening calls	
Billing information				
Plan type		Type for which customer pays (Pre/ Post; usage)		
Service commencement date			Initial charges	
Day charges	Evening charges		Night charges	
Initialization minutes			Service calls	
Voice and data product				
Voice mail messages		B/Band	Special data tariff	

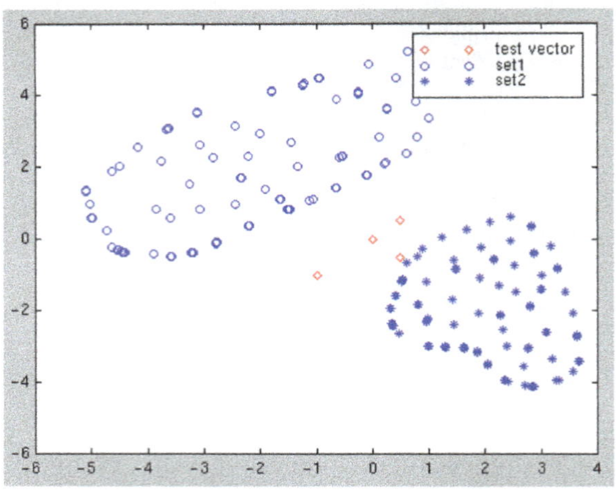

Fig. 9 The initial datasets and test vectors

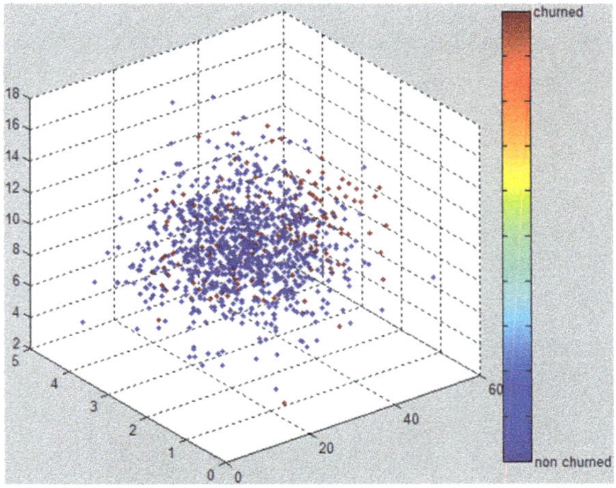

Fig. 10 Scatter plot of the customers classified by the IULDA model as churners and non-churners

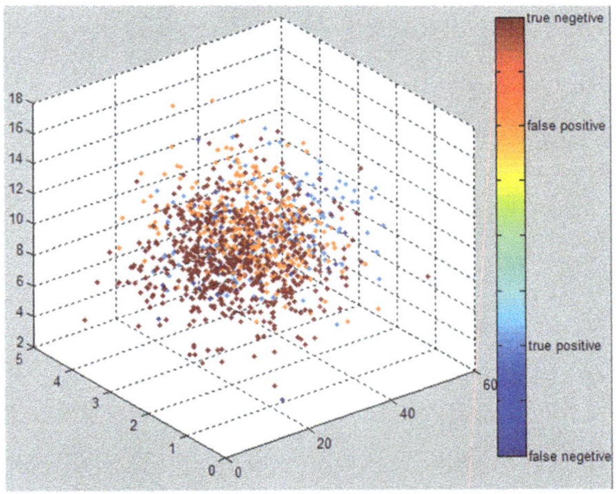

Fig. 11 Scatter plot of the test cases of IULDA model as TP, TN, FP, and FN

6 Data Results and Analysis

During this training process, feature adjustment was applied to the original data by ETL of big data technique, and the reduced relevant feature subset was obtained (Table 2).

These selected attributes were used to plot three-dimensional graph to represent the nature of customer under classifications, that is, churners and non-churners (Figs. 9, 10, 11, and 12).

The performance was measured with the techniques of sensitivity, specificity, precision, and accuracy.

This shows that

Fig. 12 Performance calculation of the proposed model

```
w =

  1.0e+004 *

  Columns 1 through 7

    0.0529   -0.0011   -0.1793    0.0001    1.0531   -0.0474   -0.0001

  Columns 8 through 14

    0.5554    0.0254   -0.0000   -0.5652    0.1830    0.0019   -0.6900

  Column 15

   -0.0120

tp =

    0.7416

fp =

    0.2584

tn =

    0.7031

fn =

    0.2969
```

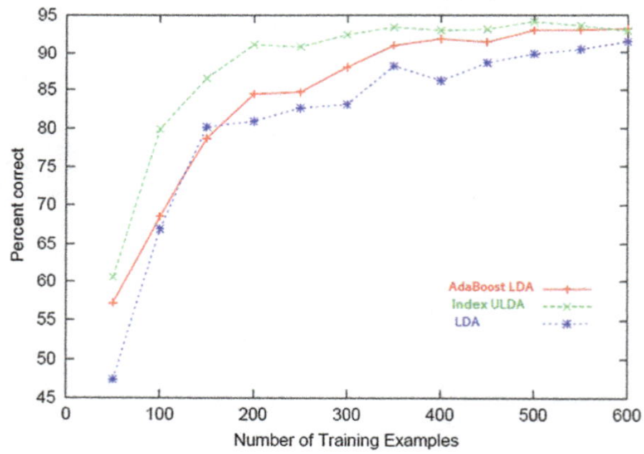

Fig. 13 Comparative plot of the training data modeled on existing LDA variants and proposed work

- 74% of the customers were positively identified by the model as churners.
- 26% of the customers were negatively identified by the model as churners, who actually did not churn.
- 70% of the customers were positively identified by the model as non-churners.
- 30% of the customers were positively identified by the model as non-churners, who actually churned (Figs. 13, 14, and Table 3).

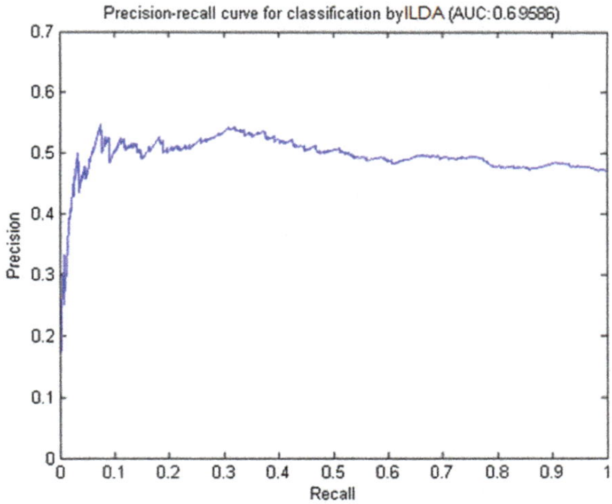

Fig. 14 Calculated area under curve (AUC) based on precision and recall values

Table 3 Work comparison with standard publications

Author	Algorithm	DOI (Publisher, Year)	Accuracy
Yaya Xie, Xiu Li	LD boosting	https://doi.org/10.1109/icmlc.2008.4620409 (ieeexplore, 2008)	63.80%
Saravanan M, V. Raaja	GraphDB	https://doi.org/10.1007/978-3-642-35527-1(springer, 2012)	72.18%
Utku Yabas, Hakki C. Cankaya	Decision tree, bagging	https://doi.org/10.1109/glocomw.2013.6825120 (ieeexplore, 2014)	ROC (0.69)
Yong S. Kim, S. Moon [10]	EDA	https://doi.org/10.1016/j.eswa.2012.04.04 (acm.dl, 2012)	MSE (0.49)
OurWork	**IULDA**	–	**72.4%**

7 Future Scope and Conclusion

Our result was compared with the existing LDA variants on data of 4500 customers about the churning behavior, and it was successfully predicted with an accuracy of **72.4% for volatile churners** identified with some willingness to churn in near future. Thus, the quality assurance parameters could be further analyzed over a sample time period taken here as 3 months to elucidate churn risk parameters with trusted model of reputation index in future.

References

1. Kumar A, Roy S, Ranjan P (2014) Dimensionality reduction for high dimensional data: a CFS and Chi Square approach. In: Proceedings of the 2014 international conference on information and communication technology for competitive strategies (ICTCS'14). ACM, New York, NY, USA. https://doi.org/10.1145/2677855.2677922
2. Glady N, Baesens B, Croux C (2009) Modeling churn using customer lifetime value elsevier editorial system for European Journal of Operational Research. EJOR-D-06–01563R2
3. Khan MR, Manoj J, Singh A, Blumenstock J (2015) Behavioral modeling for Churn prediction: early indicators and accurate predictors of custom defection and loyalty. In: 2015 IEEE International congress on big data (BigData Congress), pp 677–680
4. Brânduşoiu I, Toderean G, Beleiu H (2016) Methods for Churn prediction in the pre-paid mobile telecommunications industry. In: 2016 International conference on communications (COMM), pp 97–100
5. Zhang X, Liu Z, Yang X, Shi W, Wang Q (2010) Predicting customer churn by integrating the effect of the customer contact network. In: 2010 IEEE International conference on service operations and logistics and informatics (SOLI), pp 392–397
6. AlJazzar M (2012) A comparative study between the linear discriminant analysis and multinomial logistic regression in classification and predictive. Master thesis, Al-Azhar University, Gaza, Palestine
7. Pohar M, Blas M, Turk S (2004) Comparison of logistic regression and linear discriminant analysis, a simulation study. Metodoloski Zvezki 1(1):143–161
8. White paper on extract, transform, and load big data with Apache Hadoop Retrived on 20th Aug 2017 https://software.intel.com/sites/default/files/article/402274/etl-big-data-with-hadoop.pdf
9. Agrawal S, Bora K, Routh S (2017) Chapter 9 Machine learning approaches for supernovae classification. IGI Global. https://doi.org/10.4018/978-1-5225-2498-4.ch009
10. Kim YS, Moon S (2012) Measuring the success of retention management models built on churn prob-ability, retention probability, and expected yearly revenues. Expert Syst Appl 39:11718–11727

Effect of RC Surge Suppressor in Reduction of Over Voltages at Motor Terminal Caused by PWM-Based Inverter

Md Sajid, Amer Ali Khan, M. Suryakalavathi and B. P. Singh

Abstract Problems relating to the achievement of protection for inter-turn insulation in multi-turn coils in PWM-based induction motors have currently aroused the interest of system designers and are being studied in great detail. However, such methods are impracticable to be carried out at every installation. A system designer should have prior knowledge of the transient overvoltages which are likely to occur in a proposed installation. For this, determination of transient overvoltage is essential. Present paper deals with the investigation of transient overvoltages in 5 HP synchronous motor caused by PWM-based inverter as well as the reduction of transient overvoltages at motor terminal by connecting appropriate RC suppressors. The simulation results show the effect of placing RC suppressor in the system to reduce the overvoltages at motor end. For this, a Simulink model is developed and implemented in MATLAB.

Keywords MATLAB · PWM inverter · RC suppressor
Characteristic impedance · Reflected wave

M. Sajid (✉)
Electrical Engineering Department, Muffakham Jah College of Engineering & Technology, Hyderabad, India
e-mail: sajideed@mjcollege.ac.in

A. A. Khan
Department Electrical Engineering, Shadan College of Engineering & Technology, Hyderabad, India
e-mail: md.amerkhan@gmail.com

M. Suryakalavathi
Electrical Engineering Department, JNTUH, Hyderabad, India

B. P. Singh
Electrical Engineering Department, St. Martins Engineering College, Hyderabad, India

V. Nath and J. K. Mandal (eds.), *Nanoelectronics, Circuits and Communication Systems*, Lecture Notes in Electrical Engineering 511,
https://doi.org/10.1007/978-981-13-0776-8_19

1 Introduction

PWM-based inverters are proposed for controlling the industrial motors. VSI-based PWM inverters are more popular for medium power drives. The magnitude of output voltage and frequency can be adjusted with the help of controllable switches used in inverters. In general, IGBTs can be used for medium power applications. The specific feature of IGBT is the switching frequency that is around 20 kHz. Nevertheless, this advantage can only achieve under some circuit condition. In most of the cases, switching operation of PWM-fed induction motors leads to transient overvoltages at motor terminal. If a motor terminal experienced steep front surges, then the inter-turn insulation of stator winding may stress and cause turn insulation failure. With recent technology, SiC MOSFETs switching time is decreased. Fast switching produces a lot of advantages. However, it also creates a major problem like overvoltages at motor terminal. Experimental and theoretical investigations [1] reported that there is a severe reaction on reflect wave. It is noticed that [2] for short length cables magnitude of overvoltage is higher than long length cable. Authors also reported that filters connected at motor terminal are not able to reduce the rate of change of voltage. If modulation technique [3] is not properly adjusted, then motor side voltage rises to three to four time's system voltage. High dv/dt [4, 5] gradually deteriorates the turn insulation. Experimental [6] investigations concluded that, for long length cables, severity is reached up to 4 p.u. The accuracy [7] has been improved if inverter dead time effects on both inverter input and output voltages are considered in the proposed system. Medora et al. [8] reported that GTO-based current source inverter is not a contributor for overvoltage. Several root causes lead to a deterioration of the winding insulation. Out of these are thermal, electrical stress, and mechanical stress [9]. In case of PWM-based inverter-fed operation, the different rise times of the applied voltage by the voltage source inverter (VSI) in connection with the mismatch of motor impedance and cable impedance result to reflection of the pulse at the motor terminal and hence overvoltage. This incident additionally stresses the insulation of stator winding of motor. This problem is not found in circuit breaker switching operation. Several investigations carried out to detect problem causing this overvoltage. Nevertheless, reduction of steepness and magnitude of overvoltage with suitable suppressors could not be covered so far. The present paper objective is to present the effect of RC suppressor in the reduction of overvoltage at motor terminal caused by PWM inverter-fed operation.

2 Voltage Reflection Theory

As described in the previous section, the mismatch impedance of cable and machine as well as rise times of voltage pulse of switching device leads to transient overvoltage at the motor terminal. When an induction motor is fed by PWM-based inverter, there would be voltage reflections due to the mismatch of characteristics impedance of the motor and cable. This phenomenon is sometimes known as "Reflected Wave", "Transmission Line Effect", or "Standing Wave". In general characteristics, impedance of the cable is around 75 Ω, whereas characteristic impedance of the motor is much higher and it totally depends upon the motor power rating. This event can be explained by traveling wave theory [10]. It is well known that motor characteristic impedance is much higher than the cable impedance. Consequently, the reflection coefficient is close to one and the voltage pulse applied to the system is theoretically fully reflected [11]. Figure 1 shows the voltage reflection process.

For high frequencies, the motor acts like an open circuit. The reflection results in oscillating overvoltage with decaying amplitude. The cable length defines the propagation time of the applied voltage pulse and thus, influences the characteristic of the reflection. The reflection takes place if the propagation time is more than the rise time of the voltage pulse applied by the inverter [12]. It is well known that the transient voltage can reach up to twice, and it could be four times the DC-link voltages for fast switching. The oscillating frequency is in between 10 kHz and 10 MHz [13].

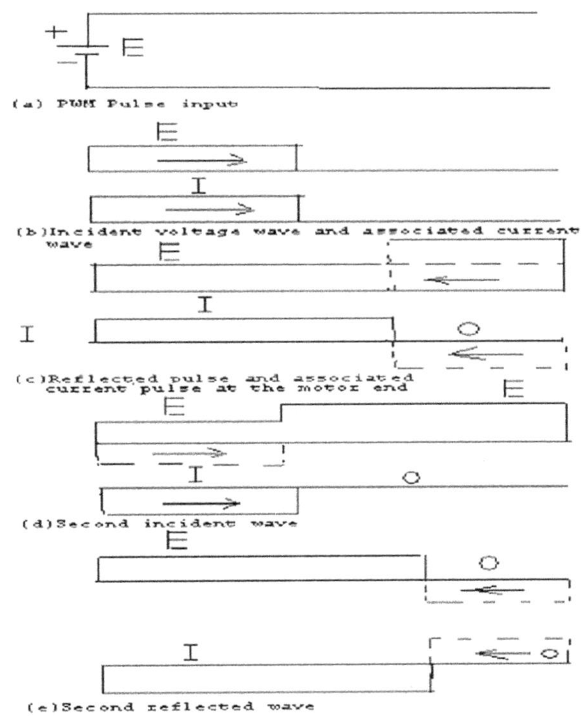

Fig. 1 Voltage reflection process

2.1 *Influence of Parameters on Transient Over Voltage*

Different parameters of the drive system influence the transient overvoltage shape. Kerkman et al. reported that [14] amplitude, rise time of the applied voltage, and length of cable are the important parameters which affect the transient overvoltage. As per Hou Yong [15], the cable's characteristic impedance also determines the shape of the overvoltage. It is well known that voltage pulse propagation time in the cable decides the oscillation frequency. Propagation time always depends upon the [16–18] length of cable used in the system. It is reported that [19] in general speed of propagation of the reflected waves in typical cables may be around 150–200 m/μs. According to the length of cable, there is an increment and decrement in oscillation frequency. It is [20] found that as the length of cable increases the magnitude of overvoltage also increases. Authors also reported that for short rise time, pulse steepness of voltage increases. Mihaila et al. [21] reported that for the first oscillation there is good agreement between computed and experimental results, but for further oscillations difference is more. This difference may be due to unexpected parameters of the equivalent circuit. Authors suggested that the impact of resistance values has an important influence on damping, and their definite values are strappingly dependent on frequency.

3 Modeling of Motor Drive System in MATLAB

Figure 2 shows the circuit model of the PWM-based inverter. Since IGBTs have good characteristics than other controllable switches, in the present work IGBTs are used as switches in the inverter circuit. Figure 3 shows the output of the IGBT-based inverter. However, the typical rise time and fall time of the inverter output is in the range of 0.05–0.06 μs. But voltage with this rise time leads to voltage reflection at switching frequency.

Figure 4 shows the model circuit of motor drive system fed by PWM-based inverter. Cable having 10 m length is connected between motor and inverter which is represented by distributed parameters. A 460 V pulse voltage is applied to the 5 HP motor through cable. As described in the previous section, pulse initiated from the cable reflected toward the motor side due to mismatch impedance of cable and motor. It results in transient overvoltage at the motor terminal which causes to stress the turn insulation. For determining the transient overvoltages caused by PWM-based inverter switching, a model circuit is simulated using MATLAB/Simulink software.

From Fig. 5, it is observed that there is an overvoltage at motor terminal. The calculated transient overvoltage at motor terminal without RC suppressor is 810 V. This voltage either stresses the weaker insulation points or deteriorates the turn insulation.

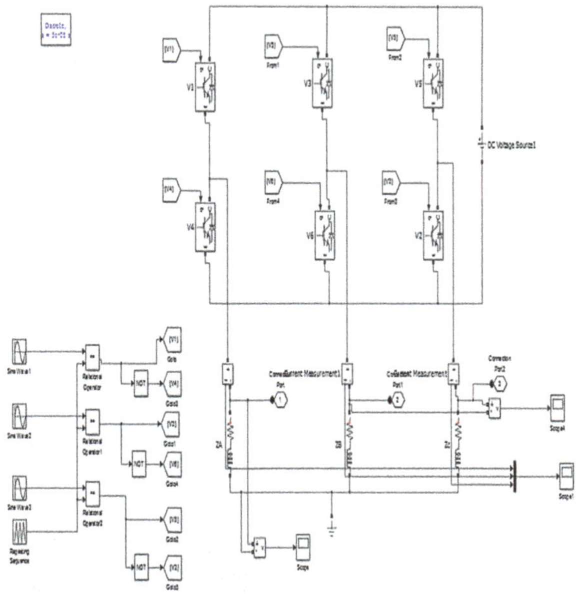

Fig. 2 IGBT-based PWM inverter

Fig. 3 Inverter output

4 Influence of RC Suppressor in Reduction of Over Voltages

In this section, influence of RC suppressor is analyzed by connecting the RC suppressor at the inverter terminal and at motor terminal. The complete motor drive system with RC surge suppressor connected at inverter terminal as shown in Fig. 6 is simulated first in MATLAB/Simulink software. The standard values for RC suppressor are given as follows:

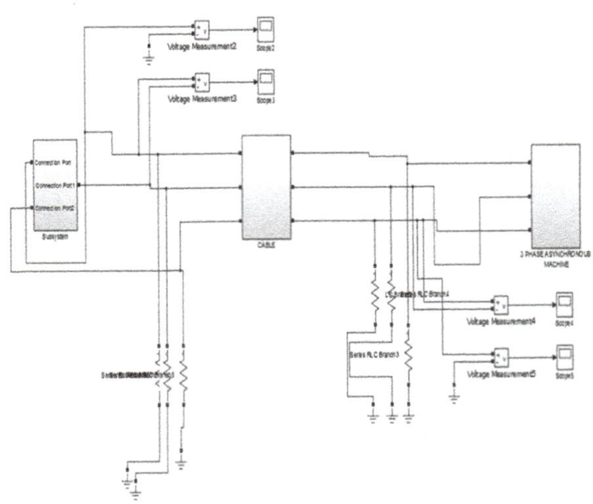

Fig. 4 Model circuit without RC suppressor

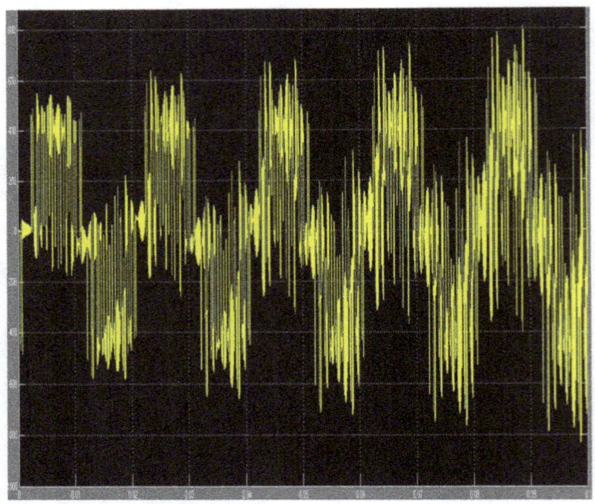

Fig. 5 Transient overvoltage at motor terminal

$$R = 250\,\Omega$$

$$C = 0.5\,\mu\text{F}$$

Figure 6 shows the model circuit with RC suppressor connected at inverter terminal and Fig. 7 shows the simulated voltages at inverter terminal.

Figure 8 shows the model circuit with RC suppressor connected at motor terminal that is simulated in MATLAB/Simulink. The simulated voltage profile at motor terminal is shown in Fig. 9.

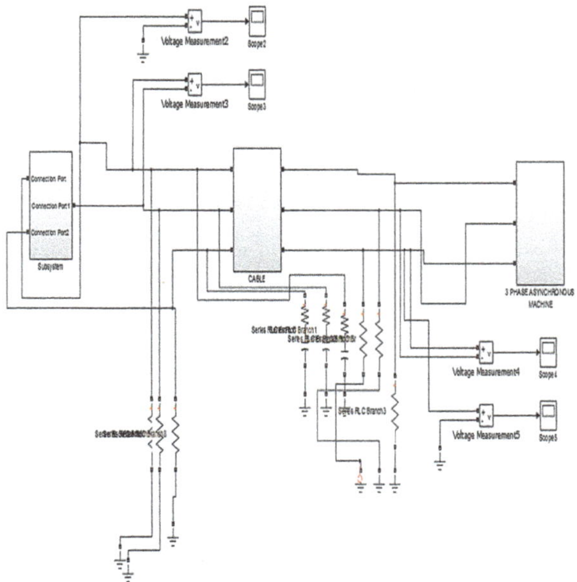

Fig. 6 Model circuit with RC suppressor connected at inverter terminal

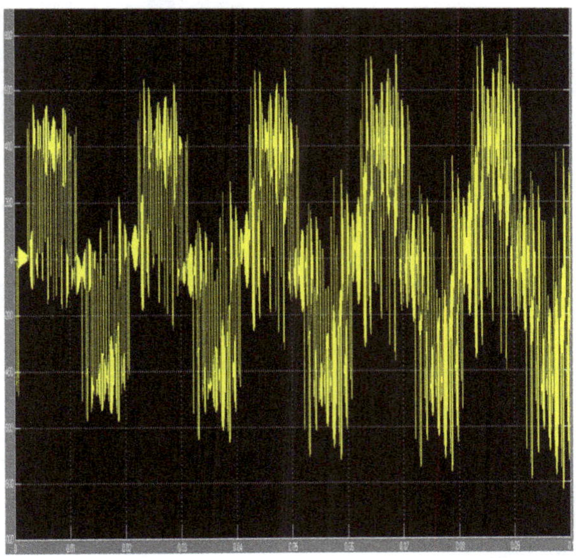

Fig. 7 Voltage profile at motor terminal with RC suppressor at inverter terminal

5 Results and Discussions

From Fig. 5, it is observed that the overvoltage may lead to the occurrence of partial discharge, and either stresses the weaker turn insulation or degrades the insulation strength. In addition to this, the wave shape is also affected. This is due to the effect

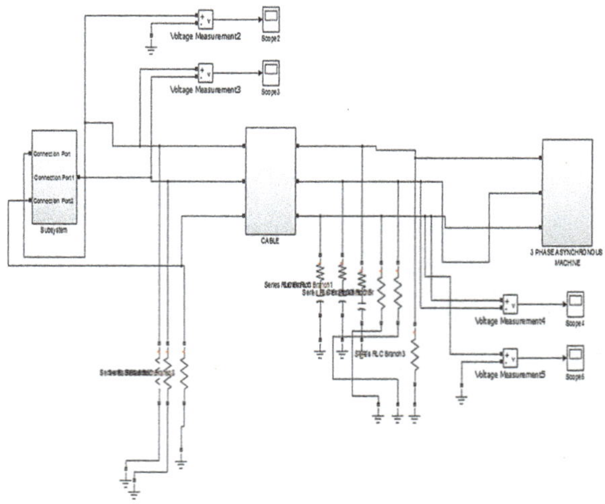

Fig. 8 Model circuit with RC suppressor connected at motor terminal

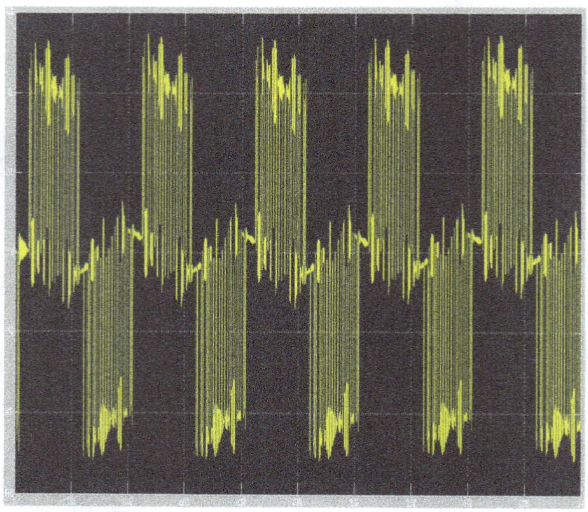

Fig. 9 Voltage profile at motor terminal with RC suppressor at motor terminal

of parameters used in the drive system. To reduce the overvoltages, a suitable RC suppressor should be added to the system as it is an impedance matching circuit. The only requirement of this circuit is to select proper values of *R* and *C* to match the load.

From Fig. 7, it has been observed that the influence of RC suppressor at inverter terminal is not much effective as it is found that placing of RC at inverter terminal is not capable to reduce the d*v*/d*t* as much as RC placed at motor terminal. It is well known that if characteristic impedance matched with the motor surge impedance then voltage reflection may not occur and consequently no overvoltages occur at the motor terminal.

From Fig. 9, it is observed that the peak voltage at motor terminal is 460 V which is in safe limit. It is found that placing the RC suppressor at motor terminal is more effective than that of placing at inverter terminal. The main reason may be that the mismatch of impedance is in between motor and cable and not in between inverter and cable.

6 Conclusion

In the present paper, MATLAB/Simulink-based program is proposed to investigate the transient overvoltage at motor terminal caused by PWM-based inverter. For this, a model is developed with and without RC suppressor. Results show that there are transient overvoltages and distorted wave at motor terminal. To reduce the severity, RC suppressor is placed at inverter terminal as well as at motor terminal. In certain investigations, RLC filter is used for reducing the overvoltage. But in this study with RC surge suppressor, overvoltage is reduced to rated value. It has been also observed from the results that placing RC suppressor at motor terminal is much effective than placing RC suppressor at inverter terminal to reduce the dv/dt. It is found that magnitude of voltage is reduced from 810 to 460 V by connecting the RC suppressor at motor terminal.

Acknowledgements Authors are grateful to MJCET, Jawaharlal Nehru Technological University Hyderabad and Shadan College of Engineering & Technology for providing opportunity to publish this paper.

References

1. Iosif V, Duchesne S, Roger D (2015) Voltage stress predetermination for long-life design of windings for electric actuators in aircrafts. In: Proceedings of IEEE conference, annual report conference on electrical insulation and dielectric phenomena, pp 318–321
2. Nussbaumer P, Zoeller C, Wolbank TM, Vogelsberger MA (2013) Transient distribution of voltages in induction machine stator windings resulting from switching of power electronics. In: Proceedings of IEEE conference on Industrial Electronics Society, IECON 2013, pp 3186–3191
3. Grubic S, Aller JM, Lu B, Habetler TG (2008) A survey on testing and monitoring methods for stator insulation systems of low-voltage induction machines focusing on turn insulation problems. IEEE Trans Ind Electron 55(12):4127–4136
4. Basavaraja B, Siva Sharma DVSS (2008) Impact of non-uniform distribution of voltage on stator winding of PWM fed motor. In: Proceedings of IEEE conference, TENCON Nov 2008, pp 1–6,19 to 21
5. Basavaraja B, Siva Sharma DVSS (2007) Modelling, simulation and experimental analysis of transient terminal overvoltage in PWM-inverter fed induction motors. In: Proceedings of IEEE conference

6. Krings A, Paulsson G, Sahlen F, Holmgren B Experimental investigation of the voltage distribution in form wound windings of large AC machines due to fast transients. In: Proceeding of IEEE, 22nd international
7. Guha A, Narayanan G (2016) Impact of dead-time on inverter input current, DC-link dynamics and light-load instability in rectifier-inverter-fed induction motor drives. In: Proceedings of IEEE conference
8. Noshirwan MK, Kusko A Voltage stress on 8000 HP induction motor due to GTO current source inverter — Case study, IECON 2010–36th annual conference on industrial electronics society, pp 2479–2484
9. Burger D, Tenbohlen S, kohler W, Ebbinghaus W (2012) Impact of multiple restrikes at vacuum circuit breaker on the EMC of medium voltage switchgear. In: Proceedings of IEEE conference, international symposium on electromagnetic compatibility, Europe, pp 1–6
10. Haq SU, Omranipour R, Luis (2014) Surge withstand capability of electrically and thermo-mechanically aged turn insulation of medium voltage form-wound AC stator coils. In: Proceedings of IEEE electrical insulation conference, Philadelphia, Pennsylvania, USA, 8–11 June 2014, pp 78–81
11. Lindell E, Liljestrand L (2016) Effect of different types of overvoltage protective devices against vacuum circuit-breaker-induced transients in cable systems. IEEE Trans Power Delivery 31(4):1571–1579
12. Rahman MF, Haider T, Haque E, Blackburn TR (1999) A study of the over-voltage stress with IGBT inverter waveforms on motor and supply cabling and their remedial measures. In: Proceedings of IEEE conference, pp. 519–525
13. Sajid M, Singh BP, Suryakalavathi M (2016) Impulse and fast surge distribution of voltage in 11 kV industrial motor. In: Proceedings of IEEE conference, international symposium on electromagnetic compatibility, Europe, pp 1–6, 2012 ICEEOT, India, pp 215–219
14. Kerkman RJ, Leggate D, Skibinski GL (1997) Interaction of drive modulation and cable parameters on AC motor transients. IEEE Transactions on industry applications, 33(3) May/ June 1997
15. Yong H, Dehong L, Jianguo J (2000) An effective method to determine voltage waveform at terminal of induction motor fed by PWM inverter. Power electronics and motion control conference, 2000. In: Proceedings of IEEE conference, IPEMC 2000
16. Niayesh K, Berth M (2002) Resonance like phenomena in motor drive circuits caused by multiple re-ignitions of vacuum circuit breakers. In: Proceedings of IEEE conference and exbhition on transmission and distribution, vol. 26–10 Oct 2002, pp. 914–918
17. Weinert H (1999) Interruption and re-ignition behavior of vacuum switching gaps with small gap lengths after high frequency current zeros. Ph.D. Dissertation, RWTH-Aachen, Germany
18. Borghetti A, Napolitano F, Nucci CA, Paolone M, Sultan M, Tripaldi N (2011) Transient recovery voltages in vacuum circuit breakers generated by the interruption of inrush currents of large motors. In: Proceedings of international conference on power systems transients, in Delft, the Netherlands, 14–17 June 2011
19. Templaar HG, Kema NV, Arnehm (1988) Determination of transient over voltages caused by switching of high voltage motors. IEEE Trans Energy Convers 3(4), Dec 1988
20. Berth M, Kung M, Limbeek EFDE (2002) Switching over voltages in motors circuits. IEEE Trans Ind Appl 37(6), Aug 7, 2002:1552–1589
21. Mihaila V, Duchesne S, Roger D (2011) A simulation method to predict the turn-to-turn voltage spikes in a PWM fed motor winding. IEEE Trans Dielectr Electr Insul 18(5), Oct 2011, pp.1609–1615

Usage of Deep Learning in Epileptic Seizure Detection Through EEG Signal

Nalini Singh and Satchidananda Dehuri

Abstract Epileptic seizure occurs due to neuronal disorder that results in abnormal pattern of brain signal. Electroencephalogram (EEG) signal represents a modest measure of electric flow in a human brain. An EEG is one of the main diagnostic tests for epilepsy. Due to the presence of seizures, normal pattern of brain waves disappears and different other brain waves can be visualized during the recording of EEG. Approximately, 1% of the total population in the world is affected by this disease. This paper is based on a systematic approach for epilepsy detection of human brain by extraction of features and classification of EEG signal. Feature extraction is completed by discrete wavelet transform (DWT) and multilayer perceptron neural network (MLPNN) with deep learning is used for classification. Experimental study of the proposed work is done through Python platform with an encouraging performance.

Keywords EEG signal · Epilepsy · DWT · MLPNN

1 Introduction

Epilepsy is chronic non-communicable neurological disorder which ranges from severe, disabling, and life-threatening to fragile ones. Here, the normal pattern of the neuronal activity creates annoyance, causing strange sensations, variance in consciousness and behavior or at times cause loss of sensations, cramps, convulsions, etc. [1]. According to the cause the types of seizure depend, any disturbance or

N. Singh (✉)
Department of Applied Electronics and Instrumentation Engineering,
Silicon Institute of Technology, Silicon Hills, Patia, Bhubaneswar 751024
Odisha, India
e-mail: nalini.singh1905@gmail.com

S. Dehuri
Department of Information and Communication Technology,
Fakir Mohan University, Vyasa Vihar, Balasore 756019, Odisha, India
e-mail: satchi.lapa@gmail.com

V. Nath and J. K. Mandal (eds.), *Nanoelectronics, Circuits and Communication Systems*, Lecture Notes in Electrical Engineering 511,
https://doi.org/10.1007/978-981-13-0776-8_20

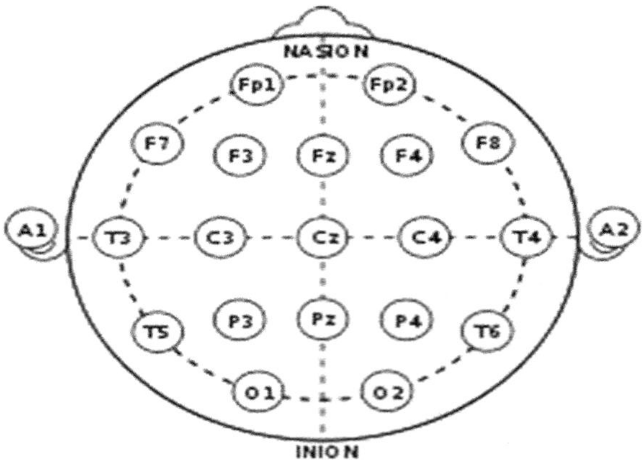

Fig. 1 The 10–20 electrode placement system

annoyance in the neuronal activity leads to seizures like abnormal brain growth, brain damage, or any illness [3]. Development of epilepsy is due to neuronal disorder in brain which changes some significant features and other factors of brain cells. Having a head injury or a single seizure due to high temperature (named febrile seizure) does not essentially mean that somebody is having epilepsy.

An internationally recognized method known as the 10–20 electrode placement system is used to place the electrodes over the scalp for both medical and experimental EEG recordings, which is schematically shown in Fig. 1.

In this analytical study, we have considered publicly available EEG dataset related to epilepsy for all experimental valuations [2]. Based on this, there are mainly two phases of the epileptic seizure detection process that are carried out. The first phase is to analyze EEG signal and convert it to a set of samples with a set of features. The second phase is to classify a ready processed data into different classes such as epilepsy or normal [3].

2 Data Description

In this paper, we have used an open-source epileptic data from Bonn University, Germany. The dataset comprises five sets of EEG data from A, B, C, D, and E. Hundred single channels are used for recording of each dataset and are recorded for the duration of 23.6 s. Five normal subjects are selected for recording of datasets A and B by means of open and closed eyes [4]. C is recorded previous to the epileptic attack, D represents beginning of epileptogenic region, and E is recorded for the duration of an occurrence of epilepsy in an epileptogenic region. For dataset recording, an amplifier system of 128-channel was used and all the datasets were recorded and finally with 173.61 Hz sampling rate, datasets are digitized. Here, every channel is having a total of 4097 samples be divided into 8 identical data frames of 512 sizes with leaving the final data. As a result, from 100 single

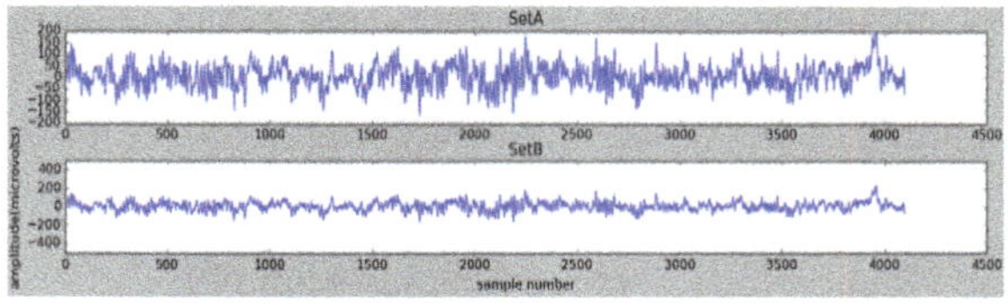

Fig. 2 Sample EEG signal from A and B

channels, for every dataset, 800 total data segments are obtained. Discrete wavelet transform (DWT) coefficients are used for statistical feature extraction of each data section intended for detection of seizure by the method of pattern recognition. Figures 3, 4, and 5 show graphs of EEG signals from sets A to E. In all, there are 500 different single-channel EEG data. In the next section, we will illustrate how to crack these signals using discrete wavelet transform [5] and prepare several statistical features and form a proper sample feature dataset (Fig. 2).

3 Feature Extraction

The purpose of this process is to eliminate the redundancy in the EEG signals by selecting the discriminative features from the raw data. The process also helps in reducing the size of the input feature vector. Here, we are using wavelet transform

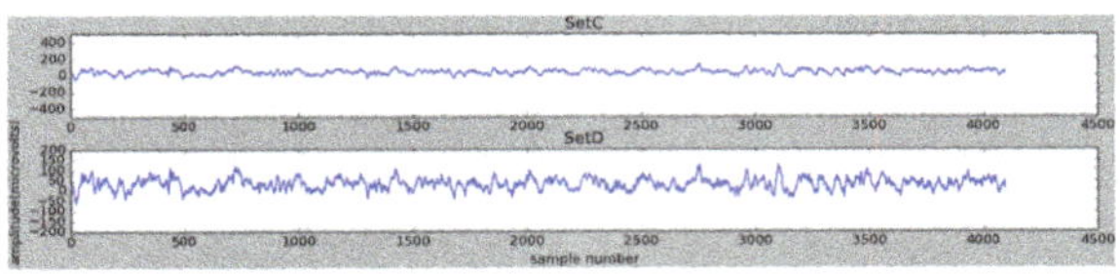

Fig. 3 Sample EEG signal from C and D

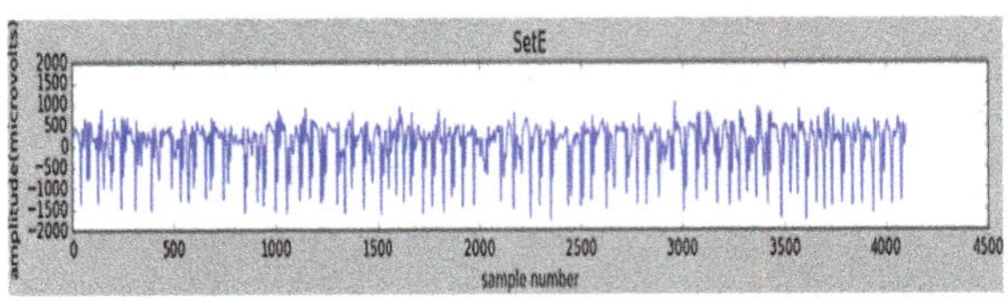

Fig. 4 Sample EEG signal from E

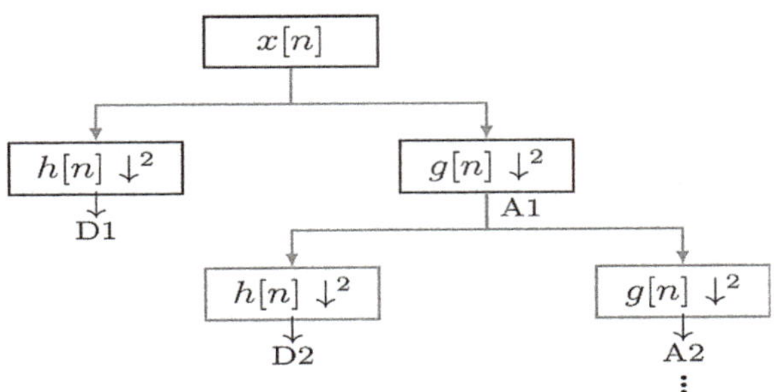

Fig. 5 DWT decomposition scheme

(WT) method for feature extraction as EEG signals have very uncertain and rapidly changing frequency [5]. The basic idea behind this technique is to use a scale for analysis. Two types of wavelet transform are there. First is continuous wavelet transform (CWT), and second is discrete wavelet transform (DWT). CWT is defined as [3]

$$\text{CWT}\ (a, b) = \int_{-\infty}^{\infty} X(t).\varphi_{a,b}^{\nabla}(t)\mathrm{d}t \tag{1}$$

where *X*(*t*) is the original signal, *a* represents the scaling factor, and *b* is the translation along the time axis. The ∇ symbol denotes complex conjugation and $\varphi_{a,b}^{\nabla}$ is computed by scaling the wavelet at time band scale *a* (as shown in Eq. 2):

$$\varphi_{a,b}^{\nabla}(t) = \frac{1}{\sqrt{|a|}}\varphi\left(\frac{t-b}{a}\right) \tag{2}$$

where $\varphi_{a,b}^{\nabla}(t)$ stands for the mother wavelet. Here, it is presumed that "*a*" and "*b*" change continuously. But the main disadvantage of CWT is that the calculation of wavelet coefficients for every possible scale can result in a large amount of data. It can surmount with the help of DWT [5]. It is almost same as CWT except that the values of *a* and *b* do not change continuously. It can be defined as (shown in Eq. 3)

$$\text{DWT} = \frac{1}{\sqrt{2^p}} \int_{-\infty}^{\infty} X(t)\varphi\left(\frac{t-2^p q}{2^p}\right)\mathrm{d}t \tag{3}$$

where *a* and *b* of CWT are replaced in DWT by 2^p and 2^q, respectively.

DWT decomposes the signal into a number of levels and every level represents a particular roughness of the signal. At each level, the signal is passed through both the high-pass filter (HPF) and low-pass filter (LPF). HPF is the mother wavelet, and LPF is the mirror version of the HPF [1, 2]. The output of each level is the downsized signal by a factor of 2. Approximation (A1) consists of low-frequency components, and details (D1) consists of high-frequency components, and the coefficients are produced by LPF and HPF, respectively. For the next level of decomposition, A1 is further decomposed and the whole process is repeated until the desired level of decomposition is achieved, which means A1 and D1 are the first split of signal *x*(*t*). Then from using A1, the next split is A2 and D2, and the process continues until the desired level is reached. Figure 6 depicts the DWT decomposition scheme of a signal of length *n* and $\downarrow^2$ shows that the signal is downsized by a factor of 2 at each level [10].

The signal is decomposed using the Harr Wavelet function up to four levels. Thus, it produces a series of wavelet coefficients like four detailed coefficients (D1, D2, D3, and D4) and an approximation signal (A4). EEG signal has been

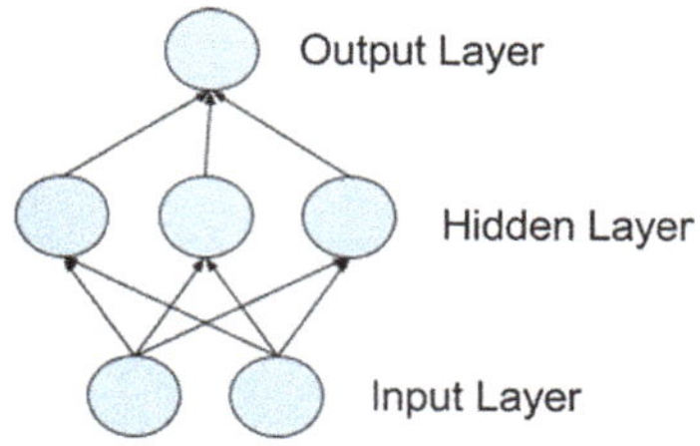

Fig. 6 Model of simple MLPNN

decomposed into four levels, and from each level, five numbers of features such as mean, median, standard deviation (SD), minimum (MIN), and maximum (MAX) of each dataset have been extracted.

4 Deep Learning Based Classification

Deep learning models are based upon the structure and function of the brain called artificial neural networks including various models. Here, very deep features of input data for every model are able to learn through extremely deep neural networks. These networks are naturally deeper than three layers. They are initialized layer-wise through unsupervised training, and after that consequently tuned in a supervised manner [7]. Here, all the features can be learned ranging from both low-level features to high-level features, while proper features are formulated in the end for pattern classification. At higher layers, deep models can potentially lead to more abstract and complex features. More abstract features are normally invariant to most local changes of the input. A number of researches say that [8, 9] deep models are able to provide improved estimate toward nonlinear functions. A computer learns from observational data using artificial neural networks (ANNs) which are biologically inspired programming platform. ANN with deep learning provides one of the best solutions to many problems in signal processing, image processing, speech recognition, and natural language processing [7].

A neural network is composed of perceptron, connected in different ways and operating with various activation functions [6]. A multilayer perceptron (MLP) is also known as an ANN and is inspired by biological nervous systems, such as the brain. With a single hidden layer, MLPNN can be represented graphically as follows.

Multilayer perceptron with backpropagation is a deep learning algorithm for supervised training [5]. The basic process is as follows:

1. A sample is presented and propagated forward through the network.
2. Output error which is the mean squared error (MSE) is found out by

$$E = \frac{1}{2}(t - y)^2$$

where t represents target value and y gives the actual network output.

3. Minimization of network error is done by stochastic gradient descent method.

Derived statistical DWT features are useful for classification. The aim of classification is to identify the epilepsy in EEG signal by linear or nonlinear mathematical approach. Here, epileptic seizure is found out by using MLPNN classifier for both the individual and joint features that are derived by using DWT with various combinations of A to E. MLPNN classifier with ADAM optimization performance is assessed with the accuracy from the derived features for detection of the epileptic seizure.

Here, we had done binary classification with one epileptic and one non-epileptic signal. In this, the combined feature extracted dataset was used. In this paper, Bonn University dataset is used. This dataset describes medical record for epileptic and non-epileptic individuals. Here, our work is based on binary classification where the occurrence of seizures represents 1 and nonappearance as 0. First two layers use rectifier ("relu") activation function, and at the output layer sigmoid activation function was used. Sigmoid function ensures network output in between 0 and 1 and also easy to map to either a probability of class 1 or snap to a hard classification of either class with a default threshold of 0.5.

4.1 Adaptive Moment Estimation (ADAM)

ADAM is a type of stochastic gradient descent (SGD) which is used for objective function minimization, written as a sum of differentiable functions [11]. In other words, SGD tries to find minima or maxima by iteration. Adaptive learning rates for every parameter are computed by ADAM. It stores an exponentially decaying average of both past squared gradients and of past gradients similar to momentum. Adam is an optimization algorithm for first-order gradient-based optimization of stochastic objective functions, based on adaptive estimates of lower order moments [12]. Implementation of this method is straightforward due to its computationally efficiency. It requires less memory and is invariant toward diagonal rescaling of the gradients. It is also suitable for problems related to large data and/or parameter. This is suitable for nonstationary objectives and problems with very noisy and/or sparse gradients. Hyperparameters typically require little tuning [12].

5 Experimental Study

Here, we are using python 2.7 for both feature extraction and classification. Here, some of the statistical features such as minimum (MIN), maximum (MAX), mean, median, and standard deviation (SD) have been extracted from datasets A, B, C, D, and E using DWT. The same procedure can be followed for all five EEG recordings to make a perfect set. So, after this level, we are ready with a sample feature dataset of order 100 by 25 matrixes of each dataset. After this, classification is also done by

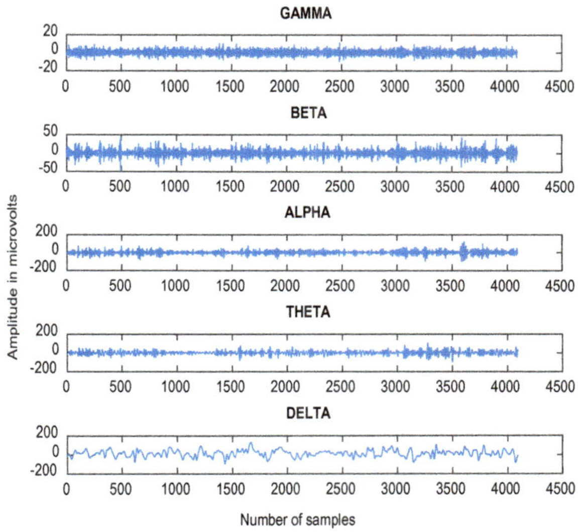

Fig. 7 Single-channel EEG from set A with four detailed and one approximation coefficient after DWT decomposition

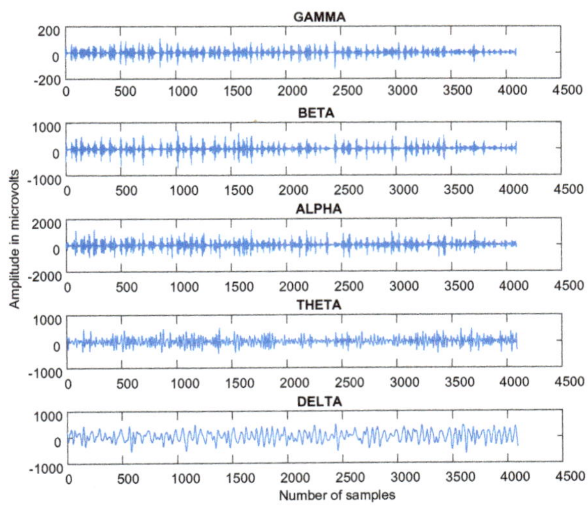

Fig. 8 Single-channel EEG from set E with four detailed and one approximation coefficient after DWT decomposition

Python 2.7. Epileptic seizure identification from EEG [5, 6] signal is possible by using MLPNN with backpropagation algorithms. This model consists of both training phase and testing phase. Here, we have used backpropagation training algorithm (Figs. 7, 8, 9, 10, and 11; Tables 1 and 2).

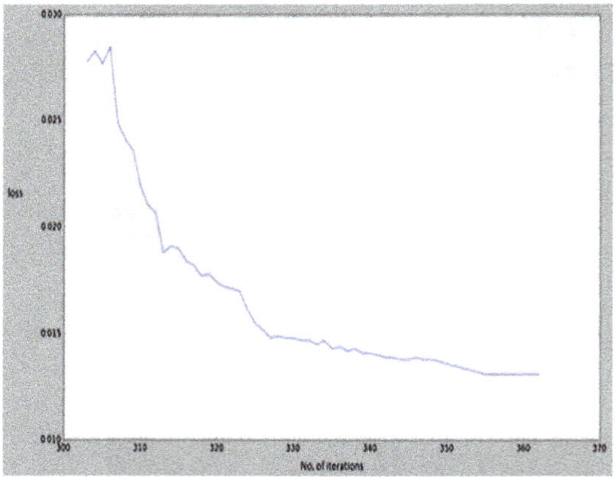

Fig. 9 Plot between loss and number of iterations with one hidden layer

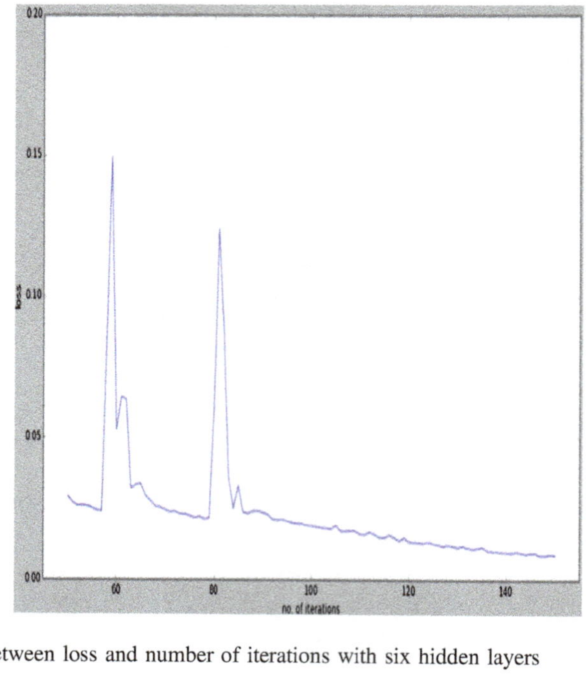

Fig. 10 Plot between loss and number of iterations with six hidden layers

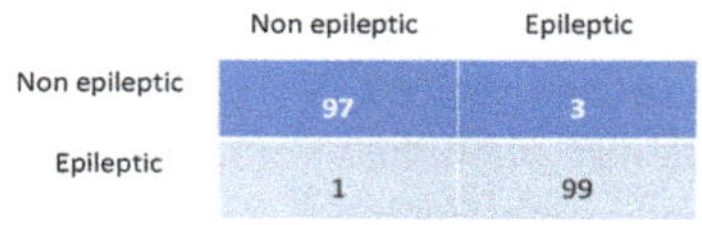

Fig. 11 Confusion matrix of two-class classification

Table 1 Feature comparison of Daubechies and Harr (D1 coefficient) of D dataset

Transform	MEAN	Standard deviation	Median	MIN	MAX
DB2 (DEB)	−24.59230	874.60371	−157.22065	−3452.53212	5650.81773
DB1 (HARR)	−24.98505	893.47791	−231.30818	−3059.52487	3992.54559

Table 2 Binary class classification accuracy used for various combinations of datasets from A to E using MLPNN

Datasets	Accuracy (%)
A-E	99.50
B-E	97.00
C-E	98.51
D-E	100
AB-E	89.00
AC-E	99.67
AD-E	99.67
BC-E	94.67
BD-E	93.33
CD-E	99.33
ABC-E	91.50
ACD-E	98.00
BCD-E	95.75
ABD-E	94.50
ABCD-E	95.60

6 Conclusion

Epilepsy detection can be done by classifying EEG signal collected from different patients in different situations. This classification can be accomplished by using different machine learning techniques. In this work, we are doing binary classification of different datasets and their accuracies have been illustrated. The basic structure of an MLPNN was used for the above, and the parameters were varied which was an attempt to establish patterns based on their change. Different combinations of activation functions were tried and compared. The relu, sigmoid, softmax, and softplus were the various activation functions used. The highest accuracy of 100% was obtained when the number of neurons was set to 40 and 8 in the first and second layers, respectively. A graph between loss and number of iteration was plotted which converge after 350 iterations. No particular pattern was observed beyond this value of epochs which could be explained by the fact that neural networks are infinitely configurable.

7 Future Work

Our future work includes implementation of other classifiers such as SVM, other advanced neural networks, and different deep learning classifiers with other optimization techniques for validating the result. Also, we go for different other features which may increase the classification accuracy.

Acknowledgements The first author would like to thank the technical support of the Department of Information and Communication Technology, Fakir Mohan University, Vyasa Vihar, Balasore.

References

1. Sharmila A, Geethanjali P (2016) DWT based detection of epileptic seizure from EEG signals using Naive Bayes and k-NN Classifiers. IEEE Accesses 4:7716–7727
2. Supriya S, Siuly S, Wang H, Cao J, Jhang Y (2016) Weighted visibility graph with complex network features in the detection of epilepsy. IEEE Accesses 4:6554–6566
3. Satapathy SK, Dehuri S, Jagadev AK (2017) ABC Optimised RBF network for classification of EEG signal for epileptic seizure identification. Egypt Inform J 18(1):55–66
4. EEG Data (2001) [Online] http://www.meb.unibonn.de/science/physik/eegdata.html
5. Gandhi T, Panigrahi BK, Anand S (2011) A comparative study of wavelet families for EEG signal classification. Neurocomputing 74(17):3051–3057
6. Pradhan N, Sadasivan PK, Arunodaya GR (1996) Detection of seizure activity in EEG by an artificial neural network: a preliminary study. Comput Biomed Res 29(4):303–313
7. Chen Y, Lin Z, Zhao X, Wang G, Gu Y (2014) Deep learning-based classification of hyperspectral data. IEEE J Sel Top Appl Earth Obs Remote Sens 7(6):2094–2107
8. Sutskever I, Hinton GE (2008) Deep, narrow sigmoid belief networks are universal approximators. Neural Comput 20(11):2629–2636
9. LeRoux N, Bengio Y (2010) Deep belief networks are compact universal approximators. Neural Comput 22(8):2192–2207
10. Qazi KI, Lam HK, Xiao B, Ouyang G, Yin X (2016) Classification of epilepsy using computational intelligence techniques. CAAI Trans Intell Technol 1:137–149
11. Kingma DP, Ba J (2015) Adam: a method for stochastic optimization. In: International conference for learning representations, pp 1–15
12. http://sebastianruder.com/optimizing-gradient-descent/index.html, 2016

An Experimental Study of Filter Bank Approach and Biogeography-Based Optimized ANN in Fingerprint Classification

Annapurna Mishra and Satchidananda Dehuri

Abstract In this paper, we have developed a robust classifier by considering the best attributes of biogeography-based optimization (BBO) and artificial neural network (ANN) for classification of fingerprints into five classes. This work involves two phases. In phase one, we extract features by using Gabor filter bank. In the second phase, we optimize the parameters of ANN by using BBO to mitigate the problem of local optima. We have conducted experimental study on a few benchmark functions for verifying the proposed model accuracy. The outcomes of experiment claim that the accuracy of our model is significantly better than other competitive models.

Keywords Classification · Gabor filter · BBO · ANN

1 Introduction

Classification of fingerprint refers to assigning a particular class to a particular fingerprint which is perhaps the most important stage in fingerprint identification system because it facilitates matching process over a large database by classifying the fingerprint images into several standardized classes. Time complexity of the identification system can be improved by having information about the class of the input fingerprint. Search space can be reduced significantly as identification now requires only searching the database to which the fingerprint belongs. Various approaches of fingerprint classification are available in literature out of which the

A. Mishra (✉)
Department of Electronics and Communication Engineering, Silicon Institute of Technology, Silicon Hills, Patia, Bhubaneswar 751024, Odisha, India
e-mail: annapurnamishra12@gmail.com

S. Dehuri
Department of Information and Communication Technology, Fakir Mohan University, Vyasa Vihar, Balasore 756019, Odisha, India
e-mail: satchi.lapa@gmail.com

V. Nath and J. K. Mandal (eds.), *Nanoelectronics, Circuits and Communication Systems*, Lecture Notes in Electrical Engineering 511,
https://doi.org/10.1007/978-981-13-0776-8_21

model-based technique proposed by Edward Henry (1900) is most famous. It uses the number and locations of singular points (cores and deltas). A set of heuristic rule is used to detect the singularities in fingerprint image and classifies them into five classes namely loops, arches, and whorl.

But it was observed that the precise location of singular points is affected by noise present in the fingerprint. In this paper, features derived from local ridge structures are used to represent a fingerprint. This representation does not use core, delta, and orientation field, explicitly. Classification is done using a biogeography model based artificial neural network. This neural network uses BBO algorithm, which is a geographical computing technique for its weight update. This technique is less sensitive toward noise since it does not depend on the detection of singular points. So this is a more robust and accurate technique for fingerprint classification [1, 2].

Some other approaches used are frequency-based and hybrid approach. In frequency-based approach, the frequency spectrum of fingerprint is used for classification [3]. Hybrid approach combines two or more different methods for classification [4, 5]. Our algorithm is based on the multichannel representation of fingerprints using a Gabor filter bank tuned in a different orientation and a multi-stage BBO–ANN classifier. The steps involved in the adopted classification algorithm [1, 3] are as follows:

- Reference point location and
- Tessellation, standard deviation calculation, and generation of feature vector.

Then, the feature vector is passed to the classifier for classification; and in this algorithm, we have used a two-stage BBO–ANN classifier.

2 Feature Extraction

Global ridge and furrow structure contain information about the class of a fingerprint. Capturing these global features will lead to an effective classification [3, 4].

The steps for feature extraction algorithm are as follows:

- Separate the region of interest from the rest of the image by detecting a valid reference point and defining a spatial tessellation around the reference point that is core point as shown in Fig. 1.

Fig. 1 Singular points of a fingerprint

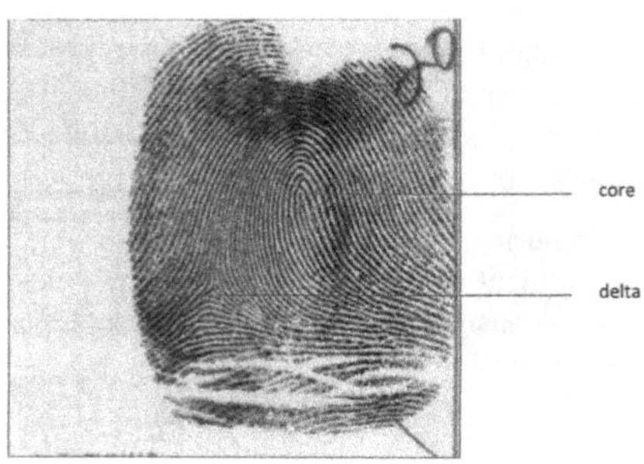

- Decompose the region of interest into a set of sub-images, which preserves global ridges and furrows structures.
- Calculate the average absolute deviation of intensity values of pixels in each sector from its mean intensity value to form the feature vector.

A fingerprint image is extracted in terms of four segmented images corresponding to four different values of θ (0°, 45°, 90°, and 135°) with respect to x-axis. Gabor filters of different orientations are used to produce different component images. Filter with 0° orientation captures the ridges having 0° orientation, i.e., parallel to x-axis and filters out (smoothes) the ridges oriented in the rest directions. Other orientation filters act in a similar manner. As per the research, the core point is defined as a point where the concave ridges have maximum curvature.

A fingerprint is characterized by the variation in its local ridge pattern and is properly captured by the standard deviation of the grayscale values. Let $C_{i\theta}$ (x,y) be the component image corresponding to direction θ for sector S_i. For $i = 0, 1, 2, \ldots, 36$ and $\theta \in [0°, 45°, 90°, 135°]$, a feature is the standard deviation $F_{i\theta}$ which is defined as

$$F_{i\theta} = \sqrt{\sum_{k_i} \left(C_{i\theta}(x, y) - M_{i\theta}\right)^2} \quad (1)$$

where k_i is the number of pixels in S_i and $M_{i\theta}$ is the mean pixel intensity in $C_{i\theta}(x, y)$. So we have a 144-dimensional feature vector.

3 Classification Using BBO–ANN

Fingerprint classification process is a difficult as there are small interclass variations and large intraclass variations among the five classes [6] as shown in Fig. 4. BBO–ANN is an artificial neural network based on evolutionary learning [7, 8]. The idea behind this is the fact that the success of an individual not only depends on his personal knowledge or skills which he has acquired through learning (concept of neural network) but it also depends on his genetic factors (concept of genetic algorithm) [9–11].

4 Integration of BBO With ANN

Here, we have used BBO algorithm to train our neural network (Fig. 2 illustrates the idea of our model). The steps involved during this process are as follows:

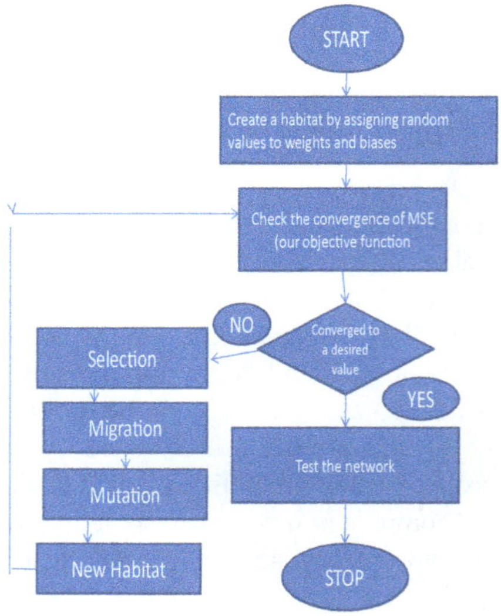

Fig. 2 Flowchart of neural network training using BBO

- A given number of habitats are generated. Each habitat displays a set of weights and biases of a multilayer perceptron neural network. Hence, a habitat corresponds to one network.
- The fitness value of the originated candidate network in step 1 is calculated. In our algorithm, MSE is used as fitness function. The aim is to reduce the difference between the actual and the estimated values.
- Then, various operations like emigration, immigration, and mutation are performed to improve the solutions.
- Some network values with higher fitness (low MSE) are kept intact in the system and passed to the next generation.
- These steps are repeated until the predetermined number of iterations is reached or the desired goal (desired MSE) is reached as given in the BBO–ANN cost function graph in Fig. 2.

Finally, the best set of parameters is assigned to the network and it is tested on a test dataset. The parameters which need to be optimized are the weights and biases associated with each layer as shown in Fig. 3. The network is constructed using each set of weights and biases, and its MSE is evaluated using the fingerprint feature dataset. The objective function of our optimization problem is the cost function, and the parameters which are to be optimized are network weights and biases.

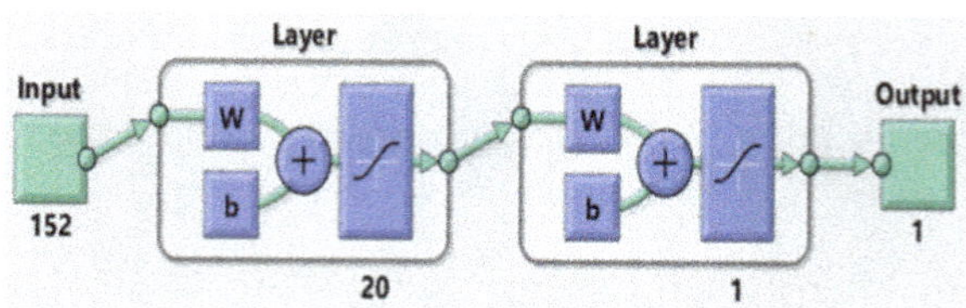

Fig. 3 Architecture of neural network

5 Experimental Works

The experimental result is given in two subsections. In subsection A, we describe the dataset and environment. The results and analysis have been discussed in subsection B.

5.1 Dataset and Environment

We report our experimental observations of fingerprint classification algorithm on NIST-9 database for the five-class classification problem as shown in Fig. 4. NIST-9 database consists of 5400 fingerprints from 2700 fingers with their natural frequency of occurrence where loops and whorls are majority. Each fingerprint is labeled as one of the five classes (R, L, W, A, and T) [2, 12]. However, we have

Fig. 4 Five fingerprint classes from NIST-9 database **a** right loop, **b** left loop, **c** whorl, **d** arch, **e** tented arch

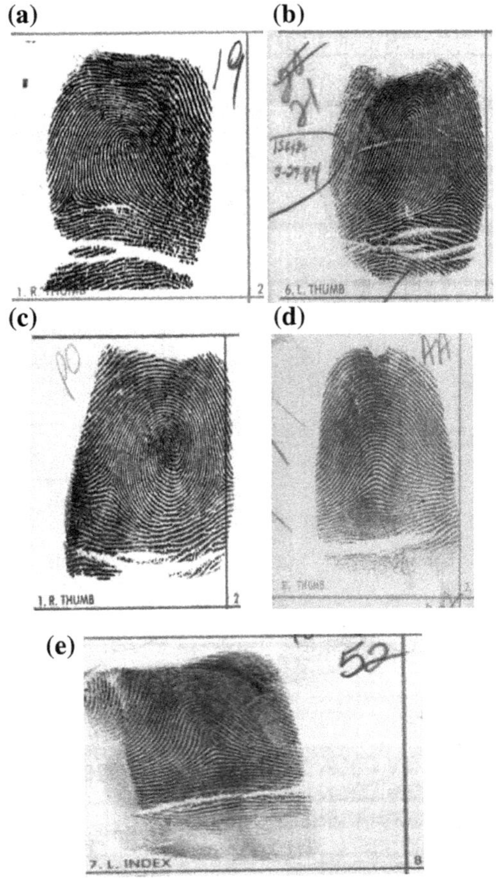

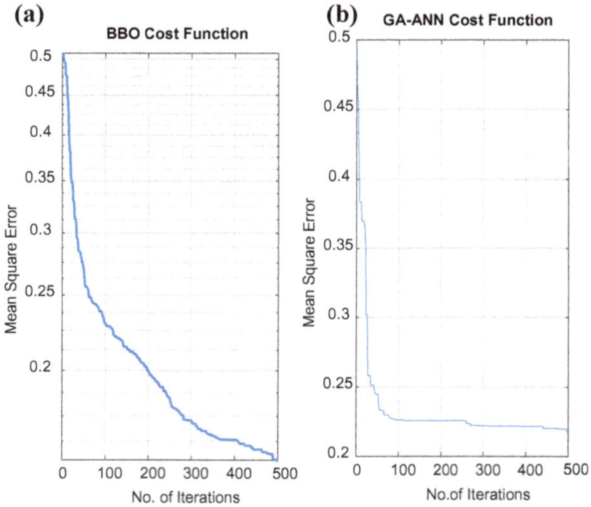

Fig. 5 Comparison of BBO–ANN with GA–ANN cost function (MSE vs. No. of iterations) in (**a**) and (**b**), respectively

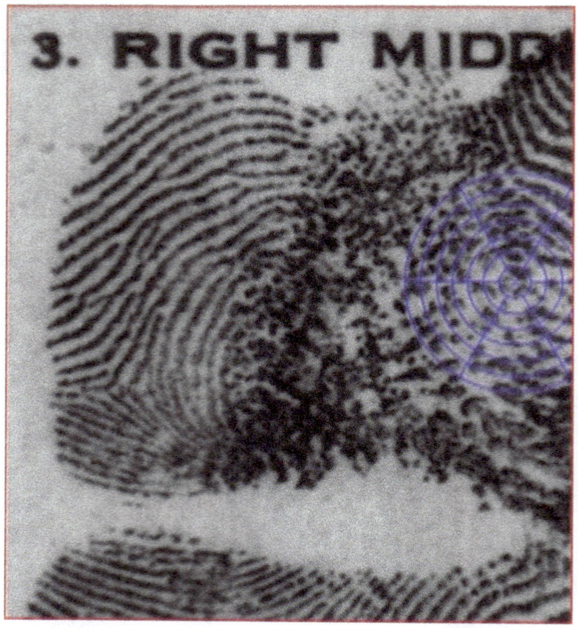

Fig. 6 Fingerprint rejected during feature extraction

used 1648 images out of which 80%, i.e., 1318 images for training the first stage classifier and 800 images to train each classifier in the second stage. For testing purpose, we used rest 20%, i.e., 330 images for testing purpose (Figs. 5 and 6).

5.2 *Results and Analysis*

The four-class classification accuracy by changing the number of hidden layer neuron is shown in Table 1. Figure 7 shows the confusion matrix for five-class classification using a two-stage classifier. Classification accuracy also changes by tuning the parameters of BBO algorithm like migration, mutation, and recombination for habitat percentage. Table 2 shows the variation of classification accuracy by changing the above parameters. Classification accuracy is reported to be increasing by incorporating rejection option in the feature extraction stage [13, 14]. If some of the images for which a valid tessellation cannot be defined are rejected, then accuracy increases.

From Table 1, it can be seen that 20 hidden neurons produce the best result. Too few number of hidden neuron leads to under-fitting in which the neural network misclassifies. Excess number of hidden neuron leads to over-fitting in which the network remembers things and it fails to generalize. Table 2 shows increase in

Table 1 Classification accuracy obtained by changing the number of hidden layer neuron

S. No.	Number of hidden layer neuron	Classification accuracy (%)
1.	16	40.12
2.	18	48.60
3.	20	55.50
4.	22	56.24

Fig. 7 Confusion matrix of five-class classification (for test samples)

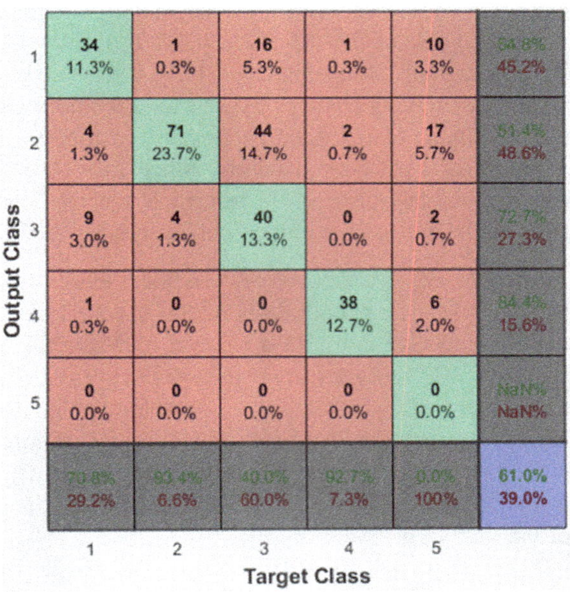

Table 2 Classification accuracy with % of rejected fingerprints

S. No.	% of rejected fingerprint	Classification accuracy (%)
1.	10	60
2.	20	64.50
3.	30	72.40

classification accuracy with increase in rejection rate. It is maximum when around 30% of fingerprints are rejected due to invalid tessellation.

A further increase in rejection rate decreases the training database of the classifier which leads to misclassification due to improper training of the network. Figure 6 shows rejected fingerprint for which a valid tessellation cannot be defined.

Here, from Fig. 5 we can see that our algorithm is more efficient in minimizing error as compared to GA–ANN algorithm. It is producing a very less MSE with better accuracy. For five-class classification, accuracy is obtained by changing the number of hidden neurons. Also, incorporating rejection option in the classifier training stage improves the performance of the classifier as shown in Figs. 5 and 6.

6 Conclusion

We have successfully classified fingerprints into five broad classes, using a multichannel filter based feature extraction algorithm and BBO which produces precise and better accuracy than some of the algorithms found in the literature on NIST 9 database. Accuracy is almost 56% with no rejection. But as compared to GA–ANN algorithm and other algorithms, the running time is slightly high like 3039.121 s, whereas in case of GA–ANN it is 895.201 s. But we can get better MMSE in BBO–ANN algorithm as compared to GA–ANN and accuracy as well. Our classifier needs less training samples than general feedforward and backpropagation neural network. A more enhanced performance can be easily expected by improving the accuracy of core point location. Since image decomposition (convolution with Gabor filter) takes almost 90 percent of the total computational time, special purpose hardware for convolution can effectively diminish the time complexity.

Future Works Our feature work includes empirical analysis of different classifiers like particle swarm optimizstion (PSO) neural network, support vector machine (SVM), and GA and PSO optimized SVM. It also includes principal component analysis (PCA)-based feature dimensionality reduction without losing significant information. Our algorithm is sensitive toward the location of core point; therefore, in our future attempt, we focus on a more robust core point estimation algorithm in order to enhance the performance of the classifier. This proposed algorithm does not handle rotation. This problem can be solved by approximating a frame of reference of the fingerprint in future.

Acknowledgements The first author would like to thank the technical support of Department of Information and Communication Technology, Fakir Mohan University, Vyasa Vihar, Balasore

References

1. Jain AK, Prabhakar S, Hong L (1998) A multichannel approach to fingerprint classification. In: Proceedings of Indian conference on computer vision, graphics, and image processing (ICVGIP), pp 153–158
2. Cho BG, Kim JS, Bae IG, Yoo KY, Byoung-Ho C, Jeung-Seop K, Jae-Hyung B, In-Gu B, Kee-Young Y (2000) Fingerprint image classification by core analysis. In: 5th International conference on signal processing proceedings, WCCC-ICSP2000, vol.3, pp 1534–1537
3. Cappelli R, Maio D, Maltoni D (2000) Combining fingerprint classifiers. In: Proceedings of first international workshop on multiple classifier systems (MCS2000), pp 351–361
4. Senior A (2001) A combination fingerprint classifier. IEEE Trans Pattern Anal Mach Intell 23:1165–1175
5. Cappelli R, Lumini A, Maio D, Maltoni D (1999) Fingerprint classification by directional image partitioning. IEEE Trans Pattern Anal Mach Intell 21(5):402–421
6. Zheng Y, Feng Ling H, Yun Xue J (2014) Eco-geography based optimization, Enhancing biogeography-based optimization with eco-geographic barriers and differentiations. Comput Oper Res 50(4):115127
7. Jain AK, Prabhakar S, Hong L (1999) A multichannel approach to fingerprint classification. IEEE Trans Pattern Anal Mach Intell 21(4):348–359
8. Samanta B, Al-Balushi KR, Al-Araimi SA (2003) Artificial neural networks and support vector machine with genetic algorithm for bearing fault detection. Eng Appl Artif Intell 16(7-8):657–665
9. Giri P, De S, Dehuri S (2016) A novel locally and globally tuned biogeography based optimization(LGBBO). In: SoCTA: soft computing: theories and applications, Springer
10. Ma H, Fei M, Ding Z, Jin J (2012) Biogeography-based optimization with ensemble of migration models for global numerical optimization. IEEE World Congress on Computational Intelligence, Brisbane, Australia
11. Li X, Wang J, Zhou J, Yin M (2011) A perturbs biogeography based optimization with mutation for global numerical optimization. Appl Math Comput 598–609
12. Jain AK, Prabhakar S, Hong L, Pankanti S (2000) Filterbank based fingerprint Matching. IEEE Trans Image Process 9(5):846–859
13. Karu K, Jain AK (1996) Fingerprint classification. Proc Pattern Recogn 29(3):389–404
14. Jain A, Hong L (1996) On-line fingerprint verification. In: Proceedings of 13th ICPR, Vienna, pp 596–600

Performance and Capacity Testing of Medium and Large Managed Plane and Control Plane Optical Networks

Sandeep Dabhade, Sumit Kumar, Shishir Kumar and K. B. Sivasubramanian

Abstract Element and network management system is used worldwide for remotely managing the network elements. With EMS/NMS, we can provision circuits, reboot network elements, provision protection rings, and provision in-band communication remotely. Due to its wide utility, performance and scalability testing of medium and large optical networks needs to be done extensively. This paper covers how performance and scalability testing of medium and large optical networks is done in both control and managed plane scenarios. Test bench has more than five thousand network elements and more than three lakhs paths. Time to retrieve nodes, optical connections, and alarms will be measured in medium and large network scenarios. Nodes are added into network management system which is mentioned in the above setup.

Keywords EMS · NMS · Control plane · OTN

1 Introduction

In optical networks domain, both managed plane and control plane networks are extensively used. In managed plane domain, nodes do not have generic MPLS routing engine (GMRE) while in control plane, nodes are GMRE enabled. Each configuration has a mesh setup of five nodes. Such multiple mesh setups are created. Figure 1 shows an example of five-node mesh setup. In control plane, path is

S. Dabhade (✉) · S. Kumar · S. Kumar · K. B. Sivasubramanian
NSM NFM-T Department, NOKIA, Bangalore, India
e-mail: sandeep.dabhade@nokia.com

S. Kumar
e-mail: sumit.2.kumar@nokia.com

S. Kumar
e-mail: Shishir.2.kumar@nokia.com

K. B. Sivasubramanian
e-mail: sivasubramanian.k_b@nokia.com

V. Nath and J. K. Mandal (eds.), *Nanoelectronics, Circuits and Communication Systems*, Lecture Notes in Electrical Engineering 511,
https://doi.org/10.1007/978-981-13-0776-8_22

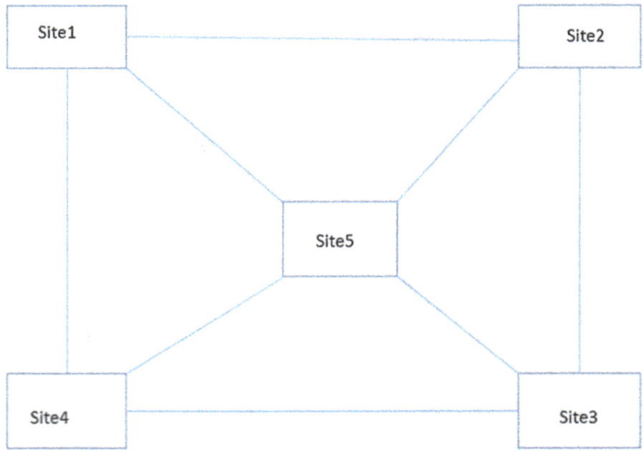

Fig. 1 Mesh setup

auto-discovered by nodes itself since they are GMRE enabled. Telecom market is moving toward control plane enabled nodes since connection provisioning becomes easier in compare to managed plane scenarios. The major advantage of control plane over managed plane network is restoration at the time of failure of path/link. Nodes used in these setups have amplifiers packs, optical transponders, and filter cards.

2 Medium Size Setup

The network management system setup has one virtual machine namely OTN. OTN, the NMS, does the functionality of FCAPS. OTN server manages nodes and end-to-end connections/paths. Medium size server has 1700 nodes and 89,000 paths. It has three network adapters managing these nodes. Figure 2 explains about OTN with three network adapters.

Table 1 indicates server configuration of medium size network.

Configuration of server is very critical in performance and capacity test scenarios.

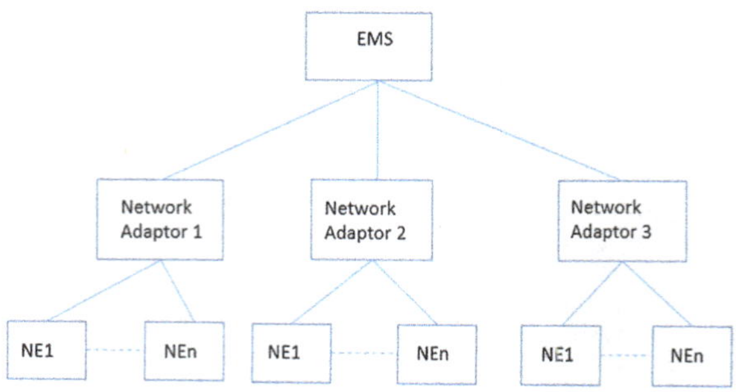

Fig. 2 OTN with network adapters

Table 1 Medium size: server configuration

Parameters	OTN (NMS) (Medium)
CPUs	8 Core
RAM	36 GB
Disk	400 GB+

Table 2 Time/performance measurements of medium size network

Operations performed from NMS on nodes	Time taken (h:min:s)
List all the physical connections (1243)	00:00:51
List all the trails (12,636)	00:01:31
List all the services (89,329)	00:01:38
List all the nodes (1727)	00:00:22
List all the alarms (29,726)	00:02:31

3 Time/Performance Measurements of Medium Size Optical Network

Take the following time measurements (Table 2).

4 Large Size Optical Network Setup and Time/Performance Measurements

The large network management system setup has three virtual machines, namely OTN and two EMS. This large EMS/NMS optical network system manages more than 5000 nodes, 100 simultaneous users, 3 lakhs connections/paths. Figure 3 explains the large network setup (Table 3).

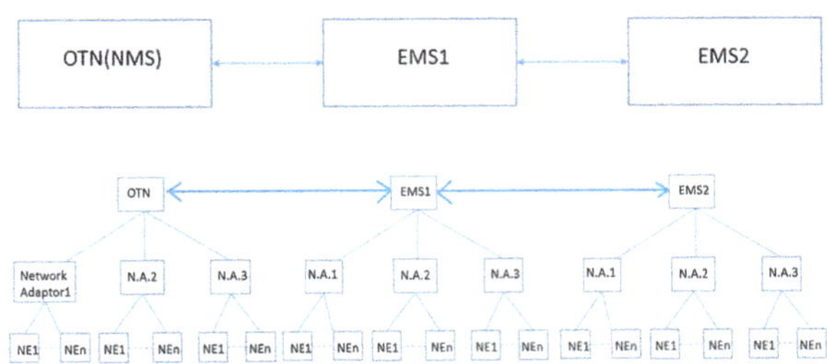

Fig. 3 Large network setup

Table 3 Time/performance measurement for large size network

Operations performed from NMS on nodes	Time taken (h:min:s)
List all the physical connections (1893)	00:01:20
List all the trails (33,592)	00:02:21
List all the services (361,113)	00:02:25
List all the nodes (12,057)	00:01:11
List all the alarms (87,956)	00:03:22
List all the control plane service (5000)	00:00:08

Table 4 Large size: server configuration

Parameters	OTN (Large)	EMS
CPUs	16 Core	8 Core
RAM	72 GB	48 GB
Disk	800 GB+	400 GB+

Table 4 indicates the server configuration for large network

Simulation tools used:

- Sikuli (GUI automation tool—for connection creation),
- Selenium (to simulate operator activities),
- JConsole (inbuilt tool to monitor different processes),
- top command is used for the memory and CPU usage, and
- MS excel used to plot the graphs.

Below are the screenshots for the memory and CPU utilization for large network scenarios.

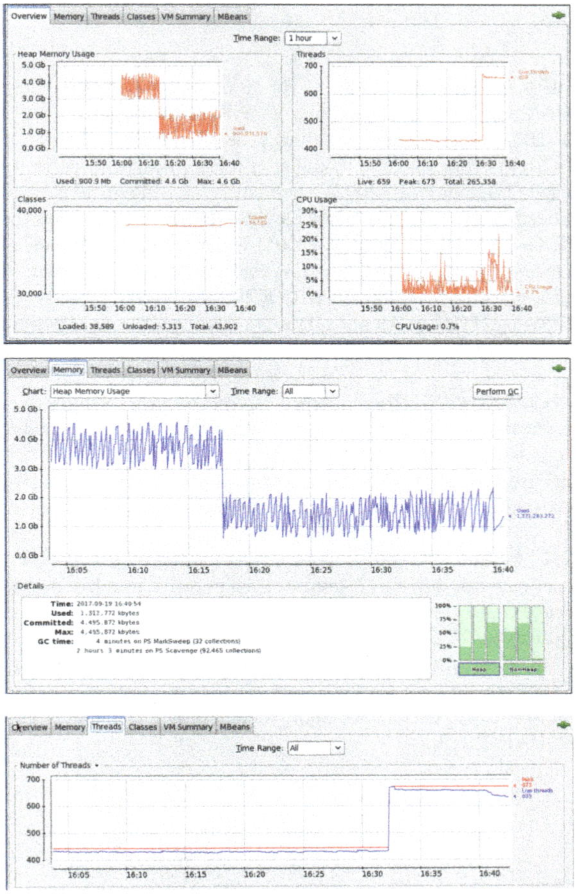

5 Conclusions

Medium and large optical network management system with thousands of nodes and more than hundred thousand connections were created and tested successfully.

It must be good to test medium and large optical network involving thousands of nodes, lakhs of connections since bandwidth requirement is growing in a nonlinear way. The time taken (performance) to retrieve the volumes of nodes, paths, equipment inventory, and alarms using network management system (NMS) is found within limits. This benchmark test is demonstrated for the satisfaction of performance and capacity results and ease of performing various operations on medium and large optical network setups.

References

1. Telco VoIP scalability test results … for 10 million subscribers. https://www.communigate.com/Papers/CommuniGate-Superdome-VoIP-Benchmark.pdf
2. Uyless B (2002) Optical networks, third generation transport systems
3. Eiselt M (2011) Programmable modulation for high-capacity networks. In: ECOC, Geneva
4. Aboul-Magd O, Jamoussi B (2002) Automatic switched optical network (ASON) architecture and its related protocols. https://shiraz.levkowetz.com/pdf/draft-ietf-ipo-ason-02.pdf
5. Chen LK, Cheung MH, Chan CK (2005) From optical performance monitoring to optical network management: research progress and challenges. http://www.lightwave.ie.cuhk.edu.hk/publication/document/Conference/ICOCN/2004/lkchen_icocn04.pdf
6. Samadi P, Wang H (2014) An optical programmable network architecture supporting iterative multicast for data-intensive applications. http://lightwave.ee.columbia.edu/files/Samadi2014a.pdf
7. Gifre L, Velasco L (2014) Experimental assessment of a high performance backend PCE for flexgrid optical network re-optimization. http://personals.ac.upc.edu/lvelasco/docs/research/2014-OFC-4.pdf
8. Okumura Yasuyuki, Kato Maki, Fujii Katsuyuki (2014) Signal shaping to achieve OOK and PSK co-existence for improved optical access network performance. Am J Netw Commun 3 (4):56–62
9. Chen LK, Li M, Liew SC (2014) Optical physical-layer network coding—another dimension to increase network capacity? http://www.lightwave.ie.cuhk.edu.hk/publication/document/Conference/IPC/2014/2014_Chen_IPC.pdf
10. Xia M, Shirazipour M, Zhang Y, Green H, Takacs A (2014) SDN and optical flow steering for network function virtualization. https://www.usenix.org/system/files/conference/ons2014/ons2014-paper-xia.pdf
11. Prabazhini M (2014) Performance improvement of optical network using dynamic bandwidth allocation. IJAICT 1(1)

Analysis of Single-Stage Three-Phase DC–AC Boost Inverter for Distributed Generation System

Prabhat Kumar Ranjan and P. R. Thakura

Abstract The increase in energy demand and reduction of greenhouse gas emission is pushing research interest towards distributed generation systems. Distributed generation systems are renewable, non-renewable energy sources and also energy storage devices. The single-stage three-phase boost inverter can provide higher value of sinusoidal AC output voltages from low-voltage DC sources without an intermediate DC–DC boost chopper. This unique property is absent in classical voltage source buck inverter which produces an instantaneous AC output voltage, which is always less than input DC voltage. Also, this topology is a suitable choice in many applications such as AC drives, uninterruptible power supplies (UPS) and distributed power generation scheme (grid integration of renewable sources). Thus, it has an advantage of boosting and inversion process in a single conversion stage, negligible shoot-through problem, reduced components, size and volume, low losses and higher efficiency, in comparison to double-stage conversion, which consists of a DC-to-DC boost converter and a conventional three-phase buck voltage source inverter in between the low-voltage DC sources and AC output. The three-phase DC-to-AC boost inverter can also be applied to storage devices such as batteries, fuel cells and supercapacitors with the grid. Working principle, operations, modelling and simulation of three-phase single-stage DC–AC boost inverter with battery as input are performed under Simulink platform of the MATLAB and different waveforms are plotted. These simulations explain the feasibility of the proposed single-stage three-phase boost inverter.

Keywords Single-stage boost inverter · Distributed generation
Conventional buck VSI

P. K. Ranjan (✉)
Department of Electrical & Electronics Engineering,
Birla Institute of Technology, Mesra, Off-Campus, Deoghar, Jharkhand, India
e-mail: prabhatrical@gmail.com

P. R. Thakura
Department of Electrical & Electronics Engineering,
Birla Institute of Technology, Mesra, Ranchi, Jharkhand, India

V. Nath and J. K. Mandal (eds.), *Nanoelectronics, Circuits and Communication Systems*, Lecture Notes in Electrical Engineering 511,
https://doi.org/10.1007/978-981-13-0776-8_23

1 Introduction

The important issue of today's society is increasing demand of electric energy and reduction of greenhouse gas emission, the depletion of fossil fuels and to protect the environment from pollution caused by conventional energy sources. Therefore, there is a need for application of renewable energy sources and storage devices based on distributed generation systems. Among these renewable energy sources, wind generators, solar panels and fuel cells are the most common and important. The powers generated from nonconventional energy sources such as solar, wind, etc. are not adequate to deliver the high power load demand. Therefore, use of the power electronics converters units becomes necessary in between the nonconventional energy sources and the load according to requirements of application area. The DC–DC boost chopper and conventional voltage source buck inverter are used in grid integration applications, which generates sinusoidal AC power output from the lower DC input and requires two-stage power conversion. For these applications, we may use single-stage boost inverter topology circuit in which generated sinusoidal AC output is greater than lower DC input voltage. The major advantages of using three-phase single-stage DC–AC boost inverter are reduced volume, weight and cost of the system and improved efficiency as compared to two-stage inverter topology configuration. In this paper, operating principle, mathematical modelling, MATLAB simulation analysis and advantages of single-stage three-phase DC–AC boost inverter are presented.

Significance of new inverter topologies enhances due to convenient and green energy conversion of renewable energy sources such as photovoltaics (PV), wind turbine systems, fuel cells, etc. Consequently, important aspects become overall reduction of converter size and passive element [1, 2]. For DC–AC conversion, most common converter topology is conventional two-level buck voltage source inverter [3], where the peak value of output AC voltage is always lower than the input DC voltage and output AC current peak is always greater than the input DC current. Due to the buck voltage nature of VSI, a DC–DC boost chopper is installed for voltage matching and maximum power point tracking (MPPT); consequently, total weight, volume, losses and hence cost of overall system will be increased. Single-stage buck and boost inverter topologies are investigated. Three-phase boost DC–AC inverters generate higher sinusoidal AC output voltages than the input one in single stage [4–6]. Due to differential output voltage, boost inverter topology has unique operating behaviour. The control of these boost inverters is usually done with the help of control scheme, which consists of inductor current control inner loop and output voltage control outer loop. To control the boost nature of the model around a particular operating point, small signal linear models have been analysed. The sliding mode controller describes variable operating point conditions with achievements of good steady-state results. However, the selection of controlling parameters related to variable switching frequency and inductance average current control is a major drawback of sliding mode control technique [7–13]. A new control strategy has been adopted for three-phase boost inverters in case of standalone distributed generation systems [14, 15]. Rule-based controller is proposed to reduce second-order harmonics current ripple component appears at the DC side of the inverter [16] Neuro-fuzzy controllers are also analysed for control of single-phase switched boost inverters [17]. So in this paper single-stage three-phase DC–AC boost inverter's both boosting and inversion characteristics are analysed simultaneously from lower DC input voltage. Finally, sinusoidal AC output voltage is obtained which is the difference of outputs of any two boost converters. The three-phase DC–AC boost inverter operation method is examined by the MATLAB simulations.

2 Topology of Proposed Three-Phase DC–AC Boost Inverter

DC-biased sinusoidal AC output voltage is obtained from the proposed three-phase DC-to-AC boost inverter. It is a cascade connection of three symmetrical DC–DC boost converters with a common DC input voltage (V_{in}). Each DC–DC boost converters have DC-biased sinusoidal AC output voltage across the output capacitor. Due to same DC biasing voltage in each DC–DC converter, DC-biased voltage is cancelled to each other in resultant output voltage. Hence, resultant output voltage of any two converters is a pure sinusoidal waveform. The load is connected differentially across all the boost converters. The basic configuration of proposed three-phase boost inverter consists of three inductors and three capacitors which are used as energy storing elements, six power semiconductor switches (MOSFETs), DC supply voltage source as input and load (R, RL). The analysis for R and RL load is done. The proposed three-phase DC–AC boost voltage source inverter circuit diagram is shown in Fig. 1. The operation of proposed DC–AC boost inverter can be explained by considering the circuit operation in the following

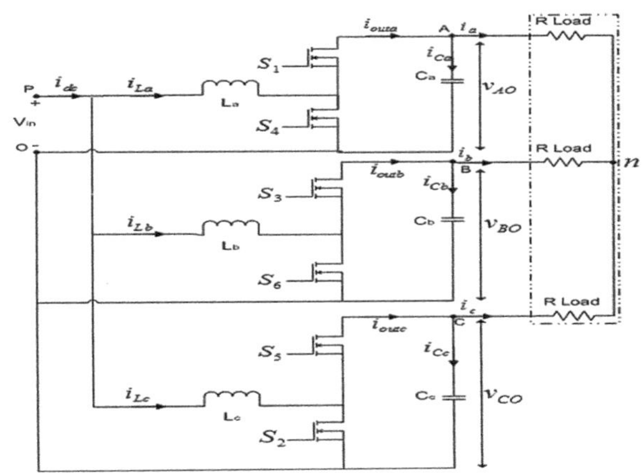

Fig. 1 Circuit diagram of proposed single-stage three-phase DC–AC boost inverter

ways—Six power semiconductor switches (MOSFETs) have different switching sequences in six modes 1, 2, 3 4, 5, and 6 (180° mode of operation).

In mode 1, only switches **S5, S6,** and **S1** conduct simultaneously and remaining three switches are in off state. Similarly in Mode 2, switches **S6, S1,** and **S2** conduct, in mode 3, switches **S1, S2,** and **S3** conduct, in mode 4, switches **S2, S3,** and **S4** conduct, in mode 5, switches **S3, S4,** and **S5** conduct and in mode 6, switches **S4, S5,** and **S6** conduct.

In each individual mode, one inductor magnetizes and stores energy in it through one switch from input supply, and consequently its current rises and remaining two inductors demagnetise and deliver energy to the capacitors and loads. Also, the remaining two capacitors discharge via the load resistance *R* and the capacitor. In this way, capacitors charge and output voltage develops across the load.

The following mathematical equations for boost inverter in case of continuous conduction mode are as follows:

$$V_{an} = K \sin \omega t + V_{\mathrm{DC}} \tag{1}$$

$$V_{bn} = K \sin(\omega t - 2\pi/3) + V_{\mathrm{DC}} \tag{2}$$

$$V_{cn} = K \sin(\omega t - 4\pi/3) + V_{DC} \tag{3}$$

$$\begin{aligned} V_{ab} &= V_{an} - V_{bn} = (K \sin(\omega t) + V_{\mathrm{DC}}) - (K \sin t(\omega t - 2\pi/3) + V_{\mathrm{DC}}) \\ &= \sqrt{3} K \sin(\omega t + \pi/6) \end{aligned} \tag{4}$$

$$V_{bc} = V_{bn} - V_{cn} = \sqrt{3} K \sin(\omega t - 5\pi/6) \tag{5}$$

$$V_{ca} = V_{cn} - V_{an} = \sqrt{3} K \sin(\omega t + 5\pi/6) \tag{6}$$

V_{in} is the common DC input voltage of all three converters. V_{an}, V_{bn}, V_{cn} are the output phase voltages, and V_{ab}, V_{bc}, V_{ca} are the line-to-line output voltages across the capacitor of the DC–DC boost converter. V_{DC} is the DC-biased voltages. *K* is the amplitude of the desired sinusoidal voltages. The outputs of all three converters are connected with load in series. The desired boost inverter sinusoidal output phase voltages are as follows:

$$V_{aN} = K \sin \omega t, \tag{7}$$

$$V_{bN} = K \sin(\omega t - 2\pi/3) \tag{8}$$

$$V_{cN} = K \sin(\omega t - 4\pi/3) \tag{9}$$

To avoid zero level crossing, the minimum DC voltage required is $V_{\mathrm{DC}} \geq V_{\mathrm{in}} + K$.

The system is analysed by two distinct sets of equations: first describing the input side and the other the output side. Both the input and output sides include DC and AC components. The generalized input and output equations are as follows:

$$V_{\text{in}}=Ri_{L_j}+L\mathrm{d}i_{L_j}/\mathrm{dt}\;+\;p_jV_{kn},\; j=a,\;b,\;c\;\&\;k=A,\;B,\;C \tag{10}$$

$$\sum_{j=a,b,c} i_{L_j}=I_{\text{in}} \tag{11}$$

$$\begin{aligned} i_j &= -C\frac{\mathrm{d}V_{\text{in}}}{\mathrm{d}t}+i_{\text{OUT}_j} \\ i_j &= -C\frac{\mathrm{d}V_{\text{in}}}{\mathrm{d}t}+p_ji_{L_j} \quad j=a,\;b,\;c \quad k=A,\;B,\;C \end{aligned} \tag{12}$$

$$\sum_{j=a,b,c} i_j=0 \tag{13}$$

where i_{L_a},i_{L_b} and i_{L_c} are the currents flowing through inductors $L_a=L_b=L_c=L$ and $R_a=R_b=R_c=R$ is the resistive part (not shown in Fig. 1) for each inductor. i_a, i_b, i_c are the currents flowing across the loads and the value of capacitance of each capacitor is C.

3 Simulation Results and Discussion

Circuit parameter specifications are as follows:

V_{in} = 100 V (input voltage); V_o = 200 V (output voltage); f_s = 1 kHz, 5 kHz (switching frequency),

Load R = 4 Ω, L = 0.75 mH, $L_a=L_b=L_c$ = 8 mH; $C_a=C_b=C_c$ = 9 μF; f_o = 50 Hz (output voltage frequency).

Figure 2 shows the MATLAB Simulink diagram of single-phase DC–AC boost inverter. Figure 3 shows the gate pulses for six power switches. Figures 4 and 5 show the output voltage waveform across the R and RL load; AC voltage is obtained with switching frequency of 5 and 1 kHz with 100 V DC source input and output frequency of 50 Hz is also obtained. Figures 6 and 7 show the output current waveform across the load. Figures 8 and 9 show the voltage across the capacitor, which is DC-biased sinusoidal in nature.

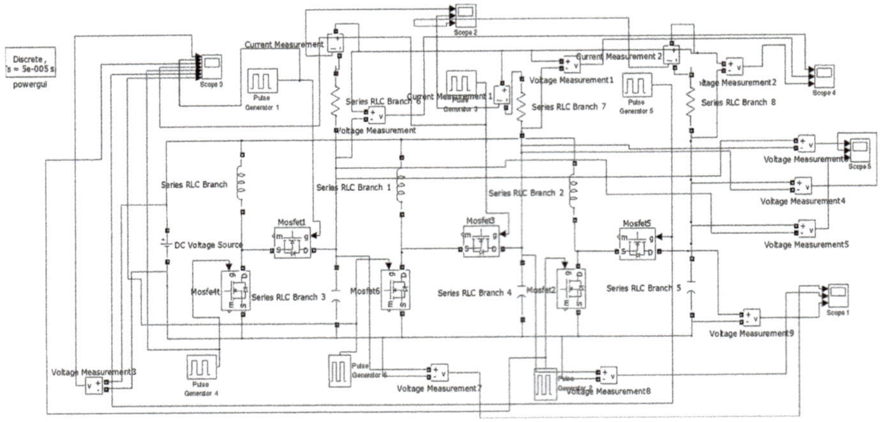

Fig. 2 MATLAB Simulink diagram of proposed single-stage three-phase DC–AC boost inverter

Fig. 3 Gate pulse voltage across switches S1, S2, S3, S4, S5, and S6

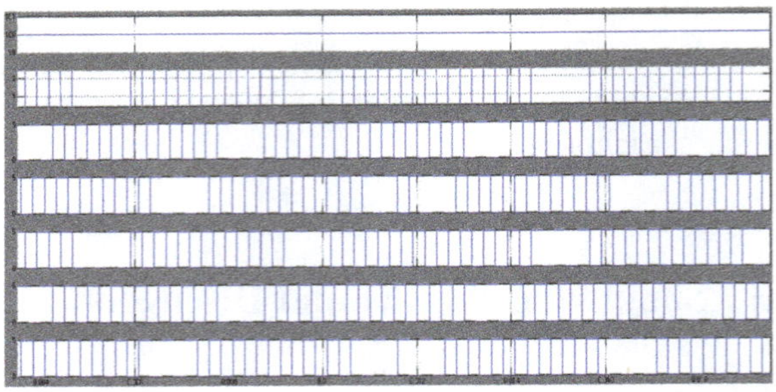

Fig. 4 Output line voltages across RL load (for 50% duty cycle and 5 kHz switching frequency)

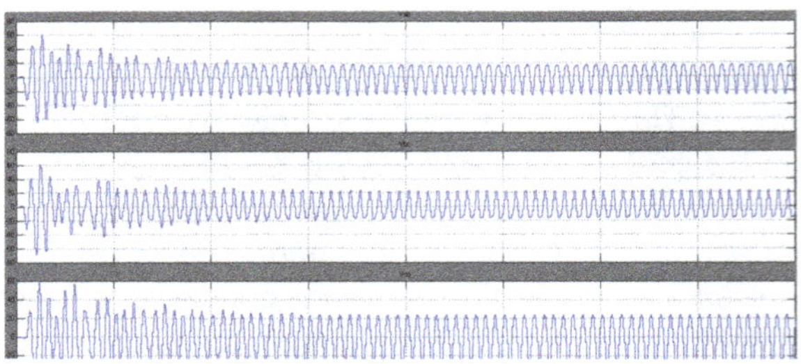

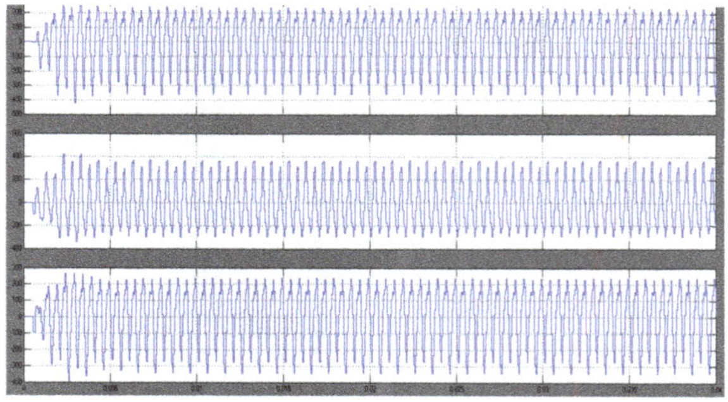

Fig. 5 Output line voltages across RL load (for 60% duty cycle and 1 kHz switching frequency)

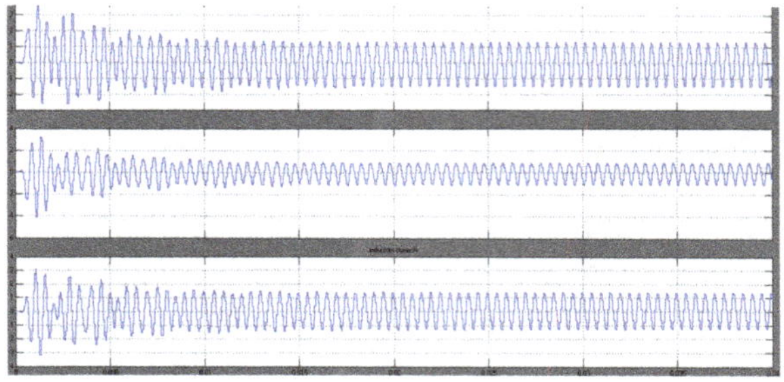

Fig. 6 Output current across $R = 4\ \Omega$, $L = 0.75$ mH load (for 5 kHz switching frequency and 50% duty cycle)

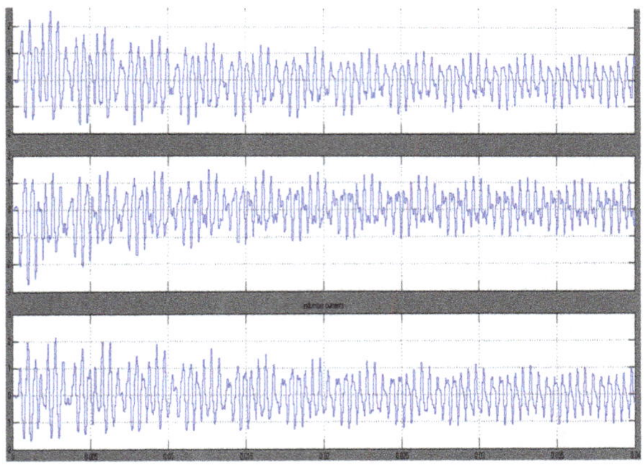

Fig. 7 Output current across $R = 4\ \Omega$ load (for 5 kHz switching frequency and 50% duty cycle)

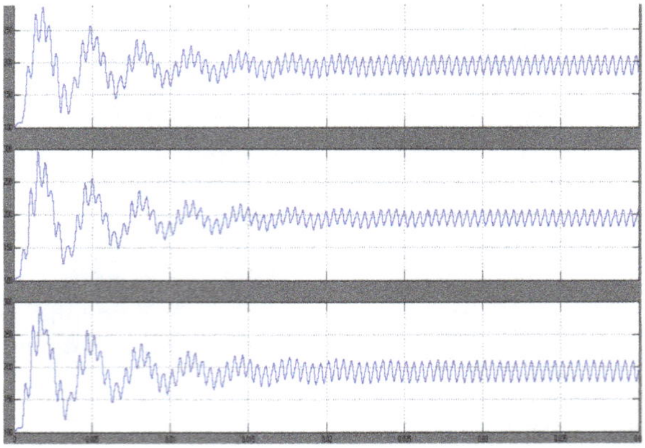

Fig. 8 Voltage across capacitors C_a, C_b, C_c (for $R = 4\ \Omega$ and $L = 0.75$ mH)

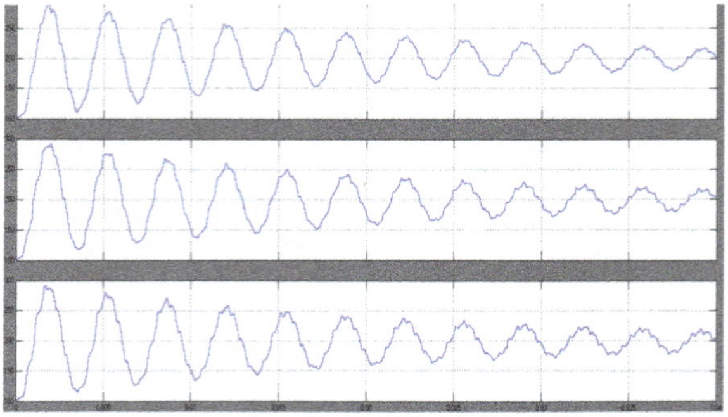

Fig. 9 Voltage across capacitors C_a, C_b, C_c (for resistive load $R = 4$ only, 5 kHz switching frequency, 50% duty cycle)

Comparison with conventional buck VSI

Conventional buck VSI	Three-phase DC–AC boost inverter
1. It requires two-stage power conversion	It requires single-stage power conversion
2. It requires extra filter design for achievement of pure sinusoidal AC output voltage	It does not require extra filter design for achievement of pure sinusoidal AC output voltage
3. Its output is always less than input	Its output is greater than input
4. Overall size of system is bulky and hence efficiency will decrease	Overall size of system is less bulky than conventional buck VSI and hence efficiency will increase

4 Conclusion

In this paper, three-phase single-stage DC–AC boost inverter topology has been presented, which always produces sinusoidal AC output voltage whose average value is greater than the input DC voltage. This configuration and unique property of three-phase DC–AC boost inverter produces the differential of DC-biased sine wave output of three separate DC–DC boost converters and differentiates it from conventional three-phase buck voltage source inverters (VSI). It may be applied as power electronics converter in the field of grid integration of distributed generation system, storage devices, electric vehicles, hybrid electric vehicles to drive the power train of vehicles and interfacing with AC motor drives. Simultaneous boosting and inversion process in one stage can be analysed from the mathematical expression and simulation results of circuit configuration, which is a clear advantage over two-stage (DC–DC boost converter and DC–AC conventional buck voltage source inverter) power conversion. The common dangerous shoot-through problem in conventional three-phase VSI will not occur in this proposed three-phase boost inverter due to inductor in the input side. The future work will be oriented towards the controller design for higher duty cycle operation, output voltage regulation, grid integration and system stability issues of three-phase boost inverter for distributed generation system.

References

1. Mazumdar P, Enjeti PN, Balog RS (2014) Analysis and design of smart PV modules. IEEE J Eng Sel Top Power Elecron 2:451–459
2. Kassakian JG, Jahns TM (2013) Evolving and emerging applications of power electronics in systems. IEEE J Eng Sel Top Power Elecron 1(2):47–58
3. Chen Y, Smedley K (2008) Three phase boost type grid connected inverter. IEEE Trans Power Electron 23(5):2301–2309
4. Do TD, Leu VQ, Choi YS, Choi HH, Jung JW (2013) An adaptive voltage control strategy of three phase inverter for standalone distributed generation systems. IEEE Trans Ind Electron 60(12):5660–5672
5. Cecati C, Aquila AD, Liserre M (2004) A novel three phase single stage distributed power inverter. IEEE Trans Power Electron 19(5):1226–1233
6. Koushki B, Khalilinia Ghaisari HJ, Nejad SMS (2008) A new three phase boost inverter: topology and controller. In: IEEE CCECE, 08, 4–7 May, 2008, pp 757–760
7. Darwish A, Holliday D, Ahmeed S, Massoud AM, Williams BW (2014) A single-stage three phase inverter based on Cuk converter for PV applications. IEEE J Eng Sel Top Power Elecron 2:797–807
8. Zhu GR, Xiao CY, Wang HR, Tan SC (2016) Closed-loop waveform control of boost inverter. IET Power Electron 9(9):1808–1818
9. Myrzik JMA (2001) Novel inverter topologies for single phase stand alone or grid connected photovoltaic and wind energy systems. In: Power electronics and drive systems, 2001. Proceedings, 2001 4th IEEE international conference on. vol 1, pp 103–108
10. Cáceres RO, Barbi I (1996) Sliding mode controller for the boost inverter. In: Proceedings of the IEEE CIEP'96, Cuernavaca, Mexico, 14–17 Oct 1996, pp 247–252
11. Vázquez N, Álvarez J, Aguilar C, Arau J (1998) Some critical aspects in sliding mode control design for the boost inverter. In: Proceedings of the IEEE CIEP'98 conference, Morelia, Mexico, 12–15 Oct 1998, pp 76–81
12. CáceresR, Barbi I (1995) A boost DC-AC converter: operation, analysis, control and experimentation. In Proceedings of the IEEE IECON'95 conference, Orlando, FL, 5–11 Nov 1995, pp 546–551

13. Cáceres RO, Barbi I (1999) A Boost dc–ac converter: analysis, design and experimentation. IEEE Trans Power Electron 14(1):134–141
14. Sanchis Gúrpide P, Alonso Sádaba O, Marroyo Palomo L, Meynard T, Lefeuvre E (2001) A new control strategy for the boost dc–ac inverter. In: Proceedings of the IEEE PESC'01 conference, Vancouver, Canada, 17–21 Jun 2001, pp 974–979
15. Sanchis P, Ursaea A, Gubia E, Marroyo L (2005) Boost dc-ac inverter: a new control strategy. Trans Power Electron IEEE 20:343–353
16. Abeywardana DBW, Hredzak B, Ageidis VG (2016) A rule based controller to mitigate DC side second order harmonic current in a single phase boost inverter. IEEE Trans Power Electron 31(2)
17. Wai RJ, Lin YF, Liu YK (2015) Design of adaptive fuzzy-neural-network control for a single-stage boost inverter. IEEE Trans Power Electron 30(12):7282–7298

Close Loop Speed Controller for Brushless DC Motor for Hybrid Electric Vehicles

Avanish Kumar and P. R. Thakura

Abstract Vehicle emissions are the great concern of the modern world and hybrid electric vehicles is one of the feasible solutions because it mitigates pollution and increases fuel efficiency. Power electronics plays an imperative role in Hybrid Electric Vehicle because it consists of converters for high-performance electric traction motor drives. It is important that the traction motor must fulfill the demands of the various driving cycles. In this chapter, a detailed performance analysis of controlling the speed of Brushless DC (BLDC) motor with or without disturbances (or load) is addressed. The chapter also involves the design and implementation of a controller which minimizes the initial overshoots. This work was conducted with the purpose of developing a three-phase inverter for BLDC motor drive system. The two different controllers are designed and simulated for closed-loop speed control of BLDC motor at different loading conditions. The simulation work is performed in MATLAB Simulink platform and results match with the theoretical one.

Keywords BLDC motor · Hybrid electric vehicles · MATLAB Simulink
PID controller · Speed control · Traction motor drive

1 Introduction

Hybrid Electric vehicles (HEVs) uprising has been feasible only after the beginning of the state-of-the-art power electronics devices and converters. The development high-performance and high power density motors also contribute significantly in the development of HEVs [1, 2]. The two major companies Toyota and Honda started the production of HEVs in the year 1997. After that continuous efforts are taken

A. Kumar (✉) · P. R. Thakura
Department of Electrical and Electronics Engineering, Birla Institute of Technology, Mesra, Ranchi 835215, India
e-mail: avanishmishra@bitmesra.ac.in

P. R. Thakura
e-mail: prthakura@bitmesra.ac.in

V. Nath and J. K. Mandal (eds.), *Nanoelectronics, Circuits and Communication Systems*, Lecture Notes in Electrical Engineering 511,
https://doi.org/10.1007/978-981-13-0776-8_24

towards the realization of high-performance, environmentally suitable HEVs around the globe. In countries like India, the HEVs are still far from the realm of automobile sector. But nowadays many efforts are being taken to bring commercially sustainable HEVs.

HEVs have two energy sources which feed the two prime movers, i.e. an electrical motor drive and the Internal Combustion Engine (ICE). The fuel feeds the ICE and the battery or fuel cell feeds the electrical motor drive. The prime movers may be connected together in series or in parallel using various possible architectures, i.e. series, parallel and series—parallel. The control strategy for each is different but the entire focus is to capitalize on the fuel economy, while meeting the demand torque and speed along the entire driving cycle [3, 4].

This chapter consists of 7 sections. Section 2 reviews different hybrid electric vehicle architecture. Section 3 introduces the types of synchronous motor, and elaborates BLDC motor. Section 4 presents power converter which is used to feed the BLDC motor also some technical background to 3-phase inverter and the gate driver circuit is discussed. Section 5 addresses the motor model and three-phase inverter model. Section 6 discusses the simulation models and results which are used to simulate the three-phase inverter with a PMBLDC motor in MATLAB Simulink and Sect. 7 presents the conclusion and future work.

2 Hybrid Electric Vehicles

In HEVs there are various power trains architectures, the most important architectures are the series, parallel and series–parallel ones.

2.1 Series Architecture

A Series Hybrid Electric Vehicles (SHEVs) as shown in Fig. 1, has power sources connected in electromechanical series. The propulsion power to the drive wheels is

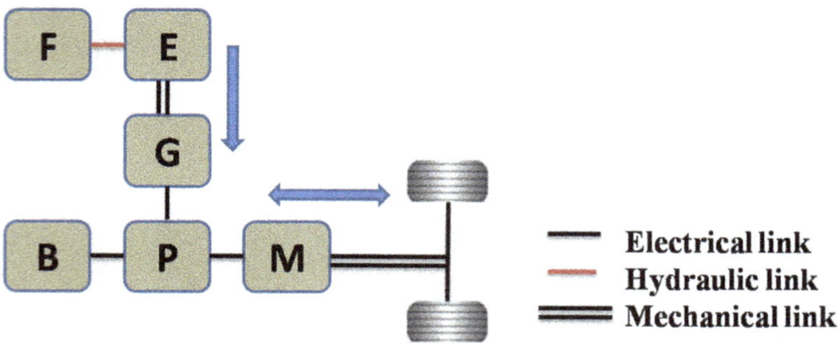

Fig. 1 SHEVs powertrain architecture. B: battery, E: ICE, F: fuel tank, G: generator, M: motor, P: power converter

provided by electrical powertrain only and an engine—generator (Genset) unit recharges the energy storage system (B) that provides energy to the electrical powertrain.

In the Genset unit, the Fuel tank (F) feeds the ICE which provides mechanical power input to the Generator (G). G converts mechanical power into electrical power and supplies to the motor (M) or to B. Merits and demerits of the series architecture may be as follows.

Merits:

(a) G and ICE are suitably designed for the average propulsion power or even less.
(b) Genset and electrical drive train are mechanically separated allowing the ICE efficiency to maximize with a significant reduction of emissions.

Demerits:

(a) It requires two electrical machines (G and M).
(b) In order to supply the peak propulsion power, M must be sized accordingly.
(c) At least two energy conversions are required, i.e. from mechanical energy to electrical energy to chemical energy inside B, and vice versa, so as to transfer generated power from ICE to wheels of the vehicles which, in turn, reduces the efficiency of the system.

2.2 Parallel Architecture

In a parallel HEVs (PHEVs) architecture, shown in Fig. 2, the conventional powertrain arrangement is connected to an electrical powertrain arrangement through a clutch that enables the vehicles to be driven by M or ICE independently or collectively. PHEVs comprise of two independent drive trains, one is electrical drivetrain and the other one is mechanical drive train, their powers are added together by using a three-way mechanical device.

Section 2.2 provides the power demanded by M to drive the vehicle. Merits and demerits of the PHEVs architecture are as follows:

Merits:

(a) In PHEVs, only one electrical machine is required;

Fig. 2 PHEVs architecture

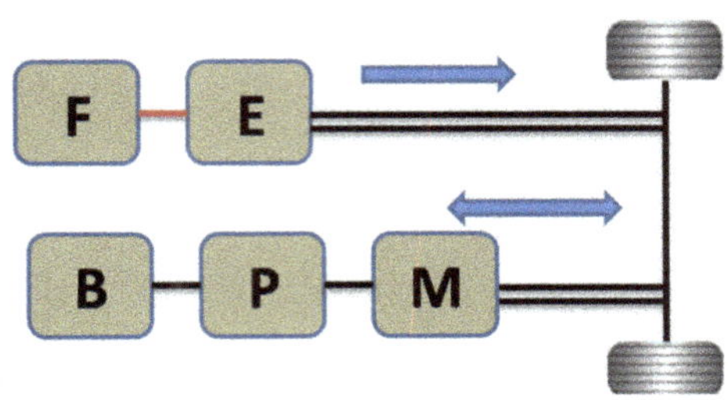

(b) M requires lower peak power in this architecture than that in SHEVs because in this architecture both ICE and M provide the propulsion power.
(c) PHEVs are more efficient than SHEVs because the generated power is transferred to wheels directly.

Demerits:

(a) A three-way mechanical device is necessary to combine ICE, M and wheels together which increases the cost.
(b) Such coupling affects the operation of ICE and imposes an unnecessary constraint on the power flow possibly turning this architecture into inferior one.

2.3 Series–Parallel Architecture

Series–parallel HEVs (SPHEVs) architecture is shown in Fig. 3. In a series–parallel configuration M, G, ICE, and the wheels of the vehicle are coupled through one or multiple planetary gear sets.

3 BLDC Motor

Permanent Magnet Brushless DC (PMBLDC)/PMBLAC motor is similar to synchronous AC machine having a permanent magnet on the rotor and winding on the stator. Motors using permanent magnets (PM) can be broadly classified as follows.

(a) Conventional DC permanent magnets motors whose armature, commutator and the brushes are the same as that of a normal DC motor except that the permanent magnets replace the field winding in the stator [5, 6].
(b) PMBLDC and Permanent Magnet synchronous motors (PMSM), the construction is similar to a synchronous motor with armature windings in the stator, but whose field winding (in the rotor) is replaced by permanent magnets and the commutation of currents in the stator phases is carried out electronically in synchronous with the rotor movement.

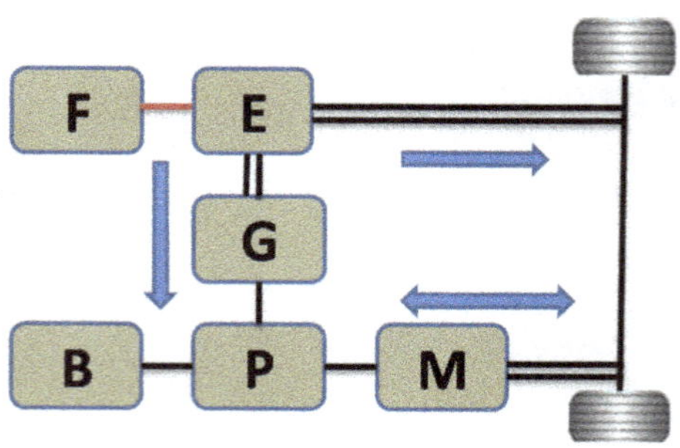

Fig. 3 SPHEVs powertrain architecture

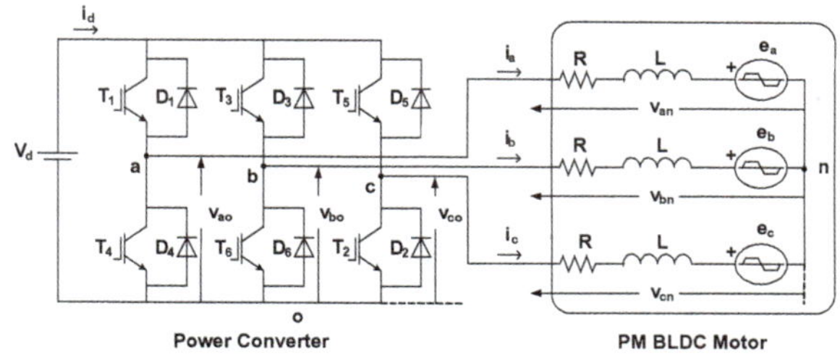

Fig. 4 Equivalent circuit of three-phase PMBLDC motor drive

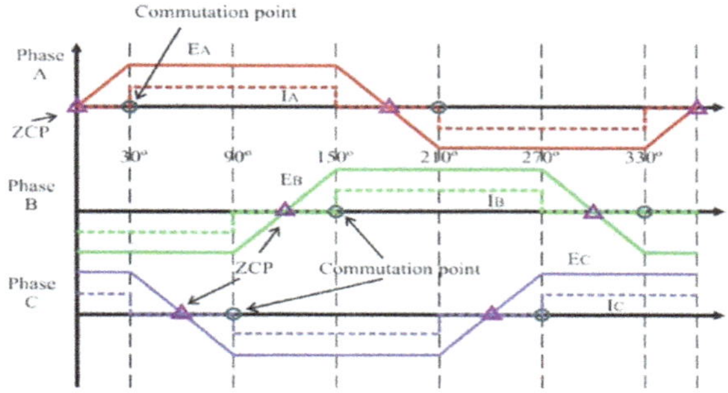

Fig. 5 Ideal back EMF voltages and three-phase currents of the PMBLDC motor

The PMBLDC motor is similar to an AC synchronous motor shown in Fig. 4 with permanent magnets on the rotor and three-phase concentrated winding on the stator. It is characterized by having a trapezoidal back EMF which is shown in Fig. 5 and driven by square-shaped currents. On the other hand, PMSM has distributed winding on the stator, having a sinusoidal back EMF and is driven with sinusoidal phase currents. The BLDC motor is usually designed in three configurations, i.e. single-phase, two-phase and three-phase BLDC motor, among these three-phase configurations are the most common. The number of power electronic switches increases with the number of phases. The stator is made up of stacked steel laminations and construction is similar to induction motor. The rotor is made up of PM pairs having north and south pole in consecutive order. Since it is a synchronous motor, the stator and rotor rotate at the same frequency. This means that there is no slip between the stator and the rotor in BLDC motor.

PMBLDC motor is best suited for drive application in HEVs because of the following reasons:

(a) PM BLDC motor has high efficiency with respect to conventional DC motor because the rotor which consists of permanent magnet does not require any power for producing excitation and also since there is no mechanical commutator is present in PMBLDC motor the frictional losses as well as heat produced by the commutator during arcing is completely eliminated.
(b) The permanent magnets are made up of very high flux density magnetic materials, which can produce very high torque for the same stator current.

(c) The characteristics of PM BLDC motor and separately excited DC motor are very similar which makes the control strategy easier as we have less number of variables to control.
(d) In the PM BLDC motor, the carbon brushes were absent so there is no wear and tear and also the operation is maintenance free. Due to inside out construction of PMBLDC motor, the stator winding produces heat on the outer periphery, the air gap and bearing of rotor do not get heated up so as like DC motor.
(e) The induced EMF in PM BLDC motor is of trapezoidal shape. The magnitude of induced EMF is constant for 120° in both half of the cycle, i.e. positive and negative half, the output power can be uniform when we excite the phases of rotor by 120° wide current. The power density of PM BLDC motor is 15% more than PMSM

4 Power Converter

The power converter device which converts DC power into AC power at desired output voltage and frequency is known as inverter. Some industrial applications of inverters include variable-speed AC drives, induction heating, UPS for computers, etc., there are mainly two types of inverters;

(a) Voltage Source Inverter (VSI): In this inverter, the DC source has negligible impedance, i.e. it has high voltage at its input terminals which shown in Fig. 6a. as the load varies output voltage remains unchanged or constant.
(b) Current Source Inverter (CSI): the CSI is fed with variable current and very high impedance DC source, for this inductor is used in series with stiff DC source which is shown in Fig. 6b. In CSI, output current remains constant even if the load changes.

For the hardware as well as MATLAB simulation Metal–Oxide Semiconductor Field Effect Transistor (MOSFET) is used as a power switch for three-phase VSI.

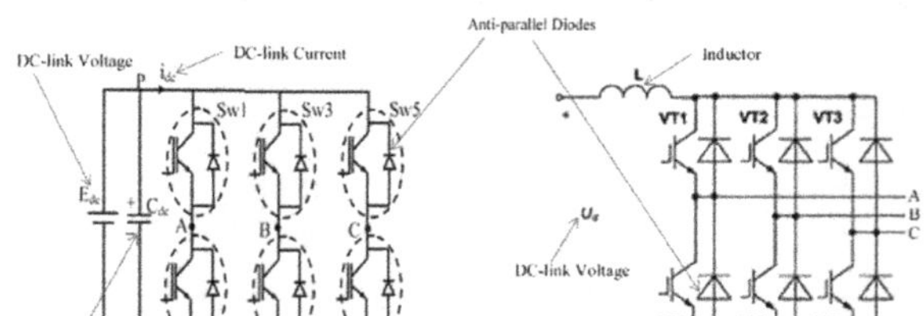

Fig. 6 Three-phase inverter **a** VSI **b** CSI

5 BLDC Motor Model

The mathematical model of PMBLDC motor is a fundamental requirement for the corresponding performance analysis and control system design. The mathematical model which was used in order to simulate the BLDC motor consisted of an electrical and a mechanical part [7, 8] (Fig. 7)

$$V_{\mathrm{AN}} = Ri_{\mathrm{A}} + L\frac{\mathrm{d}i_{\mathrm{A}}}{\mathrm{d}t} + e_{\mathrm{A}} \tag{1}$$

$$V_{\mathrm{BN}} = Ri_{\mathrm{B}} + L\frac{\mathrm{d}i_{\mathrm{B}}}{\mathrm{d}t} + e_{\mathrm{B}} \tag{2}$$

$$V_{\mathrm{CN}} = Ri_{\mathrm{C}} + L\frac{\mathrm{d}i_{\mathrm{C}}}{\mathrm{d}t} + e_{\mathrm{C}} \tag{3}$$

where V_{AN}, V_{BN}, V_{CN} denotes the phase voltage, i_{A}, i_{B}, i_{C} are the stator current, L is the stator inductance/phase, R is the stator resistance/phase and e_{A}, e_{B}, e_{C} represents the induced back EMF in each phase.

The back EMF of a three-phase BLDC motor is the function of rotor position with each phase is 120° electrical degree apart from each other and given by

$$e_{\mathrm{A}} = K_{\mathrm{e}}\omega_{\mathrm{m}}F(\theta_{\mathrm{e}}) \tag{4}$$

$$e_{\mathrm{B}} = K_{\mathrm{e}}\omega_{\mathrm{m}}F\left(\theta_{\mathrm{e}} + \frac{2\pi}{3}\right) \tag{5}$$

$$e_{\mathrm{C}} = K_{\mathrm{e}}\omega_{\mathrm{m}}F\left(\theta_{\mathrm{e}} + \frac{4\pi}{3}\right) \tag{6}$$

where K_{e} motor back EMF constant, θ_{e} is electrical rotor angle ω_{m} rotor speed and F is a trapezoidal back EMF reference function with respect to the position of rotor, with boundaries between +1 and −1.

Simulating an inverter of a BLDC motor in Simulink uses phase currents, back EMF, rotor position and DC-source voltage as inputs. For each of the different position intervals shown in Table 1, we get different voltages to the motor.

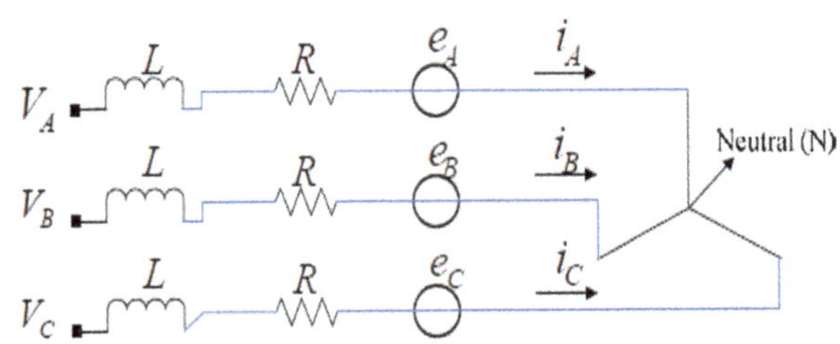

Fig. 7 Equivalent circuit of the star-connected BLDC motor

Table 1 Switching sequences

Sequence	Rotor position (Θe)	Switching (ON)	Phase current		
			A	B	C
1	0°–60°	Sw6, Sw1	Edc+	Edc−	OFF
2	60°–120°	Sw1, Sw2	Edc+	OFF	Edc−
3	120°–180°	Sw2, Sw3	OFF	Edc+	Edc−
4	180°–240°	Sw3, Sw4	Edc−	Edc+	OFF
5	240°–300°	Sw4, Sw5	Edc−	OFF	Edc+
6	300°–360°	Sw5, Sw6	OFF	Edc−	Edc+

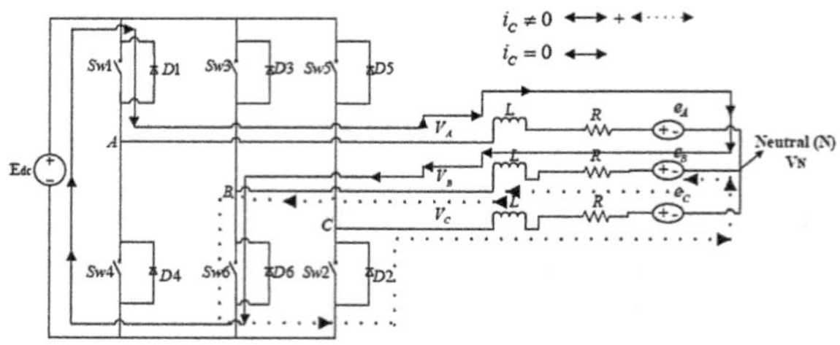

Fig. 8 Mode I (0°–60°) operation of the inverter

Rotor position (0°–60°), Sw1 and Sw6 are ON (previously Sw5 and Sw6 were ON). Figure 8 shows the current path through the three-phase inverter for the interval 0°–60°, where the path represented by bold lines with arrows represents the current through the active phase, and the dotted path represents the commutating current from the previous switching sequence.

6 MATLAB Simulation and Results

For the simulation of electrical motor drives, we have to select a proper simulation tool. In MATLAB, there is Simulink platform which is used for simulating dynamic systems. Simulink platform can also be used in the graphical environment, complex dynamic system simulations, with virtual real-time programming and a broad selection of other toolboxes (Table 2).

Figure 9 shows a block diagram of open-loop MATLAB Simulation of PMBLDC motor with three-phase Voltage Source Inverter (VSI).

The various waveform obtained in the open-loop simulation are plotted in Fig. 10a–d.

Table 2 Parameters of the motor considered for the simulation

Parameters	Value
DC voltage 'volts (V)'	310 V
Speed (rpm)	4600 rpm (481.7 rad/s)
Moment of inertia J (Kg m^2)	0.00018
Resistance R_a (Ω/phase)	1.535
Inductance L_a (mH/phase)	3.285
Torque constant K_t (N m/A)	0.49
Poles (P)	4
Frequency (Hz)	50

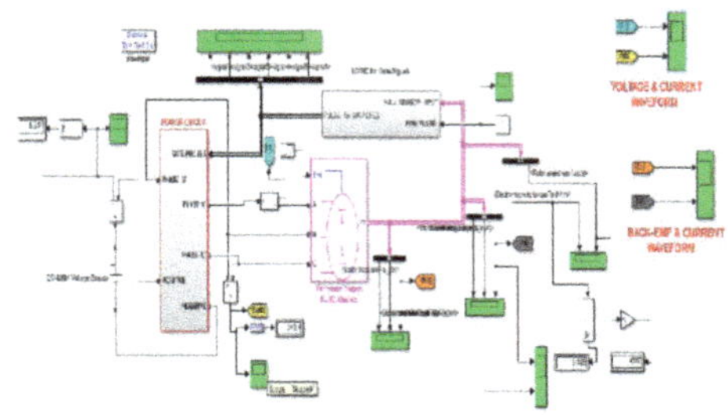

Fig. 9 MATLAB simulation of PMBLDC motor with three-phase VSI (open loop)

6.1 Controller Design

Controller is a device which monitors and alters the operating conditions of a given dynamical system. Any control method can be divided into two parts: Open-loop control and Closed-loop control. The open-loop simulation in MATLAB is already discussed. Now, we mainly focus on the closed-loop control using PID and modified PID controller and compare their Simulink results [9–11].

(a) PID Controller

Traditionally, PID control strategy has been the oldest strategy in the linear control systems. PID controllers are still used in industrial control systems due to there simplicity, reliability, robustness, and easily tuneable parameters. PID is commonly interpreted as proportional, integral and derivative controller and possesses the following properties which are given in Table 3.

The PID control structure with unity feedback is shown in Fig. 11. The role of PID controller is to calculate the error $e(t)$ by comparing the reference value and the real value, after that the variable $u(t)$ of the plant (system) is controlled tuning the proportional–integral–derivative terms.

(b) Modified PID Controller

In this controller, the only integral term is connected in the feed-forward path and proportional with derivative terms is applied to the output of the plant and is fed

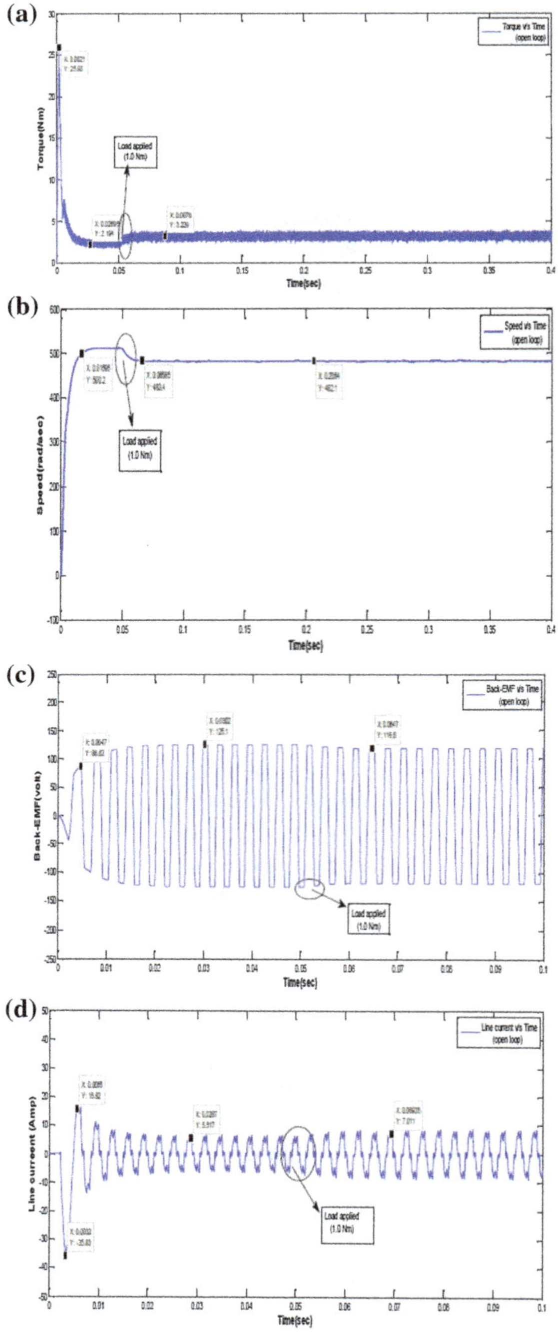

Fig. 10 **a** Torque versus time waveform for open loop (*X*-axis time; *Y*-axis toque in N m). **b** Speed versus time waveform for open loop (*X*-axis denotes time; *Y*-axis denotes 'speed (rpm)'). **c** Back EMF versus time waveform for open loop (*X*-axis time; *Y*-axis back EMF in 'volts (V)'). **d** line current versus time waveform for open loop (*X*-axis time; *Y*-axis current in 'ampere (A)')

Table 3 Properties of PID controller

Controller	Response time	Overshoot	Error
Proportional (*P*)	Small	Large	Small
Integral (*I*)	Decrease	Increase	Zero
Derivative (*D*)	Increase	Decrease	Small change

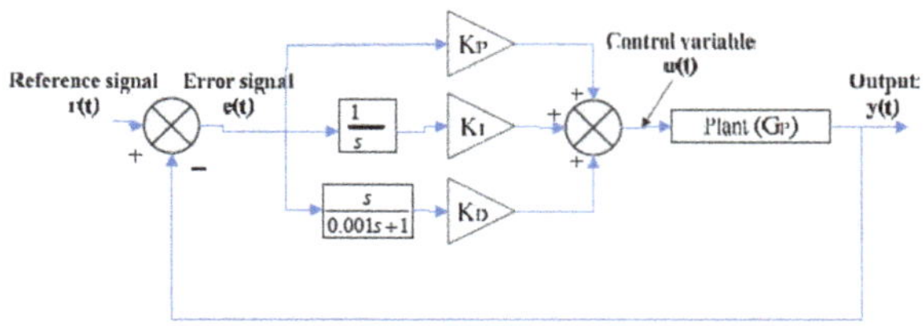

Fig. 11 Block diagram representation of a PID control system

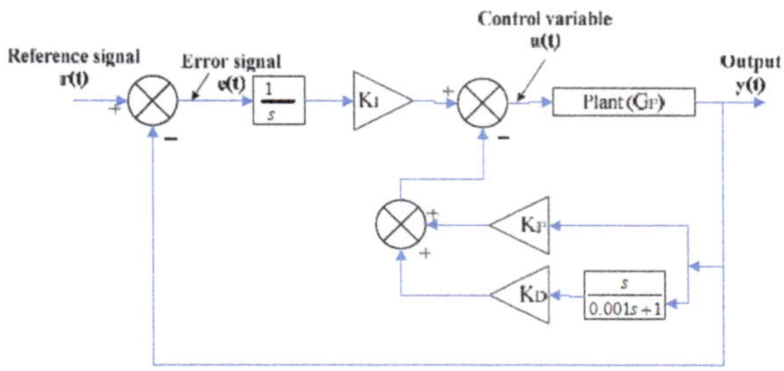

Fig. 12 Block diagram representation of modified PID control system

to the negative part of the comparator of the inner loop. This negative feedback minimizes the initial overshoot that happens in case of conventional PID.

The integral term present in the feed-forward path helps the system to reach the steady state. This type of response can be considered as little sluggish system without steep overshoots. The typical structure of modified PID control with unity feedback is given in Fig. 12.

Simulation results for closed-loop control of PMBLDC motor drive using PID and modified PID controller are given. The torque and time, speed and time, line current and time and voltage and time waveform profile obtained in the closed-loop simulation with PID and Modified PID controller is shown in Fig. 13a–d.

From the above Simulink results, it can be observed that overshoot obtained using PID controller is more compared to that of modified PID controller in transient condition during the initial time period. Both PID and modified PID controller is able to track the desired speed during the no-load and full-load conditions

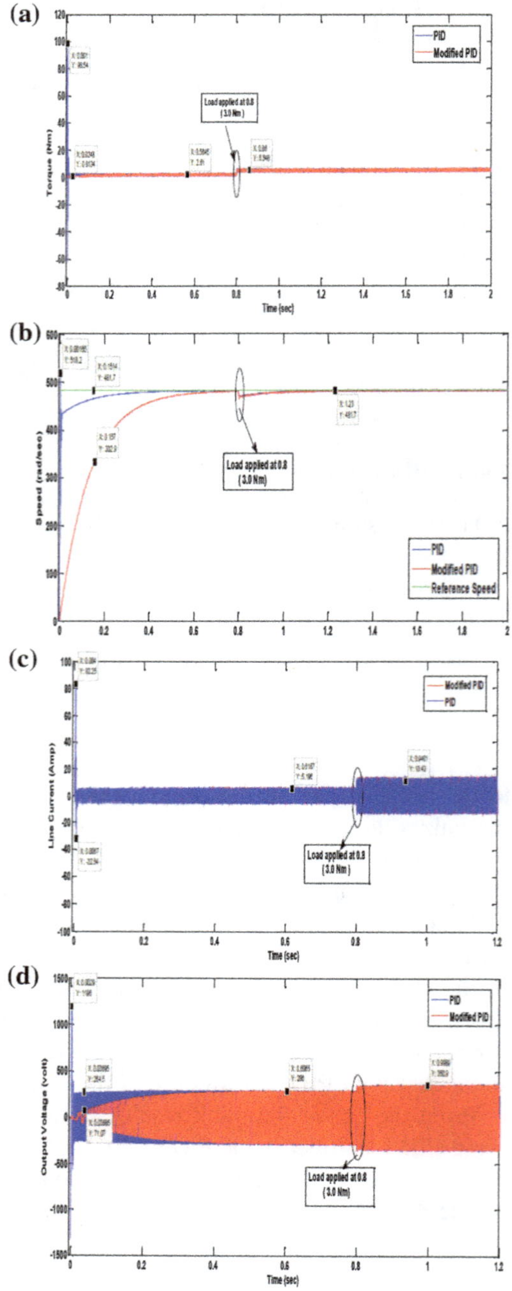

Fig. 13 **a** Torque versus time waveform for closed loop (*X*-axis time; *Y*-axis 'torque (N m)'). **b** Speed versus time waveform for closed loop (*X*-axis time; *Y*-axis 'speed (rpm)'). **C** Line current versus time waveform for closed loop (*X*-axis time; *Y*-axis 'current (A)'). **d** Output voltage versus time waveform for closed loop (*X*-axis time; *Y*-axis voltage 'volts (V)')

7 Conclusion

The control of electrical drive for the HEVs is presented in this chapter. The different vehicle architectures are being reviewed in addition to that controllers are designed and simulated for closed-loop operation of BLDC motor. The simulation result indicates that the peak overshoot gets reduced when the modified PID controller is used and gives better response in both no-load and full-load conditions compared to PID controller.

References

1. Chan CC (2007) The state of the art of electric, hybrid, and fuel cell vehicles. Proc IEEE 95(4)
2. Ehsani M, Gao Y, Butler K (1999) Application of electric peaking hybrid propulsion system to a full size passenger car with simulation design verification. IEEE Trans Veh Technol 48:1779–1787
3. Emadi A, Lee YJ, Rajashekara K (2008) Power electronics and motor drives in electric, hybrid electric, and plug-in hybrid electric vehicles. IEEE Trans. Ind Electron 55(6):2237–2245
4. Emadi A, Williamson SS, Khaligh A (2006) Power electronics intensive solutions for advanced electric, hybrid electric, and fuel cell vehicular power systems. IEEE Trans Power Electron 21(3):567–577
5. Chau K, Chan C, Liu C (2008) Overview of permanent-magnet brushless drives for electric and hybrid electric vehicles. IEEE Trans Ind Electron 55(6):2246–2257
6. Pillay P, Krishnan R (1989) Modeling, simulation, and analysis of permanent-magnet motor drives, Part II: the brushless DC motor drive. IEEE Trans Ind Appl 25(2):274–279
7. Pillay P, Krishnan R (1989) Modeling, simulation, and analysis of permanent-magnet motor drives, Part I: the permanent-magnet synchronous motor drive. IEEE Trans Ind Appl 25 (2):265–273
8. Rong J, Li Y (2011) Modeling and simulation of the double-closed loop control system of counter rotating permanent magnet brushless DC motor. In: International conference on electronic & mechanical and information technology, 12–14 Aug 2011, pp 3309–3313
9. Srikanth S, Chandra GR (2012) Modeling and PID control of the brushless DC motor with the help of genetic algorithm. In: IEEE international conference on advances in engineering, science and management (ICAESM), 30, 31 Mar 2012, pp 639–644
10. Xu C, Huang D, Huang Y, Gong S (2008) Digital PID controller for brushless DC motor based on AVR microcontroller. In: IEEE conference on mechatronics and automation, 2008, pp 247–252
11. Xia C (2012) Permanent magnet brushless DC motor drives and controls, 1st edn. Wiley, Singapore Pte. Ltd.
12. Cui S, Han S, Chan CC (2014) Overview of multi-machine drive systems for electric and hybrid electric vehicles. IEEE ITEC, pp 1–6
13. Gao DW, Mi C, Emadi A (2007) Modeling and simulation of electric and hybrid vehicles. Proc IEEE 95(4):729–745
14. Li X, Williamson SS (2007) Comparative investigation of series and parallel hybrid electric vehicle (HEV) efficiencies based on comprehensive parametric analysis. In: Proceedings of the IEEE vehicle power and propulsion conference, Arlington, TX, Sept 2007
15. Thakura PR (2008) Role and technology of electric power splitter of hybrid electric vehicle. In: IEEE vehicle power and propulsion conference (VPPC), Harbin, China, 3–5 Sept 2008

16. Avanish Kumar, Vijay Karan, Gurusivakumar G, Thakura PR (2013) Development of series hybrid electric vehicle. In: Seminar on technology solution in manufacturing 01–02 Mar 2013
17. Karthikeyan J, Sekaran D (2011) DC-DC converter CSI fed BLDC motor for defense applications. In: IEEE International conference on recent advancements in electrical, electronics and control engineering, 2011, pp 68–72
18. Al-Mashakbeh ASO (2009) Proportional integral and derivative control of brushless DC motor. Eur J Sci Res 198–203
19. Arulmozhiyal R, Kandiban R (2012) Design of adaptive fuzzy PID controller for speed control of BLDC motor. Int J Soft Comput Eng (IJSCE) 2:386–391
20. Venkataratnam K (2008) Special electrical machines, 1st edn. Universities Press (India) Private Limited

A Survey on Grid Schedulers

Ankita and Sudip Kumar Sahana

Abstract A grid is an infrastructure to meet the ongoing demands of science and engineering (Foster et al. in Int J High Perform Comput Appl 13(3):200–222, 2001) [1]. In the midst of the 1980s and the 1990s, researchers observed that parallel computing and distributed computing was not only sufficient for solving the biggest challenges of engineering problems. They needed some mechanism which could utilize the power of distributed as well as parallel computing. Grid computing (Foster and Kesselmen in The grid: blueprint for a future computing infrastructure. Morgan Kaufmann Publishers, pp 1–593, 1999; Jacob et al. in Introduction to grid computing, 1st edn., 2005) [2, 3] was the solution to their problem. But working in a grid environment is not really easy since the grid users are increasing and services are becoming commercial, so it is desirable to free the users from the load of job handling. The grid scheduler or resource broker performs the task of job handling such as resource management and fulfilling user requirements. In this chapter, an anatomy of grid schedulers has been discussed which discusses intrinsic properties of different grid schedulers.

Keywords Middleware · Globus toolkit · Resource management
Grid scheduler

1 Introduction

The notion of grid [1–4] in computer networks is very much similar to the electric grid in terms of ubiquity, but the performance of resources in grid computing may vary even though they support the same code. There are various areas where grid applications are being used such as business, research, science, medicine, and

Ankita (✉) · S. K. Sahana
Computer Science and Engineering, BIT Mesra, Ranchi, India
e-mail: sharma.ankita0211@gmail.com

S. K. Sahana
e-mail: sudipsahana@gmail.com

V. Nath and J. K. Mandal (eds.), *Nanoelectronics, Circuits and Communication Systems*, Lecture Notes in Electrical Engineering 511,
https://doi.org/10.1007/978-981-13-0776-8_25

entertainment. The network is the lifeline of any grid. It is the network which connects different distributed resources together and makes sure that a user interfaces to the resources in a uniform way. One of the most propelling forces behind grid computing is the complex problems and advances made in the area of technology. With the amalgamation of recent technology trends and advancements in the research area, it has been made possible as well as feasible to think the Grid Vision—a vision which is enough to meet the ongoing demands of science and technology.

2 Management of Resources

Proper resource management plays a crucial role in successful job completion and throughput enhancement in grid computing network. The working and control of resources are governed by the resource providers, who are the owner of these resources. There are some specific policies which actually govern the working of resources under the resource management plan. Here, it is important to consider the heterogeneity of resources as well as the various administrative policies. The above-stated factors increase the complexity of the resource management problem.

3 Related Work

A taxonomy of grid resource management system is presented in [5] for distributed computing. The next level of distributed computing is grid computing, which spans across a large geographical area and connects machines at different autonomous sites. Each site has its own security and scheduling policies and resources being heterogeneous add on to the problem of grid management system. In [6], an anatomy of resource scheduling approaches is discussed which can provide solutions to different problems such as heterogeneity of resources, site autonomy, co-allocation of resources. The taxonomy in [7] highlights the components in the grid workflow management system. The first two components are workflow design and workflow scheduling. The third important component is fault tolerance, since execution failure is very common in a dynamic grid environment. The last component is data movement, which is often required to stage input files of jobs at remote sites before the actual execution of the job begins. The survey presented in [8] discusses heterogeneous computing, which depends on two conditions, namely the mode of execution and the different type of machine models involved in computing.

The resource status information changes frequently in a dynamic grid environment. It is important that the updated resource status information is disseminated in the grid environment so that resource status database remains consistent. There are several approaches discussed in [9] for the purpose of resource status dissemination. Also, there are different ways discussed to minimize the cost of data dissemination.

4 The Grid Architecture

Grid architecture [10] is designed to meet certain requirements but there are few issues or challenges that are needed to be considered such as interoperability, scalability and extensibility, managing time, cost and computational constraint, achieving quality of service. The key issue is interoperability that is the common protocols. These protocols describe the basic means by which users and resources, communicate in a grid portal.

Basically, there are five layers [11] in grid architecture (Fig. 1).

1. Application layer
   The uppermost layer is the application layer, which consists of the user applications and the execution of these applications is dependent on the subsequent lower layers.
2. Collective services layer
   The collective services include different type of services such as scheduling services, database management, storage services, etc., available in grid computing. This layer is the heart of grid architecture. This layer focuses on the synergy of resources glued together and not on a single resource.
3. Connectivity and Resource services layer
   The connectivity and resource layer provides authentication (for verification and validation of both user and resources) and communication protocols (for data transfer between resources).

Fig. 1 Architecture of a grid

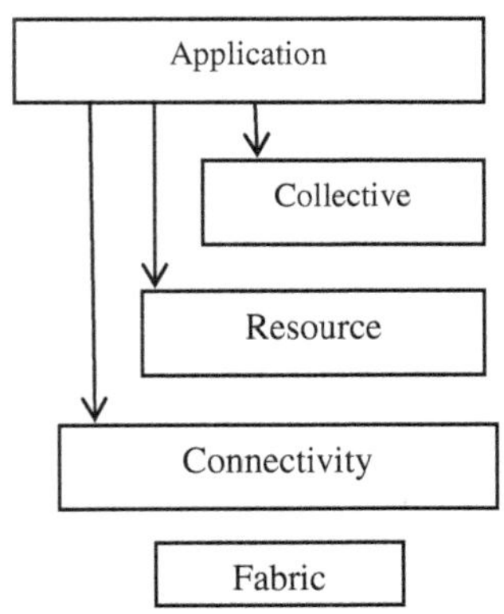

4. Fabric layer
   The fabric layer is the lowest layer in grid architecture which includes physical (servers, storage devices, printers) and logical components (data, computer programs, database directory to store well-structured data).

5 Types of Grid Systems

1. Computational Grid: A Computational grid consists of a pool of shared resources, where a large and complex task is divided into several subtasks and distributed among several processors in the pool. The answer is then sent back to the original system.
2. Network Grid: There are two important services provided by the network grid. The first service is fault tolerance and the second one is high-speed communication services. The nodes in the grid system act as a router and provide fast and secure network for communication between two endpoints.
3. Data Grid: Different services related to the handling and management of data are provided in data grid such as discovery, storage, and manipulation of data.
4. Utility Grid: In the above grids, only processing cycles or data was shared, but utility grid allows sharing of software or any special equipment.

6 Grid Middleware

Middleware [12] acts as an interface between application and components of the software and allows them to communicate. Middleware can be called as a system software. It provides various services to the application and also hides the complexity of the grid environment. Processes running on the same or different machines interact with each other with the help of a middleware. The middleware also works to enhance the security of the grid as it helps the users to validate their identity through certificate infrastructure.

7 Grid Schedulers

Grid schedulers [13] are responsible for submitting jobs to the available resources, i.e. they find the best match of jobs to the resources available in the grid (Fig. 2).

Some of the fundamental grid schedulers are discussed below:

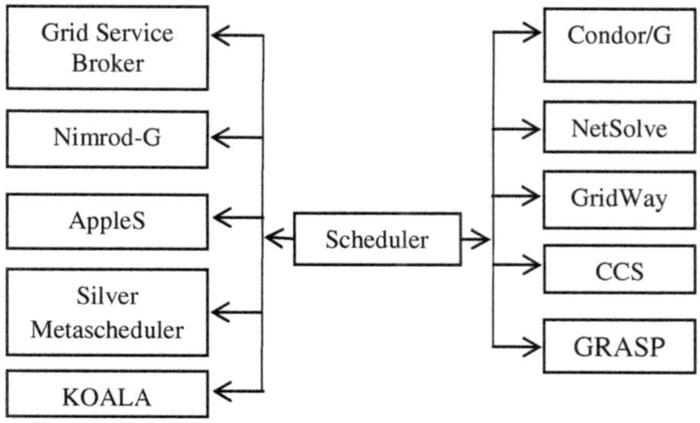

Fig. 2 Grid schedulers

7.1 Grid Service Broker

In a dynamic grid environment, job execution is a difficult task since the state of grid environment is changing continuously. Job handling is becoming a problem for users and hence service brokers are necessary to free the users from the tiresome task of job handling.

Discovery and selection of appropriate resources for a given job are one of the most important tasks in grid computing, which is handled by the grid service broker. In addition to job handling, the service broker can also handle other functions such as parameter study, rescheduling and replication.

One of the prominent tasks of service broker [14] is the fulfillment of user's requirements. There are times when the grid sites are flooded with user requests which can cause sudden failure of systems at that site. In order to handle such situations, a service broker needs to be fault tolerant.

Initially, the interface of grid portal used to be command line interface which replaced by interactive APIs for higher level utilization.

7.2 Nimrod-G

An earlier Nimrod system was developed to solve the problems or complexities encountered with parametric computing. But Nimrod was not suitable for large-scale computational grids, where resources are geographically scattered at different sites with each site having its own security and scheduling policy [15].

In order to overcome this problem, Nimrod/G [16] was developed which can perform parametric computing and can handle resource management in a large computational grid. Basically there are four main elements of Nimrod-G, namely a task-framing engine, a scheduler, a dispatcher, and resource agents. The deadline is set by the user and within the given time frame set by the user, the scheduler in

Nimrod-G tries to obtain the execution result. Nimrod-G not only optimizes time, but also tries to minimize cost, hence it an attempt towards budgetary driven grid computing.

The scheduling algorithms that are available in the Nimrod-G are: Time optimization, cost optimization, time–cost optimization and cost–time optimization.

7.3 AppLeS

The two design goals of AppLeS (Application-Level Scheduling) are to (1) study and verify adaptive scheduling as well as (2) applying the execution results to the real-time applications. Depending on the type and structure of the application, the scheduler in AppLeS [17] tries to utilize the resources available in order to meet the desired performance goals.

A scheduling agent is fitted in each of the applications that combine static and dynamic resource information as well as the user application's constraints [18].

7.4 Silver Metascheduler

Silver Metascheduler is a grid service broker, which supports advance reservation. The primary function of Silver Metascheduler [19] is to balance the workload of different resources distributed across independent administrative domains.

It is completely dependent on the local administrative site that how much a system contributes in the load sharing process. The metascheduler keeps track of all allocated loads to the systems in different sites and then distribute the workload accordingly to maintain equal or near to equal load among the active sites.

7.5 KOALA

KOALA [20] was designed to solve the problem of co-allocation. A single application may run on more than one resource of different types of multiple administrative sites managed by different site owners. This is called co-allocation.

The key issue which motivated the design of KOALA was reliability. The resources in grid are highly dynamic that is the availability of resources may change (adding a new resource or failure of a resource) at any time. So, KOALA is actually a co-allocating grid scheduler which was designed to find solutions encountered in co-allocation.

7.6 OGSI Broker

The OGSI Broker [21] facilitates automatic resource discovery and allocates the best resource to the job as per the job requirement as well as resource usage policies and also hides the system complexity.

7.7 GT Broker

GT Broker [22] is based on Globus toolkit 2, 3. It was designed to solve the problem of improper job submission. The GT Broker first examines the cluster of resources and check if the resource is sufficient for the job otherwise skips the particular cluster. GT Broker also supports dynamic requests of user during execution. Resubmission and rescheduling of jobs are also supported by GT Broker.

7.8 GridWay

The GridWay scheduler is based on Globus toolkit, which offers a common platform for integrating resources of different sites by the means of a set of policies that will govern access mechanism to shared resources other security principles.

This metascheduler provides a flexible infrastructure and resources are better utilized with higher system efficiency and application throughput. The other important features of GridWay are its fault detection mechanism and accounting properties.

7.9 GRUBER

In a resource sharing environment, there are several issues regarding the usage and access rights of resources. Resource usage policies are given by resource owners, which allow a user to use specified amounts of resource, whereas resource access policies stated by the resource owners provide special authorization privileges for a particular user to access a particular resource.

These policies can either allow or refuse the request to access a specific resource. Resource usage policies come into picture when a user has got the access rights for a particular resource, then the usage policies determine the extent or amount of the resource given to the user for its use. GRUBER [23] is a toolkit designed which facilitates selection of resources from virtual organization through automated agents.

7.10 GRASP

GRASP stands for Grid Resource Allocation System [24] is a system which is based on Open Grid Services Architecture. There are different types of services, which play a vital role in carrying out different operations in the grid.

The task of resource management in the grid is performed by GridSchedulingService. The resources are selected on the basis of the submitted job and accordingly allocated.

The next most important service is JobSubmissionService. After the allocation of resources, co-allocation takes place to balance the workload among different resources.

The jobs are also co-allocated depending on the current load of the resources. The third and last service used here is ResourceManagerService, which permits the user to execute the job on the local system (Tables 1 and 2).

Table 1 The fundamental grid schedulers, the underlying middleware and their intrinsic properties

SL. No.	Grid scheduler	Middleware	Properties
1.	Grid Service Broker	Globus Toolkit, Unicore	Supports parametric computing, fault tolerance, and support for API
2.	Nimrod-G	Globus Toolkit 2, Legion	Application-level and parametric computing
3.	AppleS	Globus Toolkit 2	Network weather service, supports master slave applications (parallel) and application-level scheduling
4.	Silver Metascheduler	Globus Toolkit	Scalability, support for advanced reservation and fault tolerance
5.	KOALA	Globus Toolkit 2,3	Co-allocation of jobs, support for parameter sweep applications
6.	OGSI Broker	Globus Toolkit 3	Automatic resource discovery, hides grid system complexity
7.	GT Broker	Globus Toolkit 2,3	Automatic staging of files, supports dynamic request for execution, reduced makespan time and rescheduling
8.	GridWay	Globus Toolkit	Support for API and job migration, fault tolerance and checkpointing
9.	GRUBER	Globus Toolkit 3,4	Designed for resource usage SLA (Service-Level Agreements), support of automated agents enable to select resources from virtual organization
10.	GRASP	OGSA	Provides support for scientific applications, parameter studies

Table 2 Some of the leading works in the area of grid brokers are given below

SL. No.	Paper	Implementation
1.	A grid service broker for scheduling distributed data-oriented applications on global grids [14]	The author has developed a broker which will run in data-intensive environment with various scheduling strategies
2.	Nimrod/G: an architecture a resource management and scheduling system in a global computational grid [16]	Nimrod/G has been developed to overcome the complexities of Nimrod and enable features such as extensibility, automatic resource discovery and interoperability
3.	KOALA: a co-allocating grid scheduler [20]	The author has discussed a co-allocating scheduler for jobs, which are having multiple parts. Also, KOALA support fault-tolerance mechanism to deal with the changeable environment
4.	Grid resource broker for scheduling component-based applications on distributed resources [22]	The author has combined two strong systems, namely ProActive and Gridbus resource broker to get more promising results
5.	GRUBER: A grid resource usage SLA broker [23]	Gruber solves the problem of resource sharing at multiple grid sites. The author has combined two strong systems, namely ProActive and Gridbus resource broker to get more promising results
6.	GRASP: a grid resource allocation system based on OGSA [24]	GRASP tries to make the process of job submission easier and convenient for the user and also provide support for MPI features
7.	A grid service broker for scheduling e-science applications on global data grids [25]	The broker has developed a model which allow access to distributed resources by performing resource discovery for a given situation

8 Conclusion

This has been proven that grid schedulers play a nontrivial role in grid computing. With this taxonomy, one can easily identify and categorize different types of grid schedulers in Grid Environment. We have discussed the important functionalities of various grid schedulers, which will help the researchers have a better understanding of these schedulers and eventually the users can select the best scheduler to run their job. The services provided by the schedulers should be available to the user in a transparent manner and also the reliability of the grid system must be enhanced by adding features like checkpointing and job migration.

References

1. Foster I, Kesselman C, Tuecke S (2001) The anatomy of the grid: enabling scalable virtual organisations. Int J High Perform Comput Appl 15(3):200–222
2. Foster I, Kesselmen C (1999) The grid: blueprint for a future computing infrastructure. Morgan Kaufmann Publishers, pp 1–593
3. Jacob B et al (2005) Introduction to grid computing, 1st edn. IBM Redbooks, International Technical Support Organisation
4. Buyya R, Venugopal S (2005) A gentle introduction to grid computing and technologies. Computer Society of India
5. Krauter K, Buyya R, Maheswaran M (2002) A taxonomy and survey of grid resource management systems for distributed computing. Softw: Practice Experience 32(2):135–164
6. Casavant TL, Kuhl JG (1988) A taxonomy of scheduling in general-purpose distributed computing systems. IEEE Trans Softw Eng 14(2):141–154
7. Yu J, Buyya R (2005) A taxonomy of scientific workflow systems for grid computing. SIGMOD Record 34(3):44–49
8. Ekmecic I, Tartalja I, Milutinovic V (1996) A survey of heterogeneous computing: concepts and systems. Proc IEEE 84(8):1127–1144
9. Maheswaran M, Krauter K (2000) A parameter-based approach to resource discovery in Grid Computing Systems. In: Proceedings 1st IEEE/ACM international workshop grid computing, 2000, pp 181–190
10. Punia A, Mittal P (2014) A review: grid computing. Int J Comput Sci Mob Comput 3(4):634–639
11. Singh MK, Pal S (2011) Five layer security architecture and policies for grid computing system. Int J Comput Sci Inf Technol 2(3):1312–1314
12. Buyya R (2007) Special issue: middleware for grid computing. Wiley Interscience
13. Rood B, Lewis MJ (2008) Scheduler on the grid via multi-state resource availability prediction. In: 9th IEEE/ACM international conference, 2008
14. Venugopal S, Buyya R, Winton L (2004) A grid service broker for scheduling distributed data-oriented applications on global grids. In: Proceedings of the second workshop on middleware for grid computing, 2004, pp 75–80
15. Abramson D, Giddy J, Kotler L (2000) High performance parametric modeling with Nimrod/g: killer application for the global grid? In: Proceedings of the 14th international parallel and distributed processing symposiums (IPDPS 2000), 2000, pp 520–528
16. Buyya R, Abramson D, Giddy J (2000) Nimrod/G: an architecture for a resource management system and scheduling system in a global computational grid. In: Proceedings of the international conference on high performance computing in Asia-Pacific Region (HPC Asia 2000), 2000
17. Berman F, Wolski R (1997) The AppleS project: a status report. In: Proceedings of the 8th NEC research symposium, May 1997
18. Casanova H, Obertelli G, Berman F, Wolski R (2000) The AppleS parameter sweep template: user level middleware for the grid. In: Proceedings of the IEEE SC 2000, international conference networking ad computing. IEEE CS Press, USA
19. Xiaohui W, Zhaohui D, Shutao Y, Chang H, Huizhen L, CSF4: A WSRF compliant meta-scheduler. Jilin University
20. Mohamed H, Epema D (2007) KOALA: a co-allocating grid scheduler. Wiley Interscience
21. Kim YS, Yu J-L, Hahm J-G, Kim J-S, Lee J-W (2004) Design and implementation of an OGSI-compliant Grid broker service. In: IEEE international symposium on cluster computing and the grid, 2004
22. Chu X, Venugopal S, Buyya R (2008) Grid resource broker for scheduling component-based applications on distributed resources. In: CyberInfrastructure technologies and applications. Nova Science Publishers

23. Dumitrescu CL, Foster I (2005) GRUBER: a grid resource usage SLA Broker. Springer, Berlin, pp 465–474
24. Kwon O-H, Hahm J, Kim S, Lee J (2004) GRASP: a grid resource allocation system based on OGSA. In: Proceedings 13th IEEE international symposium on high performance distributed computing, 2004
25. Venugopal S, Buyya R, Winton L (2005) A grid service broker for scheduling e-Science applications on global data grids. Concurrency Comput Pract Experience 18(6):685–699

Chaos-Based Colour Image Encryption Using Microcontroller ATMEGA 32

M. K. Mandal and A. K. Das

Abstract The present chapter describes the encryption algorithm of a colour image using microcontroller ATMEGA 32. Two microcontrollers are used as driver and driven systems in transmitter and receiver sides. One-dimensional logistic map is used in both the cases for generating the chaotic sequence and synchronization between them through open plus closed loop (OPCL) coupling. Under the synchronized condition, the data file of the input image is encrypted and transmitted to the receiver side. The receiver microcontroller received this encrypted data and using the reverse technique it decrypted the data and send to LCD display to reconstruct the input image. The proposed technique is verified through different images and the results confirm the validity of the algorithm.

Keywords Chaos · Synchronization · Encryption · Decryption
Microcontroller

1 Introduction

Nowadays, exchange of multimedia information like audio, text, picture and video through Internet are an integral part of our daily life. The unauthorized access and use of personal/corporate/government information are serious issues in our society. In this regard, the techniques of information security may play an important role in storage and communication of such valuable information. The cryptographic technique is one of the powerful tools for protecting valuable information by hiding

M. K. Mandal (✉)
Department of Physics, National Institute of Technology Durgapur,
Durgapur 713209, India
e-mail: mrinalkanti.mandal@phy.nitdgp.ac.in

A. K. Das
Department of Electronics and Instrumentation,
Dr. B. C. Roy Engineering College, Durgapur 713206, India
e-mail: anupkumardas29@yahoo.com

V. Nath and J. K. Mandal (eds.), *Nanoelectronics, Circuits and Communication Systems*, Lecture Notes in Electrical Engineering 511,
https://doi.org/10.1007/978-981-13-0776-8_26

its meaning from any unauthorized persons. The image encryption technique based on chaotic maps has engrossed wide-ranging attention in recent years [1]. The image data are very much confidential in case of military application, medical imaging, architectural drawings, banking sectors and so on and it should be sent through secure transmission (encrypted information) mode and legitimate receiver should decrypt it to get the original message. During the past few decades, lots of image encryption algorithms have been proposed in the literature [2–5]. Some traditional encryption algorithms like RSA, DES, AES, IDEA, etc., are not suitable for image or video data due to their bulky size with high correlation. The chaos-based image encryption algorithms are popular in the scientific community due to its simplicity in implementation, sensitivity to initial conditions and parameters, excellent security, etc. Basically, chaotic sequences are used to remove the correlation among the consecutive pixels of the image by confusing the pixel value and shifting the pixel position. The parameter values of the chaotic map and its initial conditions are different in transmitter and receiver side, therefore, it needs to synchronize for decryption process. To fulfil the need of proper encryption and decryption through chaotic synchronization, we have implemented microcontroller ATMEGA 32-based cryptosystem via synchronization of the logistic map. The used synchronization scheme is open plus closed loop (OPCL) coupling [6, 7], because it is capable of handling parameter mismatch cases. In this chapter, one-dimensional (1D) logistic map is used as driver and driven system at parameter mismatch condition for above secure image encryption–decryption technique. At synchronized condition, the pixel value of the image is encrypted with chaotic value by logical operation of driver system and send this encrypted value to the driven system, after receiving this value at the driven, the reverse way decryption has been done to retrieve the original pixel value of the image. The main feature of this paper is (i) Synchronization of 1D logistic map has been done via OPCL coupling scheme. (ii) Image encryption and decryption have been done at synchronized condition of chaos. (iii) The security verification of the system has been done through different security tests. The experimental verification of the proposed scheme has been done by microcontroller ATMEGA 32 along with some other accessories.

The rest of the paper is organized as follows: In Sect. 2, we describe the coupling theory, Sect. 3 describes proposed encryption algorithm using microcontroller, Sect. 4 describes experimental results and finally Sect. 5 gives the conclusion.

2 OPCL Coupling Theory

This section summarizes the synchronization method taken from references [6, 7], where concentration is focused on the OPCL coupling scheme. The OPCL coupling was used earlier for synchronization of identical oscillators. To describe the coupling to mismatch systems, we define a driver for n-dimensional map as

$$x_{i+1} = f(x_{i,}\mu) + \Delta f(x_i, \mu), x_i \in R^n \quad (1)$$

where $\Delta f(x_i)$ contains mismatch parameters. The system of (1) drives a response system defined as

$$X_{i+1} = f(X_{i,}\mu) + D(X_i, Ax_i). \quad (2)$$

The state variable $X_i = Ax_i$, where $A = (a_{ij})_{n\times n}$ is a real matrix, μ is the parameter and i is the number of iteration. The coupling term $D(X_i, Ax_i)$ is defined by

$$D(X_i, Ax_i) = Ax_{i+1} - f(Ax_i, \mu) + [H - \mathrm{JF}(Ax_i)](X_i - Ax_i), \quad (3)$$

where JF is the Jacobian of $f(x_{i,}\mu)$ and $H = (h_{ij})_{n\times n}$ is an arbitrary constant Hurwitz matrix, whose eigenvalues must lie inside the unit circle on the complex plane for a stable synchronization. The error signal of the coupled system can be written as $e_i = (X_i - Ax_i)$, and $f(X_{i,}\mu)$ can be written, using Taylor series expansion, by

$$f(X_{i,}\mu) = f(Ax_i, \mu) + \mathrm{JF}(Ax_i)(X_i - Ax_i) + \ldots \quad (4)$$

Taking up to the first-order terms of (4) and doing some algebraic operation we get error dynamics $e_{N+1} = H^N e_0$, where N is the number of iteration. Now, as H is the real matrix its eigenvalues are either real or complex conjugate pairs. The error is $e_N \to 0 \; as \; N \to \infty$ if the parameter of the H matrix is so preferred that its eigenvalues all lie inside a unit circle. This indicates synchronization between the driver and driven systems. In the proposed algorithm, one-dimensional logistic map is considered as the driver system as mentioned below:

$$x_{i+1} = \mu x_i(1 - x_i) + \Delta\mu x_i(1 - x_i). \quad (5)$$

The corresponding driven system according to the above OPCL coupling is given by

$$X_{i+1} = \mu X_i(1 - X_i) + a(\mu + \Delta\mu)x_i(1 - x_i) + (h - \mu + 2\mu a x_i)(X_i - aX_i) \\ - \mu a x_i(1 - a x_i). \quad (6)$$

Equations (5) and (6) are implemented in microcontroller using Proteus simulation software.

3 Proposed Algorithm

The proposed encryption and decryption techniques are implemented using microcontrollers ATMEGA 32 and the block diagram of the entire process is given in Fig. 1. In the driver and driven system, two ATMEGA32 microcontrollers are used for chaotic sequence generation and for display purpose Nokia 6610 LCD is used. Microcontroller ATMEGA32 has suitable features including 32 KB of in-system programmable flash with Read-While-Write capabilities, 1024 Bytes EEPROM, 2 KB SRAM, 32 general-purpose I/0 lines, 32 general-purpose working registers, a JTAG interface for boundary-scan on-chip debugging support and programming architecture. Here, the microcontroller is programmed with the 1D logistic map for driver and driven system illustrated in Eqs. (5) and (6) to generate the chaotic sequences under the synchronized condition for different initial conditions. The program is written in BASCOM AVR language. At first, the colour image converted the equivalent 8-bit binary value (pixel value) by the RGB LCD converter software (MCS Electronics LCD RGB-8 converter tool converts a normal windows bitmap into 256-colour RGB-coded format, it is a RLE-compressed format) and this converted value stored in driver microcontroller. The microcontrollers are programmed as per OPCL coupling mechanism so that when both the systems run simultaneously then after certain iteration they become synchronized. Under this condition, the difference between the chaotic sequence of driver and driven system will be zero, at that moment the pixel value (RLE8 conversion from image) is taken from look-up table of driver microcontroller. After synchronization, the chaotic sequence (fractional number) of the driver system will be multiplied by 1000 to get the integer decimal value and, thereafter XOR logical operation will be done with the pixel value. This new encrypted number will be sent to the driven system through Tx pin of driver microcontroller and after receiving the encrypted number the same way XOR logical operation is performed to the chaotic sequence of the driven system to get the original pixel value. The encrypted and decrypted images are displayed in Nokia 6610 LCD for visual observation. The Proteus software simulation (Proteus Design Suite 8.5 Version© Labcentre Electronics) circuit with exact pin connection is shown in Fig. 2. The detailed step-wise encryption algorithm is mentioned below:

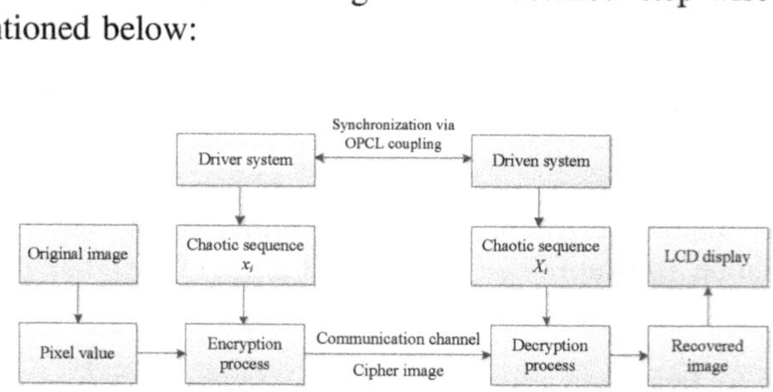

Fig. 1 The block diagram of the proposed encryption and decryption process (drawn in Microsoft Visio 2010 software)

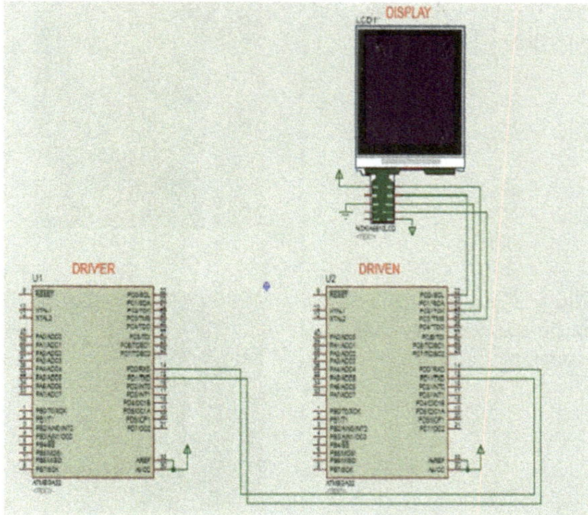

Fig. 2 Simulation circuit using Proteus software

Step-1: Take an image of dimension $m \times n$ as $I(m, n)$. The pixel values of the image are converted into decimal numbers and arranged by order from left to right and top to bottom to form a decimal number set and store in the memory.
Step-2: Start simulation to generate chaotic sequence and synchronize them.
Step-3: After synchronization, the chaotic sequence x_i is multiplied by 1000.
Step-4: Execute exclusive OR operation between the pixel value $I(i,j)$ and integer part of the chaotic sequence x_i to generate the encrypted value.
Step-5: This encrypted value will be transmitted by the driver microcontroller through Tx pin.
Step-6: Receive the encrypted value by Rx pin of the driven microcontroller.
Step-7: Generate the chaotic sequence X_i and multiplied it by 1000.
Step-8: Execute exclusive OR operation between the received encrypted data and the integer part of the chaotic sequence X_i to generate the original pixel value and store it in the memory.
Step-9: Send decrypted stored data to Nokia 6610 LCD for display the image.

4 Result and Discussion

The simulation experiment has been performed using RGB Lena image of dimension 128×128. The parameter values of the logistic map are $\mu = 3.89$, $\Delta\mu = 0.1$, $a = 1$ and $h = 0.65$. These parameter values play the role of input key of the cryptosystem. The cryptosystem has been verified by taking different images, out of those one result is included in this paper for the sake of completeness. The input Lena image (RLE converted) and its histogram are shown in Fig. 3.

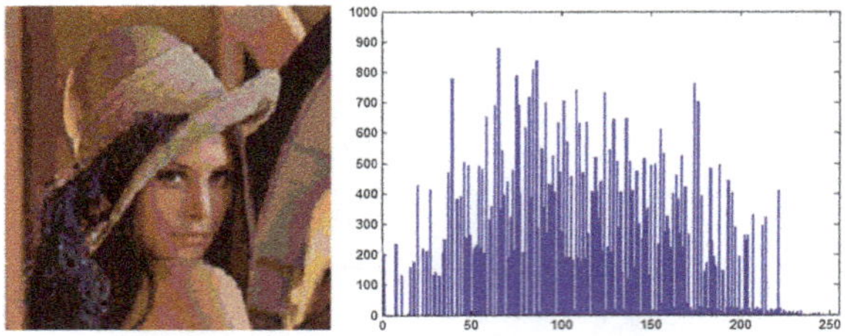

Fig. 3 Input Lena image and corresponding histogram

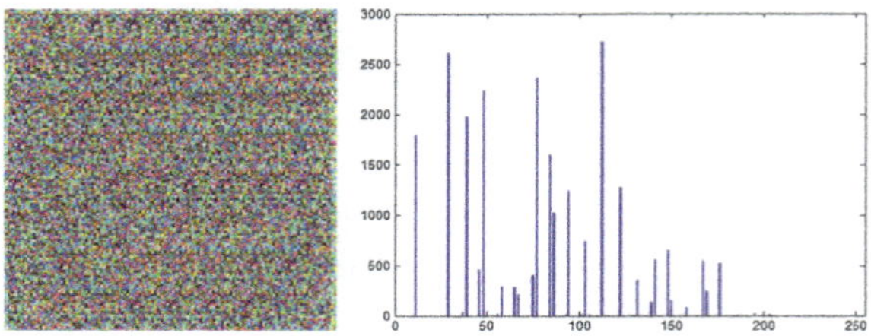

Fig. 4 Encrypted Lena image and corresponding histogram

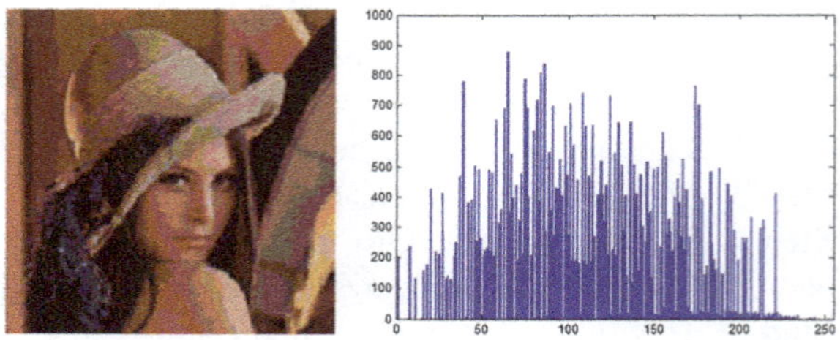

Fig. 5 Decrypted Lena image and corresponding histogram

The encrypted image and the corresponding histogram are shown in Fig. 4. The histograms of input image and its encrypted image are totally different. This result confirms the security level of the proposed cryptosystem. The decryption result is shown in Fig. 5. The decrypted figure is similar to the input figure, which confirms the validity of the decryption process via synchronization of the chaotic system using microcontroller.

5 Conclusion

In this chapter, an experimental realization of image encryption and decryption using microcontroller ATMEGA 32 has been proposed via synchronization of chaotic logistic map. This technique provides the following advantages: (i) Complete synchronization can be achieved in a precise and organized way under parameter mismatch condition. (ii) The OPCL coupling for the discrete map has been realized using a microcontroller for the implementation of chaos-based cryptosystem. (iii) The encryption and decryption process has been successfully described using different images. The proposed cryptosystem may be applicable for real-time secure communication.

References

1. Kocarev L, Lian S (2011) Chaos-based cryptography theory, algorithms and applications. Springer, February 2011
2. Yong XU, Wang H, Li Y, Pei B (2014) Image encryption based on synchronization of fractional chaotic systems. Commun Nonlinear Sci Numer Simulat 19:3735–3744
3. Xie Y, Li J, Kong Z, Zhang Y, Liao X, Liu Y (2015) Exploiting optics chaos image encryption-then transmission. J Lightwave Technol 34:5101–5109
4. Kar M, Mandal MK, Nandi D, Kumar A, Banik S (2016) Bit-plane Encrypted Image Cryptosystem using chaotic quadratic and cubic maps. IETE Tech Rev 33:651–661
5. Hou J, Xi R, Liu P, Liu T (2017) The switching fractional order chaotic system and its application to image encryption. J Automatica Sinica 4:381–388
6. Grosu I, Padmanaban E, Roy PK, Dana SK (2008) Designing coupling for synchronization and amplification of chaos. Phys Rev Lett 100:234102
7. Pal P, Debroy S, Mandal MK, Banerjee R (2015) Design of coupling for synchronization in chaotic maps. Nonlinear Dyn 79:2279–2286

An Enhanced GEAR Protocol for Wireless Sensor Networks

Rahul Priyadarshi, Surender Kumar Soni and Prashant Sharma

Abstract This chapter reflects the enhanced Geographical energy-aware routing in wireless sensor network, where the separation of nodes is totally done on the basis of their position which is further obtained by Global Positioning System. Sink node or Base Station is fixed far away from the sensed area and entry node is fixed at the central point of the sensed area. A threshold value of distance has been defined. If the obtained distance of nodes from gateway node and sink node is relatively smaller than the threshold value of distance then in that case transmission of data takes place directly. If the calculated distance is more than that of threshold value then isolate the nodes into the identical areas. Now for each area, a cluster head is nominated for data transmission. Nomination of cluster head is done on the basis of probability function and residual energy of nodes. Finally, the proposed routing protocol displays better results in terms of network lifetime, network energy consumption and packet transmission to the base station compared to the existing routing protocol.

Keywords Wireless sensor network · Nodes · Energy · Base station Cluster · Distance · Probability

R. Priyadarshi (✉) · S. K. Soni · P. Sharma
Electronics & Communication Engineering Department,
National Institute of Technology, Hamirpur (Himachal Pradesh), India
e-mail: rahul.glorious91@gmail.com

S. K. Soni
e-mail: surender.soni@gmail.com

P. Sharma
e-mail: prashantece.cet@gmail.com

V. Nath and J. K. Mandal (eds.), *Nanoelectronics, Circuits and Communication Systems*, Lecture Notes in Electrical Engineering 511,
https://doi.org/10.1007/978-981-13-0776-8_27

1 Introduction

The wireless sensor network (WSN) installed in an extensive geographical region to observe physical occurrence with satisfactory exactness and consistency. The sensor nodes can observe several individuals like metallic substances, temperature, moisture, pressure and salinity. The observed data may be efficiently used in many applications like industrial, commercial, consumer, environmental and military. WSN consists of several devices which consist of the battery to operate it. So in order to save the energy, energy efficient network should be designed [1, 2]. Because of advancements in technology, nowadays it is promising to install lakhs of sensor nodes in the particular region. It is really very difficult to replace or recharge the node battery because of deployment of a node in the area where human intervene is not possible. For that reason, a crucial issue for WSN is to restrain power outlay of nodes to elongate network lifetime. A countless number of clustering has been proposed [3, 4].

Clustering is basically a method which can manage that how network can be made more energy efficient, while communication takes place between sensor nodes and Base Station (BS). Generally, nodes cannot communicate directly with BS. First nodes have to transfer their data to the Cluster Head (CH) and once CH aggregates the data from all their respective member nodes then it transfers the accumulated data to the BS in order to make system more energy efficient. So for this purpose network area is divided into region and each region has one CH to collect the data. Sensor nodes in principal section interconnect straight away to sink node, however, nodes of other section use clustering order to communicate data to the entry node.

2 Problem Statement and Network Design

As we know that clustering protocol takes relatively a lesser amount of energy while data transmission that's what this protocol in WSN has grown rapidly in various fields of applications. Selection of CH is totally done on the basis of probability function in most of the existing protocols with some changes in probability function in most of the proposed protocol [5–7]. A Common routing topology in WSN is shown in Fig. 1.

It is not compulsory that distribution of CH be homogeneously in the entire sensor. Correspondingly certain clustering method uses imbalanced clustering. Most of the Clustering hierarchies have many advantages and disadvantages like

- Advantages
  - Any node that served as a CH in the certain round cannot be selected as the CH again
  - Because of TDMA schedule, it prevents CH from unnecessary collisions

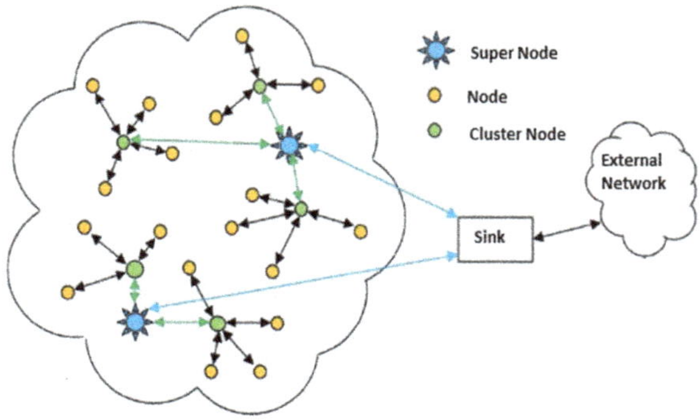

Fig. 1 Routing topology in WSN

  - Data aggregation at CH
  - CH communicate directly with sink.

- Disadvantages
  - Not applicable for large region networks
  - No real load balancing because CH is selected in terms of probabilities without energy considerations
  - Dynamic clustering brings extra overhead.

In this chapter, our main goal is to design geographical routing which is totally based on energy and meets these stipulations

- The entire network area is separated into 8 sections and an entry node is positioned in the midpoint of detecting arena that lessens the communication distance of every single node, hereafter it lessens energy ingesting and be responsible for superior network lifetime.
- Selection of CH in all 8 sections done independently from each region.

In this chapter, on considering N number of sensor nodes as of now to observe the environment where some constraints have been taken while designing the network like placement of the BS is far away from the detecting field. All the nodes are immobile in nature. An entry node is installed at the midpoint of the detecting field and after positioning entry node, it is also immobile and chargeable in nature.

Similar types of nodes that have the identical computational ability are used. An idiosyncratic identifier is allocated to every single sensor node. We are using the first order radio model while designing the network as the same principle used by the author in [8, 9]. This designed model characterizes the nodes energy debauchery for transferring, accruing and receiving the data. Extra energy is debauched at the transmitter end compared to receiver end because transmitter end needs extra energy for amplification. An energy radio communication model has been shown in Fig. 2. The given below Eq. (1) and (2) are mainly used for transmission and receiving energy calculation for k number of bit.

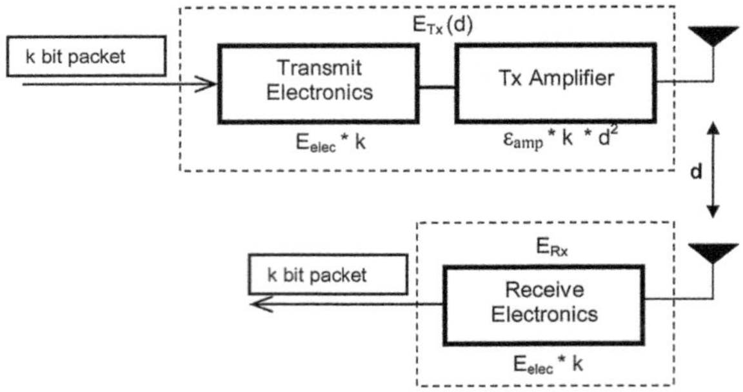

Fig. 2 An energy radio communication model

$$E_{TX}(k,d) = \begin{cases} kE_{elec} + k\varepsilon_{fs}d^2 & d < d_0 \\ kE_{elec} + k\varepsilon_{amp}d^4 & d \geq d_0 \end{cases} \quad (1)$$

$$E_{RX}(k) = kE_{elec} \quad (2)$$

where

- E_{TX} is the energy transmission and E_{RX} is the and energy reception of k number of bits towards distance d.
- E_{elec} is the electronic energy compulsory for modulation, filtering, and coding.
- E_{DA} is the energy compulsory for data accumulation.
- ε_{fs} and ε_{amp} are the amplification energy.

3 Proposed Routing Protocol

This segment reflects the feature of the suggested protocol. Sensor nodes generally have an excessive amount of data which is further processed to BS. As a result, an instinctive process of accumulating the data into a minor set of significant statistics is mandatory [10–13]. For enhancing the network lifetime and throughput, placement of the entry node which is chargeable in nature is taken into consideration at the midpoint of detecting field. The entry node or one can say gateway node gathers data from network area as well as from the CH. Afterwards, collects data and communicates to the BS. In fact, recharging of entry node is economical compared to the worth of a sensor node. Designed network model has different phases, which are explained below.

3.1 Initial Phase

In this section, similar types of sensor nodes are used for random deployment in the network area. Initially, BS broadcasts the "START" message in the entire area to get the location of sensor nodes and their residual energy as well as their idiosyncratic identifier [14]. Afterwards, BS determines the distance of every single node which is further saved to the data table. The data tables consist of location of every sensor nodes, an idiosyncratic identifier of every sensor nodes, residual energy of every sensor nodes as well as distance from BS and entry node.

3.2 Setup Phase

In this segment, the detecting field is separated into eight subparts based on the position of the nodes in the network region. In first subparts, as the distance from BS is very less so they do straight away communication with BS for data communication [15]. Similarly in second subparts, nodes which are near to the entry node, straight away transmitting the data to the entry node which collects the data and direct to the BS. These two subparts, i.e. first and second subparts are referred as non-clustered subparts or regions and all other subparts are equally divided into two parts which are referred as cluster region.

3.3 Selection of Cluster Head

In the beginning, BS split the detecting field into eight subparts and selection of CH is done in each Subparts. Also, initial stage of first round where all the nodes have equal chance to become CH because each node has the same energy in starting phase. At that moment Selection of CH is totally on the basis of probability function and residual energy of nodes. The selection of CH is determined by choice confirmed by the node with generating a random number between 0 and 1. A threshold value *T*(*n*) is defined to compare generated number. If the generated number is below *T*(*n*), then that particular node will become CH. The threshold *T*(*n*) is calculated as by the given below equation

$$T(n) = \begin{cases} \frac{P}{1-P(r \bmod 1/P)} \times \frac{E_{\text{remain}}}{E_{\text{initial}}} & \text{if } n \in G \\ 0 & \text{otherwise} \end{cases}$$

where

- P is the percentage of choosing cluster heads.
- r is the current round.
- G is set of all the sensor nodes that have not been cluster heads in $1/P$ rounds.

- E_{remain} is the remaining energy of node.
- $E_{Initial}$ is the initial energy of node.

3.4 Scheduling and Steady-State Phase

Each CH assigns a time slot for their respective member nodes based on the Time Division Multiple Access scheduling. Member nodes must have to communicate with their CH in the assigned time period only or else go to sleep mode. Therefore, energy debauchery for each node gets drops. In the steady state, all the nodes transfer their data to the CH. CH accumulates data from its member nodes and advancing to the entry node, entry nodes collect the data received by CH and advancing it to the BS.

4 Simulation and Result Analysis

The performance of proposed protocol is compared with existing geographical energy-aware routing (GEAR) protocol in WSN. The simulations have been done using MATLAB. Table 1 shows the simulation parameters and its value.

Figure 3 reflects the graph between numbers of alive node versus a number of rounds. The proposed GEAR protocol has better lifetime comparing to existing GEAR protocol as it is clearly shown from below figure. It is happening because the network has been divided into eight logical parts and the way we are choosing CH. Existing protocol's node die in 1108th round but proposed protocol's node die in 1611th round which is definitely showing longer network lifetime.

Figure 4 describes the performance of proposed GEAR protocol with existing LEACH protocol. The graph has been plotted between throughputs versus a

Table 1 Simulation parameters

Simulation variable	Set up values
Sensor area	200 m × 200 m
No. of sensor nodes	500 nodes
Initial energy of nodes	0.1 J
Base station location	(60 m, 80 m)
Entry node location	(50 m, 50 m)
Number of rounds	2000
E_{elec}	40 nJ/bit
ε_{fs}	8 pJ/bit/m^2
ε_{amp}	0.001 pJ/bit/m^4
E_{DA}	5.5 nJ/bit
k	4000 bits

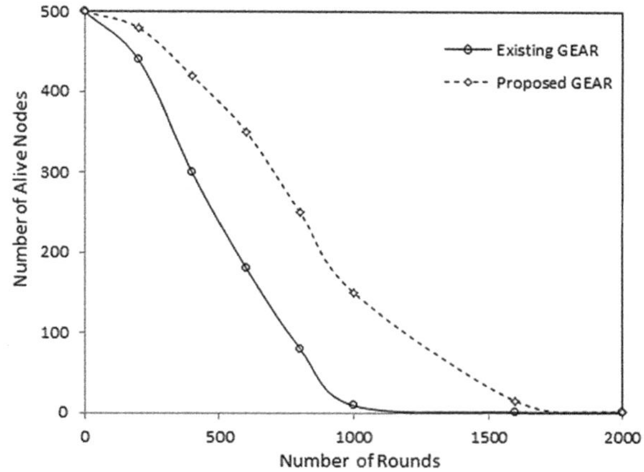

Fig. 3 Number of alive nodes

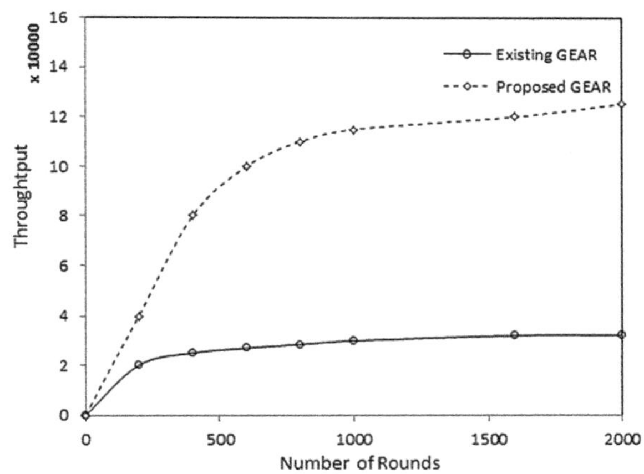

Fig. 4 Throughput evaluation

number of rounds. It illustrates the total number of a data packet that is transmitted to the BS which is more in case of proposed GEAR protocol compared to the existing GEAR protocol.

Figure 5 reflects the graph between Residual energy versus Number of rounds. Initially, each node has 0.5 J energy and total number of node is taken as 500. So the total energy of network is 50 J. Proposed GEAR protocol confirms superior results in comparison of existing protocol in reference to energy consumption.

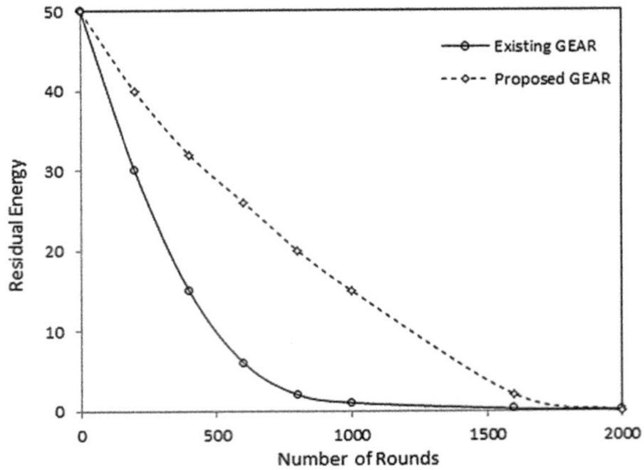

Fig. 5 Residual energy

5 Conclusion

Energy efficient GEAR to diminish the energy ingestion of sensor network has been proposed by taking consideration into entry node as well as a selection of CH is totally done on the basis of residual energy of the node and probability function. In this chapter, the network is divided into eight regions where two regions use direct communication and another six region use clustering order for communication. This displays the improved circulation of nodes in the network. The outcome displays the improved performance as compared to the existing protocol. In future, one can modify this problem statement to implement various link metrics that like ETX link metrics [16, 17].

References

1. Heinzelman WR, Chandrakasan A, Balakrishnan H (2000, January) Energy-efficient communication protocol for wireless microsensor networks. In: Proceedings of the 33rd annual Hawaii international conference on system sciences, 2000, pp 10. IEEE
2. Hussain S, Matin AW (2005) 2-tier hierarchical cluster based routing in wireless sensor networks. Tech Rep Jodrey Sch Comput Sci TR-2005-11, November, 2005
3. Ye M et al (2005) EECS: an energy efficient clustering scheme in wireless sensor network. In: Performance, computing and communications conference, 2005, IPCCC 2005, 24th IEEE International, IEEE 2005
4. Li C et al (2005) An energy efficient unequal clustering mechanism for wireless sensor network. In: IEEE international conference on Mobile Adhoc and sensor system conference, 2005. IEEE
5. Mahmood D, Javaid N, Mahmood S, Qureshi S, Memon AM, Zaman T (2013, October) MODLEACH: a variant of LEACH for WSNs. In: 2013 Eighth international conference on broadband and wireless computing, communication and applications (BWCCA), pp 158–163. IEEE

6. Barati H, Movaghar A, Rahmani AM (2015) EACHP: energy aware clustering hierarchy protocol for large scale wireless sensor networks. Wireless Pers Commun 85(3):765–789
7. Fu C, Jiang Z, Wei WEI, Wei A (2013) An energy balanced algorithm of LEACH protocol in WSN. Int J Comput Sci 10(1):354–359
8. Qing L, Zhu Q, Wang M (2006) Design of a distributed energy efficient clustering algorithm for heterogeneous wireless sensor networks. Comput Commun 29(12):2230–2237
9. Iqbal A, Akbar M, Javaid N, Bouk SH, Ilahi M, Khan RD (2013) Advanced LEACH: a static clustering-based heteroneous routing protocol for WSNs. arXiv preprint arXiv:1306.1146
10. Kotobelli E, Zanaj E, Alinci M, Bumçi E, Banushi M (2015) A modified clustering algorithm in WSN
11. Manap Z, Ali BM, Ng CK, Noordin NK, Sali A (2013) A review on hierarchical routing protocols for wireless sensor networks. Wireless Pers Commun 72(2):1077–1104
12. Min R, Bhardwaj M, Cho S, Shih E, Sinha A, Wang A, Chandrakasan A (2001) Lowpower wireless sensor networks. VLSI Design-2001, Invited Paper, Bangalore
13. Nam C-S, Jeong H-J, Shin D-R (2008) The adaptive cluster head selection in wireless sensor networks. In: IEEE international workshop on semantic computing and applications
14. Estrin D, Culler D, Pister K, Sukhatme G (2002) Connecting the physical world with pervasive networks. In: IEEE pervasive computing, pp 59–69, January–March 2002
15. Ye W, Heidemann J, Estrin D (2004) Medium access control with coordinated adaptive sleeping for wireless sensor networks. IEEE/ACM Trans Netw 12(3):493–506
16. Javaid N, Ullah M, Djouani K (2011) Identifying design requirements for wireless routing link metrics. In: Global telecommunications conference (GLOBECOM2011), 2011 IEEE, vol 1, no 5, pp 5–9 Dec 2011
17. Pan J, Cai L, Hou YT, Shi Y, Shen SX (2005) Optimal basestation locations in two-tiered wireless sensor networks. IEEE Trans Mob Comput (TMC) 4(5):458–473

A Novel Equivalent Circuit Approach for Modal Analysis of MEMS Cantilever Beam

C. Kavitha and M. Ganesh Madhan

Abstract This chapter presents a novel equivalent circuit modeling approach for the modal analysis of a MEMS cantilever beam. The modal solution is implemented in the circuit model, which provides the spatial displacement in steady-state and dynamic conditions. The sinusoidal excitation is employed to study the transient solution of mode shapes. As a test case, the steady-state displacement along the length of 40 μm cantilever is determined under different modes and compared with the results obtained by MATLAB simulation, which indicates a good agreement.

Keywords Cantilever beam · Equivalent circuit · MEMS · Modal analysis
Parallel resonator

1 Introduction

Modal analysis is a dynamic characterization of the cantilever beam, which is used to identify the natural frequencies and mode shapes. In order to increase the efficiency of the system, and to achieve deflections with low power consumption, geometrical features of the beams can be optimized. The applications include mass detection, Atomic Force Microscopy (AFM) [1], Chemical sensor [2] and to determine the vibration characteristics (natural frequencies and mode shapes) [3]. Cantilever structures can also be used as a microsensor, microactuator, microgripper, micromanipulators in different areas of industries and scientific MEMS applications. The references, Takao and Yukio [4] and Yalcin [5] report the extraction of the modal circuit parameters of the vibrator and capacitive micromachined ultrasonic transducers (CMUTs) using modal analysis based on finite

C. Kavitha (✉) · M. G. Madhan
Department of Electronics Engineering, Madras Institute of Technology,
Anna University, Chennai 600044, India
e-mail: kaviphd2011@yahoo.co.in

M. G. Madhan
e-mail: mganesh@annauniv.edu

V. Nath and J. K. Mandal (eds.), *Nanoelectronics, Circuits and Communication Systems*, Lecture Notes in Electrical Engineering 511,
https://doi.org/10.1007/978-981-13-0776-8_28

element model. Burak et al. [6] developed an experimental setup, which includes base excitation of polysilicon microcantilevers for different geometries and pressures. The modal characteristics were determined from the matrix decomposition methods which were obtained from the linear time domain modal analysis [7]. Bor Tsuen Wang and Deng-Kai Cheng [8] have developed the modal parameter extraction method for free vibration response.

The lumped equivalent circuit model for a single-cell circular capacitive micromachined ultrasonic transducer (CMUT) was presented by Mao et al. [9]. The derivation of equivalent circuit model is based on modal analysis techniques, harmonic modes, and cross-modal coupling between different Eigenmodes. However, discretizing the beam using large matrices is not a convenient scheme for implementation. Further, the computational complexity should be minimal. If the cantilever beam is discretized to '*n*' elements, '*n*' degrees of freedom are considered for each mode. Eigenvalue analysis is normally used to predict the mode shapes by modal superposition. If the number of elements increases, development of circuit for modal analysis becomes complex [10]. From the literature, it is observed that modal analysis of MEMS cantilever has been mostly carried out using FEM analysis tools such as ANSYS, ABAQUS, COMSOL, etc. Electrically equivalent circuit has been largely used for single cantilever without spatial variations. To our knowledge, only two reports focus on equivalent circuit approach for modal analysis [9, 10]. They have used a modal model with mode superposition theorem. Even though Thomas and Nenad's [10] scheme uses equivalent circuit, it also requires MATLAB tool for time domain calculations. In the case of Mao et al., their analysis is restricted to frequency domain alone.

Hence, there is a need to develop a relatively simple model with a lesser number of segments with reasonable accuracy for modal analysis of MEMS cantilever. It is also desirable to provide steady-state and transient analysis by the same circuit. An electrical equivalent circuit modeling approach for MEMS cantilever is discussed in this chapter. The detailed modal analysis of the cantilever beam under steady-state and transient condition is reported. Further, this model is simple when compared to the other existing equivalent circuit models. This approach is easy to implement without considering Eigenvalue analysis and decoupling into '*n*' degrees of freedom, for each mode. The displacement at a different position, mode shapes of different modes are obtained using circuit simulator PSpice© and compared with numerical results using MATLAB.

2 Modeling of Cantilever Beam for Modal Analysis

An Euler–Bernoulli uniform fixed-free beam, which represents the MEMS cantilever under transverse vibration condition is shown in Fig. 1.

A 40 µm rectangular MEMS cantilever beam is considered for the analysis of different modes. The beam is clamped at one end and is free at the other end. In this approach, the beam is discretized into '*n*' number of elements and '*n* + 1' number

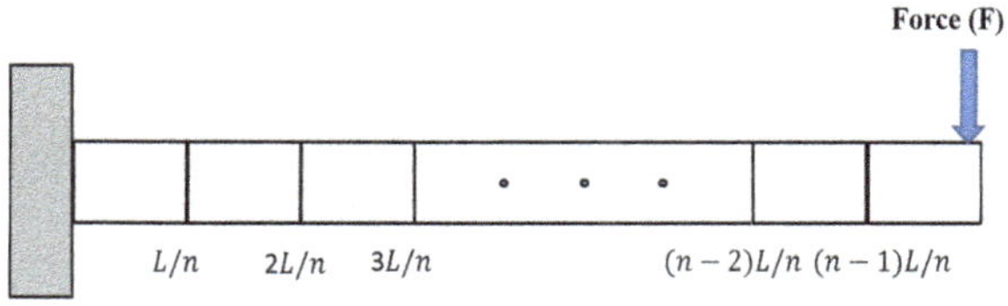

Fig. 1 Discretized cantilever beam structure with '*n*' elements and '*n* + 1' nodes

Table 1 Specifications of a silicon cantilever beam [10]

Parameters	Values
Length of the cantilever (L)	40 μm
Width of the cantilever (w)	4 μm
Thickness of the cantilever (h)	2 μm
Young's Modulus of silicon (E)	150 GPa
Density of silicon (ρ)	2330 kg/m^3
Poisson's ratio (v)	0.17

of nodes is obtained. In this work, the transverse bending modes of the cantilever beam are considered without damping. The specifications for a typical silicon MEMS cantilever beam considered for the analysis are given in Table 1.

The spring constant of the cantilever beam is given as

$$K = \frac{3EI}{L^3} \tag{1}$$

where the moment of Inertia (I) is calculated as

$$I = \frac{wh^3}{12} \tag{2}$$

and the mass of the beam is represented as

$$m = \rho V = \rho wLh \tag{3}$$

The dynamic beam equation is usually given as

$$\frac{\partial^2}{\partial x^2}\left(EI\frac{\partial^2 v}{\partial x^2}\right) + \rho A\frac{\partial^2 v}{\partial t^2} = F \tag{4}$$

where area of cantilever beam is represented as $A = wh$. Under free vibration of beam, transverse force $F = 0$. By applying the method of separation of variables, the solution becomes

$$v(x,t) = v(x)v(t) \tag{5}$$

The detailed derivation is given in [11]. The natural frequencies of different modes can be estimated from Euler–Bernoulli beam theory as

$$\omega_k = \beta_k^2 \sqrt{\frac{EI}{\rho A}} \tag{6}$$

where number of modes k = 1, 2, 3… n. Applying the boundary condition at the fixed end leads to zero displacement and zero slope. The free end cannot have a bending moment or shearing force. The general solution of the Eq. (4) is

$$\begin{aligned} v(x) = {} & C_1 \cosh(\beta x) + C_2 \sinh(\beta x) \\ & + C_3 \cos(\beta x) + C_4 \sin(\beta x) \end{aligned} \tag{7}$$

where C_1, C_2, C_3, C_4 are constants, which are obtained by using the boundary conditions. The solution for different modes is as follows:

$$v_k(x) = C_1 \left\{ \begin{array}{l} [\cosh(\beta_k x) - \cos(\beta_k x)] \\ + \left(\begin{array}{c} \left(\dfrac{\cos(\beta_k L) + \cosh(\beta_k L)}{\sin(\beta_k L) + \sinh(\beta_k L)} \right) \\ (\sin(\beta_k x) - \sinh(\beta_k x)) \end{array} \right) \end{array} \right\} \tag{8}$$

where C_1 = 1 is assumed for the mode shapes. The nontrivial solution of cantilever beam [10] is

$$\cos(\beta_k L)\cosh(\beta_k L) = -1 \tag{9}$$

Solution is obtained numerically in MATLAB, for the first six modes and values are found as $\beta_k L$ equal to 1.875, 4.694, 7.855, 10.996, 14.137 and 17.279. The frequencies of different modes of the cantilever are calculated using

$$f_k = \frac{\omega_k}{2\pi} \tag{10}$$

A detailed comparison with the literature is provided in Table 2 for highlighting the novelty of the present work.

Table 2 Comparison of literature studies

References	Technique	Inference
Ibarra et al. 2015 [12]	ANSYS and analytical methods for steel microcantilever	Modal frequencies are determined and maximum displacement is found
Liangca et al. 2015 [13]	ANSYS and analytical methods for copper and silicon microcantilever	Modal frequencies are determined
Thomas and Nenad 2013 [10]	Analytical method, finite element analysis and equivalent circuit modeling for silicon microcantilever	Modal frequencies are determined and discretized beam model is used to find the displacement along each position of the beam
Proposed method	Equivalent circuit modeling for silicon microcantilever	A novel equivalent circuit model with reduced complexity is developed for the first time. This model requires less computation time. A simple model can provide transient and frequency response

3 Circuit Modeling for Modal Characteristics

3.1 Methodology

The electrical equivalent circuit of cantilever beam, which is a parallel resonator, considered in this work, produces tip displacement of the cantilever. The normalized displacement at various positions of cantilever is obtained from the circuit comprising of polynomial voltage source model. Finally, the product of tip displacement and normalized spatial displacement gives the modal displacement at different positions and time.

The analysis of the cantilever beam involves simultaneous evaluation of transient behavior and spatial displacement as shown in Fig. 2. For evaluation of time-dependent displacement, a force is applied at the tip of the beam. The

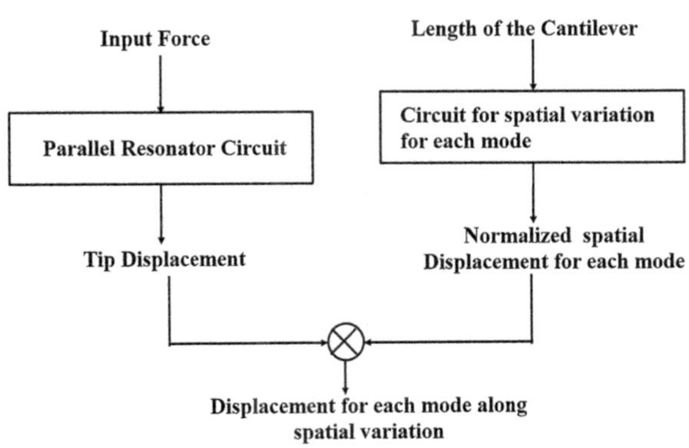

Fig. 2 Circuit implementation for the methodology

normalized displacement with respect to the cantilever tip is represented as a voltage source for each node (or) n. This parameter is evaluated for various position of the cantilever. The product of the displacement evaluated by the resonator circuit and the normalized spatial displacement provides the time-dependent displacement for various modes at various locations of the cantilever.

3.2 Circuit Implementation for the Methodology

The equivalent circuit of the cantilever is represented as a parallel resonator using force–current analogy, which is shown in Fig. 3.

The spring, mass, and damper are modeled as inductance, capacitance and resistance respectively. The equivalent equation for a cantilever is represented as [14–16]

$$C\frac{\mathrm{d}^2\varphi}{\mathrm{d}t^2} + \frac{1}{R}\frac{\mathrm{d}\varphi}{\mathrm{d}t} + \frac{1}{L}\varphi = I_\mathrm{F} \tag{11}$$

where capacitance, flux linkage, resistance, inductance and external force are represented as C, φ, R, L and I_F respectively. The damping medium is neglected and the other circuit element values of inductance and capacitance are evaluated from (1) and (3) as 0.7456 nF and 53.3 mH, respectively, for the cantilever considered in this work. The output of the parallel resonator is obtained as the flux linkage and the displacement is obtained by the product of current in inductance and the corresponding inductance value $v(t) = i_L L$. This displacement (V_D) is evaluated as a voltage drop due to the current-controlled current source (CCCS) in the circuit.

In this work, we have considered 4 elements and 5 nodes for a 40 μm beam. Each node is modeled as a DC voltage source corresponding to the distinct position in the beam and the displacement at different positions for one mode is determined as per (8), which is modeled as voltage-controlled voltage source (VCVS-i.e., $E1$, $E2$, … En) and shown in Fig. 4.

This model has to be implemented for all the elements. The displacement at various positions and times $V(x, t)$ is evaluated using a VCVS (second-order polynomial), which is dependent on $V(x)$ and $V(t)$ as shown in Fig. 5.

Figures 3, 4, and 5 have to be combined and used for different modes by changing the $\beta_k L$ values and the corresponding modal frequencies. This approach provides the transient spatial displacement under different modes.

Fig. 3 Parallel resonator circuit model for cantilever tip displacement

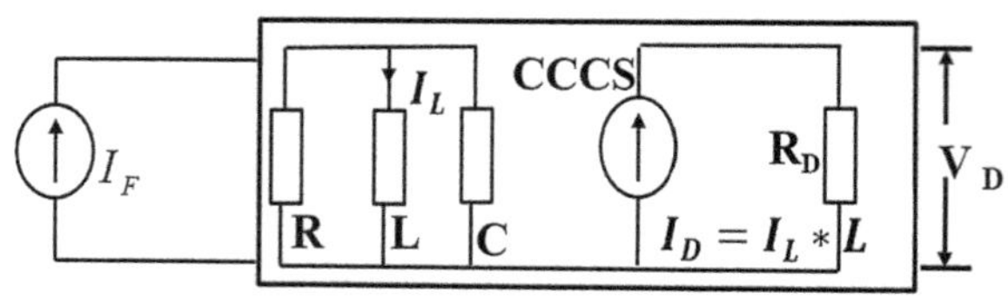

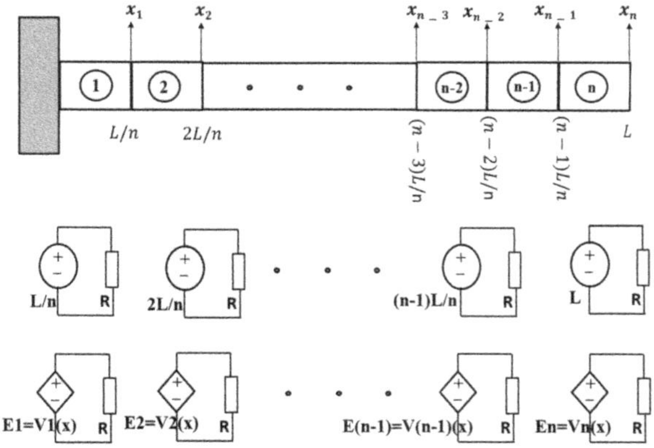

Fig. 4 Equivalent circuit model for the segmented cantilever beam

Fig. 5 Displacement at different positions for various modes

4 Simulation Results

The analysis is carried out under steady-state and transient conditions by simulating the complete electrical equivalent circuit in PSpice©. The spatial variation (modes) with respect to the corresponding frequency is also determined for sinusoidal input, without damping and bias voltage.

4.1 *Steady-State Response*

The steady-state analysis is performed by calculating the displacement at a different position due to force equivalent current applied to the cantilever beam. The displacement at various positions ($V(x)$) as per (8) is determined.

The displacement at various positions, obtained using PSpice and MATLAB are compared, for first three modes and given in Fig. 6a.

The results exactly match with each other thereby validating our simulation approach. The displacement is calculated for various positions of cantilever along the length of the beam, which is independent of time. By applying a constant external DC current value equivalent to force of 7.31434 nN at the tip of the beam, the corresponding displacement is found as 389.854 pm (≈0.4 nm). By varying the position of the cantilever beam, displacement $V(x)$ is determined for six modes and

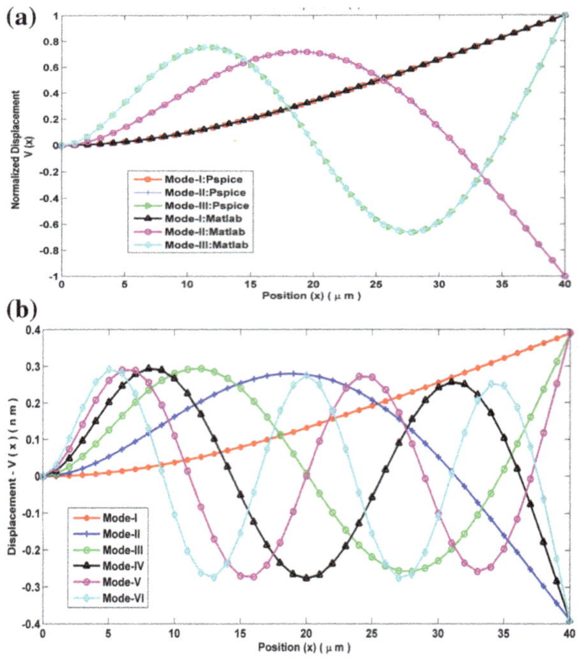

Fig. 6 Displacement $V(x)$ at various positions of the beam for six different modes **a** comparison of three modes with PSpice and MATLAB **b** position-dependent displacement for six modes

Table 3 Values of displacement versus force at 7.3143 nN for three modes

Displacement $V(x)$ at I_F = 7. 3143 nN values				
Modes	X = 10 μm (pm)	X = 20 μm (pm)	X = 30 μm (pm)	X = 40 μm (pm)
Mode-I	37.9235	132.352	256.404	389.825
Mode-II	162.666	278.2275	52.646	−389.8185
Mode-III	282.451	7.642	−226.655	389.9445

shown in Fig. 6b. The values of displacement for 7.3143 nN force for three modes are obtained and given in Table 3.

4.2 *Transient Response*

A sinusoidal external input force of 7.31434 nN is applied at the corresponding modal frequencies of 1.62, 10.15, and 28.43 MHz and the resonator response (tip displacement), for first three modes are shown in Fig. 7.

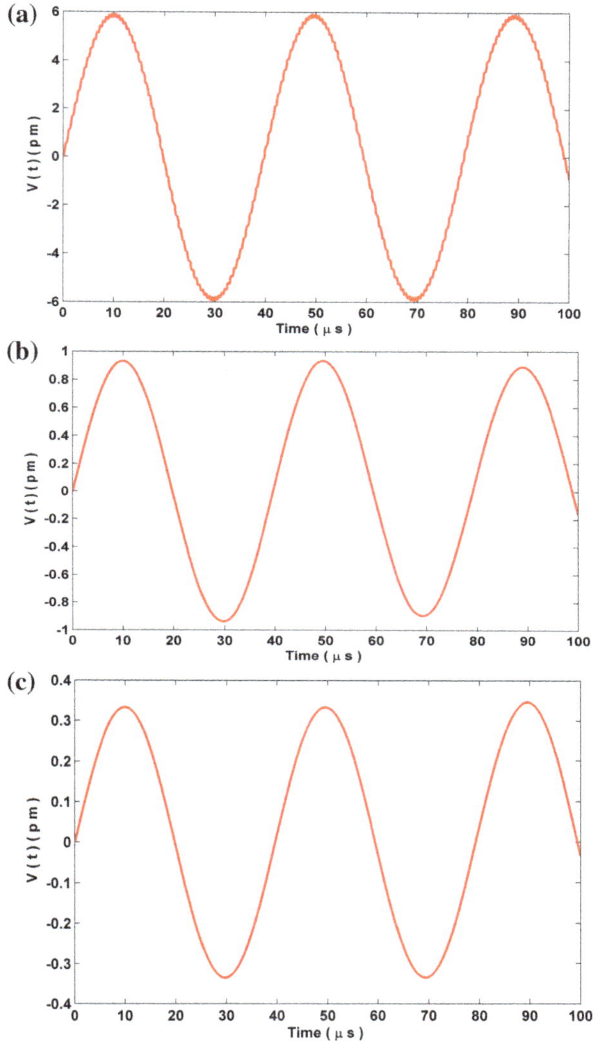

Fig. 7 Displacement *V* (*t*) with the corresponding modal frequency **a** Mode-I **b** Mode-II **c** Mode-III

Each mode displacement is measured at various times under different positions of the cantilever beam. Each mode vibrates with a different frequency and produces in phase or out of phase displacement depending on the position of the beam. The beam motion is complex and the displacement is shown at various positions of the cantilever (10, 20, 30, 40 μm) in Fig. 8.

The beam moves up and down from the rest position due to the applied external force. The first mode transient response shows that the displacement is increasing with the position of the beam and all are in phase with each other as shown in Fig. 8a. In the second mode, beam starts moving upward up to the half of the length and starts decreasing and moves in the opposite direction. The obtained beam

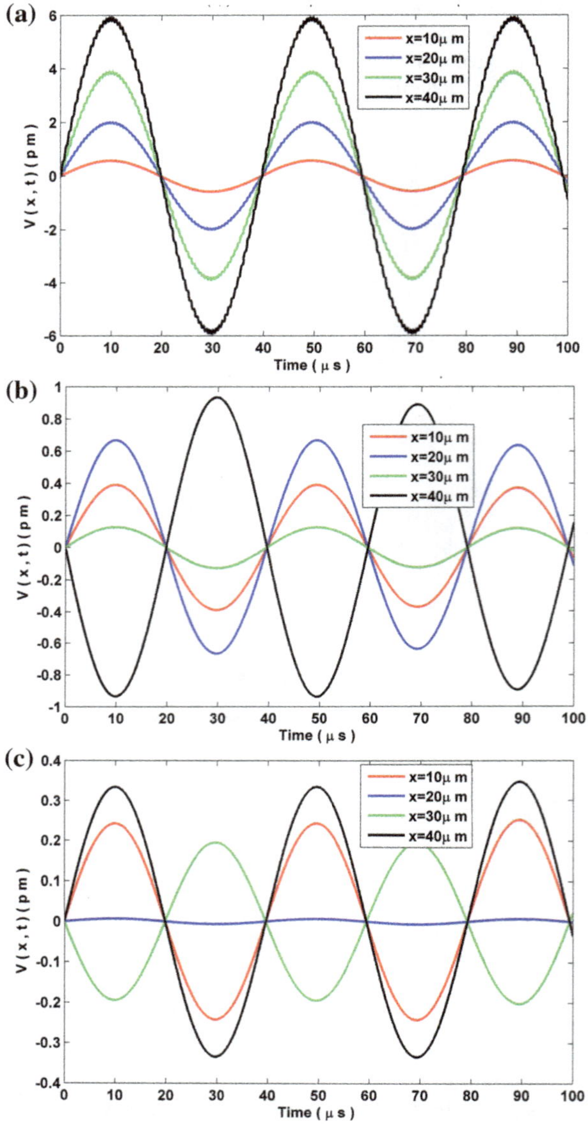

Fig. 8 Displacement $V(x,t)$ with the corresponding modal frequency **a** Mode-I **b** Mode-II **c** Mode-III

Table 4 Maximum displacement values at different positions for various modes

Positions (μm)	Mode-I (pm)	Mode-II (pm)	Mode-III (pm)
10	0.575	0.391	0.242
20	2.01	0.67	0.00655
30	3.91	0.1265	−0.194
40	5.95	−0.936	0.334

displacement at the middle and tip of the beam is out of phase, as shown in Fig. 8b. These observations are in agreement with the Ref. [10]. In the third mode, mid of the beam produces almost zero displacement and at the position of 10 and 30 μm, the beam produces out of phase displacement. The maximum displacement is measured at 10, 20, 30, and 40 μm and given in Table 4. The magnitude of displacement decreases for increasing modes.

Each mode is activated with its corresponding natural frequency with a sinusoidal input. The displacement is measured along the length of the beam during the transient. The movement of the beam is calculated with respect to rest position. The first three modes of vibration are plotted for various time scales and is given in Fig. 9. The natural frequency increases for each mode. Each characteristic mode vibrates with a different shape and frequency. For higher modes, the beam motion also becomes complex and exhibits the behavior as reported in Ref. [11].

5 Conclusion

Modal analysis of the MEMS cantilever beam is examined with a novel electrical equivalent circuit approach and simulating using PSpice©. The steady-state and transient responses have been obtained by varying the input force and identifying the displacement along the length of the beam. The modal beam displacement decreases with the increasing the number of modes as per the theory. The advantage of this model is simple implementation compared with Eigenanalysis, matrix decoupling of equations to '*n*' degrees of freedom methods.

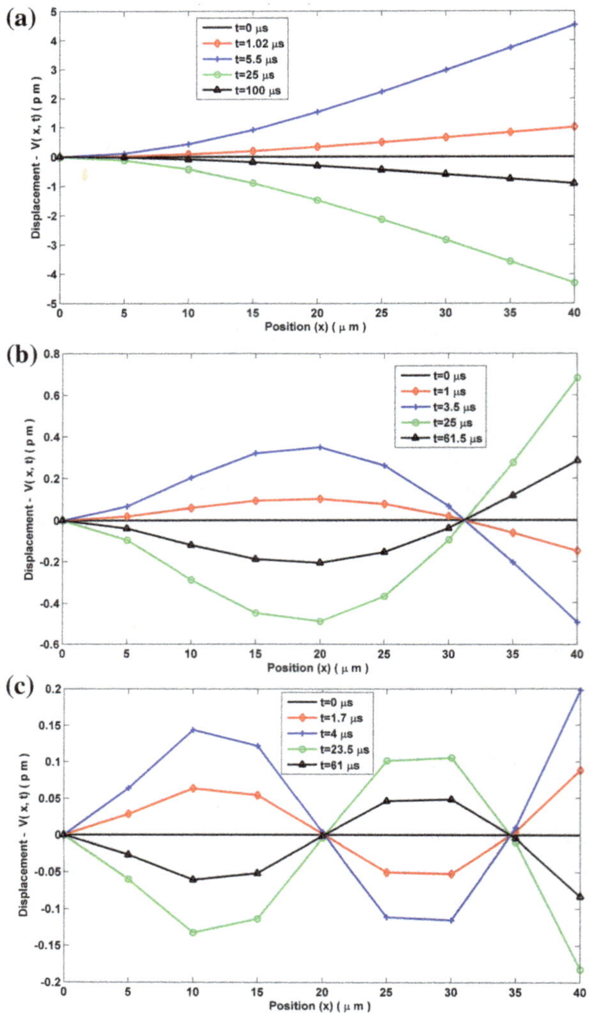

Fig. 9 Displacement for first three modes at various times along the length of the beam **a** Mode-I **b** Mode-II **c** Mode-III

References

1. Lobontiu N, Garcia E (2004) Two microcantilever designs: lumped-parameter model for static and modal analysis. J MicroElectroMechanical Syst 13(1)
2. Gibson RF, Liu Z, Srinivasan N (2005) Influence of bending-extension coupling on the modal frequencies of nonsymmetrically laminated MEMS microcantilevers. J MEMS 14(6)
3. Chatrathi K, Packirisamy M, Stiharu I, Nerguizian (2006) Effect of curvature on dynamic behavior of cantilever MEMS. IEEE ISIE 2006
4. Tsuchiya T, Kagawa Y (2004) On the passive vibration damping by piezoelectric transducers with inductive loading. In: Systems modeling and simulation: theory and applications. AsiaSim 2004. Lecture notes in computer science, vol 3398. Springer, Berlin, Heidelberg

5. Yamaner FY (2006) Finite element and equivalent circuit modeling of capacitive micromachined ultrasonic transducer (CMUT). Dissertation of Master of Science, Sabanci University Summer 2006
6. Ozdoganlar OB, Hansche BD, Carne TG (2005) Experimental modal analysis for microelectromechanical systems. Exp Mech 45:498–506, December 2005
7. Liang Z, Inman DJ (1990) Matrix decomposition methods in experimental modal analysis. J Vib Acoust 112:410–413
8. Wang B-T, Cheng D-K (2008) Modal analysis of mdof system by using free vibration response data only. J Sound Vib 311:737–755
9. Mao SP, Rottenberg X, Rochus V, Czarnecki P, Helin P, Severi S, Nauwelaers B, Tilman HAC (2017) Modal analysis based equivalent circuit model and its verification for a single cMUT cell. J Micromech Microeng 27
10. Jones TB, Nenadic NG (2013) Electromechanics and MEMS Ist edn. Cambridge University Press
11. Whitney S (1999) Vibrations of cantilever beams: deflection, frequency, and research uses
12. Ibarra-Villegas FJ, Ortega-Cisneros S, Moreno-Villalobos P, Sandoval-Ibarra F, del Valle-Padilla JL (2015) Analysis of MEMS structures to identify their frequency response oriented to acoustic applications. Superficies y Vacío 28(1):12–17, marzo de 2015
13. Xiong L, Zhou Q, Wu Y, Chen P (2015) New laser excitation method for modal analysis of microstructure. Mech Syst Sig Process 50–51:227–234, January 2015. https://doi.org/10.1016/j.ymssp.2014.05.012
14. Veijola T, Kuisma H, Lahdenpera J, Ryhanen T (1995) Equivalent-circuit model of the squeezed gas film in a silicon accelerometer. Sens Actuators A 48:239–248
15. Kavitha C, Madhan MG (2016) Study of squeeze film damping characteristics under different gas mediums in a capacitive MEMS accelerometer. J Braz Soc Mech Sci Eng 38:241–252
16. Kavitha C, Madhan MG (2017) An analysis of electrostatically actuated microvibrating structures incorporating squeezed film damping effect using an electrical equivalent circuit. J Braz Soc Mech Sci Eng 39:925–936

Energy-Efficient Data Delivery in Green Cloud Networks

M. Karuppasamy and S. P. Balakannan

Abstract Nowadays due to the increase in the cloud computing services, the energy consumption has drawn huge focus in the high-performance computing research community. With increasing utilization of Cloud network infrastructure, the energy consumption has drastically increased to the cloud networks. This has become a critical issue. At the time of data delivery in the cloud, network components consume enormous amounts of energy and connecting to sky-scraping operational expenses and carbon footsteps to the environment. The energy problem in the cloud networks is the key difficulty of relation with network data delivery. Also, it is the main objective to be solved in the cloud computing networks design. The energy-efficient related work in cloud networks combines with characteristics of the network and analyzes the approach energy consumption. The energy-efficient strategies from the aspect of cloud networks node and cloud networks are discussed from the following aspects of data size, data delivery time, bandwidth size and the delivery path between clients to and the server. The aim is to achieve an energy-efficient proportional green cloud network to provide on-demand quality of services (QOS) with energy-efficient data delivery algorithm.

Keywords Cloud computing · Green computing · Environment
Virtualization · Energy

1 Introduction

In cloud computing, a number of computers, servers, and communication devices are interconnected that provide on-demand resources as per client request. Distributed, parallel, and grid computing collectively emerged as cloud computing.

M. Karuppasamy · S. P. Balakannan (✉)
Kalasalingam Academy of Research and Education, Krishnankoil, Tamil Nadu, India
e-mail: balakannansp@gmail.com

M. Karuppasamy
e-mail: karuppasamy.m1987@gmail.com

V. Nath and J. K. Mandal (eds.), *Nanoelectronics, Circuits and Communication Systems*, Lecture Notes in Electrical Engineering 511,
https://doi.org/10.1007/978-981-13-0776-8_29

Cloud Computing is the up-and-coming technology, where the services are provided on demand over the Internet [1]. Established Information Communication Technology industry from side to side serving developers and companies to triumph over is lacking in hardware capacity similarly CPU, Memory, and Storage by allows the user to access on-demand resources from side to side the Internet. The cloud computing on-demand service models are like Infrastructure as a Service (IaaS), Platform as a Service (PaaS), and Software as a Service (SaaS) [2]. However, a reliable and efficient resource allocation has been a challenge in cloud computing that ensures the quality of service to users. Owing to emerging field cloud computing is facing many problems such as bandwidth, execution delay, response time as well as the security-related issues [3].

Cloud computing network spends more bandwidth for the delivery rate. This can be a high cost for smaller applications but can be considerably elevated for the data-intensive applications. To delivering the demanding and multipart of data, the network requires sufficient bandwidth for delivery data [4]. In cloud computing, we have focused simply on the energy consumed within the information hub, however not targeted that transporting information between information centers and residential computers will consume even larger amounts of energy than storing it [5]. In cloud computing, network infrastructure consumes more power to transfer the information. So shipping dominates the overall energy utilization to the highest degree increases the electricity. Particularly, energy for transfer may be as little as 10% and 25% at stumpy usage tiers for personal and public garage services correspondingly, and almost 60% and 90% at high utilization degrees [6].

The energy-efficient data delivery needs are common to many other application environments such as gird, parallel, distributed, and cloud computing. In cloud network, data delivery is associated with a certain predefined set of QoS. Cloud network data delivery is a narrative set of connections information distribution system [7]. The set up data deliveries built on the model of information delivery from source to destination. Cloud network data delivery is designed to provide guaranteed and accurate delivery of data over cloud computing network environments [8].

QoS is the ability to make available different precedences to various applications, users, or else data flows, otherwise to assurance a convinced intensity of the performance. A QoS criterion is various and is extremely reliant (throughput, delay, jitter, loss rate) of the application [9]. There is a most important problem connected with the management of cloud services that result in a catastrophe. With the increasing trend of the cloud services, it becomes more difficult to investigate the QoS for cloud [10]. The main challenge is to solve scalability and the dependability issues in managing the QoS in Cloud Computing.

1.1 Related Work

An important research attempt has been found in latest existence on managing records shipping, direction failures, congestion, and strengthen green routing in the cloud network. In AODV, source nodes generate the packets and route the packets to the destination node. If any malfunction occurs in route stale and it affects in any one of the node, it delivers the packets by using the packet copy transmission [11]. The backup routing scheme is based on AODV. In AODV-BBS, every node continues to neighborhood hop for locating the packets. The numerous continuation of different paths is hard, luxurious, and sustained, which turns over to reduce the network efficiency [12].

When the site visitors increase at the time of the routing, then the protocol uses a number of routing paths at the same time, the event of congestion occurrence or the data flow direction is failure and due to these reasons the network performance needs to be improved [13]. The multipath routing is used to route the data packets and break up site visitors and it mitigates congestion, if any nodes in congestion the adaptive routing protocol is used to route the packet to mitigate congestion. But these problems affect the communication throughput [14].

A two-hop routing protocol is designed for if any node desires to broadcast the information, it extremely hold direction-finding table, that incorporates path of hop remoteness nodes. If any path is established, information packets are introduced; in any other case, spanking new path detection method will be initiated. If route malfunction arises in-between nodes, packet buffering begins in physical memory module [15].

Existing pull-based information delivery approaches are often classified on many proportions. The primary proportion is that the objective of the matter. The target is subsequently to optimize the utility; by a utility, we tend to represent a number of clients specified maneuver to live the worth of an entity to a client, supported a metric corresponding to information recency are significance to the client [16]. The second proportion is that the constraints of the issue. Constraints area unit precincts, e.g., bandwidth, to that the model ought to adhere. A third proportion is when substance area unit invigorated, whichever within the background, and on-demand once clients request them, or a few arrangements of the two [17].

1.2 Proposed Method

1.2.1 Green Cloud Computing

Green cloud is environmentally responsible and eco-friendly use of computing and its resources. The core green clouds are virtualization, power optimization, efficient resource utilization. Energy consumption will come down during the computing resources for the duration of hit the highest point operation and accumulate the energy for the period of idle operation, and also shrink hurtful possessions of computing resources and shrink the computing wastes [18].

1.3 Green Cloud Network Virtualization

Green Cloud Network assigns the control along with tracking of a whole computing network as it comes under the solitary organizational being from a software-based administrator's console. Green Cloud Network virtualization is considered for the public network optimization of bandwidth, elasticity, scalability, consistency, and protection [19]. It automates numerous network organizational responsibilities, which essentially envelops up a network's true difficulty. Each and every network server along with services are well thought-out to one group of resources, which may possibly be used not including to think about to the hardware components. Green Cloud Network virtualizations are increased in usage of unpredictable network hardware components [20]. Green Cloud Network virtualization involves data transfer rate into independent channels, it may be assigned, or reassigned, for separate servers or network devices on on-demand. The Green Cloud networking is a term which is mainly used to trim down power utilization and improve the efficient makes use of network resources in cloud networking environments [21].

1.4 Utilization-Based Energy Model

In a typically virtualized network, there exists a multi-layered hardware stack, which consists of physical devices, hubs, switch, bridge, gateways, network interface cards, and modems.

The hardware framework in the virtualized network for the convergence of representation, the energy components as $N = \{\text{Hubs, Switch, Bridge, Gateways, Network Interface Cards, Modems}\}$ typically, the energy consumption of a network is formatted as

$$\text{ENC} = \text{Estatic} + \mathring{a}(\text{DNEn} + \text{UNEn}) \tag{1}$$

where ENC is Energy utilized by the network components Estatic is the fixed energy consumption when there is no workload on it. UNE is the utilization of each kind of network components DNE is dynamic energy coefficient. When a network is virtualized, its energy model can be rewritten as

$$\text{ENC} = \text{Estatic} + \mathring{a}\,\text{M}\,\text{ENCi} \tag{2}$$

where ENCi is the dynamic energy consumption of NCi (network components), M is the active network components (NC) number on the network. As the network components (NC) cannot be connected by hardware energy meter, their actual energy consumption ENCi should be measured in an indirect way. The most mentioned per network components energy model is as follows:

$$\text{ENCi} = \frac{\text{Estatic}}{M} + Wi * \mathring{a}(\text{DNEn} + \text{UNEn}) \tag{3}$$

1.5 System Architecture

In this green cloud network architecture, S represents source node, D represents Destination node, R represents Router, and L represents a link between each node. In each path created from sources to destination, bandwidth allocation to each path is based on the energy-efficient data delivery algorithm. After successful allocation, the network components which are not used or idle can be switched off to save the energy from switched off and unutilized network components (Fig. 1).

1.6 Network Energy Reduction Scheme

Network energy reduction can be done by sinking for how long time of network components is utilized, decrease the operational data transfer rate, or still switched off the network components. Network energy reduction scheme, is approximately divided into time, data rate (bandwidth), network components [22]. Time domain

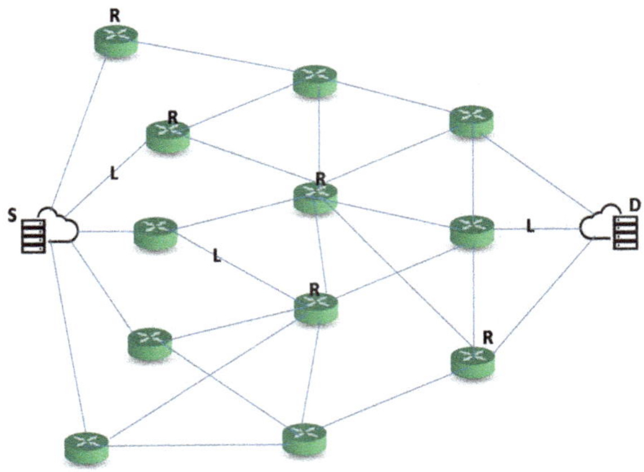

Fig. 1 Green cloud network

solutions in cloud network temporally switched off network components within a cloud network, when in particular time there is no data transfer in the downlink and signal-free symbols, turning off network components; Bandwidth reduction procedure embraces the bandwidth through the downlink transmit consignment [23]. In order to maintaining the equal energy spectral density, lesser bandwidth requires not as much of radiated power. If the downlink transfer is stumpy, the channel bandwidth can be shrinked with the purpose of less power utilization. There are mainly two ways to reduce the bandwidth: minimize the bandwidth but maintain the delivery service frequency; or minimize bandwidth and modify the delivery service regularity [24]. Reducing the number of network components is the most frequently used energy reduction method in the cloud network. Network Energy Reduction mechanism is used, when the data transfer loading unit is low, the unutilized network components are switched off [25].

1.7 Energy-Efficient Data Delivery Algorithm

In cloud network communications phase, one of the major troubles is the energy consumption. The energy-efficient data delivery algorithm used to increase the bandwidth utilization and decrease the unutilized network components in the green cloud network. The main objective of green cloud network consider with high data rate (bandwidth) for the application and continuously monitoring the network components for energy saving in the cloud network.

1. To identify resources and available resources
2. To identify sources and destination
3. To identify the shortest path
4. To find arrival task size
5. To identify the particular resources for task work processing time
6. To assign the task to the particular resources using a fitness function {Bandwidth} > $\sum$ Bandwidth (requirement for individual tasks)
7. To reallocate the resources according to the task processing time
8. To unutilized resources to be switched off
9. To repeat the step-1 to step-9 for further allocation.

2 Simulation Results

The simulation is done with a normal network and green cloud network designed with 200 systems and 25 routers connected across (Figs. 2, 3 and Tables 1, 2).

Fig. 2 Normal networks and power utilization

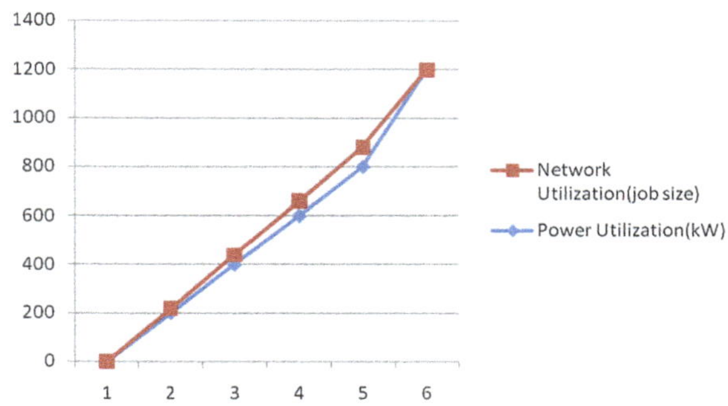

Fig. 3 Green cloud network and power utilization

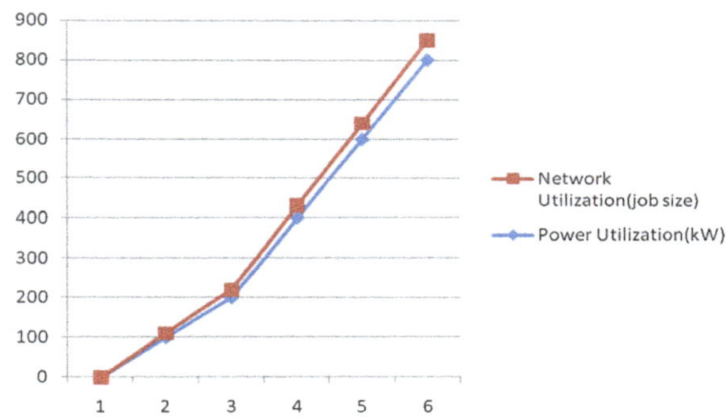

Table 1 Network energy consumption

Network components name	No. of network components	Energy (in terms of power consumption approximately)
Router	25	15.4 * 25 = 385 W (in h) * 24 (h) = 9240 W * 365 (days) = 3372600 W (year)

Table 2 Green cloud network energy consumption

Network components name	No. of network components	No. of network components kept switched off	Energy (in terms of power consumption approximately)	
			On stage	Off stage (saving per h)
Router	25	10	10 * 15.4 = 154 W	10 * 15.4 = 154 W

3 Conclusion

The energy-efficient data delivery has achieved the objective of minimizing the no. of active cloud network components, humanizing the cloud network components consumption and reduced power consumption. The energy-efficient data deliveries have to be achieved with the least amount of active cloud network components as well as by means of switching off the inactive cloud network components. An energy-efficient data delivery algorithm employed for resources allocation in the energetic cloud environment to reduce energy consumption but not in a straight line to decrease the number of cloud network utilization. These research efforts are done in the direction of energy-aware resources allocation as cloud network environment have become energy hungry and a major source of CO_2 emissions. The most important endeavor is to decrease the energy consumption cloud network exclusive of degrading network performance and reduces the CO_2 emission. In this work area, various open issues are also discussed which can be taken up for further research.

References

1. Buyya R (2013) Introduction to the IEEE transactions on cloud computing. IEEE Trans Cloud Comput 1(1), January–June 2013
2. Chiang Y-J, Ouyang Y-C, Hsu CHR (2015) An efficient green control algorithm in cloud computing for cost optimization. IEEE Trans Cloud Comput 3(2), April/June 2015
3. Beloglazov A, Abawajy J, Buyya R (2011) Energy-aware resource allocation heuristics for efficient management of data centers for cloud computing 0167-739x/$ © 2011 Elsevier B.V. All Rights Reserved. https://doi.org/10.1016/j.future.2011.04.017
4. Hassan MM, Kamruzzaman SM, Alamri A, Almogren A, Alelaiwi A, Alnuem M, Islam M, Razzaque A (2014) Design of an energy-efficient and reliable data delivery mechanism for mobile ad hoc networks: a cross-layer approach. Concurrency Comput Pract Experience Concurrency Comput Pract Exper (2014) Published Online in Wiley Online Library (wileyonlinelibrary.com). https://doi.org/10.1002/cpe.3309
5. Cordeschi N, Amendola D, De Rango F, Baccarelli E (2014) Networking-computing resource allocation for hard real-time green cloud applications 978-1-4799-6606-6/14/$31.00 ©2014 IEEE
6. Xiong N, Han W, Vandenberg A (2011) Green cloud computing schemes based on networks: a survey. Published in IET communications received on 9th April 2011 revised on 19th July 2011. https://doi.org/10.1049/iet-com.2011.0293
7. Muhammad A, Nasser N (2015) SLA-based optimization of energy efficiency for green cloud computing. 978-1-4799-5952-5/15/$31.00 ©2015 IEEE
8. Boru D, Kliazovich D, Granelli F, Bouvry P, Zomaya AY Energy-efficient data replication in cloud computing datacenters. Cluster Comput https://doi.org/10.1007/s10586-014-0404-x. Received: 18 Mar 2014. Revised: 4 July 2014. Accepted: 24 Sept 2014 © Springer Science + Business Media New York 2015
9. Ardagna D, Casale G, Ciavotta M, Pérez JF, Wang W (2014) Quality-of-service in cloud computing: modeling techniques and their applications Ardagna et al. J Internet Serv Appl. http://www.jisajournal.com/content/5/1/11

10. Garg SK, Versteeg S, Buyya R (2013) A framework for ranking of cloud computing services. Future Gener Comput Syst 29:1012–1023 (2013) © 2012 Elsevier B.V. All rights reserved. https://doi.org/10.1016/j.future.2012.06.006
11. Huang T-C, Huang S-Y, Tang L (2010) AODV-based backup routing scheme in mobile ad hoc networks. In: Proceedings of the 2010 international conference on communications and mobile computing—vol 03, CMC'10. IEEE Computer Society, Washington, DC, USA, 2010, pp 254–258
12. Lee SJ, Gerla M (2001) Split multipath routing with maximally disjoint paths in ad hoc networks. In: IEEE international conference on communications (ICC), vol 10, Helsinki, Finland, pp 3201–3205
13. Ela Kumar RNR (2011) THR: a two-hop look ahead with packet buffering protocol for MANET. Int J Inf Technol Knowl Manag 4:109–112
14. Bright L, Raschid L (2002) Using latency-recency profiles for data delivery on the web. In: Proceedings of very large data bases
15. Cherniack M, Galvez E, Franklin M, Zdonik S (2003) Profile-driven cache management. In: Proceedings of 19th international conference on data engineering (ICDE)
16. Kim D, Jung J, Lee H, Yi Y (2015) Energy-efficient sensing data delivery for low power environmental sensors
17. Andres F, Boulos J The data delivery system. NACSIS R&D Center 3-29-1 Otsuka, Bunkyo-ku Tokyo//2 Japan member of IFATEC/Euriware, 12-14 rue du Fort de St-Cyr Montigny Le Bretonneux
18. Valentini GL, Khan SU, Bouvry P (2013) Energy-efficient resource utilization in cloud computing
19. Jain N, Choudhary S (2016) Overview of virtualization in cloud computing. Colossal Data Analysis and Networking (CDAN), https://doi.org/10.1109/cdan.2016.7570950 IEEE
20. Franco Callegati Walter Cerroni Chiara Contoli Giuliano Santandrea Performance of Network Virtualization in cloud computing infrastructures: The OpenStack case. https://doi.org/10.1109/cloudnet.2014.6968981978-1-4799-2730-2
21. Fabio Baroncelli Barbara Martini Piero Castoldi. Network virtualization for cloud computing. https://doi.org/10.1007/s12243-010-0194-y. Springer 0003-4347
22. Chen T, Yang Y, Kim H, Horneman K (2011) Network energy saving technologies for green wireless access networks. 1536-1284/11/$25.00 © 2011 IEEE IEEE Wireless Communications, October 2011
23. More NS, Ingle RB (2017) Challenges in green computing for energy saving techniques. 978-1-5090-3404-8/17/$31.00 ©2017 IEEE
24. Sun X, Ansari N (2017) Green cloudlet network: a distributed green mobile cloud network. 0890-8044/17/$25.00 © 2017 IEEE IEEE Network, January/February 2017
25. Lin T, Alpcan T, Hinton K (2017) A game-theoretic analysis of energy efficiency and performance for cloud computing in communication networks. IEEE Syst J 11(2), June 2017

Electroencephalogram (EEG) Signal Analysis for Diagnosis of Major Depressive Disorder (MDD): A Review

Shalini Mahato and Sanchita Paul

Abstract Depression or Major Depressive Disorder (MDD) is a psychiatric disorder. It is the major contributor to overall global burden of disease. Any deterioration in brain functioning is reflected in Electroencephalogram (EEG) signals. EEG signals are highly complex, nonstationary and nonlinear. So, it is very difficult to analyze visually and identify changes in the waveform in order to classify MDD and normal subjects. Thus, computerized quantitative EEG is used for the analysis of signals. Support Vector Machine (SVM) using band power feature reported an accuracy of 98.33% and using Kernel Eigen-Filter-Bank Common Spatial Patterns (KEFB-CSP) gave an accuracy of 91.67% [1, 2]. Logistic Regression (LR) using band power feature reported an accuracy of 98.33%, using 4 nonlinear features combination provided an accuracy of 90%, using wavelet transform provided an accuracy of 87.5% and using only alpha power gave an accuracy of 73.3% [1, 3, 4]. Naïve Bayesian (NB) using band power feature provided an accuracy of 96.8% [1]. Artificial Neural Network using Relative Wavelet Energy (RWE) reported an accuracy of 98.11%, using power spectrum feature gave accuracy of 84% and using Lep–Ziv complexity accuracy of 60–80% was reported [5, 6, 7]. Linear Discriminant Analysis (LDA) reported an accuracy of 91.2% using SASI (Spectral Asymmetry Index) and DFA (Detrended Fluctuation Analysis) [8]. Decision Tree provided an accuracy of 80% using EEG band power as feature [9]. The study reveals that, in general, high classification accuracy is achieved by SVM, LR and ANN and highest classification accuracy of 98.33% is achieved by SVM. Highest accuracy is achieved by SVM because it is more robust and computationally more efficient due to maximal margin gap between separating hyper planes and kernel trick. The study gives some ideas which could be helpful for guiding and improving future researches. Since linear and nonlinear method for feature extraction are both equally efficient. So any of the linear/nonlinear feature can be used. For feature

S. Mahato (✉) · S. Paul
Department of Computer Science and Engineering,
Birla Institute of Technology, Mesra, Ranchi 835215, India
e-mail: swarup.shalini@gmail.com

S. Paul
e-mail: sanchita07@gmail.com

V. Nath and J. K. Mandal (eds.), *Nanoelectronics, Circuits and Communication Systems*, Lecture Notes in Electrical Engineering 511,
https://doi.org/10.1007/978-981-13-0776-8_30

selection and reduction, Genetic Algorithm (GA), Principal Components Analysis (PCA) and Linear Discriminant Analysis (LDA) can be used. Since SVM, LR and ANN give high classification accuracy and any of them can be used or any hybrid technique like GA-SVM, GA-ANN can be used. This chapter compares various EEG signal analysis techniques, compares their accuracy and methodology used and finally recommends the most suitable technique based on the accuracy for detection of depression. The chapter also summarizes some of the key finding related to EEG features based on present state of art. These results could be helpful for guiding and improving the future research in this area.

Keywords Major Depressive Disorder (MDD) · Electroencephalogram (EEG) Biomarker · Quantitative EEG · Band power · Spectral Asymmetry Index (SASI) Relative Wavelet Energy (RWE) · Higuchi's Fractal Dimension (HFD)

1 Introduction

According to World Health Organization (WHO), depression is the major cause of disability globally with respect to Years Lived with Disability (YLDs). Around 322 million people are suffering from depression globally. Females (5.1%) are more affected by depression as compared to males (3.6%). Rate of increase of number of people suffering from depression is 18.4% from 2005 to 2015 [10]. It is predicted by WHO that depression will be the leading cause of disease burden globally by 2030 [9].

A person suffering from MDD is not interested in usual activities and pleasure along with four additional symptoms of depression for at least continuous 2 weeks [11]. A person affected from depression suffers from setback at occupational, social, and personal life. At worse, a person suffering from depression can even commit suicide. Only less than 50% of the depression patients get treatment because of incorrect diagnosis, insufficient resources, lack of trained medical care providers and societal disgrace associated with mental disorder. At times, it happens that a person suffering from depression is not detected and unaffected person is wrongly identified and prescribed antidepressant.

The major reason behind wrong diagnosis is because of absence of any accepted biomarkers for MDD or any other psychiatric disorder. Diagnosis of depression is done on the basis of accepted depression classification criteria like Diagnostic and Statistical Manual of Mental Disorders-Fourth Edition (DSM-IV), Beck Depression Inventory (BDI), and Hamilton Depression Rating Scale (HAM-D), which depends on some questionnaire asked and observed behavior during the interview session between MDD patient and health practitioner.

No objective criteria exist for the assessment of depression presently. Analysis on the basis of objective criteria would be helpful in detection of depression in early stages so that it can be prevented from becoming severe and reaching untreatable state.

Change in bioelectrical activity of the brain is expected whenever there is any deterioration in brain functioning. This change can be analyzed by brain screening, which can be done with the help of EEG. EEG is cost-effective and easily available. It also provides high temporal resolution as the time scale of EEG is in milliseconds which is same as the time frame of neural activity.

Since EEG signals are highly complex, nonstationary and nonlinear, manually interpreting the change in the signal for classifying normal and depression subjects is a complex task. So computer-aided signal processing is needed to automatically classify depression and normal subjects. Extensive research work has been done in this field for distinguishing between MDD and normal subjects.

The study helps in identifying certain key ideas for future work. Both linear and non linear method for feature extraction can be used as they are equally efficient. The study reveals that highest classification was achieved using band power and RWE. So either of them can be used or a combination of both can be used. GA, PCA, LDA all are efficient for feature selection and reduction, so any of them can be used. Since high classification accuracy is achieved by SVM, LR and ANN any of them can be used or any hybrid technique like GA-SVM, GA-ANN can be used.

2 Quantitative EEG

First human EEG was recorded in 1920s by Hans Berger, a German neuroanatomist and psychiatrist. With the development of digital computer technology, it was possible to access and quantify correctly the exact values of many more EEG variable compared to that obtained by visual inspection of EEG signals. This lead to the development of Quantitative EEG (QEEG) It provides additional measurement or topographic display of voltage or frequency of EEG signal. It mathematically processes EEG signal to format or domain such that specific waveform component is highlighted, which gives information or numerical values related to EEG data which can be used for comparison or subsequent evaluation. QEEG technique which deals with topographic display of voltage or frequency over entire scalp is called EEG Brain Mapping. QEEG involves factors like amplitude of signal, frequency, waveform, hemispheric reactivity.

3 Methodology

EEG signal consists of information about brain activity from different regions of brain with the help of electrodes. These EEG signals may be contaminated with various kinds of artifacts like eye blinks, movements, muscular activity, and electrical noise from power line. These artifacts need to be removed with the help of software or by visual detection by experts.

After the removal of artifacts, feature extraction is carried out from the signals which are used to differentiate between depression and normal subjects.

Basically two types of methods are used for feature extraction: linear and nonlinear. In linear analysis, frequency and time domain approaches are used. Mathematical theory of dynamical nonlinear systems forms the basis of nonlinear EEG analysis.

Feature extracted with linear methods are inter-hemispheric asymmetry, band power, EEG measurements (amplitude, frequency and power), Wavelet Transform (WT), Fast Fourier Transform (FFT), Auto Regressive (AR) model, and so on. Feature extracted with nonlinear methods are such as Higuchi's Fractal Dimension (HFD), Approximate Entropy (ApEn), Lyapunov Exponent (LE), Correlation Dimension (CD), Detrended Fluctuation Analysis (DFA), and so on.

Since a large number of features are extracted from each of the channels, using all the features for classification would lead to high computational cost. Some of the researchers prefer to decrease this cost by using some of significant features and not all features. This would substantially reduce the computational cost. Some of the feature selection and reduction techniques used are Genetic Algorithm (GA), Principal Components Analysis (PCA), and Linear Discriminant Analysis (LDA).

Further, the classification is done with the help of different classifier like LR, SVM, NB, Back Propagation-Artificial Neural Network (BP ANN), LDA, self-organizing competitive network, k-means clustering, k-Nearest Neighbor (KNN), Kernel Principal Component Analysis (KPCA), Decision Tree (DT), Multilayered Perceptron Neural Network (MLPNN), Probablistic Neural Network (PNN) (Fig. 1).

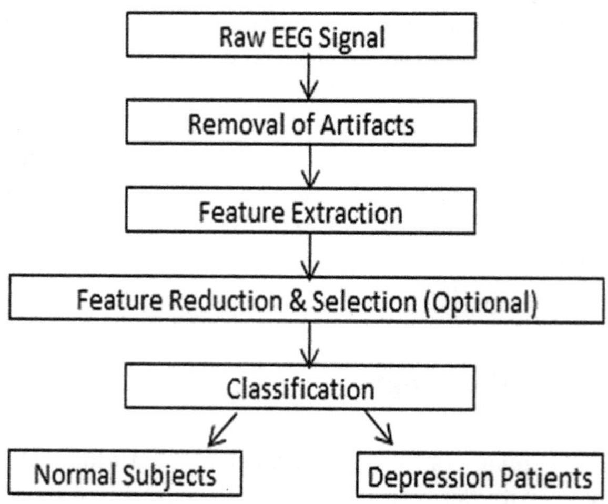

Fig. 1 Stages of EEG signal analysis for detection of depression

4 Literature Review

Classification accuracy depends to a great extent on the EEG feature used and classification techniques used. Extensive research has been done using different EEG features and different classifiers to improve classification accuracy between depression patients and normal subjects. Some of the findings related to EEG features based on present state of the art are discussed in the table.

5 Discussion

Table 1 depicts highest classification accuracy of 98.4%, which was achieved by SVM classifier using all signal power and alpha asymmetry features. LR's classification accuracy was found to be higher than LDA and KNN [3]. True detection rate of HFD (94%) was reported to be higher than SASI (88%) [12]. CD had shown the highest classification accuracy of (83.3%) among DFA, CD, HFD, LE along with LR classifier [3]. It was reported that combination of DFA, HFD, and CD with LR classifier provided high classification accuracy of (90%) (Tables 2 and 3).

It can be observed from the above analysis that there are a number of conflicting results, which are mainly due to the following reasons:

(i) Lack of standardized common dataset on MDD
(ii) Different conditions during EEG recording (EO, EC, falling sleep, emotional challenges, etc.) and different environment (heat, temperature, noise, illumination, etc.)
(iii) Different channels are used for signal analysis
(iv) EEG recording length and length of epoch are different
(v) Different procedures for artifacts removal
(vi) Different or incomparable measures used.

6 Conclusion

From the above discussion, it is evident that EEG signals are efficient and reliable in discriminating between MDD patient and normal subjects. But the major barrier in finding standardized biomarkers is the absence of standardized common dataset, techniques, and methodology used. Thus, these issues need to be solved in order to establish clinically useful and authentic biomarker.

In this review chapter, several papers relating to EEG signal analysis and depression has been summarized which would be beneficial in guiding and improving the future research work in this area such that higher accuracy, sensitivity and specificity can be achieved. This would help in the development of

Table 1 Summary of findings related to EEG signal's features

Feature	Description	Findings
Band power (linear method)	EEG signal can be split into frequency bands: delta (0.5–4 Hz), theta (4–8 Hz), alpha (8–13 Hz), and beta (13–30 Hz). Band power is the average power contained in a particular frequency band	**Alpha power**: Exceptionally high alpha activity was found in the resting state in the posterior region of the brain during depression [13, 2] but conflicting results were also found as per [4] which reported lower alpha signal power in all parts of the brain during resting state. Low left frontal alpha activity during emotional challenge (approach and withdrawal) in MDD patients [2]. Alpha power attains the highest classification accuracy for discriminating between normal subjects and MDD as compared to other bands [3, 14]. During emotional challenge (approach and withdrawal), EEG alpha asymmetry was a stronger indicator of depression than in resting state alpha asymmetry [2]. Alpha asymmetry was found to be decreased remarkably in MDD especially in women when neuro-feedback treatment was done [15]. Occipital alpha activity was found to be increased in depression patient. Right parieto-temporal region showed comparatively higher alpha activity for depression patients [13] **Beta power**: Found to be higher in depression patients in central, temporal and parietal region of brain [13, 8]. Beta power positively correlates with stress symptoms. For depression patients beta values increased only for left hemisphere [13] **Theta power**: It is found to be less in all parts of brain in depression patient [8] which was contradictory to the finding of increase of theta activity in occipital and parietal region [13]. For depression patients, there was increase in theta values in both left and right hemisphere [13] **Delta power**: It is slightly more in temporal, central and parietal region of the brain in depression [9]. But significantly higher temporal delta activity was found in Acute Unstable Depression Syndrome (AUDS) patients than MDD patients. Compared to MDD, Acute Unstable Depressive Syndrome (AUDS) had higher delta activity in right hemisphere [16] Classification accuracy higher than alpha band power was achieved, when all the bands were analyzed together [3, 14]. Increase in the current source density maximal

(continued)

Table 1 (continued)

Feature	Description	Findings
		in the right hemisphere and highest in the alpha (55.6%) then theta (30.6%) band in parietal and frontal region in depression patients [17]
SASI (linear method)	SASI is calculated as the difference between especially selected higher and lower frequency band further from which central band is eliminated	SASI values were positive for depression patients and negative for normal subjects. It is positively correlated with Hamilton Depression Scale (HAM-D) scores [7, 18]. Pz channel provided maximum difference in SASI in differentiating normal subjects and MDD patients. Using SASI, no statistically significant difference in MDD and normal subjects with respect to hemispheric asymmetry was found. SASI classifying accuracy were highest in parietal brain area [12]
HFD (nonlinear method)	Higuchi's algorithm is used to find the fractal dimension in time domain directly. Fractal dimension determines the degree of irregularity in discrete time sequence	Value of HFD increases with depression in all EEG channel. Statistically significant difference between MDD patients and normal subjects for HFD in temporal, parietal and occipital channel [19]. Using HFD, no statistically significant difference was found in MDD and normal subjects with respect to hemispheric asymmetry. HFD classifying accuracy highest in parietal brain area [12]
RWE (nonlinear method)	RWE is the relative energy related with different frequency band. It reveals the degree of disorderness in different frequency bands	Normal subjects reported higher value of RWE in gamma, beta, alpha, and theta bands as compared to depression patients. Depression patients reported higher RWE value in delta band [20]. Higher Wavelet Entropy (WE) (mean WE = 1.327) was found in depression patients than those of normal subjects (mean WE = 1.266). For depression patients, resting state WE (mean WE = 1.349) was found to be higher than in mental arithmetic state (mean WE = 1.243) [21]. Using RWE, it was found that right hemisphere has higher energy distribution in depressed patients than normal subjects [22]
ApEn (nonlinear method)	ApEn is the complexity measure which quantifies regularity and complexity in time series data. ApEn value is positively correlated with unpredictability and randomness of the signal	ApEn was much lower in MDD patients as compared to normal subjects [23]
DFA (nonlinear method)	DFA analyzes the deviation of the signal with respect to local signal's tendency in data window	In channels O2 and Pz, the difference between MDD patients and normal subjects were found to be a maximum for DFA. DFA values are significantly lower for depressive group [18]

Table 2 Summary of previous work on EEG signal analysis for detection of depression

S. No.	Ref.	Methodology	Features used	Analysis method	Accuracy
1.	[8]	Power of band and alpha inter-hemispheric asymmetry used as feature. Z-score standardization was done. Most significant feature selected using Receiver Operating Characteristics (ROC). LR, SVM and NB used as classifier	EEG band power, Inter-hemispheric asymmetry	LR, SVM, NB	98.33% (LR), 96.8% (NB), 98.4% (SVM)
2.	[4]	Power spectrum used as feature and ANN is used as classifier	Power spectrum, FFT	ANN	84%
3.	[12]	Linear SASI and nonlinear HFD used separately as feature and t-test is used for analysis	SASI, HFD	t-test	85% (for both SASI and HFD)
4.	[22]	High-frequency noise is removed using Total Variation Filtering (TVF). Classification of frequency bands was done using 6-level multi-resolution decomposition method for Discrete Wavelet Transform (DWT). RWE used for feature extraction in different decomposition level. ANN used as classifier	RWE, WE	ANN	98.11%
5.	[3]	4 nonlinear feature and 4 band power used for feature extraction. Feature selection done using GA. KNN, LDA, and LR used as classifier using both set of features separately	EEG band power, 4 nonlinear features (DFA, HFD, CD and LE)	LDA, LR, KNN	90% (LR + 4 nonlinear features), 80% (KNN + 4 nonlinear features), 86.6% (LDA + 4 nonlinear features), 70% (KNN + alpha band power), 73.3% (LR and LDA + alpha band power)
6.	[14]	Band power is used for feature extraction. LDA and GA for feature reduction and selection. DT used as classifier	EEG band power	DT	80%
7.	[24]	Cross-correlation is used for feature extraction for which 5 features extracted (Peak Value (PV), instant at which peak occurs (IP), Equivalent Width (EW), Centroid (C) and Mean Square	Cross-correlation	SVM	95.5%

(continued)

Table 2 (continued)

S. No.	Ref.	Methodology	Features used	Analysis method	Accuracy
		Abscissa (MSA)) and 5 D feature vector is formed. SVM is for classifying epilepsy patients and normal subjects			
8.	[6]	The author has proposed a new method: Kernel Eigen-Filter-Bank Common Spatial Patterns (KEFB-CSP) for feature extraction. KPCA is used for feature reduction. LDA, KNN and SVM used as classifier	Kernel Eigen-Filter-Bank Common Spatial Patterns (KEFB-CSP)	SVM	91.67% (KEFB-CSP + SVM)
9.	[20]	WT used for feature extraction. Most significant features are selected using ROC. LR used as classifier to predict the response of antidepressant's treatment outcome in MDD patients	Wavelet transform	LR	87.5%
10.	[18]	Linear SASI and nonlinear DFA used for feature extraction in Pz channel. LDA and Mann–Whitney statistical test was used as the classifier. Performance of SASI and DFA are compared and combination of SASI and DFA Also used as feature	DFA, SASI	LDA, Mann–Whitney statistical test	76.5% (SASI), 70.6% (DFA), 91.2% (SASI + DFA)
11.	[5]	Classification of schizophrenic, depression and normal subjects in 4 different states (eyes closed and 3 cognitive tasks) is done using Lep–Ziv complexity for feature extraction and BP ANN as classifier	Lep–Ziv complexity	BP ANN	60–80%

Table 3 Summary of previous works related to EEG

S. No.	Ref.	Methodology	Features used	Analysis method	Accuracy
1	[25]	DWT used for splitting signal into sub-band. Set of statistical features were used as feature. PCA, ICA and LDA used for feature reduction. SVM used for classifying individual as healthy or having epileptic seizure	Set of statistical features (Each sub-band's standard deviation, absolute values mean, wavelet average power, adjacent sub-band's ratio of absolute mean)	SVM + LDA, SVM + PCA, SVM + ICA	100% (SVM + LDA), 99.5% (SVM + ICA), 98.75% (SVM + PCA), 98% (SVM)
2	[26]	Four feature extraction methods are used: (AR) model parameters, band power, fractal dimension and wavelet energy. Bidirectional search and plus-L minus-R techniques are used to select most informative channel. GA used for feature selection. LDA and SVM used for classifying schizophrenic and normal subjects	(AR) model parameters, band power, fractal dimension and wavelet energy	SVM + bidirectional search, LDA + bidirectional search, SVM + plus-L minus-R, LDA + plus-L minus-R	99.38% (SVM + bidirectional search), 84.62% (LDA + bidirectional search), 99.54% (SVM + plus-L minus-R), 88.32% (SVM + plus-L minus-R)
3	[1]	Band power was used as feature. PCA used for dimension reduction. BP ANN, self-organizing competitive network for classifying subject to normal or schizophrenic	Band power	BP ANN, self-organizing competitive network	60–80% (BP ANN), 40–60% (Self-organizing competitive network)
4	[27]	DWT is used for splitting into sub bands. Statistical features and frequency domain features are used. GA used for feature reduction. k-means clustering used for identifying abnormal brain's EEG signal	Statistical feature (Mean, Median, Mode, Standard deviation, Maximum, Minimum) Frequency domain features (Band power, FD and Energy)	k-means clustering	Highest 83.3% (alpha band feature + k-means clustering)
5	[28]	DWT is used for splitting into sub bands. Statistical feature of each sub-band used as input to SVM classifier which detects seizure signal	Statistical feature such as variance, energy, maximum and minimum sample value in power spectral density	SVM	98.75%

(continued)

Table 3 (continued)

S. No.	Ref.	Methodology	Features used	Analysis method	Accuracy
6	[29]	ICA with 5 different methods (Linear trends, Spectral pattern, Extreme values, Kurtosis, Data Improbability) was used for artifact removal. DWT done for decomposition of signal. Power spectrum, variance and mean of the signal used for feature extraction. MLPNN, PNN and SVM used to classify cursor movements	Power spectrum, variance and mean of the signal	Multilayered perceptron neural network (MLPNN), Probablistic neural network (PNN), SVM	88.75% (MLPNN), 83.75% (PNN), 82.2% (SVM)
7	[23]	Entropy and Standard deviation used as input feature to NB, ANN and SVM (using linear, quadratic polynomial, and Gaussian radial basis function) for classifying epileptic patients	Entropy, Standard deviation	MLPNN, SVM (Radial Basis Function)	100% (MLPNN and SVM)
8	[30]	Feature extracted with power spectral density and input to ANN for epileptic seizure detection	Power spectral density	ANN	98.8%
9	[31]	Approximate Entropy (ApEn) and Auto Mutual Information (AMI) was used for feature extraction-test is used for identifying Alzheimer's disease patients	Approximate Entropy (ApEn) and Auto Mutual Information (AMI)	t-test,	80%
10	[32]	Autoregressive (AR) parameter are used as input features to ANN for identifying psychiatric disorder(obsessive compulsive disorder and schizophrenia	Autoregressive coefficients	ANN	88.89%

computer-aided system for identifying depression patients. This system could be of great help in identifying depression patients in early stages preventing it from becoming severe and untreatable which in terms would improve quality of life of the patient and substantially decrease the cost associated with medical care.

References

1. Li Y, Fan F (2005) Classification of schizophrenia and depression by EEG with ANNs. In: 27th annual international conference of engineering in medicine and biology society, IEEE-EMB, 2005, pp 1–6
2. Stewart JL, Coan JA, Towers DN, Allen JJB (2014) Resting and task-elicited prefrontal EEG alpha asymmetry in depression: support for the capability model. Psychophysiology, 446–455
3. Hosseinifarda B, Moradia MH, Rostami R (2013) Classifying depression patients and normal subjects using machine learning techniques and nonlinear features from EEG signal. Comput Methods Programs Biomed 109:339–345
4. Mantri S, Agrawal P, Patil D, Wadhai V (2015) Non invasive EEG signal processing framework for real time depression analysis. In: SAI intelligent systems conference, pp 518–521
5. Fan F, Li Y, Qiu Y, Zhu Y (2005) Use of ANN and complexity measures in cognitive EEG discrimination. In: Engineering in medicine and biology society, 27th annual international conference, IEEE-EMBS, 2005, pp 1–6
6. Liao S, Wu C, Huang H, Cheng W, Liu Y (2017) Major depression detection from EEG signals using kernel eigen-filter-bank common spatial patterns. Sensors, pp 1–14
7. Hinrikus H, Suhhova A, Bachmann M, Aadamsoo K, Vohma U, Lass J, Tuulik V (2009) Electroencephalographic spectral asymmetry index for detection of depression. Med Biomed Eng Comput, pp 1291–1299
8. Mumtaz W, Xia L, Ali SSA, Yasin MMAM, Hussain M, Malik AS (2017) Electroencephalogram (EEG)-based computer-aided technique to diagnose major depressive disorder (MDD). Biomed Signal Process Control 31:108–115
9. World Health Organization (2017) Depression and other common mental disorders global health estimates. WHO Document Production Services, Geneva, Switzerland
10. World Health Organization (2011) Global burden of mental disorders and the need for a comprehensive, coordinated response from health and social sectors at the country level. Report by the Secretariat. EB 130/9
11. American Psychiatric Association (1994) Diagnostic and statistical manual of mental disorders, 4th edn. American Psychiatric Association Washington, DC, pp 339–345
12. Bachmann M, Lass J, Suhhova A, Hinrikus H (2013) Spectral asymmetry and Higuchi's Fractal dimension measures of depression electroencephalogram. Comput Math Methods Med 2013:1–8
13. Grin-Yatsenko VA, Baas I, Ponomarev VA, Kropotov JD (2010) Independent component approach to the analysis of EEG recordings at early stages of depressive disorders. Clin Neurophysiol 121:281–289
14. Mohammadi M et al (2015) Data mining EEG signals in depression for their diagnostic value. BMC Med Inf Decis Mak, pp 108–123
15. Bruder GE, Stewart JW, Hellerstein D, Alvarenga JE, Alschuler D, McGratha PJ (2012) Abnormal functional brain asymmetry in depression: evidence of biologic commonality between major depression and dysthymia. Psychiatry Res, pp 250–254

16. Bjork MH, Sand T, Bråthen G, Linaker OM, Morken G, Nilsen BM, Vaaler AE (2008) Quantitative EEG findings in patients with acute, brief depression combined with other fluctuating psychiatric symptoms: a controlled study from an acute psychiatric department. BMC Psychiatry, 2008, pp 1–6
17. Ricardo-Garcell J (2009) EEG sources in a group of patients with major depressive disorders. Int J Psychophysiol 71:70–74
18. Bachmann M, Lass J, Hinrikus H (2017) Single channel EEG analysis for detection of depression. Biomed Signal Process Control 31:391–397
19. Subasi A (2007) EEG signal classification using wavelet feature extraction and a mixture of expert model. Expert Syst Appl 32:1084–1093
20. Mumtaz W, Xia L, Yasin MAM, Ali SSA, Malik AS (2016) A wavelet-based technique to predict treatment outcome for major depressive disorder. PLoS ONE, pp 1–6
21. Li Y, Li Y, Tong S, Tang Y, Zhu Y (2007) More normal EEGs of depression patients during mental arithmetic than rest. In: Joint meeting of the 6th international symposium on noninvasive functional source imaging of the brain and heart and the international conference on functional biomedical imaging, 2007, pp 165–168
22. Puthankattil SD, Joseph PK (2012) Classification of EEG signals in normal and depression conditions by ANN using RWE and signal entropy. J Mech Med Biol 12:1240019–1240032
23. Sood M, Bhooshan SV (2014) Automatic processing of EEG signals for Seizure detection using soft computing techniques. In: IEEE international conference on recent advances and innovations in engineering, 2014, pp 1–6
24. Chandaka S, Chatterjee A, Munshi S (2009) Cross-correlation aided support vector machine classifier for classification of EEG signals. Expert Syst Appl 36:1329–1336
25. Subasi A, Gursoy MI (2010) EEG signal classification using PCA, ICA, LDA and support vector machines. Expert Syst Appl 37:8569–8666
26. Sabetia M, Boostani R, Katebi SD, Price GW (2007) Selection of relevant features for EEG signal classification of schizophrenic patients. Biomed Signal Process Control 2:122–134
27. Kalaivani M, Kalaivani V, Devi AV (2014) Analysis of EEG signal for the detection of brain abnormalities. In: IJCA proceedings on international conference on simulations in computing nexus pp 1–6
28. Kumar RSS, Jose JP (2011) Seizure detection in EEG using time frequency analysis and SVM. In: International conference on emerging trends in electrical and computer technology (ICETECT), IEEE, pp 1–6
29. Kousarrizi MRN (2009) Feature extraction and classification of EEG signals using wavelet transform, SVM and artificial neural networks for brain computer interfaces. In: International joint conference on bioinformatics, systems biology and intelligent computing, IEEE, 2009, pp 352–355
30. Tzallas AT, Tsipouras MG, Fotiadis DI (2007) The use of time-frequency distributions for epileptic seizure detection in EEG recordings. In: Proceedings of the 29th annual international conference of the IEEE EMBS, 2007, pp 3–6
31. Abásolo D, Hornero R, Escudero J, Gomez C, Garcia M, Lopez M (2006) Approximate entropy and auto mutual information analysis of the electroencephalogram in Alzheimer's disease patients. In: IET 3rd international conference on advances in medical, signal and information processing, IEEE, 2006, pp 1–6
32. Tsoi C, So DSC, Sergejew A (1993) Classification of electroencephalogram using artificial neural networks. In: 7th NIPS conference conference: advances in neural information processing systems 6, 1993, pp 1151–1158

Numerical Simulation of High-Temperature VCSEL Operation and Its Impact on Digital Optical Link Performance

K. Murali Krishna and M. Ganesh Madhan

Abstract An 863 nm, bottom-emitting VCSEL diode is simulated under a wide range of operating temperatures. Based on this VCSEL, a digital optical link comprising of Graded-index Multi-Mode Fiber (MMF) and a Si-based P-type/intrinsic/N-type (PIN) photodetector is numerically evaluated. The working temperature of the laser diode is changed from 20 to 120 °C and its impact on the MMF link is evaluated for 1.25 and 2.5 Gbps. The link performance is evaluated by determining Q factor and BER. For a data rate of 1.25 Gbps, the maximum distance is found to be 0.64 km at 120 °C and 0.3425 km for 2.5 Gbps at 100 °C, for a BER = 10^{-12}, as per IEEE 802.3.z standard.

Keywords Thermal effects · Rate equations · Transfer function
Intermodal dispersion · Gigabit Ethernet

1 Introduction

Vertical-Cavity Surface-Emitting Lasers (VCSEL) is characterized by high beam quality, low beam divergence, long lifetime, and short resonator round-trip time. Hence, they find applications in optical fiber communication, gas sensing and in miniature optical clocks. VCSELs are available in the form of single mode, multimode and arrays for high-power output for different applications. The existing optical fiber infrastructures are many years old, and replacing them with new ones are costly as well as time consuming mainly in dense urban areas. Hence, interconnecting 50/125-μm and 62.5/125-μm multimode fibers with VCSEL provide a cost-effective solution for short-range telecom and datacom applications. For the past two decades, 850 nm lasers are used for data transmission over multi-mode

K. Murali Krishna (✉) · M. Ganesh Madhan
Department of ECE, MIT Campus Anna University, Chennai, India
e-mail: kmuralikrishnamk@gmail.com

M. Ganesh Madhan
e-mail: mganesh@annauniv.edu

V. Nath and J. K. Mandal (eds.), *Nanoelectronics, Circuits and Communication Systems*, Lecture Notes in Electrical Engineering 511,
https://doi.org/10.1007/978-981-13-0776-8_31

fibers (MMF) covering several hundred meters in Gigabit Ethernet application (IEEE 802.3z).

In general, it is a challenging task to transmit error-free data with high-speed VCSEL, at high temperature since, the nature of the materials with which VCSELs are developed favor normally lower temperatures only. Also, the usage of this technology becomes indispensable in the areas of data center, airborne and light-weight communications.

Recently, Surface relief VCSEL [1] based on gain-cavity mode detuning were demonstrated from 22 to 90 °C. The mode selectivity is done based on shallow surface relief.

For high-speed and temperature performance, KL Chi [2] proposed p-type-doped VCSEL structure which showed significant improvement in performance operating at 25 and 85 °C. An 980 nm VCSEL with bandwidth of 23 GHz at 85 °C and 25.6 GHz at 25 °C was reported [3]. The key features of the literature [3] are the development of improved VCSEL using 5 μm oxide-aperture diameter along with studies on energy efficiency and temperature stability. Based on Differential Mode Delay (DMD) plot and VCSEL spectrum, Effective Bandwidth (EB) of optical link was evaluated without actual measurement [4]. To mitigate the Chromatic Dispersion (CD) effect in short distance VCSEL based MMF transmissions, MMF jumpers are introduced, which carry the desired DMD feature and also provides flexibility to do CD compensation [5]. An 850 nm oxide VCSEL was investigated for power–current, relative intensity noise, eye and BER characteristics over a wide temperature range from −40 to 125 °C, for automotive applications [6]. However, the link lengths are in the order of meters only. With the application of finite difference technique, a flip-chip-bonded 850 nm AlGaAs/GaAs implant-apertured VCSEL was analyzed for thermal characteristics [7]. A 2.5 Gbit/s transmission over 250 m graded index MMF link with a BER of less than 10^{-11}, was reported [8]. The link includes a GaAs-based VCSEL operating between −20 and 100 °C. A bias-free transmission of 2.5 Gbit/s data rate with $2^{31} - 1$ PRBS signals using selectively oxidized GaAs VCSEL source and a graded index 50 μm core diameter MMF, was demonstrated [9]. The performance of a high-speed VCSEL whose modulation bandwidth is above 13 GHz was reported [10]. The working of DFB-SMF link at various temperatures has also been studied [11]. From the literature, it is observed that very high data rate transmission through VCSEL has been achieved using either single-mode fibers or incorporating special arrangements. However, studies on the 850 nm band VCSEL at high operating temperatures for Gigabit MMF links have not been reported extensively. Also, maintaining the operating temperature of VCSEL is not cost-effective and ambient temperature may also vary over a wide range. Since many buildings and academic campuses have already installed MMFs, it becomes essential to evaluate the performance of low-cost VCSELs over wide operating temperature for Gigabit Ethernet applications. Conventionally, link evaluation papers do not incorporate detailed device models, for e.g. Ref. [12], uses a simple expression for representing modulated optical pulse, whereas this work includes a detailed rate equation for dynamically generating the optical pulse under gigabit modulation. Previously, we have reported a link analysis for an 850 nm VCSEL [13] with a different structure, which had a maximum operating temperature of 43 °C only. The work of P.V. Mena et al. [14] deals with different VCSEL laser models for experimental devices. We have considered an 863 nm VCSEL model and evaluated the device response under static and dynamic conditions. Further, we verified our results with experimental values provided in Ref. [14], and found to agree well. Based on this model, we have investigated two different data transmission for multi-mode fiber optical link under a temperature range of 20–120 °C. Even though many authors have discussed about VCSEL thermal dependence, their impact on BER and Q factor of a MMF link is not reported in detail. The proposed work differs from Ref. [8], due to the choice of bottom-emitting VCSEL as laser transmitter, while the authors of Ref. [8] have considered a selectively oxidized GaAs VCSEL. The optical power output of the device considered in this work is higher compared to the literature, in the temperature range of 20–100 °C. A detailed comparison with the literature is provided in Table (1) for highlighting the novelty of the present work. The impact of thermal effects of VCSEL on the data link is analyzed by determining the eye diagram and BER. The maximum possible link distance for various temperatures is evaluated for a maximum BER of 10^{-12}, as per IEEE 802.3.z standard. Hence, we feel that this work will be useful for the implementation of low cost, uncooled VCSEL-based transmitters for short distance MMF links. The present work is done using Matrix Laboratory software.

2 System Model

The optical link is established by the following components, namely VCSEL diode, multi-mode fiber, and PIN photodetector-based receiver. The eye diagram analysis is used to evaluate the link performance. The simulation is carried out for various temperatures from 20 to 120 °C, for data rates 1.25 and 2.5 Gbps, to determine the thermal effects on link performance. The modeling of VCSEL diode is discussed first and then the multimode fiber. A PIN diode provides the received signal. This approach follows our previous work [13].

2.1 VCSEL Transmitter

The structure of the VCSEL diode is explained in Ref. [15], with an emission wavelength of 863 nm. This diode has potential to operate until 120 °C, in continuous wave mode.

The temperature effects can be explained using the rate equations as given in [14]. They are

Table 1 Literature review

S. No.	Authors	Ref. No.	Technique	Inference
1.	L. Xiang et al.	[1]	Gain cavity mode detuning in shallow etched surface relief VCSEL structure	VCSEL operation at high temperature was realized. Fiber link performance was not studied
2.	KL. Chi et al.	[2]	Asymmetric 3λ/2 cavity length; p-type doping in the active layer of VCSEL	VCSEL characteristics were studied while fiber link was not incorporated
3.	X. Chen et al.	[5]	VCSEL-based MMF link with MMF jumper	VCSEL-MMF link was demonstrated with fixed laser parameters and for a short link distance of 200 m
4.	K. Nishikata et al.	[6]	characteristics and data transmission of 850 nm and 1300 nm oxide VCSEL	High temperature, short wavelength VCSEL characteristic was obtained without fiber link. For long-wavelength VCSEL, SMF link data transmission was carried out
5.	P. Schnitzer et al.	[9]	Bias-free transmission using high-performance selectively oxidized GaAs VCSEL	VCSEL-MMF link for specific laser condition and fixed fiber distance of 500 m was investigated
6.	K. Murali Krishna et al.	[13]	InGaAs-GaAs-based VCSEL with MMF link	Impact of thermal effects on VCSEL-MMF link performance was carried out
7.	K. Murali Krishna and M. Ganesh Madhan	Present work	GaAs-AlGaAs-based VCSEL with MMF link	For the first time, a high-temperature, low-cost VCSEL operation is simulated without any feedback and thermal control system for already installed MMF links inside buildings

$$\frac{\partial N}{dt} = \frac{\eta_i(I - I_{off}(T))}{q} - \frac{N}{\tau_n} - \frac{G_o(N - N_o)S}{1 + \varepsilon S} \tag{1}$$

$$\frac{\partial S}{dt} = -\frac{S}{\tau_p} + \frac{\beta N}{\tau_n} + \frac{G_o(N - N_o)S}{1 + \varepsilon S} \tag{2}$$

$$T = T_o + (IV - P_o)R_{th} - T_{th}\frac{\partial T}{\partial t} \tag{3}$$

The laser parameters mentioned in the above equations and their values are given in [14, 16].

2.2 MMF channel

The approximated equation for the impulse response of a MMF by following Ref. [12] can be written as

$$h_{mmf}(t) = \frac{1}{\sqrt{2\pi\sigma^2}}\exp(\frac{-(t-\tau)^2}{2\sigma^2}) \quad (4)$$

where the terms have their usual meaning as given in [13].

3 System Simulation Results

3.1 VCSEL diode DC characteristics

The rate equations are solved using an ODE solver available in the Matrix Laboratory tool, for different temperatures from 20 to 120 °C. The work reported herein is based on VCSEL model, which is derived from the experimental characteristics of an 863 nm bottom-emitting VCSEL, by P.V. Mena et al. [14]. We use the same device model and the simulation results of our calculation represented in Fig. 1a matches well with the VCSEL power–current (P-I) characteristics reported in [14]. Further, transient conditions for the same device are evaluated by H. Kanj [17]. We also compared our results as displayed in Fig. 1b and found to match well with [17]. As the DC and transient characteristics are coinciding with the experimental results, we proceed to evaluate the link performance.

3.2 Gigabit Ethernet Link

The multi-mode optical fiber waveguide is modeled based on the Eq. (6), and the optical pulse is attenuated and dispersed as it propagates through the fiber. Equation (7) provides the time domain version of the optical power output from the multi-mode graded index fiber, by convolving the input power with the MMF transfer function. The receiver is modeled as a fourth-order low-pass filter with cut off frequency at 75% of the data rate. Receiver noise sources such as thermal and shot noise along with relative intensity noise (RIN) of the light source are included in the model [12, 18].

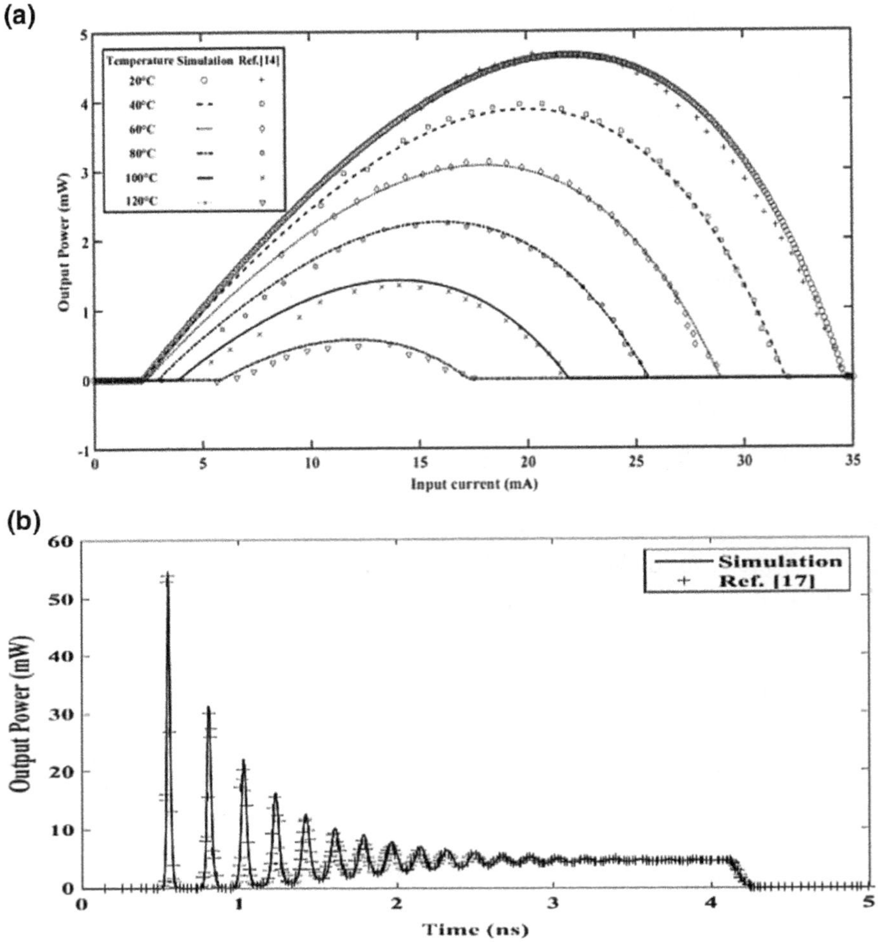

Fig. 1 **a** P-I characteristics of VCSEL diode **b** transient characteristics of VCSEL output power at 20 °C

3.2.1 1.25 Gigabit transmission:

An 8-bit sequence consisting of alternate zeros and ones, which indicate a worst-case scenario at a data rate of 1.25 Gbps is generated and provided as an input current to the VCSEL. The device is biased at 6 mA and the pulse amplitude is kept as 4 mA. These conditions are fixed based on the threshold current and linear operating region, obtained from the P-I characteristics at a specific temperature. An operating temperature of 120 °C and fiber lengths of 0.25, 0.64 and 1 km are considered for this study. By solving the rate equations under square current pulses (data) with appropriate initial conditions, the laser output is obtained. Figures 2a–c represents the input current (data), VCSEL output power and detector output respectively.

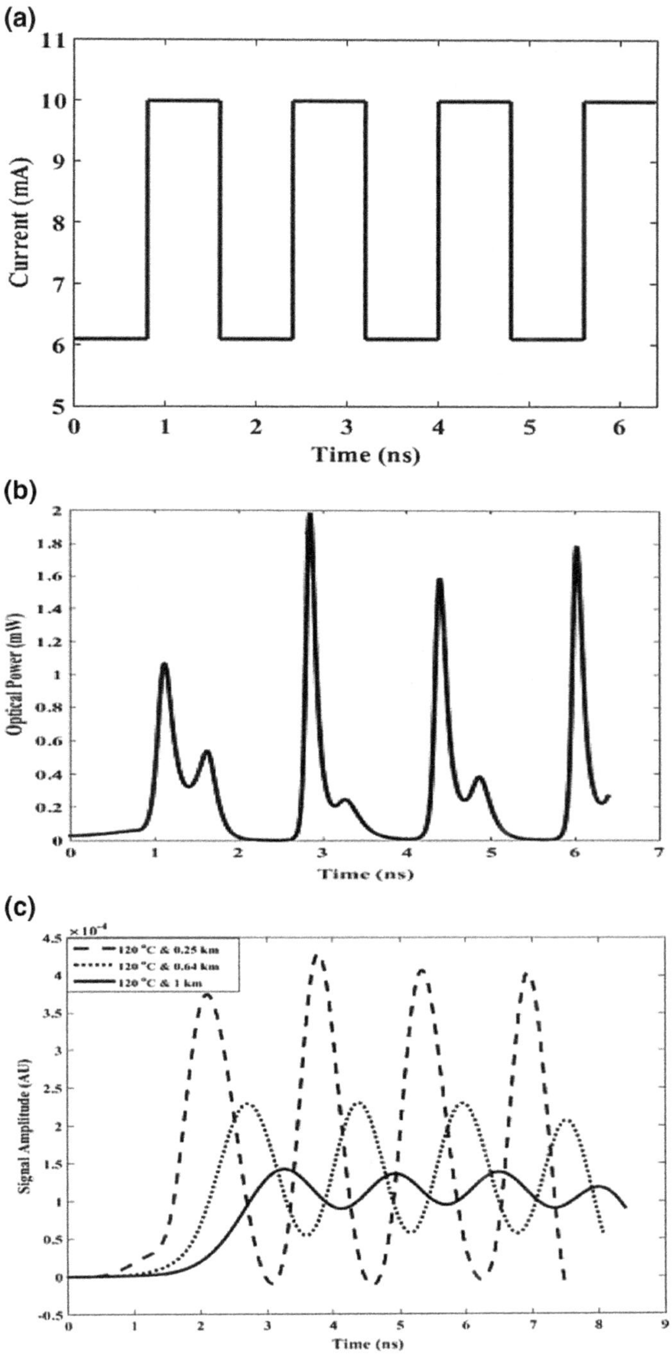

Fig. 2 **a** Current sequence **b** VCSEL power output **c** PIN diode output at 120 °C

The MMF lengths are varied to show the effect of pulse broadening. As fiber length increases, pulse spreading predominates. Moreover, a reduction in the envelope amplitude along with an increase in dc offset is observed with increased distance. As seen from the P-I characteristics, the optical power output reduces with increasing temperature. Hence, a combination of reduced power and increased dispersion leads to the behavior shown in Fig. 2c, where 120 °C and 1 km results indicate a poor signal quality.

The 1.25 Gbps link is simulated for various temperatures of VCSEL and performance are measured with the aid of valid eye openings. To achieve a BER = 10^{-12}, at 20, 60 and 120 °C temperatures, the maximum possible transmission distances are evaluated and corresponding eye diagrams are shown in Figs. 3a–c. The eye height measurements show a decrease in amplitude as temperature and fiber length increases. This is due to the fact that increasing temperature will reduce VCSEL power and increase in fiber length makes intermodal dispersion and attenuation more pronounced. For maintaining a BER of 10^{-12}, at 20 °C, the maximum possible distance is found as 0.77 km. However, the link distance decreases for higher operating temperatures.

The result in Fig. (4) shows that the amplitude of the received signal envelope reduces, leading to a decrease in modulation depth. This arises due to the filtering effect of intermodal dispersion in MMF.

The Q factor and BER are computed for various MMF distances and VCSEL temperatures [19]. For BER $\leq 10^{-12}$, the performance metrics are calculated using the eye diagrams. The graph in Fig. 5a indicates a gradual decrease in the link performance as temperature and MMF link distance increases. For a Q factor around 17 dB, the maximum transmission distances are 0.77 and 0.64 km for 20 and 120 °C respectively. Similarly, for obtaining a specific value of BER, higher operating temperature restricts the link distance, as observed in Fig. 5b.

The maximum link distance to a BER of 10^{-12} corresponding to 120 °C is found to be 640 m.

3.2.2 2.5 Gigabit transmission:

A pulse of 6 mA amplitude and DC biased at 4.2 mA is given as the input current to the diode. Figures 6a–c represent the input current, VCSEL output power at 100 °C and the detector outputs at 0.15, 0.3425 and 0.525 km. The output optical pulse in this case is narrower than that in the previous one, as expected. The maximum operating temperature is restricted to 100 °C, in this case, due to thermal effect in VCSEL and pulse broadening effect in MMF link. The received signals are shown in Fig. 6c.

The eye diagrams corresponding to BER = 10^{-12} at temperatures of 20, 60 and 100 °C for 2.5 Gbps link are evaluated and shown in Figs. 7a–c. Figure (8) shows a reduction in the envelope amplitude with respect to distance. This arises due to the filtering effect of MMF, which increases with distance.

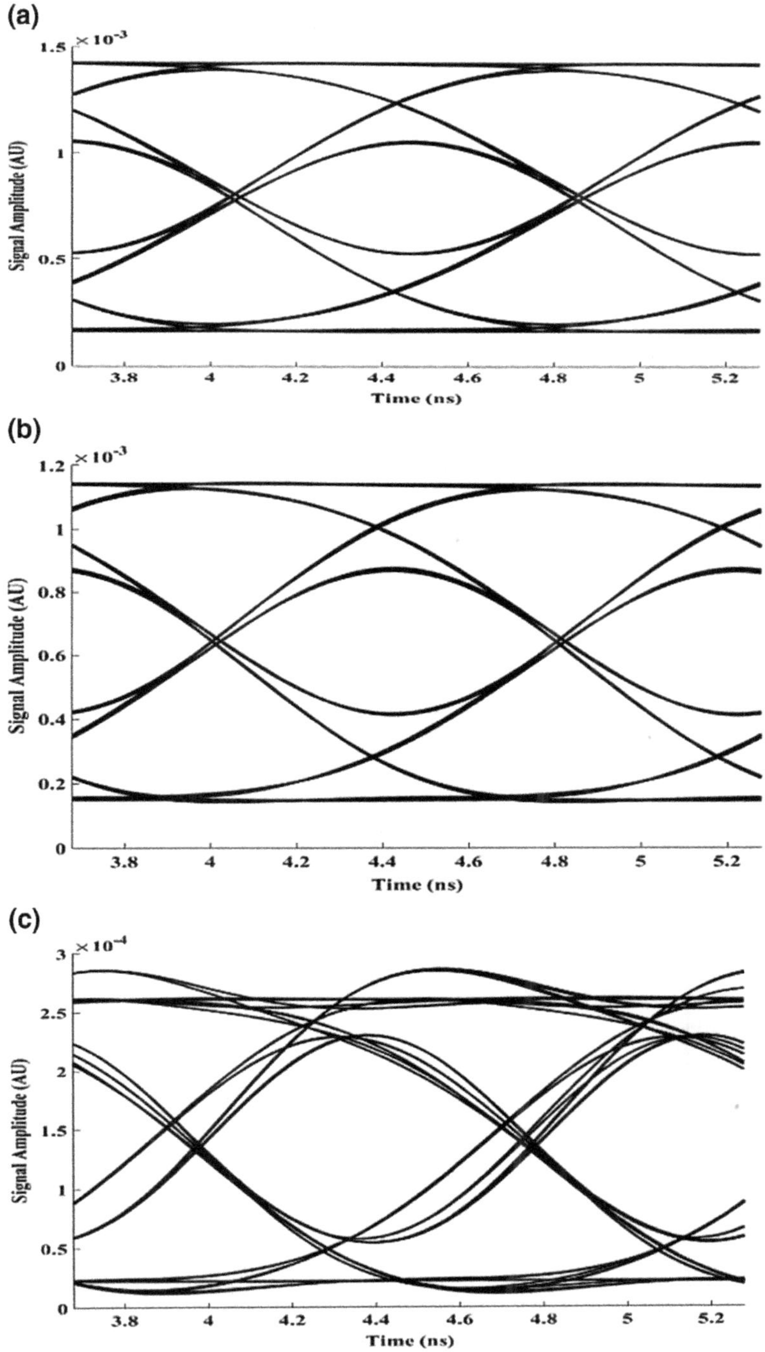

Fig. 3 The temperature and maximum link distance for the eye diagrams **a** 20 °C, 0.77 km **b** 60 °C, 0.74 km **c** 120 °C, 0.64 km

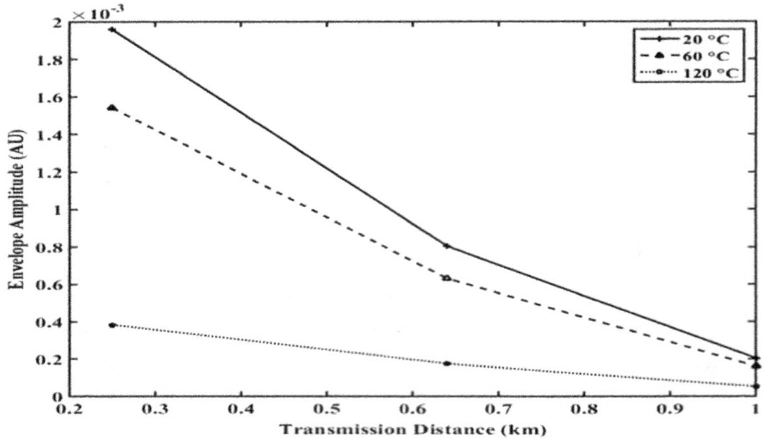

Fig. 4 Envelope amplitude for 1.25 Gb/s data rate versus fiber length

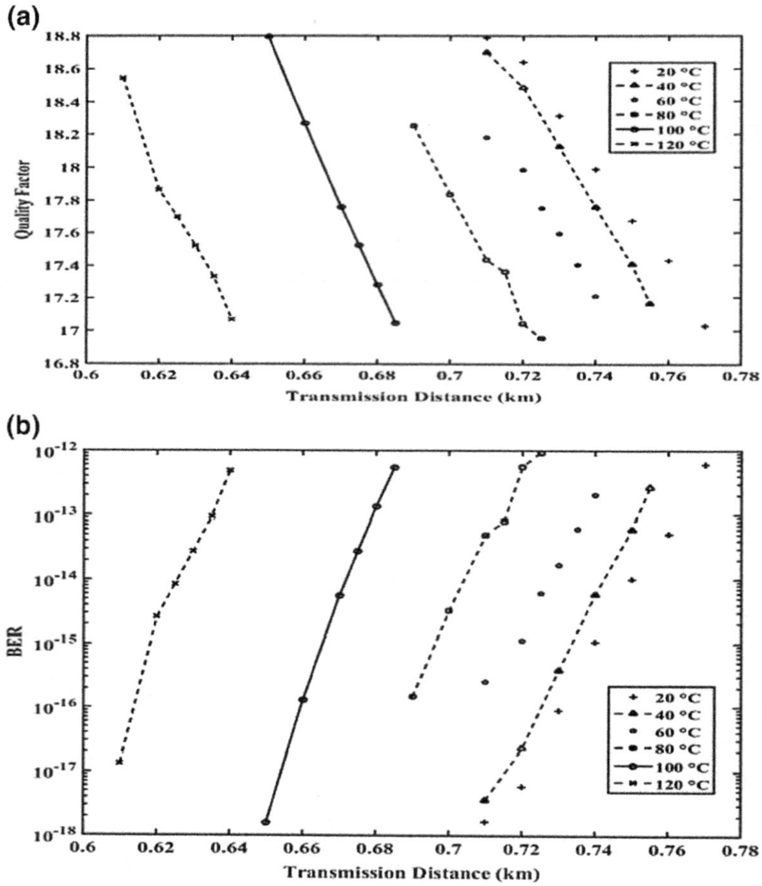

Fig. 5 **a** Q factor **b** BER versus MMF link distance

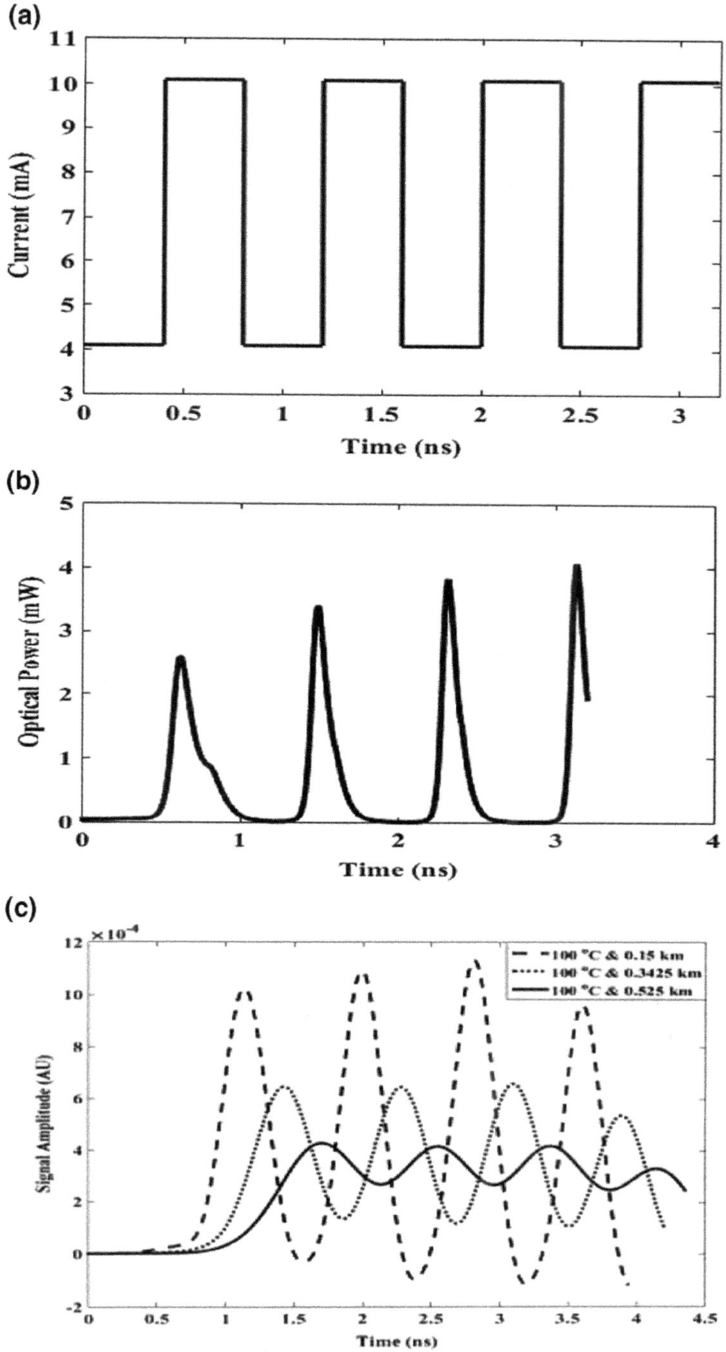

Fig. 6 **a** Current sequence **b** VCSEL power output **c** PIN diode output at 100 °C

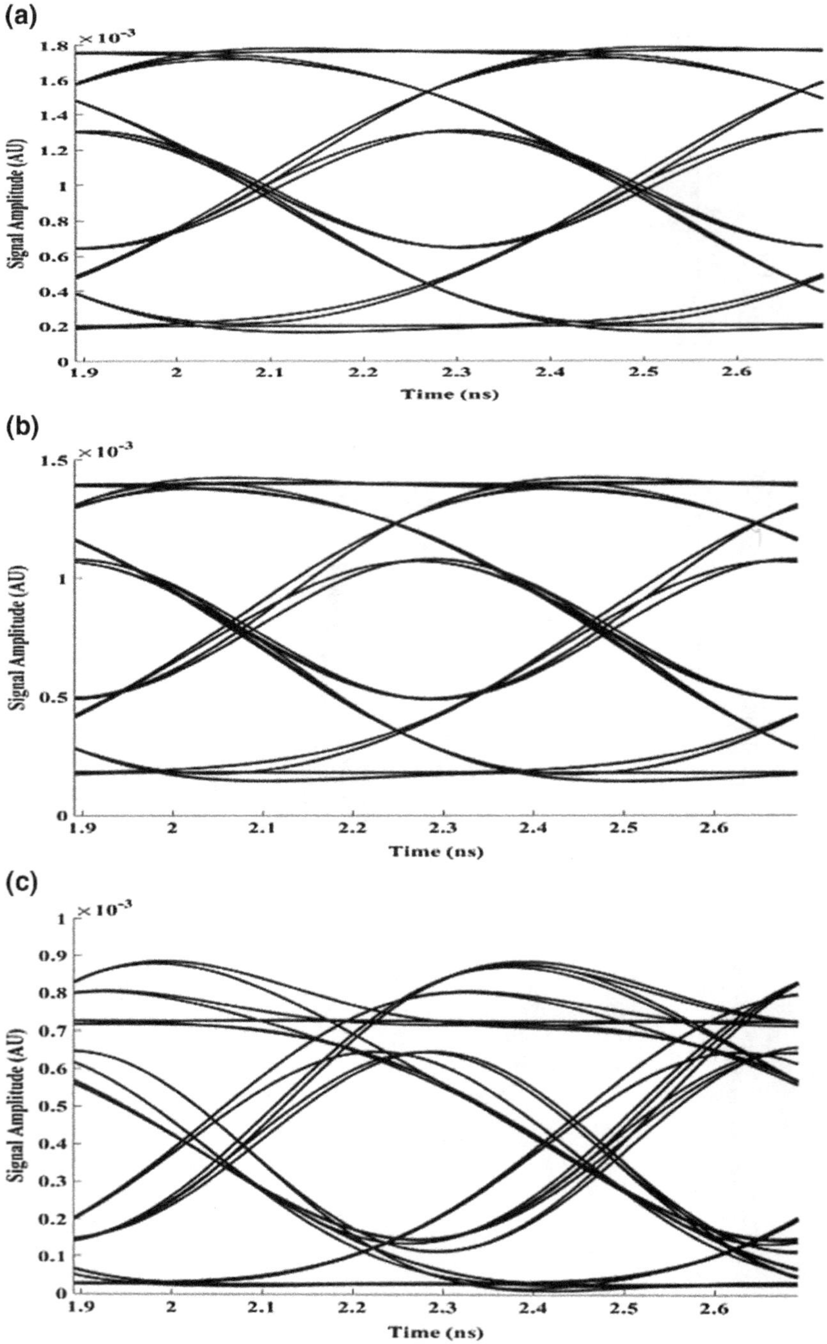

Fig. 7 The temperature and maximum link distance for the eye diagrams **a** 20 °C, 0.405 km **b** 60 °C, 0.39 km **c** 100 °C, 0.3425 km

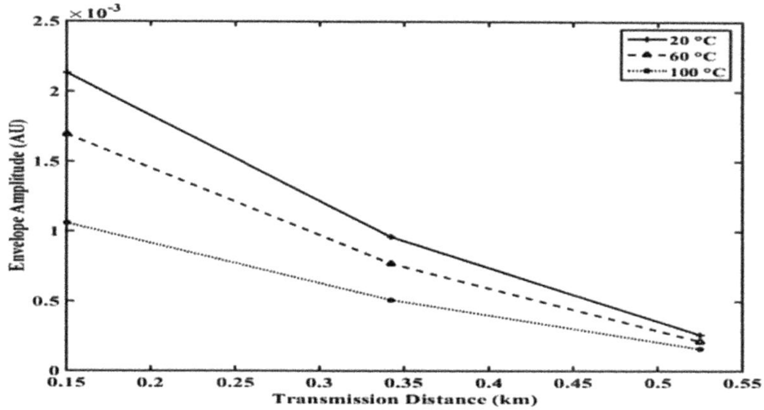

Fig. 8 Envelope amplitude versus MMF link distance

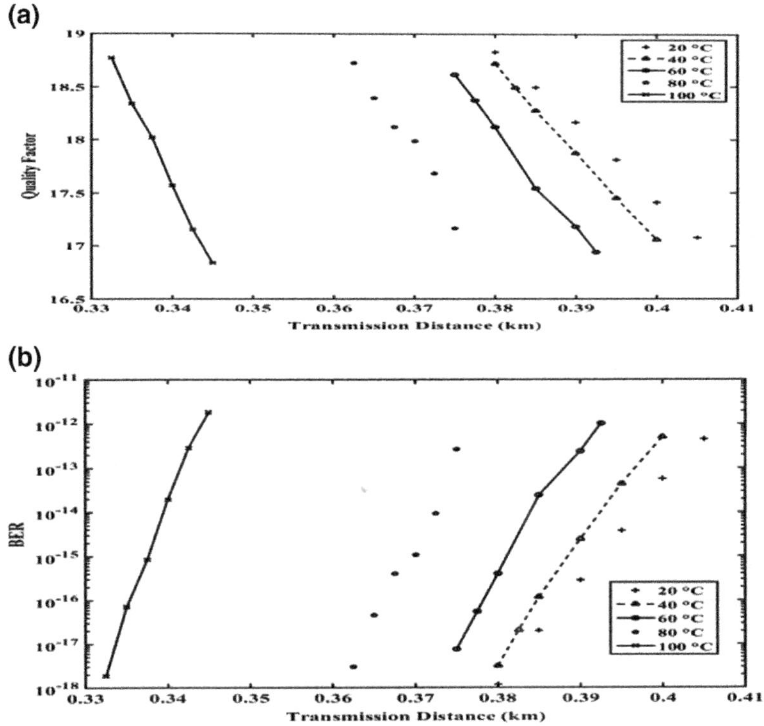

Fig. 9 **a** Q factor **b** BER versus MMF link distance

Table 2 Comparison of the results obtained with the available literature

S. No.	Parameter	Authors [Ref. No.]	As obtained by the reference	As obtained by present work
1.	Threshold current, I_{th} Max. output power, P_{out}	L. Xiang et al. [1]	I_{th} = 2.15 mA at 80 °C P_{out} = 0.7 mW at 80 °C	I_{th} = 2.9 mA at 80 °C P_{out} = 2.258 mW at 80 °C
2.	Oxide relief diameter (or) Aperture diameter, W_o Slope efficiency, dL/dI Max. output power, P_{out}	KL. Chi et al. [2]	P-doping W_o = 6–7 μm dL/dI (at 4 mA) = 0.55 W/A, P_{out} = 5.5 mW at 25 °C dL/dI (at 4 mA) = 0.46 W/A P_{out} = 2.9 mW at 85 °C	W_o = 16 μm dL/dI (at 4 mA) = 0.36 W/A, P_{out} = 4.46 mW at 25 °C dL/dI (at 4 mA) = 0.28 W/A P_{out} = 2.05 mW at 85 °C
3.	BER and link distance, d	P. Schnitzer et al. [9]	For 2.5 Gbps, BER of 10^{-11}, at 27 °C, d = 500 m	For 2.5 Gbps, BER of 10^{-11}, at 20 °C, d = 420 m
4.	BER and link distance, d	K. Murali Krishna and M. Ganesh Madhan [13]	For 1.25 Gbps, BER = 10^{-13}, at 40 °C, d = 770 m For 2.5 Gbps, BER of 10^{-13}, at 40 °C, d = 411 m	For 1.25 Gbps, BER = 10^{-13}, at 40 °C, d = 752 m For 2.5 Gbps, BER of 10^{-13}, at 40 °C, d = 397 m

The Q factor and BER for different link distances and ambient temperatures are shown in Figs. 9a, b. Beyond 100 °C, the eye measurement is not feasible due to reduced optical power and hence at this data rate, calculations are carried out till a link length of 0.3425 km at 100 °C. In, a similar work carried out by Mederer et al. [8], the bias current and modulation amplitude are fixed. Hence, for a fixed distance, at all operating temperatures, BER $\leq 10^{-11}$. In our case, the maximum linear region of operation for all temperatures is utilized in the simulation. Hence maximum transmission distance for achieving BER of 10^{-12}, for each temperature variation. Also, for a BER of 10^{-12}, the MMF link distance at 100 °C VCSEL operation is less than 250 m for Ref. [8], while for our case it is around 342.5 m. This is due to the difference in the VCSEL structures.

It is evident from Figs. (5 and 9) that, as data rate increases from 1.25 to 2.5 Gbps, the transmission distances are nearly halved. Further, the performance of the link also deteriorates with increasing operating temperature of the VCSEL.

The maximum transmission distance for 2.5 Gbps data, is found to be 342.5 m, for the entire range of temperature till 100 °C. This satisfies the BER recommendation for GigE application specified by IEEE 802.3.z standard.

Table (2) illustrates a comparison of the results obtained in the present work with literature, which indicates the suitability of simple, uncooled 863 nm GaAs-AlGaAs VCSEL for already installed MMF links for in building applications.

4 Conclusion

In this paper, a VCSEL-based MMF link and PIN diode is simulated for Gigabit Ethernet application. The effect of laser temperature and fiber dispersion are evaluated by numerical simulation. From this study, the optimum link length up to which 1.25 and 2.5 Gigabit data can be sent with BER $\leq 10^{-12}$, is identified for various operating temperatures. For short-range Gigabit data transmission using uncooled, cost-effective VCSEL diode transmitters, this analysis can be useful.

Acknowledgments The authors gratefully acknowledge DST, New Delhi for providing financial support to carry out this research work under Promotion of University Research and Scientific Excellence (PURSE) II scheme. K. Murali Krishna is thankful for the award of DST-PURSE fellowship.

References

1. Xiang L et al (2017) Stable single-mode operation of 894.6 nm VCSEL at high temperatures for Cs atomic sensing. Chin Phys B 2017 26(7):074209 (1–4)
2. Chi KL et al (2016) 850 nm VCSELs with P-type δ-doping in the active layers for improved high-speed and high temperature performance. IEEE J Quantum Electron 52(11):2400607 (1–7)
3. Moser P et al (2015) Impact of the oxide-aperture diameter on the energy efficiency, bandwidth, and temperature stability of 980-nm VCSELs. J Lightwave Technol 33(4): 825–831
4. Molin D, Bigot-Astruc M, Sillard P (2014) New MMF and VCSEL metrics for system reach predictions. In: Conference and exhibition on optical fiber communications, pp 1–3
5. Chen X et al (2013) Chromatic dispersion compensated 25 Gb/s multimode VCSEL transmission around 850 nm with a MMF jumper. In: Conference and exposition on optical fiber communication and conference on national fiber optic engineers, pp 1–3
6. Nishikata K et al (2005) Wide temperature operation of 850 nm VCSEL and isolator-free operation of 1300 nm VCSEL for a variety of applications. Vertical-Cavity Surf-Emitting Lasers IX, Proc SPIE 5737:8–19
7. Mehandru R et al (2002) Finite difference analysis of thermal characteristics of CW operation 850 nm lateral current injection and implant apertured VCSEL with flip-chip bond design. Solid State Electron 46(5):699–704
8. Mederer F et al (2000) 2.5 Gbit/s data transmission with singlemode GaAs VCSELs over −20 to 100 °C temperature range. Electron Lett 36(3):236–238

9. Schnitzer P et al (1998) Bias-free 2.5 Gbit/s data transmission using polyimide passivated GaAs VCSELs. Electron Lett 34(6):573–575
10. Nazaruk DE et al (2014) Single-mode temperature and polarization-stable high-speed 850 nm vertical cavity surface emitting lasers. In: 16th Russian youth conference on physics and astronomy, pp 1–6
11. Koussalya B, Ganesh Madhan M (2010) Simulation of thermal effects in laser diode and its impact on high speed fiber optic link. J High Speed Netw 17(4):175–184
12. Yuen R, Fernando XN, Krishnan S (2004) Radio over multimode fiber for wireless access. In: Canadian conference on electrical and computer engineering, vol 3, pp 1715–1718, 2004
13. Murali Krishna K, Ganesh Madhan M (2016) Performance analysis of a low cost VCSEL transmitter based multimode fiber optic link for gigabit ethernet application. In: Proceedings of international conference on communication and signal processing, pp 270–274, 2016
14. Mena PV et al (1999) A simple rate-equation-based thermal VCSEL model. J Lightwave Technol 17(5):865–872
15. Ohiso Y et al (1996) Flip-chip bonded 0.85-μm bottom emitting vertical-cavity laser array on an AlGaAs substrate. IEEE Photonics Technol Lett 8(9):1115–1117
16. Qi C, Shi X, Wang G (2011) High-order circuit-level thermal model of vertical-cavity surface-emitting lasers. IET Optoelectron 5(1):19–27
17. Kanj H (2003) Circuit-level modeling of laser diodes. M.S., North Carolina State University, Raleigh, North Carolina, United States
18. Senior JM, Jamro MY (2009) Optical fiber communications: principles and practice 3rd edn. Pearson Education Limited
19. Agrawal GP (2010) Fiber-optic communication systems 4th edn. Wiley

Smart Activity Sequence Generator in Wearable IoT

Jatindeep Singh, Punit Mishra, Satyajit Mohapatra, Hari Shanker Gupta and Nihar Mohapatra

Abstract Sensors in activity based computing enable continuous monitoring of numerous physiological signals when attached to the human body. This finds wide application in areas of activity monitoring, bio-medical rehabilitation, and fitness tracking. Primary challenges in embedded application development for smart wearable include high energy efficiency and user compatibility. Existing algorithms and applications are still unable to fully utilize the true power of the data being collected. They provide lot of descriptive data analytics but lack in predictive analysis. Energy efficiency of computing as predicted by Koomey's is expected to strike the second law of thermodynamics based on Launder's Limit within few decades. In this work an energy efficient computing technique for next generation mobile applications is developed. Proposed Artificial Intelligence based energy-efficient embedded algorithm that provide personalized training sequence recommendation in order to achieve desired calorie goals. Suggested training sequence of 6 activities fall under high, medium and low calorie burn with achieved median for 234C:535C:688C respectively. The crux of this implementation is Calorie Matrix Regeneration via state feedback technique using Markov Decision Process (MDP) and Genetic Algorithm (GA). Number of generations required by the GA to reach a suboptimal solution is optimized. While Machine learning algorithms are written in C/C++ for effective embedded implementation, certain computationally expensive modules like MDP and GA are coded in Python with proposed IoT cloud based implementation thereby improving battery efficiency to

J. Singh · P. Mishra
Cognizant Technology Solutions, Noida 201301, Uttar Pradesh, India
e-mail: jsiitgn@gmail.com

P. Mishra
e-mail: punit.mishra@cognizant.com

H. S. Gupta (✉)
Space Applications Centre, Jodhpur Tekra, Ahmedabad 380015, India
e-mail: hari@sac.isro.gov.in

S. Mohapatra · N. Mohapatra
Department of Electrical Engineering, IIT Gandhinagar, Gandhinagar 382355, India
e-mail: nihar@iitgn.ac.in

V. Nath and J. K. Mandal (eds.), *Nanoelectronics, Circuits and Communication Systems*, Lecture Notes in Electrical Engineering 511,
https://doi.org/10.1007/978-981-13-0776-8_32

12–16 h. This implementation is first of its kind and a step ahead of available state of the art fitness training algorithms/applications.

Keywords Smart activity trainer · Activities of daily living (ADL) Markov decision process (MDP) · Genetic algorithm (GA) · Internet of things (IOT) · Human activity recognition (HAR) · Smart wearables

1 Introduction

Smart wearable provide wide application in the area of health monitoring, human activity recognition (HAR), bio-medical rehabilitation and fitness training [1]. The services featured by wearable have the potential to tremendously grow in the near future. As per the cisco global mobile data traffic forecast [2], there were 97 million wearable devices in the world in 2015. The North American and Asia-Pacific region are highest in its rankings encompassing of about 70% of the global market. The survey further estimates 601 million smart wearable globally by 2020, growing (@5× w.r.t 2015) at CAGR of 44%. This prediction indicate Asia-Pacific region to dominate regional share of wearable ~32% share by year 2020. Several factors that inhibit uptake of wearable include functionality, battery life, connectivity, data security and high production costs. Primary challenges in embedded application development for smart wearable devices include energy efficient high performance computing for advance functionalities and end user coupled integration [3, 4]. Traditional wearable and even smart phones enable to capture individual user data. In spite of the abundance of available data from these sensors, existing algorithms and applications are still unable to fully utilize the true power of the data being collected. A look at dashboards of popular fitness apps reveals a lot of descriptive data analytics but a significant lack in predictive analysis. Some of them describe the whole activity of a user and offers the user for setting up a new calorie goal. Other apps on the other hand has predefined fitness badges one can earn. By running mere collaborative filtering algorithm on user database enhance the user experience by recommending him badges he may try to earn during a week depending on the user need to burn calorie.

Personalized data provides much more insights into how much calories an individual has spent while being involved in different day to day activities. Hence different individual tend to have a totally different calorie matrix. Fitness training sequence recommendation based on these values provides a real time hypothesis testing environment, dynamically creating new solutions to the existing problem of calorie burn maximization. Conventional applications tend to use average data available on health forums while providing periodic or time based recommendations or use mere collaborative filtering approach on user database rather than personalized health recommendations required for quick recovery. We move a step ahead from collaborative filtering and offer personalized exercising schedule for the user hereby unleashing the true power of data at hand. IoT integration enable access data of thousands of individual, harnessing the power of data analytics and algorithm development. The paper is further organized into five sections. Complete description of the system is provided in Sect. 2. Section 3 comprises of the implementation details of developed algorithm. The achieved results are summarized in Sect. 4 followed by concluding remarks in Sect. 5.

2 System Description

In this work, an Artificial Intelligence based energy-efficient embedded algorithm that provide personalized training sequence recommendation in order to achieve desired calorie goals is developed. The designed personalized activity sequence training system is shown in Fig. 1. It primarily comprises of multi-sensor activity recognition, calorie matrix development module and AI based Markov Decision Process (MDP) and Genetic Algorithm (GA). The crux of this implementation is calorie matrix regeneration via state feedback technique using MDP and GA. The sensors within smart phone is used for activity recognition [5–7]. Fitness wearables come with the advantage of being with the user most of the day and therefore require battery life of more than 12 h [8]. It is difficult to define the exact energy an algorithm will spent as different systems have different architectures for performing similar computations. To get a general sense of energy being consumed Big-O, Small-O notation which is typically used for defining algorithmic efficiency [9] is used. Our algorithm is of Floyd Warshall type with computation of order $\Theta(V^3)$, hence the energy spent of the form $\Theta(wV^3)$. Performing such computations on hardware would affect battery life. Hence certain computationally expensive algorithms like the MDP and GA are moved to cloud.

2.1 *Multi-sensor Human Activity Recognition Process*

Inertial sensors in modern smart phones can potentially be exploited for activity recognition along with their connectivity and communication resources [10–12].

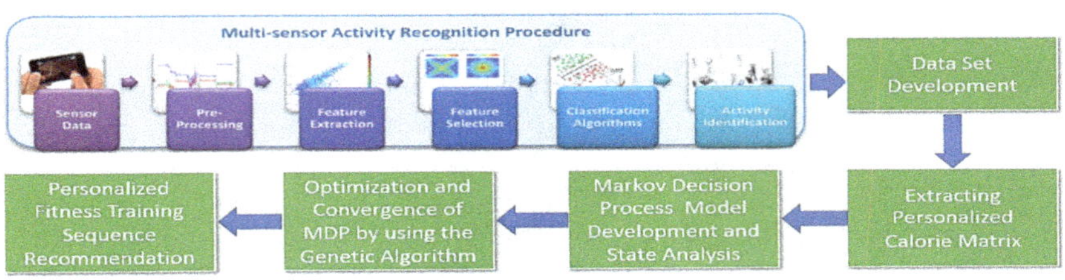

Fig. 1 A block diagram representation of the proposed personalized fitness sequence recommendation system

Smart wearables use the real-time information about user activity to provide fitness training suggestions. Data from a set of experiments [13] is used to train and test all of the multiclass support vector machine (SVM) learning classifier. 70% of data is used for training while rest 30% for testing purposes. Feature extraction processes raw data into data frame format that is suitable for being processes by machine learning algorithm like SVM [14, 15] for classification. Prior to machine learning algorithm, there are 561 correlated variables. This severely increase the computation done by machine learning algorithm on hardware. Hence we reduce the data by performing a principal component analysis [16]. We train the principal component analyzer with available historical data. All new occurrences of incoming data is then filtered through this trained PCA. The PCA further determines the multi-colinearity between the various columns and develops uncorrelated columns, thereby selecting only those features that are not correlated during feature selection. SVM classification implementations produces one of the 12 output states as the recognized activity.

2.2 *Re-development of Calorie and State Probability Matrix*

Before implementing the AI algorithm, data set is re developed. Raw data as well as processed output of model prediction and machine learning is continuously logged from the first day of use. Raw tri-axial linear acceleration and angular velocity signals from Samsung Galaxy S-II with embedded sensors provide information about energy and calorie consumption in activity and transition [13]. Energy variable from the historically logged dataset of the user individual is used for developing personalized calorie matrix as shown in Table 1. Letters A–L represent the 12 Activities of Daily Living (ADL) considered for this work include A:

Table 1 Personalized calorie matrix development

	A	B	C	D	E	F	G	H	I	J	K	L
A	41	10	5	8	2	8	8	–	–	–	9	–
B	10	102	11	9	3	9	9	–	–	–	10	–
C	5	11	67	7	1	7	7	–	–	–	8	–
D	8	9	7	15	8	9	–	6	8	–	–	–
E	2	3	1	8	29	9	6	–	–	–	9	–
F	8	9	7	9	9	11	–	–	–	8	–	9
G	8	9	7	0	8	9	58	6	8	–	–	–
H	2	3	1	8	0	9	6	58	–	–	9	–
I	8	9	7	9	9	0	–		80	8	–	9
J	8	9	7	0	8	9	–	6	8	80	–	9
K	8	9	7	9	9	0	–	–	–	8	93	9
L	2	3	1	8	0	9	6	–	–	–	9	93

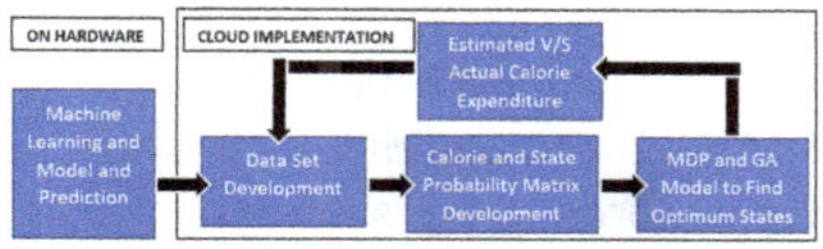

Fig. 2 The development process of calorie and state probability matrix

Walking, B: Walking Upstairs, C: Walking Downstairs, D: Sitting, E: Standing, F: Laying, G: Stand-Sit, H: Sit-Stand, I: Sit-Lay, J: Lay To Sit, K: Stand To Lay, and L: Lay To Stand.

Table 1 gives provides a sample personalized calorie matrix created from raw data captured from the user. Every user spends different calories while performing different activities. The energy used in the activities is captured by the accelerometer and gyroscope in different form and a total of 159 such variables [13] are processed. Now it is assumed that a person does an activity for 10 min and a 1 min buffer period is maintained for switching activities. Hence diagonal of this matrix has large values as they talk about calorie burnt while performing a particular activity for 10 min whereas the non-diagonals would be smaller values corresponding to calories spent during 1 min during transition from one state to another. It is observed that some of the non-diagonal elements are unavailable as these are not valid training sequences. Since different individual tend to have a totally different calorie matrix, personalized data provides much deeper insights into how much calories an individual has spent while being involved in various different day to day activities.

Fitness training sequence recommendation based on these values provides a real time hypothesis testing environment, dynamically creating new solutions to the existing problem of calorie burn maximization. Our primary target is to develop an app which takes input from the machine learning algorithms and prediction model and predicts six sequential activity states that provide high, medium or low calorie burn recommendation sequence to the user on request. This requires selecting 6 states from 12^6 possible combinations which pose a huge computational challenge. This calorie matrix plays an important role in developing Markov Decision model and find sub optimal solution to the problem, thereby developing a personalized recommendation for each individual based upon his/her reading. This helps in development of personalized calorie state matrix for AI learning via MDP and GA (Fig. 2).

3 System Implementation

Mobile sensing platforms evolved in an iterative process after several real world deployments. While hardware platform-1 provides wireless multimodal sensing, Hardware Platform-2 sends data to local storage for computations. It is shown that Platform-2 is much more battery efficient [17]. This provides the basic motivation for cloud based implementation of AI algorithms like MDP and GA. Moreover, fitness wearable have advantage of being with the user most of the day and therefore require battery life of more than 12 h [8]. It is difficult to define exact energy that an algorithm will spent as different systems have different architectures for performing the similar computation. To get a general sense of energy being consumed, Big-O, Small-O notation typically used for defining algorithmic efficiency [9] is used. Our algorithm is of Floyd Warshall type with computation of order $\Theta(V^3)$, hence the energy spent of order $\Theta(wV^3)$. Performing such computations on hardware would affect battery life. Hence computationally expensive algorithm like MDP and GA are shifted to cloud. Machine learning algorithms are coded in C/C++ for effective embedded implementation, while computationally expensive modules are coded in python for further cloud implementation.

3.1 The Markov Decision Process and Model Development

Sequential decision making can be divided into two parts continuous decision making dealt in economics and engineering subjects like control systems and discrete decision making-dealt in artificial intelligence and operation research. MDP finds application in various spheres including budgeting, asset acquisition with concave cost, asset acquisition with lagged information process, demand management [18–21], storage problems, shortest paths, dynamic assignment, pricing, information collection, technology switching dynamic programming can be used to solve MDP by using backward recursion method or by forward dynamic programming [22–24]. During Markov Formulation the transition from one state to another is assumed a stochastic process. Finite planning horizon restricts to six sequential predictions [25]. Each activity stage is referred as a state and is available at discrete intervals of time, denoted by S_t where $t \in \{1, 2, 3, 4, 5, 6\}$ involving a calorie value for being in the stage. Every state movement is referred an action state a_t and in total 12×11 actions are available. The value function for a state is given as $V(S_t) = C(S_t, a_t) + \sum V(S_{t+n})$, where $V(S_t)$ represents total calorie spent by the person if the 6 sequential predictions are followed and $C(S_t, a_t)$ represents the total calorie spent for being in a particular state and calorie spent on the action taken.

3.2 The Genetic Algorithm

GA is search technique to find an exact or approximate solution for various search and optimization problems. For finding a solution by mere brute force approach the equation for $V(S_t)$ above requires the algorithm to manually search within $\sim 12^6$ combinations, which makes it highly computationally expensive for battery limited smart phones and wearable devices [25–28]. Hence, GA optimization technique is used for solving the above mentioned problem and thereby limiting the computational cost involved in such complex processes [25, 28–30]. The GA is described in Fig. 3.

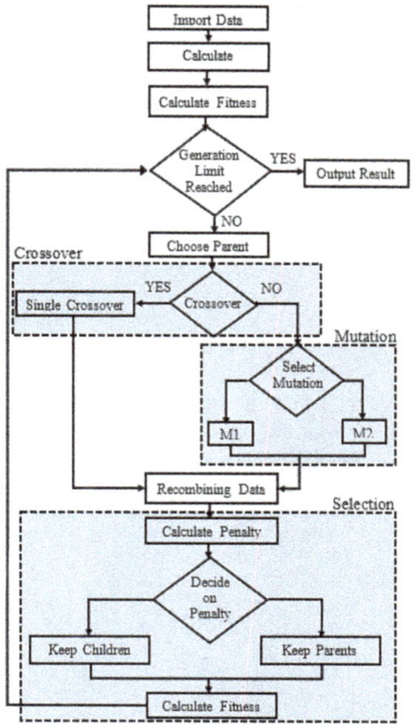

Fig. 3 The genetic algorithm sequence for selecting states

4 Results and Discussion

Matlab and Python are used as simulation tools. Machine learning algorithms are written in C for effective embedded implementation and computationally expensive modules like MDP and GA are coded in Python with proposed cloud based implementation. Figure 4 gives a fractional representation of instances from 12 ADL activities in the obtained dataset. A vast difference in proportion of translational activities like lay-sit as compared with respect to normal activities like walking is observed as translational activities require very less time as compared to normal activity.

Out of 561 variables Principal Component Analysis is used to reduce the number of features in hand and remove multi co-linearity among different variables. 12 ADL activities are color coded and plotted against PCA components. Distinct

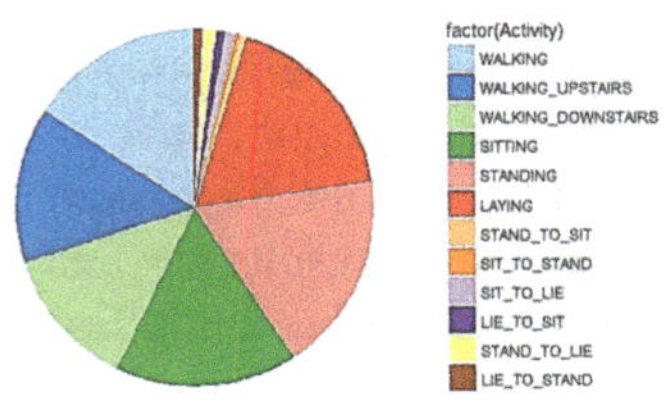

Fig. 4 Pie representation of instances from 12 ADL activities in the dataset

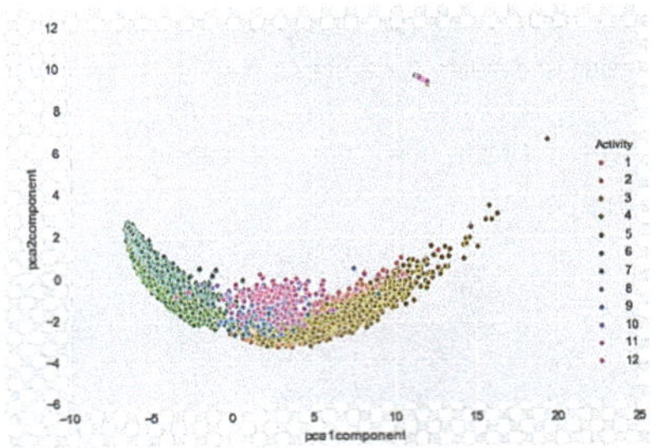

Fig. 5 2-D view of principal component analysis plot for 12 ADL activities

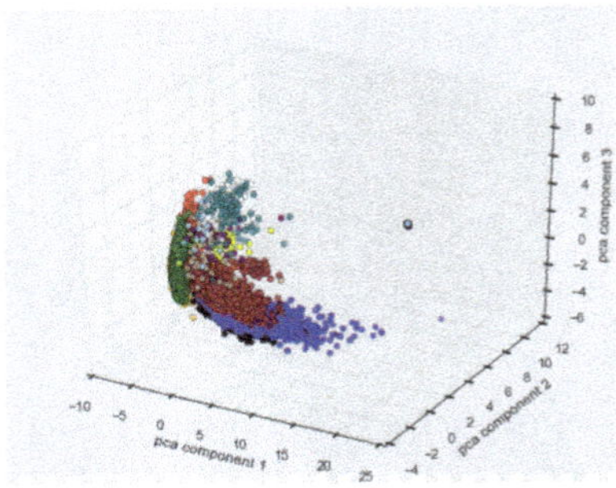

Fig. 6 3-D view of principal component analysis plot for 12 ADL activities

boundaries in principal component analysis are visible from the 2-D PCA (Fig. 5) and 3-D plots (Fig. 6). This makes it suitable for input to the SVM classification.

Figure 7 represents graphical node version of the results predicted by our algorithm. The six sequential states act as nodes which a person performing high, medium or low calorie burn exercises subsequently visited. As shown in Fig. 7, a person burning high calorie would perform the recommended exercise schedule as follows 2, 11, 6, 9, 10, 3 which translates into walking down the slope for 10 min, followed by Laying to stand exercise for 10 min, following which he spends 10 min on squats, then 10 min of lay to sit exercise, followed by stand to lay exercise for 10 min and last exercise would be to sit for 10 min for cooling down and relax.

A glimpse of the results is shown in Fig. 8. It consists of 6 columns. The transit columns represent the path that a person should take and the value columns represent the estimated calories a person would burn following the exercise routine recommended by the algorithm. Each row bears final results from one of the 1000 results iteration run during generation testing. The following result is shown for generation limit of 10.

For finding the optimal generation limit performed for mutation and crossover functions, the algorithm is run with different number of generation. To decrease the variance of result, the system is run keeping generation limit as constant over 1000 iterations and results our findings is shown in Fig. 9. B, R, G represent generation limits of 20, 40 and 200 respectively.

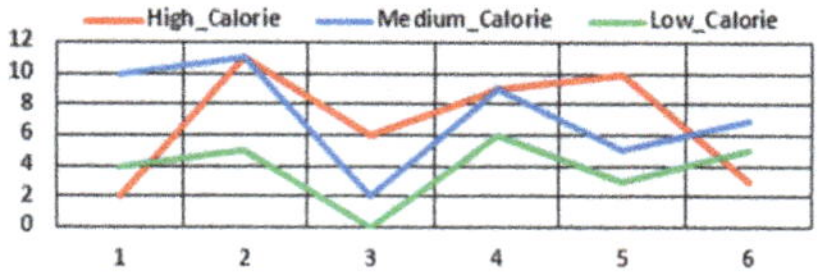

Fig. 7 Recommended sequential node switching of 12 ADL activity states

HIGH TRANSIST	HIGH VAL	MEDIUM TRANSIT	MED VAL	LOW TRANSIT	LOW VAL
[2 11 6 9 10 3]	670	[10 11 2 9 5 7]	480	[9 5 4 5 9 3]	261
[8 4 9 11 7 3]	687	[2 0 11 4 1 9]	481	[5 4 3 1 6 5]	270
[10 2 1 1 2 10]	663	[10 3 0 2 11 1]	479	[5 4 7 3 0 4]	258
[1 6 7 4 5 9]	616	[3 9 4 0 7 10]	477	[0 4 6 5 0 4]	236
[0 10 6 8 3 4]	676	[3 5 8 7 2 6]	482	[4 5 0 6 3 5]	199
[4 8 2 9 6 0]	632	[4 0 6 10 5 3]	478	[5 0 3 3 0 5]	181
[7 8 11 9 3 2]	635	[9 10 6 2 5 7]	474	[0 3 4 4 0 3]	225
[0 10 9 1 4 11]	576	[0 11 3 10 7 2]	481	[3 4 0 4 1 3]	255
[9 5 7 4 8 10]	709	[8 4 10 0 5 1]	479	[2 3 4 3 5 2]	243
[8 3 11 4 0 9]	630	[4 2 3 5 1 9]	472	[6 5 3 5 0 4]	202
[6 9 3 2 11 10]	676	[8 5 3 8 11 4]	478	[5 0 4 6 5 4]	213
[2 10 11 11 1 6]	628	[1 4 6 11 8 5]	475	[1 3 5 4 0 5]	246
[1 6 2 5 9 0]	631	[6 1 11 8 7 2]	477	[4 0 5 9 6 2]	273
[5 4 7 9 10 3]	668	[2 9 8 6 0 11]	483	[0 4 6 7 3 4]	260
[5 4 11 6 10 2]	689	[0 4 2 7 3 1]	471	[3 4 5 1 5 3]	227
[7 2 8 5 9 6]	611	[3 0 11 5 9 6]	473	[5 2 4 3 7 5]	222
[9 6 8 2 7 4]	687	[7 1 6 3 2 10]	480	[4 5 3 8 3 0]	234

Fig. 8 View of result table for personalized activity sequence recommendation

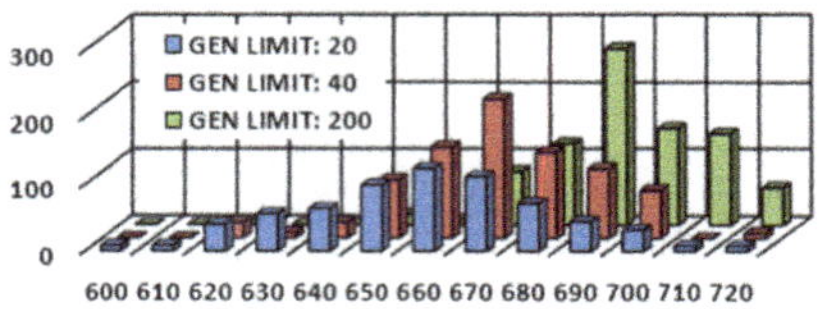

Fig. 9 Performance of genetic algorithm over varying number of generations

As is evident from Fig. 9, an increase in mean of values of high calorie sequential activity calculated, though the max value remains the same. Thus indicating the reporting of higher calorie values when more number of generations exist. Though the global maxima is achieved in all the cases and the kernel density estimation spread moves towards higher values with increase in number of generations. Thus there exists a trade-off between number of computations and the quality of solutions.

5 Conclusion

In this work, an energy efficient computing technique for next generation mobile applications is developed. Artificial Intelligence based embedded algorithm that provide personalized training sequence recommendation in order to achieve desired calorie goals is proposed. The crux of this implementation is Calorie Matrix Regeneration via state feedback technique using MDP and GA. Machine learning algorithms are written in C for effective embedded implementation and computationally expensive modules like MDP and GA are coded in Python with proposed cloud implementation for extending battery life significantly. This work finds potential application in the field of fitness training and biomedical rehabilitation. This implementation is first of its kind and a step ahead of available state of the art fitness training algorithms/applications.

References

1. Smart wearable: reflection & orientation, DG connect services, European Commission, Content & Technology, CNECT, Brussels, 2016
2. http://www.cisco.com/c/en/us/solutions/collateral/service-rovider/visual-networking-index-vni/mobile-white-paper-c11-520862.html (weblink)
3. The challenges of wearable electronics, TE Connectivity Limited, 2015
4. Verma P, Rajnish R, Fatima S (2017) Challenges: wearable computing for internet of things. Int J Sci Res
5. Molinero AR, Martinez DP et al (2007) Detection of gait parameters, bradykinesia and falls in patients with Parkinson's disease by using a unique triaxial accelerometer. World Parkinson Congress, Glasgow
6. Mannini A, Sabatini AM (2010) Machine learning methods for classifying human physical activity from on-body accelerometers. Sensors
7. Allen FR, Ambikairajah E et al (2006) Classification of a known sequence of motions and postures from accelerometry data using adapted Gaussian mixture models. Physiol Meas 27:935
8. Carroll A, Heiser G (2010) An analysis of power consumption in a smartphone. In: USENIXATC proceedings of USENIX conference
9. Demaine ED, Lynch J, Mirano GJ, Tyagi N (2016) Energy-efficient algorithms. In: Proceedings of the 2016 ACM Conference on Innovations in Theoretical Computer Science (pp 321–332). ACM
10. Anjum A, Ilyas MU (2013) Activity recognition using smartphone sensors. In: Consumer communication and networking conference, IEEE
11. Ronao CA, Cho SB (2016) Human activity recognition with smart phone sensors using deep learning neural networks. ESA, Elsevier
12. Capela NA, Lemaire ED et al (2016) Evaluation of a smart phone human activity recognition application with able-bodied and stroke participants. J NeuroEng Rehabil
13. Anguita D, Ghio A, Oneto L et al (2013) A public domain dataset for human activity recognition using smartphone dataset. In: ESANN
14. Ravi N, Mysore P et al (2005) Activity recognition from accelerometer data. In: Proceedings of innovative applications of artificial intelligence
15. Kwapisz JR, Weiss GM, Moore SA (2011) Activity recognition using cell phone accelerometers. SIGKDD Explor Newsl 12(2)
16. Vasan KK, Surendiran B (2016) Dimensionality reduction using principal component analysis for network intrusion detection. Elsevier
17. Choudhury T, Consolvo S et al (2008) The mobile sensing platform: an embedded activity recognition system, an Intel research. IEEE
18. Powell WB (2007) Approximate dynamic programming: solving the curses of dimensionality, vol 703. John Wiley & Sons
19. Li Y, Cao F (2011) Infinite horizon gradient estimation for semi Markov decision process. In: 8th Asian control conference, IEEE
20. Rout RR, Krishna MS, Gupta S (2016) Markov decision process-based switching algorithm for sustainable rechargeable wireless sensor networks. IEEE Sens J

21. Garcia MG, Ruiz J, Ledesma S et al (2010) Combination of acceleration procedures for solving stochastic shortest path Markov decision processes. In: Intelligent systems and knowledge engineering, IEEE
22. Theodoridis S, Kourtoumbas K (2004) Pattern recognition, 2nd edn, p 582
23. Gendreau M, Laprte G, Potvin J-Y (1994) Metaheuristics for the vehicle routing problem. Manag Sci 40:1276–1290
24. Laporte G, Gendreuau M, Potvin J-Y, Semet F (2000) Classical and modern heuristics for the vehicle routing problem. Int Trans Oper Res 7:285–300
25. Moon S, Bawane N (2015) Optimal feature selection by genetic algorithm for classification using neural network. IRJET. ISSN: 2395-0056
26. Pei M et al (1995) Genetic algorithm for classification and feature extraction. In: Classification society of North America, USA, 22–25, 1995
27. Patriarche J, Manduca A, Erickson B (2003) Improved classification accuracy by feature extraction using genetic algorithms. In: Proceedings of SPIE, USA
28. A deeper lay than Moore's? The Economist online, web link: https://www.economist.com/blogs/dailychart/2011/10/computing-power
29. Koomey JG et al (2009) Assessing trends in the electrical efficiency of computation over time. IEEE Ann Hist Comput
30. Bennett CH (2015) Notes on Landauer's principle, reversible computation and Maxwell's demon. Stud Hist Philos Mod Phys

Hand Gesture Recognition: A Survey

Shamama Anwar, Subham Kumar Sinha, Snehanshu Vivek and Vishal Ashank

Abstract A human–computer interaction is generally limited to taking input from the user using handheld devices like keyboard, mouse, or scanners. With the advancement in computers, the user interaction approaches have also advanced. Direct use of hands as an input device is an attractive method for providing natural Human–Computer Interaction. It is also helpful for people who use sign language. The chapter aims to study the existing methods for Hand Gesture Recognition and provide a comparative analysis of the same. The entire process of hand gesture recognition is divided into three phases: hand detection, hand tracking, and recognition. The chapter includes a review of the different methods used for the hand gesture recognition. The recognition phase is classified based on the way the input is received as glove based or vision based. For recognition, various methods like Feature extraction, Hidden Markov Model (HMM), Principal Component Analysis (PCA) are compared along with the reported accuracy.

Keywords Computer vision · Hand gesture · Hidden Markov model
Principal component analysis

1 Introduction

Communication among people varies from being verbal to use of body language or gestures. Gesture forms an important means of communication. People tend to involuntarily use hand movements (termed as gestures) when they normally talk or even during telephonic conversations. Hand gestures provide a separate complementary modality to speech for expressing one's ideas giving emphasis on certain points. Humans can conveniently interact with computing devices by using hand

---

S. Anwar (✉) · S. K. Sinha · S. Vivek · V. Ashank
Department of Computer Science and Engineering, Birla Institute
of Technology, Mesra, Ranchi 835215, India
e-mail: shamama@bitmesra.ac.in

V. Nath and J. K. Mandal (eds.), *Nanoelectronics, Circuits and Communication Systems*, Lecture Notes in Electrical Engineering 511,
https://doi.org/10.1007/978-981-13-0776-8_33

gestures. The method is more suitable than using other input devices but the major challenge is how to make hand gestures understood by the computing devices.

For this purpose, hand gesture recognition systems have evolved. The entire hand gesture recognition process can be divided into three phases: Hand Detection, Tracking, and Recognition. In the first phase, a video input is given to the system, which is then divided into frames (images). The aim of this phase is to recognize the object of interest (i.e., the hand) from the frames. This phase may require some form of preprocessing like noise removal, background subtraction, etc. Once the hand is isolated from the frames, the tracking is done in subsequent frames to detect the motion. There are various existing models to aid this process. The chapter presents some prominently used methods.

The approaches used for hand detection can be mainly divided into "Data-Glove based" and "Vision Based" approaches based on the way the input is taken by the system. The Data-Glove-based methods require the use of sensor devices for recognizing finger and hand movement, which then needs to be represented in an appropriate form for further computations. The sensors aid the collection of hand configuration and movement data. However, the devices are quite expensive and bring much cumbersome experience to the users. In contrast, the vision-based methods acquire the input by means of a camera. This method of input is more convenient and portable as well. Any handheld or stationed device can be used for acquiring the input. These systems need to be background invariant, lighting insensitive, person and camera independent to achieve real-time performance, which is a challenge. Moreover, such systems must be optimized to meet the requirements, including accuracy and robustness. The purpose of this chapter is to present a review of Hand Gesture Recognition techniques for human–computer interaction, consolidating the various available approaches, pointing out their general advantages and disadvantages along with their reported accuracy.

1.1 Glove Based

In glove-based recognition system, a glove with sensors is provided that detects the finger and hand movement. The type of sensors used in these gloves varies from flex sensors to LED sensors. The positioning of sensors also varies across different models [1]. Some system use gloves with sensors on fingertips, while others prefer gloves with sensors at the finger joints (Fig. 1).

1.2 Vision Based

In vision-based gesture recognition, the movement of the hand is recorded by a camera. The video is decomposed into a set of images (frames). Some preprocessing may be required to isolate hand from other body parts and to eliminate the

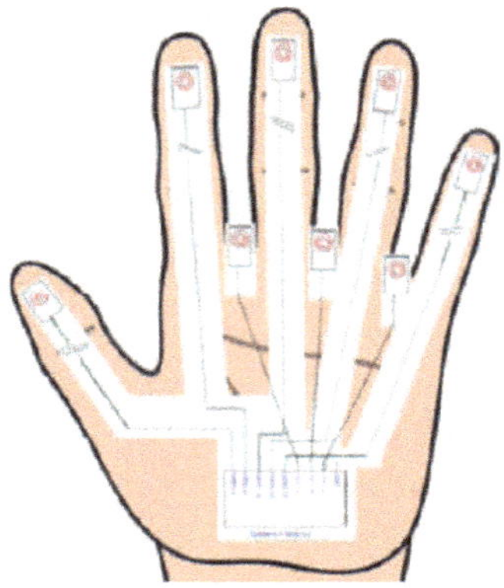

Fig. 1 Conceptual view of the gesture recognition glove consisting of sensors [1]

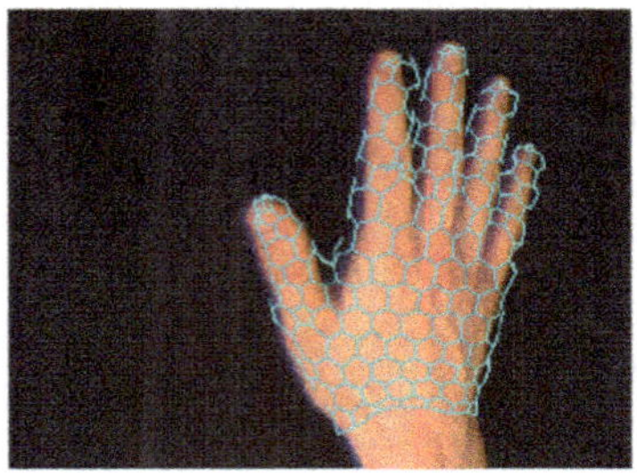

Fig. 2 Snapshot of 3D tracker [2]

Fig. 3 Real-time hand tracking using color glove [3]

background. The approaches also differ based on the background elimination techniques used. Simple background subtraction can be used if the background is static. But for real-time tracking, the background is not static. So, these implementations require a more complex background elimination technique. After the background has been eliminated, hand recognition is performed. The common approaches used for hand detection in vision-based recognition are skin color detection and 3D hand model approach [2, 3]. A description of the techniques is included in the subsequent section (Figs. 2 and 3).

2 Gesture Recognition Techniques

There are various gesture tracking techniques available. Some are glove-based recognition while others are based on vision. Some efficient algorithms exploit the advantages of both these methods. Once the hand is detected from an input frame, its movement is tracked for further recognition. There are various approaches for the same. A simple approach for recognition is the template matching technique. This method requires creating a template of predefined actions. Few researchers experimentally determined the number of templates required for a certain gesture and maintained a database of the same [4]. They also used linear regression to calculate the exact number of templates to be used on a certain gesture based on the average time the gesture was performed. The experiments were conducted on hand gestures taken on a fixed background. Hand pose recognition in the cluttered background has more applicability in real-life tracking. To achieve this, many techniques have been combined by Stenger [5] in their proposed system. The color model is initialized and updated by a frontal face detector. Hand locations and scale are hypothesized efficiently using cumulative likelihood maps, and the hand pose is estimated by normalized template matching. The system eliminated the need for background subtraction and the method was efficient enough to detect the hand in each frame independently. A drawback for template-based methods is the need to maintain a template set for the recognizable gestures.

Other than template-based matching, feature extraction-based methods have also been used. A method to recognize the unknown input gestures by using Hidden Markov Models (HMMs) was proposed by Chen et al. [6]. Since the variation of the hand gestures is usually large, the transition between states is necessary in each gesture for an effective hand tracking. The experiments in the paper recognized a single action in a stationary background. Hence, the system had a smaller search region for tracking. Addition of a new gesture required retraining the HMM for the new gesture. Repeated experiments could recognize 20 different gestures, and the recognizing rate is above 90% [6].

Another method based on feature extraction was implemented to recognize American Sign Language and Arabic numbers. The method used stereo color image sequences in HMMs. The system has three stages: preprocessing, feature extraction, and classification. In preprocessing stage, color and 3D depth maps were used to detect and track the hand. In the second stage, 3D combined features of location, orientation, and velocity with respect to Cartesian and Polar systems were used. Additionally, *k*-means clustering was also employed for HMM. In the final stage, the hand gesture path was recognized using Left-Right Banded topology (LRB). This system successfully recognized isolated hand gestures with 98.33% recognition rate [7]. But methods based on feature extraction are found to be computationally expensive.

Methods based on active shape model have also gained popularity. In [8], an active statistical model for hand gesture extraction and recognition is applied. After the hand contours are found out by a real-time segmenting and tracking system, a set of feature points (landmarks) were marked out automatically and manually along the contour. Mean shape, eigenvalues, and eigenvectors are computed out and composed the active shape model. When the model parameter is adjusted continually, various shape contours are generated to match the hand edges extracted from the original images. The gesture is finally recognized after well matching.

A method using Principal Component Analysis (PCA), which used skin color detection (vision based) for hand recognition was also designed, which was tested in the controlled background and in different lightning conditions. The database collected in the ideal conditions has proved to be the most efficient database in terms of accuracy and gives 100% accuracy. When the lightning conditions are changed, the accuracy decreases as compared to the previous one. The system shows 91.43% accuracy with low brightness images [9]. But the model was not capable of working with the images containing hands of other than skin color. The proposed model does not evaluate the images clicked in other light colors, where the hand gestures have been clicked and the model works only with a static gesture. But there might be miss-recognitions in case the background has elements that resemble the human skin [10].

3 Comparison of the Methods

Based on the study of the different techniques, a comparison table is provided (Table 1) that lists the advantages, disadvantages, and accuracy of the different methods reviewed.

Table 1 Comparison of different techniques for gesture recognition

Method	Advantages	Disadvantages	Accuracy
Template matching	1. Simple 2. Accurate for small gesture set	1. Efficiency decreases for large posture sets	98.80% [11]
Feature extraction	1. Works well for large gesture set	1. Computationally expensive	90% [6]
Active shape model	1. More suitable for real-time recognition 2. Also efficient in posture, face recognition	1. More accurate for open hand gestures, face recognition	88% [12]
Principal component analysis	1. Better recognition accuracy	1. Depends on lighting condition and illumination 2. Needs retraining	91–100% based on lighting factor [9]

4 Conclusion

Based on the review of the different techniques involved in hand gesture recognition, it is observed that the two major ways a human–computer interaction system can take input is, either glove-based or vision-based method. The glove-based method although is more accurate but the cost of such systems is generally high due to the need of a sensor-enabled glove. An additional hardware component (glove) is needed for implementing such system. The user comfort is also compromised as these methods require a certain restriction on the hand anatomy and hence the portability of such systems is less. In contrast, the vision-based methods are portable and generally does not require any specific or special hardware for implementation. A similar study on the various recognition techniques reveal the pros and cons of the different methods used. Template matching-based methods are simple and accurate for a small set of gestures or postures. It requires maintaining a large set of databases and may not be feasible if the applicability of such systems is on a large scale.

The feature extraction-based methods and active shape model methods are more suitable for real-time recognition and are generally vision based as well. These methods along with PCA and HMM needs more training to adapt the system for more accurate recognition. The chances of misrecognition are higher in real-time HMM-based methods due to real-time moving background.

References

1. Abhishek KS, Qubeley LCF, Ho D (2016) Glove-based hand gesture recognition sign language translator using capacitive touch sensor. In: IEEE international conference on electron devices and solid-state circuits (EDSSC)
2. Garg P, Aggarwal N, Sofat S (2009) Vision based hand gesture recognition. Int J Comput Electr Autom Control Inf Eng 3(1):186–191
3. Wang RY, Popovic J (2009) Real-time hand-tracking with a color glove. ACM Trans Graph 28(3)
4. Carrera KCP, Erise APR, Abrena EMV, Colot SJS, Telentino RE (2014) Application of template matching algorithm for dynamic gesture recognition of American sign language finger spelling and hand gesture. Asia Pac J Multidiscip Res 2(4):154–158
5. Stenger B (2006) Template-based hand pose recognition using multiple cues. In: Asian conference on computer vision
6. Chen F-S, Fu C-M, Huang C-L (2003) Hand gesture recognition using a real-time tracking method and hidden Markov models. Image Vis Comput 21(8):745–758
7. Elmezain M, Al-Hamadi A, Pathan SS, Michaelis B (2009) Spatio-temporal feature extraction-based hand gesture recognition for isolated American sign language and Arabic numbers. In: 6th international symposium on image and signal processing and analysis
8. Liu N, Lovell BC (2005) Hand gesture extraction by active shape models. In: Digital image computing: techniques and applications
9. Ahuja MK, Singh A (2015) Hand gesture recognition using PCA. Int J Comput Sci Eng Technol 5(7):267–271

10. Bansal M, Saxena S, Desale D, Jadhav D (2011) Dynamic gesture recognition using hidden Markov model in static background. Int J Comput Sci Issues 8(6):391–398
11. Azad R, Azad B, Kazeroni IT (2013) Real-time and robust method for hand gesture recognition system based on cross-correlation coefficient. Adv Comput Sci Int J 2(5):121–125
12. Edwards GJ, Cootes TF, Taylor CJ (1998) Face recognition using active appearance models. In: Computer vision—ECCV, LNCS

Design of Smart Embedded System for Agricultural Update Using Internet of Things

Qaiser Razi and Vijay Nath

Abstract In a developing nation like India, most of the population is dependent on agriculture. 60–70% population [1] depends only on agriculture and related sectors. Despite technological development in the entire field, now full attention is required in agriculture system to enhance the production and quality of crops. The production of crops plays a major role in the enhancement of GDP of the country. Quality of food gives the healthy life to the people. Not only human life depends on the agriculture animals and other creatures too depends on agriculture. There are so many factors that can generate good quality of crops, e.g., good quality seed, fertilizers, soils, and water and weather. As the farmers of our country are not so much literate and aware of the technology, as a result, the crop production is not up to the mark. If the farmers can embrace new technology the condition would be different and the agriculture sector can become the major employment source. Application of IoTs can help the farmer to increase the crop yield and there by minimizing the hard work done by them. In this chapter, the main objective is to let the farmers know about their necessary items such as good quality seeds, fertilizers, soils, water, weather etc. They can also know about the market value and portability of their crop. For a good yield setup, how the farmer will know how much water is required, when it is required and how much fertilizers are required, which type of crops they can plant, etc. These objectives can be achieved with the support of advanced technologies like sensors and GSM. IoTs will help in predicting the crop yield, soil quality, soil temperature, water level, and proper timing of crop to be delivered. Therefore, a system needs to be developed to enhance the farming procedure and reducing human effort and thereby increased farming output.

Keywords IoTs · GSM · Sensors · GDP

Q. Razi (✉) · V. Nath
VLSI Design Group, Department of Electronics and Communication Engineering,
Birla Institute of Technology Mesra, Ranchi 835215, Jharkhand, India
e-mail: qaiser.razi@gmail.com

V. Nath
e-mail: vijaynath@bitmesra.ac.in

V. Nath and J. K. Mandal (eds.), *Nanoelectronics, Circuits and Communication Systems*, Lecture Notes in Electrical Engineering 511,
https://doi.org/10.1007/978-981-13-0776-8_34

1 Introduction

With the increasing population in India, there will be huge requirement of food. More than 70% [2] of the population depends on agriculture. Agriculture plays a major role in the economy of the country. Farmers need to perform lots of task like seeding, weeding, fertilizing, watering, and crop cutting to grow certain crops. These tasks requires decision-making if a farmer by mistake added some different fertilizers then crop may be damaged or added more water than required then also crop gets damaged and may effect the crop production. One of the major issues faced by the farmers during crop production is that they have to keep a regular watch on their agricultural land if there are animals or not. Smart agriculture helps to deal with many of the issues stated above and thus increasing the crop yield.

2 Methodology

Smart Soil Management plays an important role in the production of any crop. One should have sound knowledge about the property of soil that which soil is suitable for what type of crop. If in any land cultivation is done then it remains fallow land or barrel land for entire season if no cultivation is done. So if farmers have knowledge about which crop can be grown after the cultivation of one crop they can make their life different and it is a boost the economy of our country. But unfortunately, farmers are not so much educated and technology updated, so IoTs will help them to come up with this problem. Smart Climate Management plays an important role in the production of any crop. But nowadays climate is changing drastically due to which the production of the crop is disturbed. In a study, it is found that agriculture production will decrease 10–20% by 2050 [1] because of climate change. If an IoTs-based system is developed that warns the farmers about sudden change in climate that will be very useful for them as they can protect their crop from sudden damage. Smart Water Management deals with the amount and quality of water which is very much important for the production of the crop. Crop water requirement depends on several factors like crop type, season, and climate and growth stage of the crop. Farmers should have sound knowledge of these factors but they are not so much aware of these factors. So IoTs-based system will help them to deal with the water management that is when and how much water is required. Smart Crop Management deals with the growth of the crop. During the growth of the crop, it requires right amount of nutrients that is fertilizers and adequate protection from pests and diseases. Right quality, right amount, right type and right time fertilizers are required to the crops for better production. This is not possible for the farmers to keep a eye on all these factors at a time so development of IoTs based system will help them in dealing with fertilizers and pest issue. Smart Delivery of the crops means delivering the crop to the market smartly that means according to the need the crop is delivered. Sometimes, it may happen that in

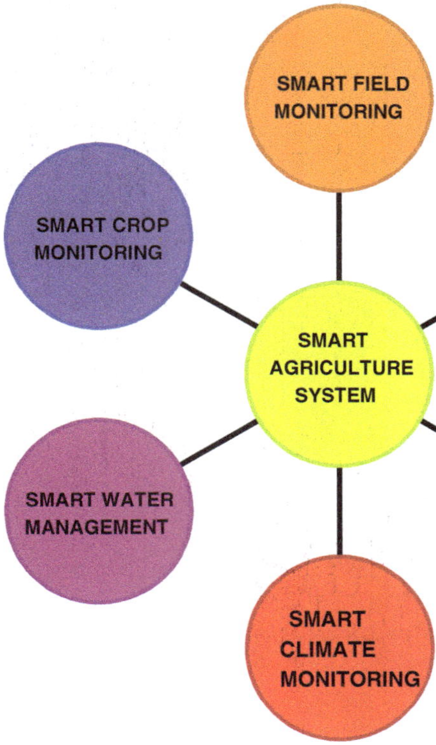

Fig. 1 Smart grid for agriculture update

certain area one crop is available in the market in large quantity due to which the price is down and the farmers may face loss. And there is a shortage of certain crop so to keep a balance of the crop availability system based on IoTs will help the farmers to deliver the crop to the market according to the requirement. Field monitoring means keeping an eye on the agricultural field. This is not possible for the farmers to regularly keep looking to their agricultural land from some damage by animals or human beings. So IoTs-based system will help them without going into the field they can monitor their field (Fig. 1).

Block Diagram (Fig. 2):
Sensors Application in Smart Agriculture System:

Sensors can play the most important role in smart agriculture system. Sensors are the device that can sense the condition and send the related information to the users. In making agriculture smart, we can use lot of sensors like soil moisture sensors, water-level sensors, temperature sensors, humidity sensors, movement sensors.

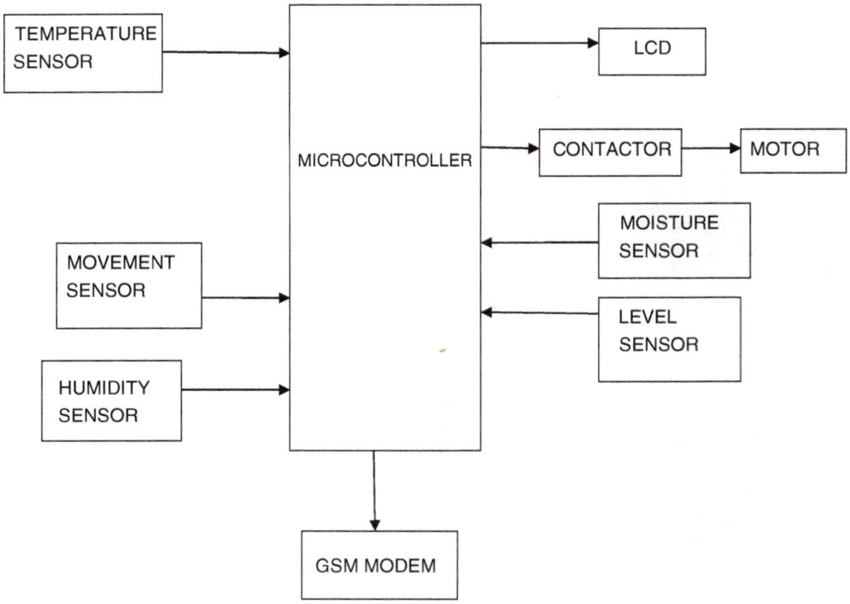

Fig. 2 Block diagram of microcontroller-based smart agriculture system

Fig. 3 Soil moisture sensor [3]

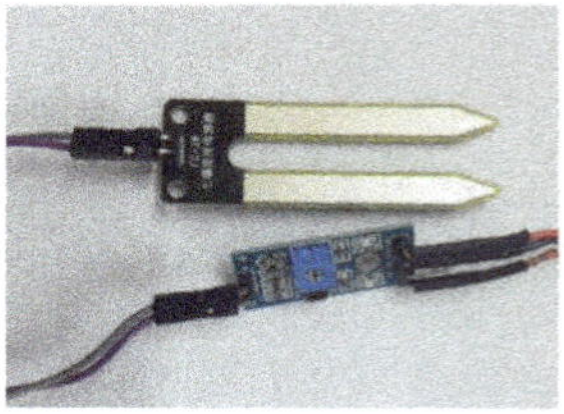

Soil Moisture Sensor (Fig. 3):

Soil moisture sensor works by measuring the dielectric constant of the soil. It helps to manage irrigation efficiently. This sensor consists of two electrodes that are inserted in the soil, which senses the moisture content of the soil and send the information if the moisture content is good or not according to that motor is switched on or off. Soil moisture sensors are of extremely low cost, water proof, small in size and output voltage is proportional to moisture level.

Flowchart (Fig. 4):
Water-Level Sensor (Fig. 5):

Sensors that detect the level of water, consist of wire which senses the level, it can be easily calibrated to nearly any range and any fluid in the field with the use of two buttons. One button records the minimum fluid level (0 V level), and the other button records the maximum fluid level (3 V level). After being calibrated, the

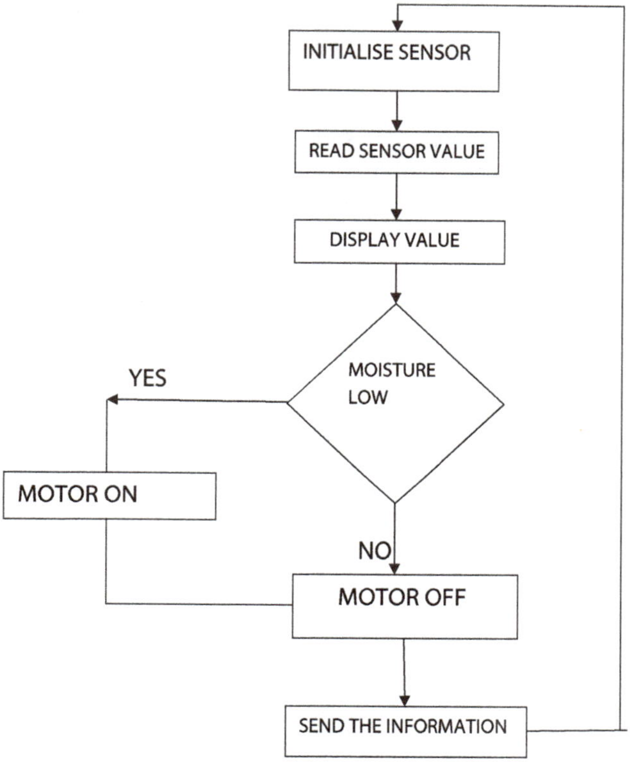

Fig. 4 Flowchart of soil moisture sensor

Fig. 5 Water-level sensor [4]

Fig. 6 Temperature sensor [5]

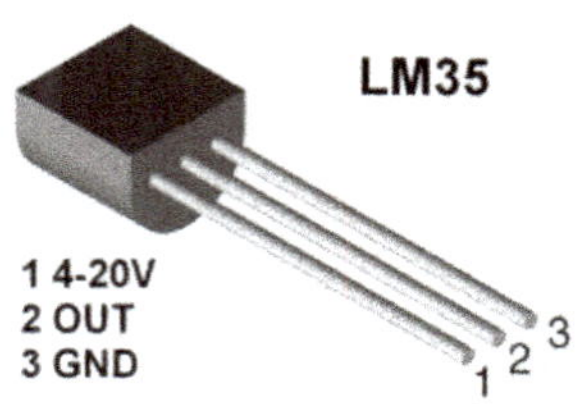

sensor will return a value of 0–3 V linear with the liquid level. Water-level sensors are compact, low cost, easy to use and take accurate reading in less than a second.

Temperature Sensors (Fig. 6):

Soil temperature sensor comes in variety of design using thermistors, thermocouple. The electricity transmitted from sensors to data logger can be converted to temperature value in Celsius, Kelvin, or Fahrenheit. Temperature sensors are of low cost, small in size, measures precisely and output voltage is linear to temperature.

Flowchart (Fig. 7):
PIR Sensors (Fig. 8):

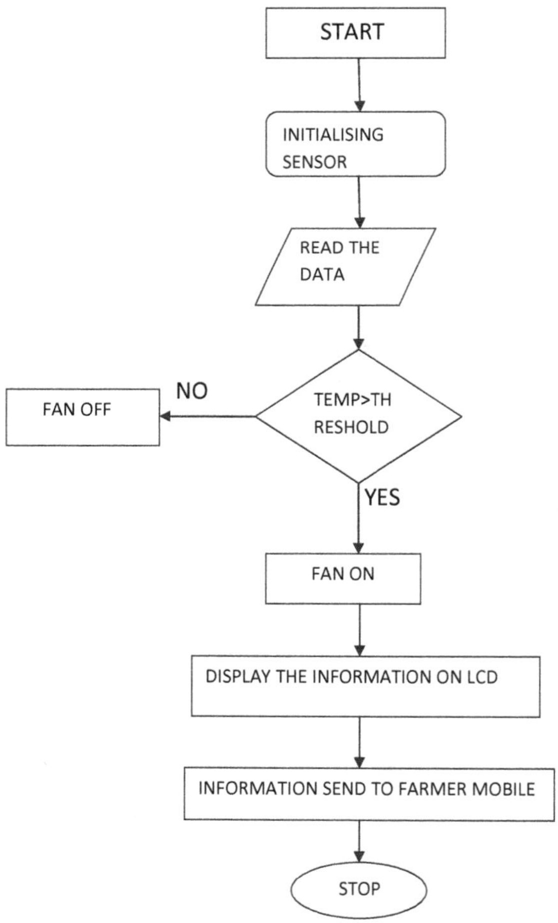

Fig. 7 Flowchart of temperature sensors

Fig. 8 PIR sensor [5]

The entire object emits heat energy in the form of radiation. It is invisible to human eye since it radiates infrared wavelength. PIR sensors detect infrared radiation emitted or reflected from object. It is to detect the movement of animals, humans being or any other objects. They are commonly used in the form of alarm or automatic lighting system. When any object passes that point the temperature there rises and thus the sensor converts the resulting change in output voltage and triggers the detection.

Flowchart (Fig. 9):

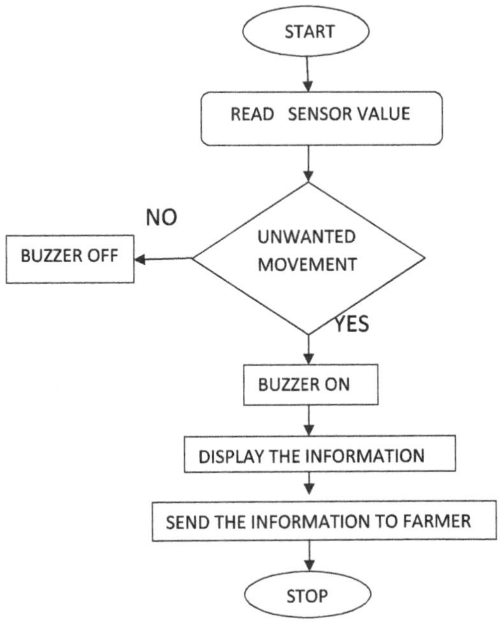

Fig. 9 Flowchart of temperature sensor

Fig. 10 16 × 2 LCD displays [6]

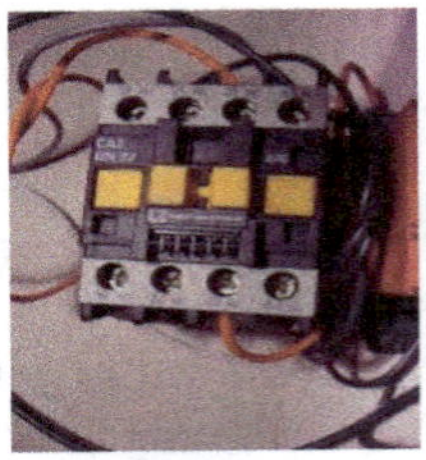

Fig. 11 AC contactor for pump application [7]

Liquid Crystal Display (Fig. 10):

LCD is a device that is used to display the required data on it. To display any character or numbers it is first interfaced with microcontroller and then the microcontroller will send the required ASCII value to the data bus of LCD. It plays an important role in making agriculture smart by letting the farmers know the condition of their agriculture land by displaying all the characteristics like temperature, soil moisture, humidity, and many more.

Contractor (Fig. 11):

Contactor is similar to relays but have higher current ratings. Contactor is an electrically controlled switch used for switching. It is connected to high-current load devices such as pumps or motor. It is connected to the microcontroller directly when the sensors senses the condition and sends it to the microcontroller. If microcontroller sends the information that soil moisture is low or water content is low accordingly contactor acts and switch on the motor.

GSM (Fig. 12):

GSM plays an important role in making agriculture smart. Nowadays everyone is having mobile phone. So if GSM modem is interfaced with all the sensors used and the microcontroller used then it will read the output sent by the sensors and sends it to the required users. This will be very helpful for the farmers as they can know the condition of their farm land without being physically present there.

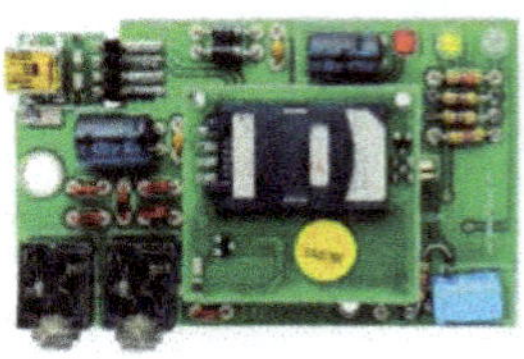

Fig. 12 GSM modem [8]

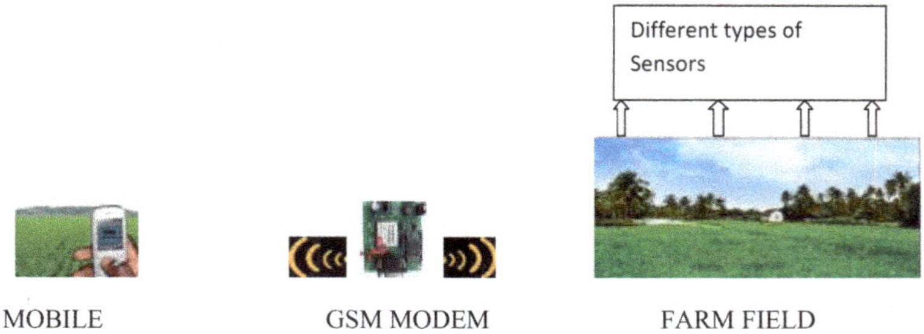

Fig. 13 Real-time implementation of e-farming [3]

Working and Result:

When the soil moisture receives the power it starts sensing the moisture of the soil and the analog value received is converted into digital by using ADC and the digital value is sent to the microcontroller. Now, if the soil moisture is less than the threshold value, it is displayed to the LCD and the information is also given to the farmers mobile using GSM modem and the action is taken that is motor is on. And if the moisture is in the limit or over the threshold value similarly the information is displayed on the LCD. When the water-level sensor receives the power it also starts sensing the water content in the agricultural land and accordingly it sends the information to the microcontroller and depending on the output of the sensor the action will be taken and the information will be displayed on the LCD and farmers mobile. As the power is received by PIR sensor it starts sensing the movement in the field by heat radiation and sends the information to the microcontroller. If there is no movement, then it will send 0 to the microcontroller and if there is movement in the field then it will sends 1 and accordingly it will be displayed on the LCD and the buzzer will start buzzing. Similarly, when the temperature sensor receives the power it will start sensing the soil temperature and if the temperature is greater than the threshold value or less than the threshold or the tem it will be displayed on the LCD (Fig. 13).

3 Conclusion

The smart agriculture system using IoT has been designed and implemented in this chapter. In India, farmers are not to much aware of the technology if they are provided with proper training and a smart mobile in hand they can perform many of their agriculture works from anywhere and anytime. Thus, application of IoTs in agriculture helps farmers to stay connected with their farm. It also helps in reducing human effort and increasing the productivity of the crops. Thus with IoTs and fully equipped software agriculture industry can become major source of economy and make India better in coming days.

Acknowledgments I would like to thanks Prof. S. Pal, Head, Department of Electronics and Communication and M. K. Mishra, Vice Chancellor, Birla Institute of Technology, Mesra, Ranchi for providing infrastructure to carry out these results.

References

1. Central Statistics Office, Directorate of Economics and Statistics, Ministry of Agriculture, Govt of India
2. Govt. of India. Pocket book on agricultural statistics-2014
3. Tyagi V, Kumar R, Fartyals G, Garg A (2017) IOT based agriculture system. Int J Eng Sci Comput 7(5)
4. Rukhmode S, Vyavhare G, Banot S, Narad A, Tugnayat RM (2017) IOT based agriculture monitoring system using wemos. Int Conf Emanations Mod Eng Sci Manag 5(3)
5. Suma N, Samson SR, Saranya S, Shanmugapriya G, Subhashri R (2017) IOT based smart agriculture monitoring system. Int J Recent Innov Trends Comput Commun 5(2):177–181
6. Roopashree V, Deekshith N, Soumya HV, Varshitha P. IOT in Indian agriculture using wireless sensor networks
7. https://en.wikipedia.org/wiki/Contactor
8. Suresh R, Gopinath S, Govindaraju K, Devika T, SuthanthiraVanitha N (2014) GSM based automation irrigation control. Int J Adv Res Comput Commun Eng 3(2)

Study and Design of Smart Embedded System for Train Track Monitoring Using IoTs

Sabiha Fatma and Vijay Nath

Abstract Trains are a key and cheaper mode of transportation systems in most developed countries around the world. Nowadays, train derailment in India is a major problem, thousands of peoples are dying due to derailing accidents. Other major problems are collision. It is also often occurring and results in severe damage to life and property. This chapter presents a thought of smart train track monitoring system and collision using IoT methods. This chapter presents various innovative ideas to control and detect the cause of derailed systems such as the strength of track, vibration measurement, load measurement, track alignment, conditioning of track. In this proposed smart system, track conditions will be monitored by IoTs systems with the help of different smart sensors pasted in track and information of track will be delivered to nearby substations, drivers, and central railway monitoring office. Before passing of the trains, the track will be scanned by systems and after the green signal of the system, train will be allowed to run on the particular track. The proposed methods can reduce the accidents.

Keywords IoT · Derail · Alignment

1 Introduction

As the railway transportation network is considered to be the safest and easiest network, but nowadays, it is not that much safer as a lot of accidents occur due to collision, wrong signaling, natural disaster, rail crossing, derail, and fault in track. Various technologies are also developed to avoid these accidents such as vibration detection, ultrasonic sensor detection technique, detection using a piezoelectric

S. Fatma (✉) · V. Nath
VLSI Design Group, Department of Electronics & Communication Engineering, Birla Institute of Technology Mesra, Ranchi 835215, JH, India
e-mail: Sabihafatma01@gmail.com

V. Nath
e-mail: vijaynath@bitmesra.ac.in

V. Nath and J. K. Mandal (eds.), *Nanoelectronics, Circuits and Communication Systems*, Lecture Notes in Electrical Engineering 511,
https://doi.org/10.1007/978-981-13-0776-8_35

transducer, optical fiber detection, etc. We can detect these faults using IoT server which is discussed in this chapter.

One of the methods we can use in monitoring of track can be optical fiber detection technique. In the optical fiber detection technique, optical fiber should be laid under the track and optical signal is transmitted through the fiber at one end, and on another end, there is a detector which detects the signal. If there is any fault or break in the track the signal will not be properly transmitted at the receiver end. In this way, the fault is detected and then the signal is transmitted to the substation. The most easiest way of determining the optical power is OTDR (optical time domain reflectometer). The fiber used here will be single mode (SM) fiber. In Fig. 1, it shows the process involved in detecting fault using optical fiber.

This is the most accurate way of detecting a fault, but practically it is difficult to implement. As the length of the track is large and fiber is also costly. So, this method cannot be implemented.

Major problem in train accident is the derailment. As the transportation of train always depends on the railway track only so, if there is any fault in the track it leads to severe damage to life and property. Derailment occurs when the train runs off the track but it is not necessary, derailment also occurs due to broken rails, collision of train with another object, mechanical failure of wheel, excessive gauge widening, improper operational control [1]. Most of the accidents are caused because of this derailment, which is sometimes not easily identified. Keeping all the points in mind the proposed smart embedded systems are designed to monitor the track and scan it before the train crosses the railway track and if there is any fault in track the sensor detects it and sends the information to nearby station master of the railway through ZigBee technology (Fig. 2).

1.1 Types of Train Accident Occurring

i. **Collision**

There are so many types of collisions observed such as: head-on collision, rear-end collision. There are so many causes of collisions such as natural disaster, improper handling of train crossing, collision with any undoable object, improper signaling, adverse weather condition, high speed of train, etc. When trains are passing through any tunnel there are chances of collision with any undoable objects.

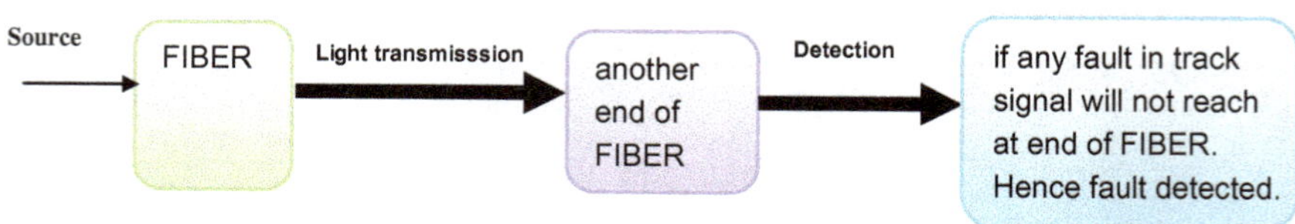

Fig. 1 Detection of fault using fiber

Fig. 2 Train derailment from the track [2]

ii. **Derail**
There are many causes for the derailment of train. In which some major are: Broken rails, mechanical failure of wheel, improper operational control, excessive gauge widening, collision with any object, harsh train handling, rail defects at bolted joints, open hydraulic lock and connection, explosives inside the track track mishap due to soil erosion, track strength and capacity, dynamic effect of track–vehicle interaction. Every year many accidents are occurring because of derail but in the year 2016 one biggest railway accident in Indore–Patna express due to derailment.

iii. **Crossing Accident**
A large number of accidents also occur at the railway crossing due to human error and improper signaling. It can be reduced by monitoring and controlling of track circuit.

iv. **Faulty Equipment and Improper Maintenance of Train**

v. **Fire Accident**

Among all types of accident, derailment is the major cause of accident according to a survey. A chart of percentage of accident in previous 2 years is shown in Fig. 3, in which it is clear that accident due to derailment is increasing every year.

Accident occurring nowadays is mostly because derailment. Derailment occurs when the train runs off the track. But it is not necessary usually, derailment occurs because of broken rails, collision of train with another object, mechanical failure of wheel, excessive gauge widening, improper operational control [3]. Figure 4 shows the derailed because of the mechanical failure of the wheel.

Except these bearing failure and track geometry are also causes of derailment. But broken rails and wields are major causes of derailment. Track geometry includes train alignment, wide gauge and elevation which are responsible for 7.3%

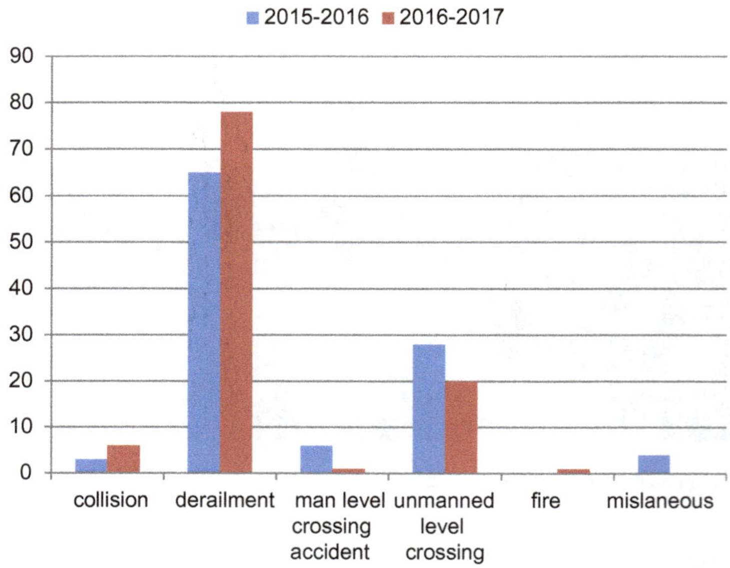

Fig. 3 Types of accident occurring in previous years

Fig. 4 Derailment in Australia

of derailment and bearing failure was responsible for 5.9% according to a survey which is shown in Fig. 5.

There are various causes of derailment. Some of the major ones are discussed below.

Broken Rails

The rail running surface should be continuous and of proper geometric layout. If the surface of rail is disrupted, i.e., surface is broken or there is any crack, then it may cause a serious accident. Due to broken rails derailment occurs. It occurs because of

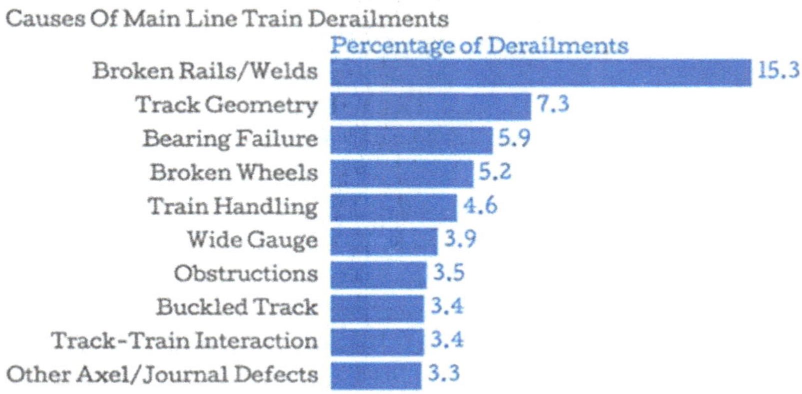

Fig. 5 DATA: federal railroad administrator [1]

Fig. 6 Broken rails [1]

hydrogen inclusion. It can be detected by track electrically detecting technique. It operates by passing a current through one or both of tracks, so if there is any crack present or rail is broken it detects it easily [1] (Fig. 6).

Gauge Widening

The distance between the inner sides of the two parallel rails that makes a single rail line is the rail gauge. Gauge widening mechanism is gradual and slow, but it cannot be ignored because sometimes derailment also occurs due to excessive gauge widening. Factors affecting this are poor maintenance of running gauge, excess speed, excessive load, etc. The dimension of wheel and edge of the track should have some slope so that train may run round in large radius curvature.

Defective Wheels
It is the most common failure where there is a failure of plain bearing, running gear such as wheelsets, suspension may fail due to the deficient lubricant. Failure of wheel is also due to the metallurgical crack propagation. Various technologies are developed to reduce the failure by eliminating the plain bearing and intervention.

Improper Operational Control
When the train is running at a very high speed and any object come in contact then it becomes very difficult to handle. Because of high speed, train also derailed from the track which causes serious damage. Recently Amtrak train was derailed on May 2015 as it was traveling at twice its speed limit. This can be avoided by advance safety technology (AST) using computers. The rail route, Amtrak left track, was not governed by AST so accident occurred. In this technology, we meant to prevent high-speed derailments.

Track Strength and Capacity
Track should have a sufficient vertical strength to withstand the maximum vehicle loads. Track quality is not simply a geometric property. The bearing strength and rigidity of individual components also play an important role.

2 Methodology

In the implementation of smart circuit using IoT to monitor and detect fault in track we require sensor, IoT, a processor named as ARM processor. ARM is based on RISC (reduced instruction set control) architecture which is developed by a British company. RISC based means the processor requires a few transistors than the typical processor in average computer and due to which its cost is reduced. It is a widely used processor and consists of desirable traits for light, portable battery-powered devices—including smartphones. In the family of ARM7, we are using LPC2148 IC. LPC2148 is the widely used IC from ARM-7 family. It has inbuilt many peripherals, so this IC is more reliable and efficient.

In Fig. 7, it will be clear how IoT will be interfaced [4]. IoT helps to connect with physical objects around us. Electronic devices like microcontroller behave like real objects and start communicating. Basically, the fundamental goal of IoT is to connect everything around us and enable effective communication between them with less human involvement.

2.1 Pin-Configuration of LPC2148

A circuit is designed for detecting and monitoring the track. There is transmitting unit and receiving unit in the circuit(Fig. 8). The transmitter unit, consist of microcontroller, IR (infrared) sensor, ZigBee device, motor, fault detector, alarm,

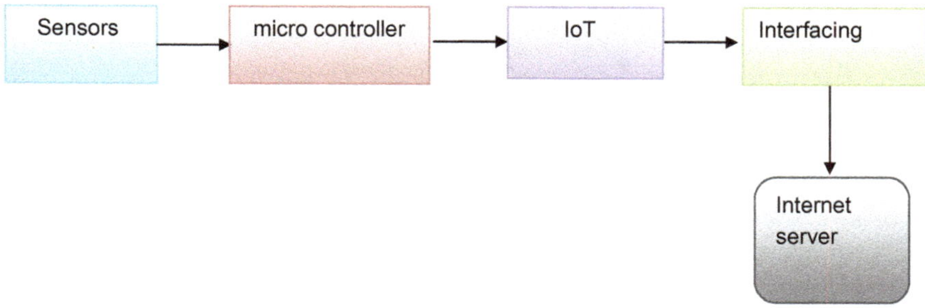

Fig. 7 Proposed architecture [5]

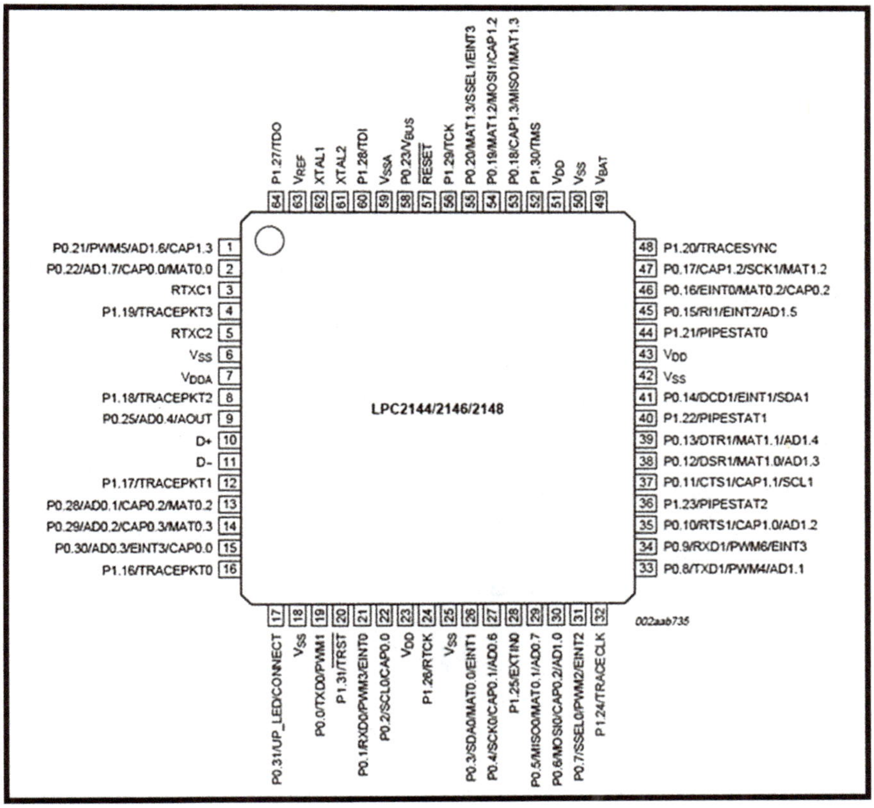

Fig. 8 Pin configuration of IC LPC2144

power supply, etc. Before passing of the train, track is scanned using this circuit and if there is any fault in the track it is detected by the sensor. The sensor will then send the information to the controller and controller then send the message to the signal

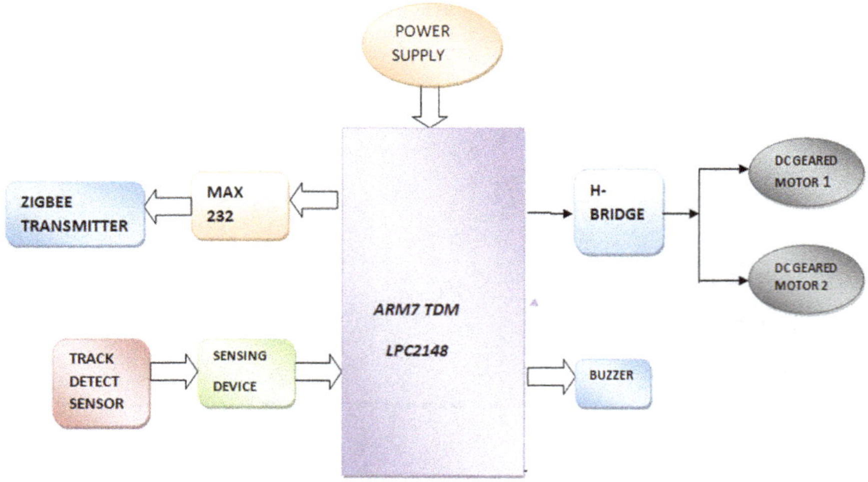

Fig. 9 Transmitting part of circuit [6, 7]

controller room through ZigBee device. The transmitter unit of the circuit is shown in Fig. 9. There is a receiver unit which is connected near the station control room, from where the signal is received through ZigBee about the fault in track. In this way, we can continuously monitor the track. Block diagram of receiving unit is shown in Fig. 10.

Technology used for communication. There are various technologies through which we can communicate. Such as: ZigBee technology, Wi-Fi, Bluetooth, etc. Here, we are using ZigBee technology for communication [9].

ZigBee which is a high-level communication protocol is based on IEEE.802.15.4.2003 standard and it can be used as wireless network WPANs such as wireless headphones with a mobile in a short range. ZigBee communication is used for control and sensor network and handles the devices at very low data rate. The ZigBee WPANs works at various ranges of frequencies, out of which 250 kps is best for two ways transmission of data between controller and sensor. ZigBee communication standard defines basically 2 layers, (1) physical layer (2) MAC (media access control). In ZigBee technology, data can be transmitted over long distance through an intermediate device and create a mesh network. ZigBee technology is a simpler, less expensive, and secure network. The ZigBee Alliance is a group of companies that maintain and publish the ZigBee standard [2]. ZigBee is a

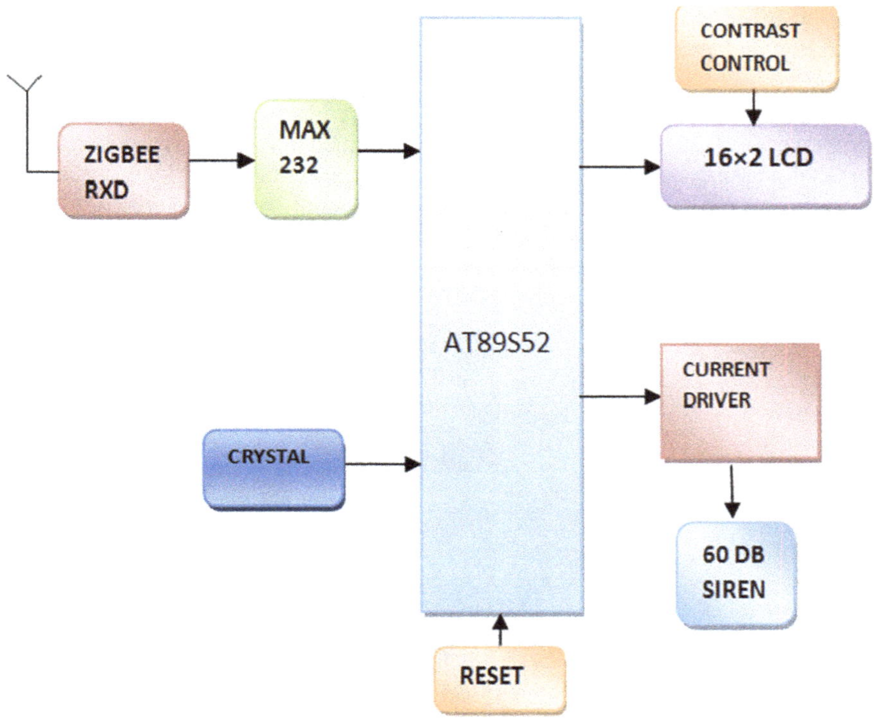

Fig. 10 Receiving part of circuit [8]

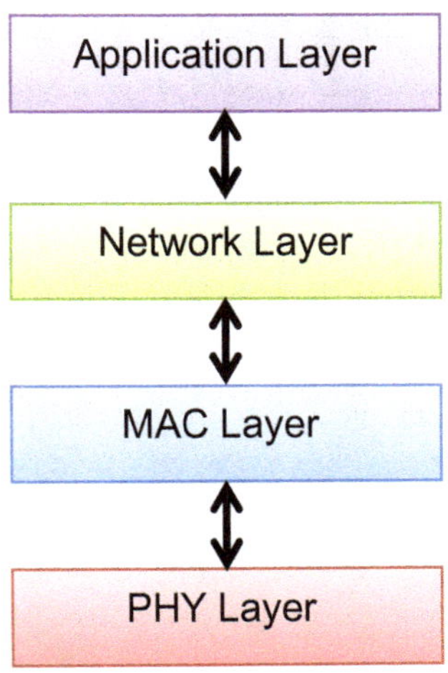

Fig. 11 Layers of ZigBee

standard-based wireless ad hoc network that consumes less power, high security, reliability, and cost efficient [10].

ZigBee follows the OSI reference model. Its layer is defined by IEEE standard which consists of physical layer, media access layer, application and network layer comes under ZigBee. It is an IEEE-based specification which is used to create personal area network (PAN). Layers are shown in Fig. 11 [10].

ZigBee technology has various appliances. Some of the applications are shown in Fig. 12.

Fig. 12 Application of ZigBee

Fig. 13 IR sensor

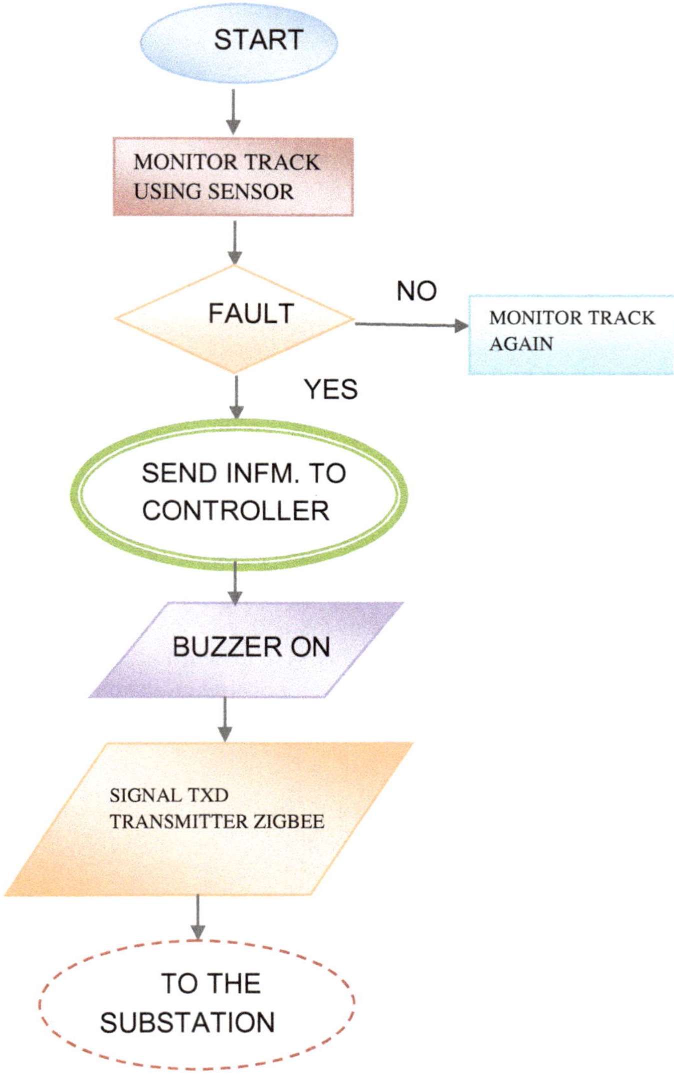

Fig. 14 Process involved in detecting fault

Sensor used here is the IR sensor. This IR sensor will sense the faults in the railway track and hence send the signal to the microcontroller.

Infrared sensors sense certain characteristics of its surrounding by detecting Infrared radiation. Infrared energy radiation lies in the invisible region. It is not visible with human eye, but we can feel it on skin as it emits lots of infrared, i.e., heat.

IR sensor is shown in Fig. 13.

3 Working Procedure

Transmitter section of the robot will be moving on the track for controlling system interfaced where a sensing device used to detect any crack in the track. Here, IR sensor is used to follow line of sight, i.e., detect the crack. It gives the information to the main controller if track is faulty anywhere. This same information is also transferred to the remote station through ZigBee. From the remote station, necessary action should be implemented. Flowchart in Fig. 14 explains the working procedure of the circuit [8, 6].

4 Conclusion

This paper consists of a design of embedded system using IoTs for monitoring tracks and detecting the cause of train accidents and also tells about the cracks and any defects in the track. With the help of this system, we can prevent many train accidents as circuit will predict earlier about the defects. Before passing the train, track will be scanned by systems and after green signal of the system train will be allowed to run on particular track. From the above-proposed methods can reduce the accident. Thus, implementation of this system will reduce the train accident. Fault in the track can also be detected by the optical fiber, which is discussed in the chapter, but the cost of installation is too high that is why its implementation is difficult.

Acknowledgments I would like to thank Prof. S. Pal, Head, Department of Electronics and Communication engineering and Prof. M. K. Mishra, Vice Chancellor, Birla Institute of Technology, Ranchi for providing infrastructure to carry out this result.

References

1. Track monitoring journals Scientific American, a Division of Nature American, INC
2. Singh M, Singh S, Jaiswal J, Hempshall J (2006) Autonomous rail track inspection using vision based system. In: IEEE international conference on computational intelligence for homeland security and personal safety, Oct 2006, pp 56–59
3. Zheng S, An X, Chai X, Li L (2012) Railway track gauge inspection method based on computer vision. In: IEEE international conference on mechatronics and automation, pp 1292–1296
4. Aboelela E, Edberg W, Papakonstantinou C, Vokkarane V (2006) Wireless sensor network based model for secure railway operations. In: Performance, computing, and communication
5. IoT for embedded systems: the new industrial revolution. Retrieved from http://www.micrium.com/iot/overview
6. Railway track Monitoring ANd accident Avoidance. https://www.ijarcce.com/upload/2016/april-16/IJARCCE%20242.pdf

7. Wu H, Wilson N (2006) Railway vehicle derailment and prevention. In: Handbook of railway vehicle dynamics
8. International Journal & Magazine of Engineering, Technology, Management and Research. A Peer Reviewed Open Access International Journal. Vol. 2 ISSN NO-2348-4845 (2015)
9. Hodge VJ, O'Keefe S, Weeks M, Moulds A (2015) Wireless network for condition monitoring in railway industry. IEEE Trans 16(3) (June)
10. Zigbee technology. https://www.elprocus.com/what-is-zigbee-technology-architecture-and-its-applications/
11. American experience: native Americans and the transcontinental railroad
12. Sambamurthy N et al (2013) Prevention of train accidents using wireless sensor networks. Int J Eng Res Appl 3(6):1592–1597. ISSN: 2248-9622. www.ijera.com (Nov–Dec)
13. Shuaib K, Boulmalf M, Sallabi F, Lakas A (2006) Introduction to ZigBee. Co-existence of ZigBee and WLAN, a performance study. In: IFIP international conference on wireless and optical communications networks, p 5
14. Culler D, Estrin D, Srivastava M (2004) Overview of sensor networks. Computer 37(8): 41–49. (Aug)
15. Hayre HS (1974) Automatic railroad track inspection. IEEE Trans Ind Appl IA-10(3): 380–384 (May)
16. Al-Nuaimy W, Eriksen A, Gasgoyne J (2004) Train-mounted gpr for high-speed rail trackbed inspection. In: Tenth international conference on ground penetrating radar, 21–24 June 2004
17. Navya K, Murthy MBR (2013) A ZigBee based system. Int J Eng Res Appl 3(5):483–486. www.ijera.com (Sept–Oct)
18. International Journal of Engineering Science Invention Research

Study and Design of Smart Embedded System for Smart City Using Internet of Things

Abhinav Kumar and Vijay Nath

Abstract Smart city is a way of developing the new areas in which it aims to integrate either city features or assets like smart homes, smart parking, waste management, traffic control system, and other community services. Each smart city is equipped with various kinds of electronic devices which are either directly or indirectly connected with IoT. This paper provides an inclusive idea of the concept of the smart city with the help of IoT. In the present day scenario, IoT plays an important role and their capabilities are merging into or apply to the different parts of smart cities for better results. The potential application of smart city pops up frequently in the context of urban development with respect to the technology involved in it. Meanwhile, there are some challenges all over the world and its implementation technique are thoroughly expressed in this chapter.

Keywords Internet of Things (IoT) · Smart city · Electronic devices

1 Introduction

The first question which arises in our mind is "what is a smart city". There are many definitions for smart cities. One such useful definition from the ITU is, a smart city is a modified or technically enhanced city that uses set of protocols to improve living standards by providing basic amenities to each and every resident in a better way. The concept of smart cities varies from place to place, depending on the level of development.

Smart cities offer various advantages over the traditional ones like good governance, efficient energy and waste management, providing quality education, etc.

A. Kumar (✉) · V. Nath
VLSI Design Group, Department of Electronics & Communication Engineering, Birla Institute of Technology Mesra, Ranchi 835215, Jharkhand, India
e-mail: abhinav.rahul99@gmail.com

V. Nath
e-mail: vijaynath@bitmesra.ac.in

V. Nath and J. K. Mandal (eds.), *Nanoelectronics, Circuits and Communication Systems*, Lecture Notes in Electrical Engineering 511,
https://doi.org/10.1007/978-981-13-0776-8_36

These features can be enhanced by incorporating with IoT which makes every unit to be connected to the Internet through a unique address, thus making it quickly and easily accessible even from remote location.

This chapter provides an insight into the concept of automated smart homes, which comprises of all the advanced facilities a smart city should possess, thus overcoming the problems of general homes. Various researchers have analyzed a lot over this concept.

2 Methodology

The concept of smart cities comprises of following things like good governance, responsible citizens, efficient energy and waste management, latest technology, modern infrastructure, better mobility and security, automated home, quality education, advanced parking, healthcare systems, etc. Figure 1 depicts the smart city features.

Good governance [1] aims at achieving greater efficiency and continuous improvement through innovation by using recent technology to improve standards of living and it also changes the ways of delivering the public services through the help of e-governance. Efficient energy management [2] provides a cost-effective and sustainable energy management in which renewable energy production and consumption are integrated and coordinated through energy services. It includes power generation through smart power grids, thus storing and consuming it in an efficient manner. Figure 2 shows a smart energy grid.

Latest technology [3] covers various issues like efficient use of energy and water, properly illuminated streets ensuring the safety of public, etc. It is also capable of self-learning by using previous experiences to improve performance by anticipating, thinking, and reasoning about the steps need to be taken in future. Thus, it has the ability to self-generate and self-sustain. Modern infrastructure [4] refers to a truly smart system, which includes sensing technologies that are embedded in infrastructure and the equipment which interacts with it. These sensors are linked to a communication channel which allows real-time data acquisition and analysis. In the present sensing environment, it is capable to respond in real time according to user's needs. Better mobility [5] is defined as information and communication technology-based system that provides cities' inhabitants with higher control to their access to transit as well as the more efficient use of their time. It improves the efficiency by acquiring traffic information in real-time scenario, taking required measures to reduce pollution (noise) and increasing safety level during traveling. This, in turn, implies, constructing a network or a system that collects and analyzes the information from companies and public institutions operating in the smart city. This creates a system of five layers which is illustrated in Fig. 3.

Smart homes [6] deals with additional facilities required in any home like multi-level security, equipping it with latest technologies, providing all luxuries and comforts, and many more which is either automated or a function of real-time

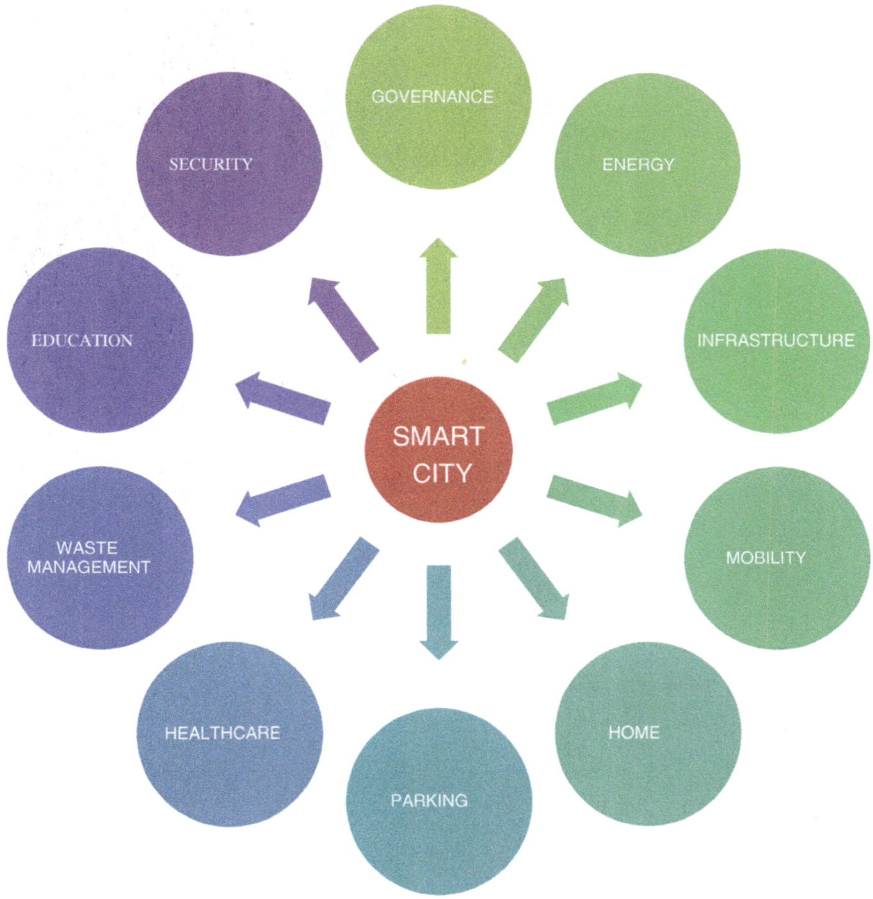

Fig. 1 Concept of smart cities

monitoring system. All these services must be eco-friendly and compatible with the real world. It makes use of information technology to share information by connecting too many subsystems, thus optimizing total home performance. Figure 4 shows the smart home with its associated features.

Smart systems are designed to deal with complex situations by predicting and being capable of deciding along with interacting with the environment. They may also be energy autonomous and networked. Better health care provides citizens with the ease of treatment round the clock, even at remote locations or in absence of doctors. Implementing with IoT offers online medical services and maintaining a digital record. Efficient waste management includes observation, detection, controls, and analysis is performed, which are based on pollution or waste caused by a human being. It requires a sudden action to be taken else it will have adverse effects on nature.

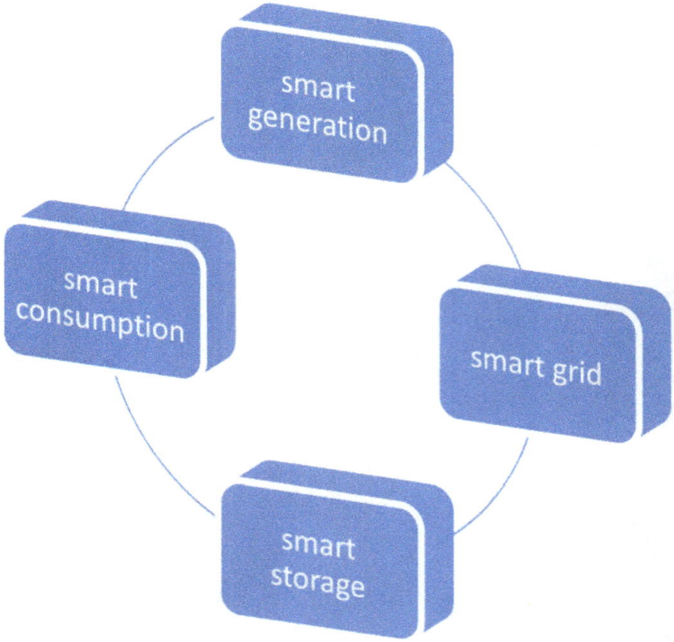

Fig. 2 Model of the smart energy grid

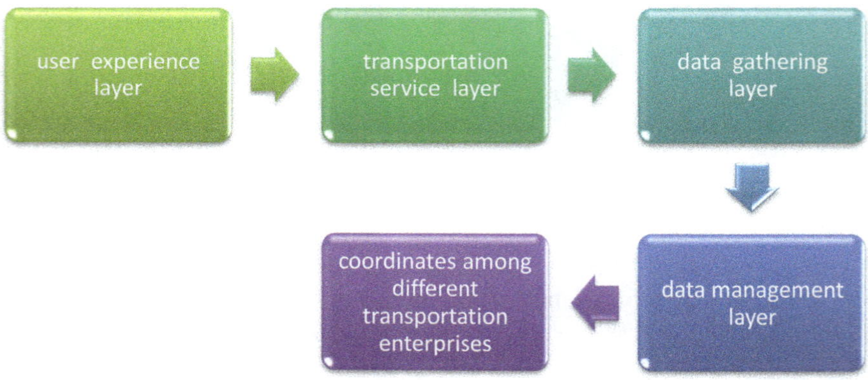

Fig. 3 Layers of smart mobility

Figure 5 shows the application of IoT in the implementation of a smart city.

In this chapter, a detailed overview on smart home is explained. Basically, efficient waste management, smart energy grid, smart parking, and better security, all these features are essential factors which make any city, smart in comparison with other city. And all these factors are associated with Internet by any means of

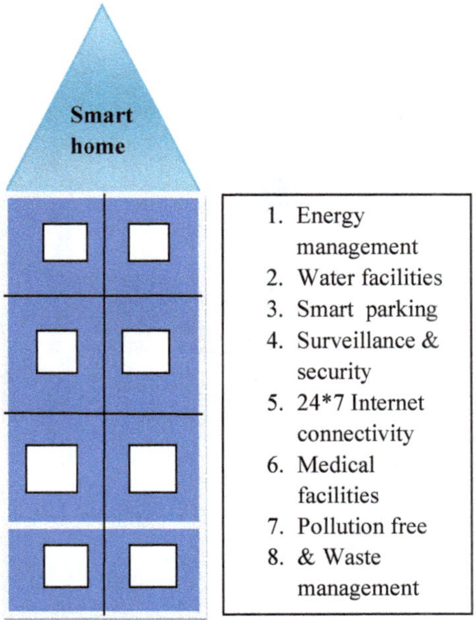

Fig. 4 Features of smart home

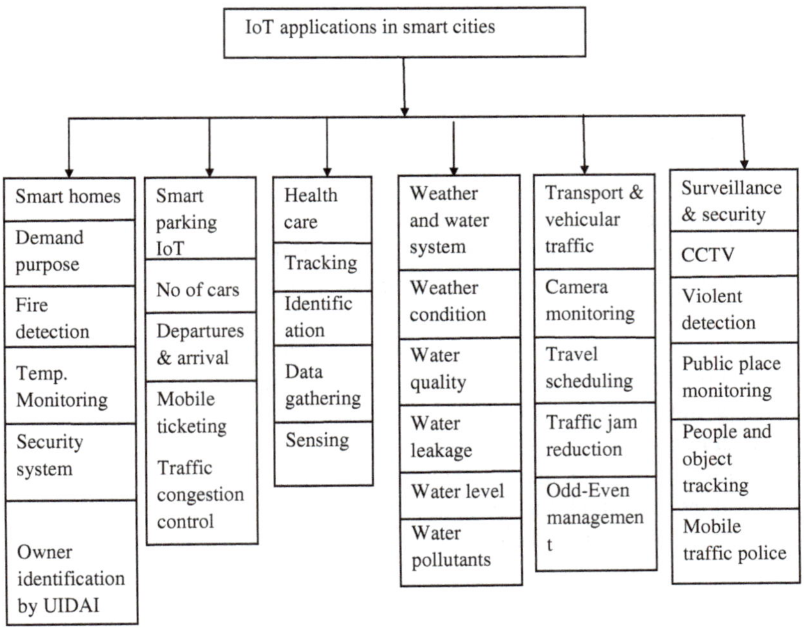

Fig. 5 IoT applications in smart city concept

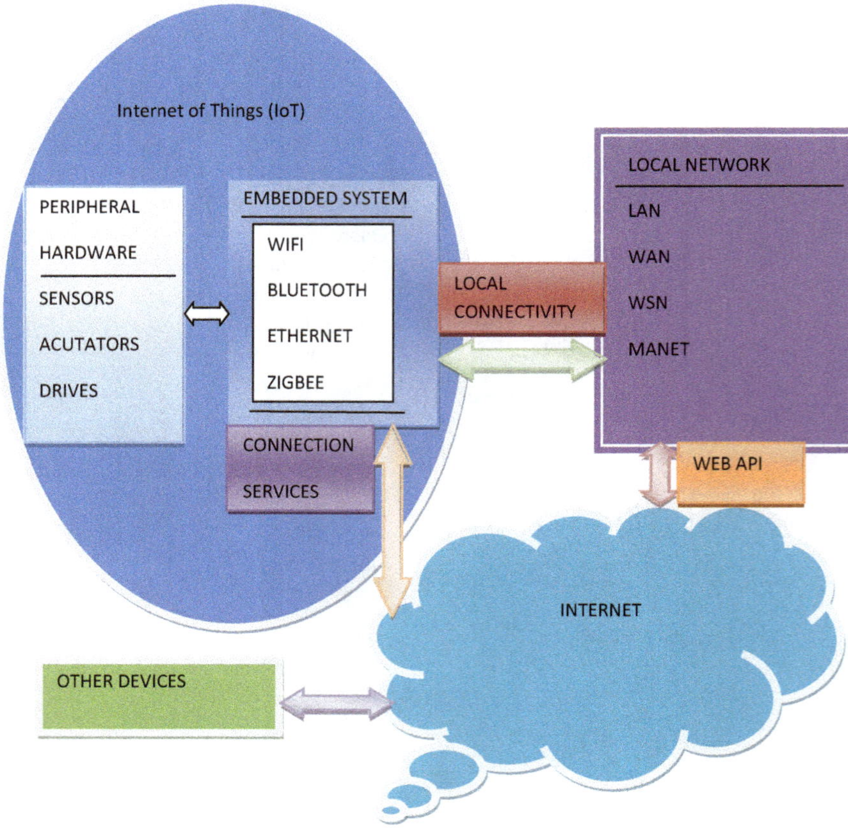

Fig. 6 IoT architecture

information technology like smartphones, a server, etc., which are either directly or indirectly attached with their respective server mechanism and further it is connected with the internet of things. A block diagram of their interconnectivity is shown in Fig. 6.

A smart home is a part of the living environment that is equipped with many hardware and software components interconnected together to improve the quality of life for its inhabitants. It comprises a lot of things which make them really smart or technically enhanced in compare to a general home. There are various regions or area such as energy, waste management, water supply, internet services, etc., which can be modified and make a home smart. Figure. 7 shows a smart home which consists of user, Internet, and IoT server is illustrated below. User refers to human being that uses all available facilities present in the real world, arranges them in a proper manner, observes and controls them wisely.

IoT server is an interface between different applications that is sensed by sensors and users via Internet. IoT server basically uses microcontroller services to collect

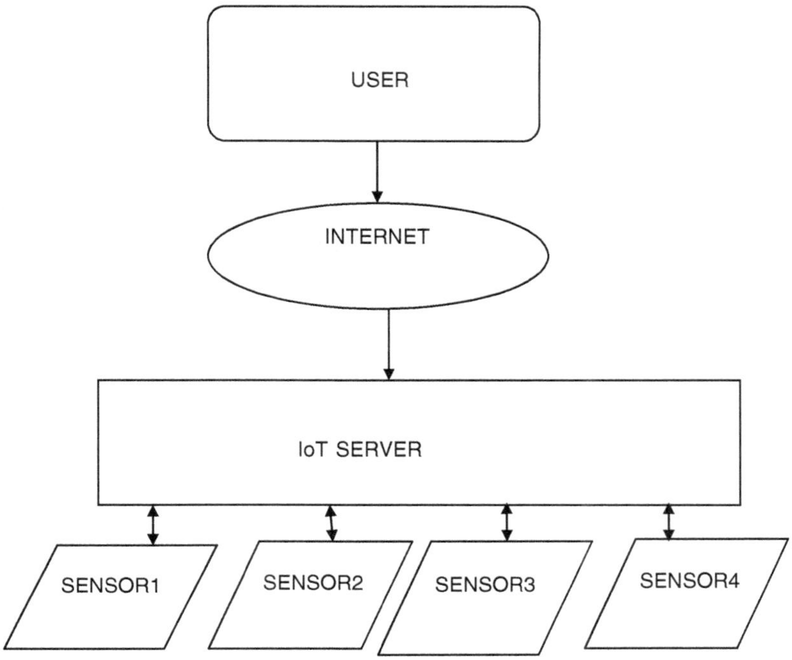

Fig. 7 Smart homes

input signal from various home-based applications and by the help of any IoT wireless communications platform like Raspberry, ZigBee [7], IoT (ESp8266), and WSN (wireless sensor network) it produces a suitable output which is under observation and control of user via Internet. Figure 8 shows the functions of IoT server.

Smart home [8] has the capability to monitor and regulate various abnormal conditions like detecting the fire and smoke, intensity of lights, speed of fan and activate alarms in an emergency situation. This is done with the help of several sensors like gas, PIR, IR, current, RFID, LDR, etc. First, the PIC controller connects to the Internet through Wi-Fi module ESP8266 which uses specific commands. All the components of IoT server are communicated with each other by using the wireless protocol (Fig. 9).

On establishing a secure connection, it initializes reading the parameters of various sensors. All sensors have a specific threshold value. The sensor data acts as an input, which is sent to web server along with temporarily storing it in the cloud. The data can be accessed from any location round the clock. If any parameter of the sensor, i.e. set value exceeds the given threshold value, then actuation comes into the picture to control many parameters like motion detection or gas leakage, thus notifying the web server which tends to control it remotely via Internet. In the same manner, all applications are observed and controlled frequently. For example,

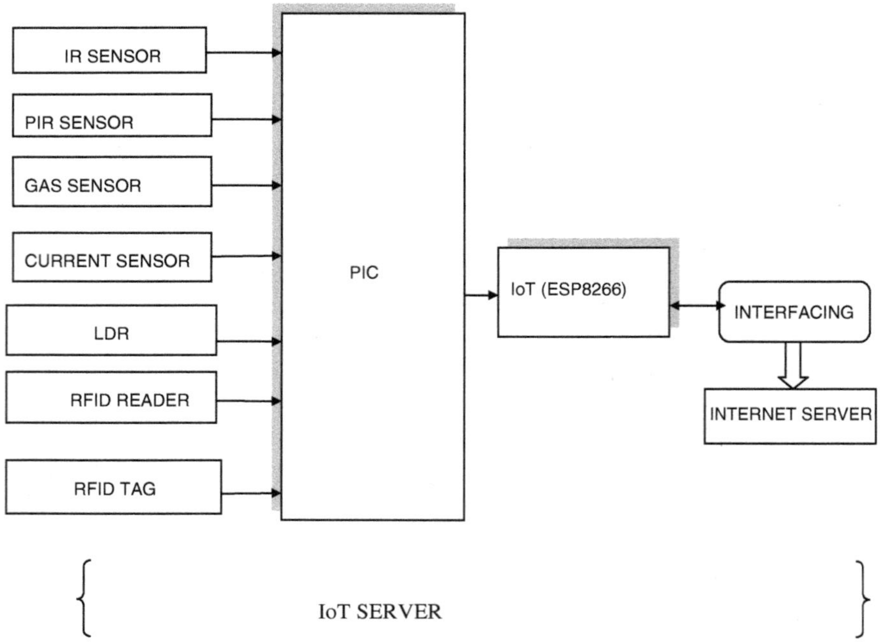

Fig. 8 IoT server

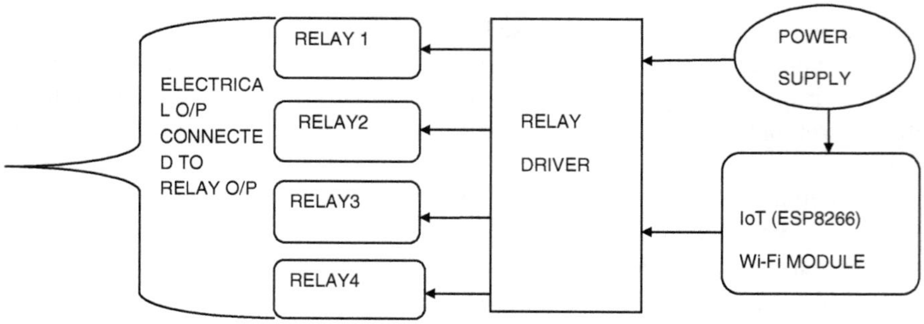

Fig. 9 Process description of ESP8266

electrical appliances like bulb, fan, alarms, surveillance system, etc., should be designed in such a way that it works in both, automated and optimized manner, by monitoring them with the help of an IoT server. Since each sensor has a unique IP address assigned to them, they can be controlled with the help of Internet, thus making it more reliable, flexible, and easily accessible.

3 Software Design

For a webpage to be designed, HTML (Hypertext Markup Language), which is mainly used to change the display. It is standard markup language for creating web pages and web applications. The documents make use of special tags or codes to write plain text files, which is further used by a web browser to interpret and display the information on the screen. These files have an extension of .htm or .html.

4 Output Analysis

A webpage has been created to continuously monitor the home appliances through internet and it is enabled by ESP8266Wi-Fi module. The sensors such as gas, PIR, LDR, and current sense the data, it will be displayed on the webpage and the user can control these data (Fig. 10).

4.1 Smart City Challenges and Concerns

There are some challenges when we form a smart city which are listed as shown in (Fig. 11).

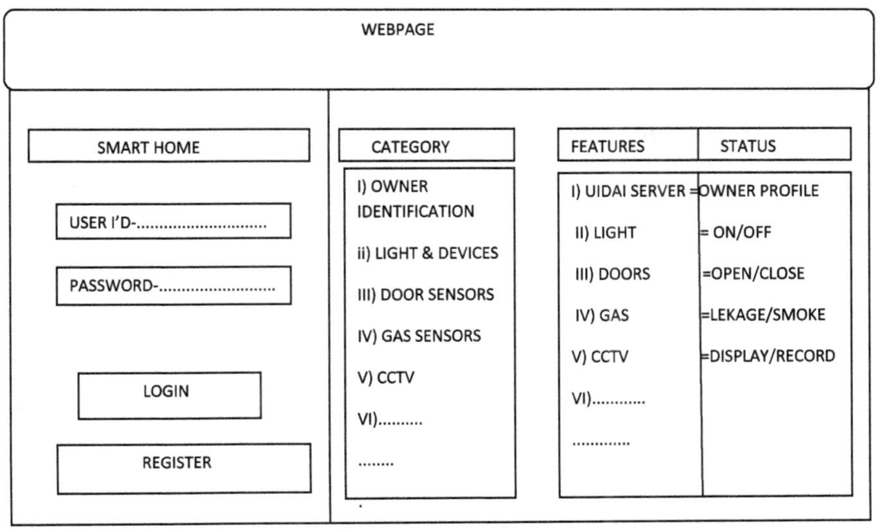

Fig. 10 Webpage of smart home

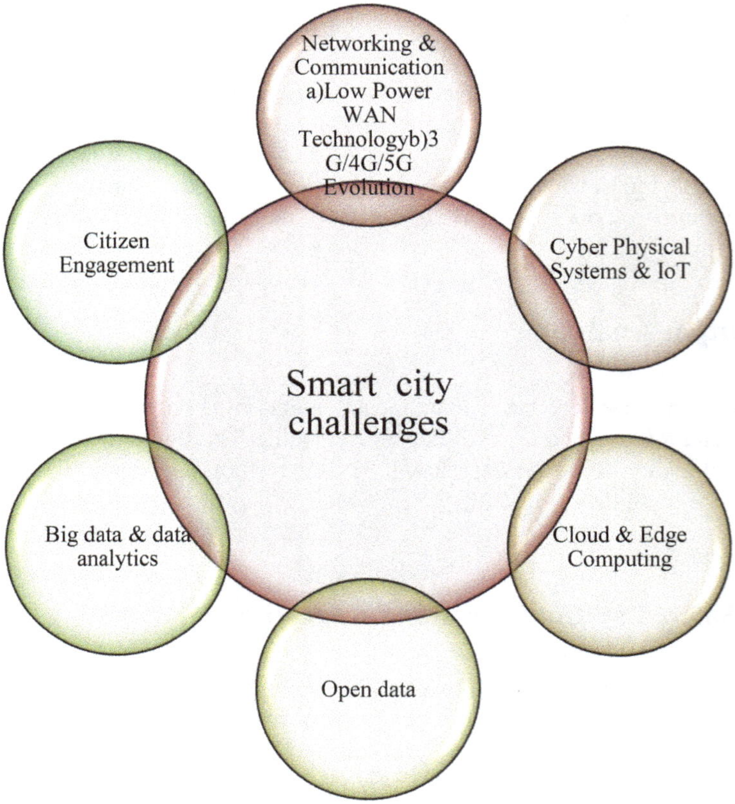

Fig. 11 Smart city challenges

4.2 IoT Challenges

See Fig. 12.

5 Conclusion

Smart city using IoT has been designed and implemented in this paper. In India, citizens are not so much aware of the technology. The Indian government has launched a project of smart city mission which comprises all basics concept of the smart city which is listed above. Smart city is an equal participation of both government and citizen. Making of a smart city is first initiated at our home with the help of smart electronics gadget and also by a proper waste management which is obtained from various household activities. The IoT uses the Internet to merge

Fig. 12 Major challenges of an IoT

various heterogeneous things with the help of IoTs various feature of smart cities can be observed and controlled from anywhere and anytime. Application of IoTs in smart cities provides an accurate and better result to improve standards of living and makes India better in coming days.

Acknowledgments I would like to thank Dr. S. Pal, HOD, Department of Electronics and Communication and M. K. Mishra, Vice Chancellor, Birla Institute of Technology, Mesra, Ranchi for their support, guidance, and sharing their experience to carry out my paperwork and to explore new ideas in my field of study.

References

1. Zhu C, Leung VCM, Shu L, Ngai ECH (2015) Green internet of things for smart world. IEEE Access 3:2151–2162
2. Zanella A, Bui N, Castellani A, Vangelista L, Zorzi M (2014) Internet of things for smart cities. IEEE Internet Things J 1:22–32
3. Kyriazis D, Varvarigou T, White D, Rossi A, Cooper J (2013) Sustainable smart city IoT applications: Heat and electricity management amp; Eco-conscious cruise control for public transportation. In: Proceedings of the 2013 IEEE 14th international symposium on "a world of wireless, mobile and multimedia networks" (Wow Mom), Madrid, Spain, 4–7 June 2013, pp 1–5
4. Smart mobility diagnosis of the present situation in Mexico pdf
5. Ei-Basioni BMM, SMA Kader, Fakhreldin MA (2016) Smart home design using wireless sensor network and biometric technologies, vol 2, Issue 3, March 2016
6. Wiu Z-L, Saito N (2013) The smart home. In: Proceedings IEEE, vol. 101, no. 11, Nov 2013
7. Han D-M, Lim J-H (2013) Design and implementation of smart home energy management system based on Zigbee. IEEE Trans Consum Electron 56(3) (Aug 2013)
8. http://smartcities.ieee.org/

Study and Design of Smart Embedded System for Remote Health Monitoring Using Internet of Things

Satvika Anand and Vijay Nath

Abstract Internet of Things (IoTs) plays a significant role in directly integrating physical world with computer-based systems by sensing and controlling various features, even remotely, which leads to a reduction in human intervention to a larger extent. This results in improved efficiency, better accuracy and easily accessible data. This chapter proposes the study and implementation of remotely accessible health monitoring system to monitor blood pressure level with IoT. It enables the user to improve health-related risks by collecting vital signs, i.e. temperature, blood pressure, sugar level, etc., of the patients and communicating these information to detector in real-time scenario. After proper examination of the current status and full medical information, it also aids in transmitting the accurate decision taken by doctor and precautionary measures required to be taken by the patients. In this manner, patients are constantly under the observation of doctor, even being physically apart with each other. Thus, IoT serves as a catalyst for health care by creating a brilliant and efficient network.

1 Introduction

Remote health monitoring system includes various components like acquiring data, storing it either in analog or digital form, accessing it easily along with supervising medical data and visualizing. It has several advantages over the traditional ones like allowing physicians to monitor the patients even being apart, improved quality of health care, a time-efficient and more reliable approach. Integration of IoT in this approach makes every unit to be connected to the Internet by providing unique addresses, thus leading to further increase of intelligence and flexibility. It makes

S. Anand (✉) · V. Nath
VLSI Design Group, Department of Electronics and Communication Engineering, Birla Institute of Technology, Mesra, Ranchi 835215, Jharkhand, India
e-mail: satvikaanand25@gmail.com

V. Nath
e-mail: vijaynath@bitmesra.ac.in

V. Nath and J. K. Mandal (eds.), *Nanoelectronics, Circuits and Communication Systems*, Lecture Notes in Electrical Engineering 511,
https://doi.org/10.1007/978-981-13-0776-8_37

use of smart sensors for accurately measuring, monitoring and analysing various health indicators like blood pressure, blood glucose level, heart rate, etc. This paper proposes an IoT-based remote health monitoring system to monitor blood pressure level by constantly tracking the parameters and sending alerts to the doctor if it deviates from the standard ranges, thus providing fast and easy healthcare service. Various researchers have devised and analysed similar systems at different levels. Smart healthcare systems have been analysed by focussing on the architecture of wearable devices and their interfaces in [1]. In [2], the patient is monitored continuously with the help of mobile phones. A gateway with the help of sensor nodes is devised for monitoring health parameters by using IoT is given in [3].

2 Methodology

Procedure Figure 1 shows the general procedure to monitor patients remotely with the help of various interfaces.
Here, doctor accesses patients' health status with the help of suitable interface and after proper diagnosis, it is capable of transmitting the required actions taken by a doctor to the patients. A common problem, i.e. interoperability [1] occurs with use of various interfaces is due to their different communication standards. To overcome this challenge, M2M (machine to machine) technology [2] is used which refers to direct communication between devices using any communication channel, either wired or wireless.

The vital health conditions monitored under different specializations are detailed in (Fig. 2), which are as follows:

- Cardiology—Cardiologist deals with the heart disorders and other circulatory system-related diseases like hypertension, arrhythmias, coronary artery disease, cardiac arrest, congenital heart defects, etc.
- Hepatology—Hepatologist deals with the problems of liver, gallbladder, pancreas, etc., along with their management.
- Rheumatology—Rheumatologist deals with problems of joints, soft tissues, autoimmune diseases, vasculitides, and heritable connective tissue disorders.
- Neurology—Neurologist deals with various disorders of the nervous system like epilepsy, stroke, neuromuscular, movement disorders, etc.

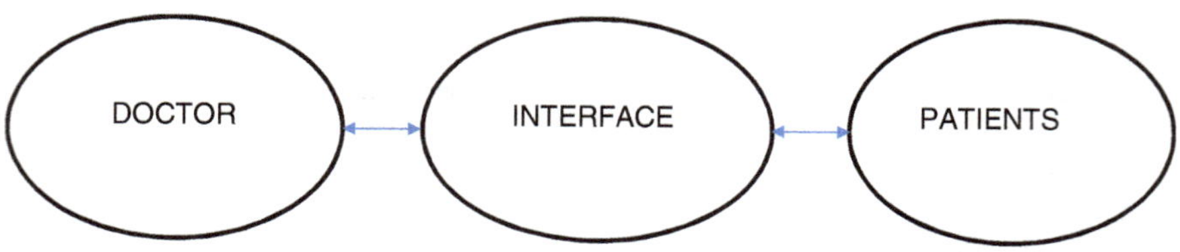

Fig. 1 Functional diagram of remote health monitoring system

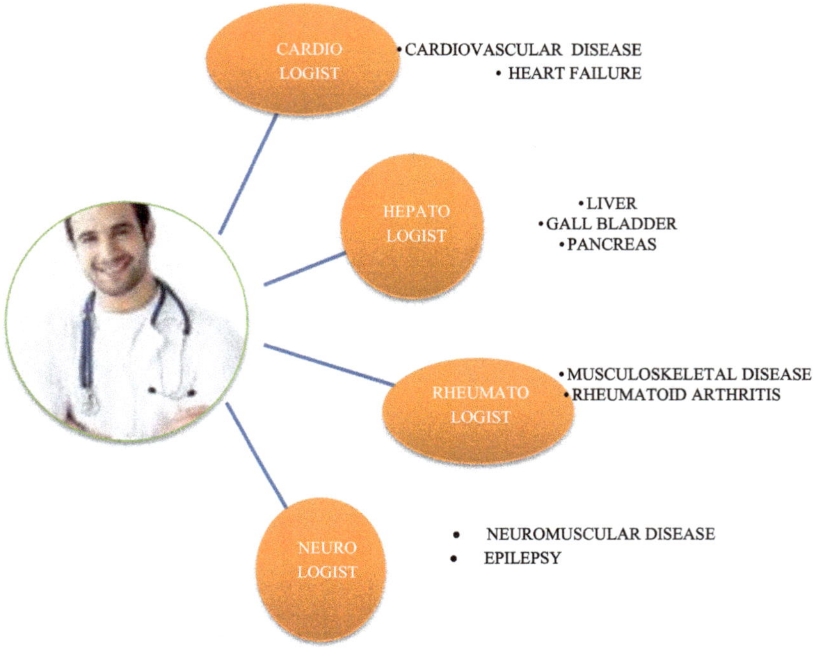

Fig. 2 Functions of various specialists

Fig. 3 Block diagram of Raspberry Pi

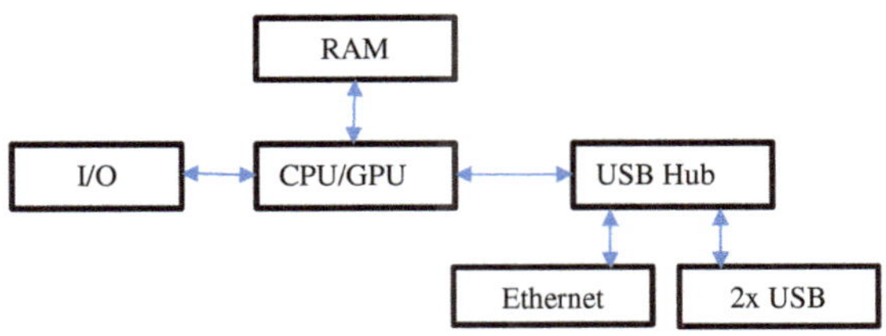

The various interfaces required are as follows:

- Raspberry Pi—It is a small single board computer which connects to various peripherals through pins and connectors (Fig. 3).
- ZigBee—It is an IEEE 802.15.4-based specification used to create personal area networks [4]. It is a wireless ad hoc network with low power, low data rate and close proximity (Figs. 4 and 5).

In this approach, the vital data is collected from the sensor and status of the monitoring system is checked whether it is ready or not to accept data. If not, then system error is generated. If yes, the connection status of remote monitoring system is checked. Again, logical decision 'no' results in a system error. For 'true' condition, crucial information of patients is sent to remotely located server for which

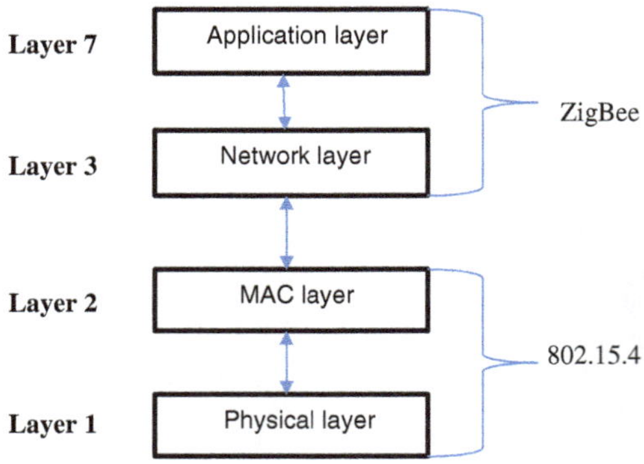

Fig. 4 Block diagram of ZigBee

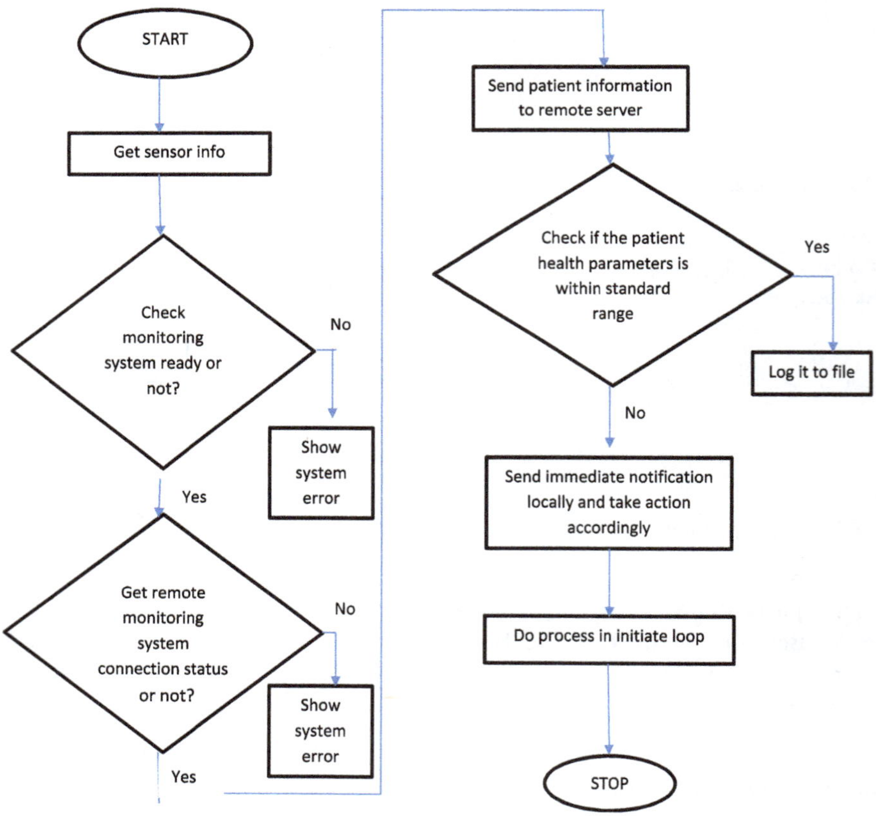

Fig. 5 Generalized flowchart

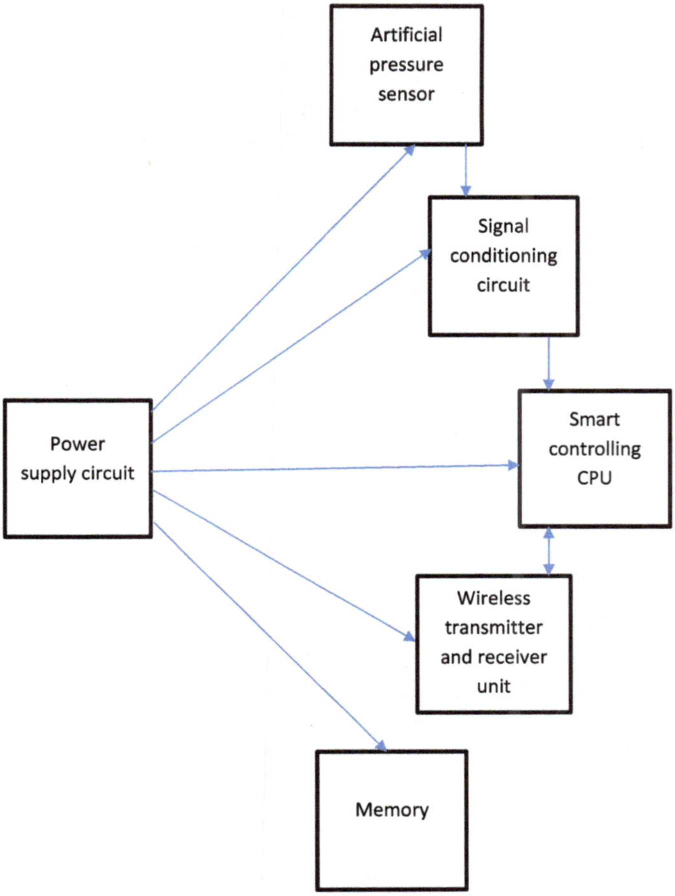

Fig. 6 Functional diagram of blood pressure monitoring system

data integrity and security is to be maintained. There, received data is compared with standard range to see if patients' health parameters lie within that range or not. If not, then immediately notification is sent to doctors else it is logged into file. This loop is repeated many times until parameters fall within the range.

The proposed system continuously monitors the blood pressure working on the principle of oscillometric method. Both systolic and diastolic pressure can be evaluated (Fig. 6).

The block diagram of the proposed scheme is detailed below:

- Artificial pressure sensor.
- Signal conditioning circuit: It is used to manipulate the output of the sensor with the help of various electronic circuits like amplifiers, filters, attenuators, converters, etc.

- Smart controlling CPU: It is an intelligent processor, being capable of programmed, which accepts input from sensors and gives the output.
- Wireless transmitter and receiver unit.
- Memory: It stores all the data and instructions required for processing.
- Power supply unit: It supplies power to all the components.

The system comprises of various hardware units like sensors, a portable and a relocatable unit along with a management services.

Sensors help in sensing the pressure of blood in terms of either high pressure or low pressure. A portable unit, which is basically a wireless system, requires an Internet connection for the transmission of health parameters. Relocatable unit helps in relocating the patients along with basic facilities. Management service provides the information regarding location of nearest hospital or doctor with the help of Global Positioning System, thus making it easy to respond quickly, even by monitoring several patients at the same time.

3 Conclusion

A smart embedded system for monitoring the blood pressure of patients remotely has been designed and implemented in this study. It helps in diagnosing the patients in real-time scenario and taking proper precautionary measures. Also, since the use of IoT makes the proposed system more reliable and convenient, offers high speed along with data accessibility ensuring data security at low cost. Thus, shifting of healthcare services from clinic centric to patient centric will make a huge difference in this vast world of medicare in terms of overall health benefits and outcomes.

Acknowledgments I would like to express my indebtedness and sincere gratitude to the Hon'ble Vice Chancellor, Dr. M. K. Mishra as well as HOD of ECE Dept., Dr. S. K. Pal, BIT Mesra for their encouraging support and valuable guidance during the course of this work.

References

1. Custodio V, Herrera FJ, Lpez G, Moreno JI (2012) A review on architectures and communications technologies for wearable health-monitoring systems. Sensors 12(10):13907
2. Pigadas V, Doukas C, Plagianakos VP, Maglogiannis I (2011) Enabling constant monitoring of chronic patient using android smart phones. In: Proceedings of the 4th international conference on pervasive technologies related to assistive environments, ser. PETRA '11. ACM, New York, NY, USA, pp 63:1–63:2
3. Ramirez-Ramirez R, Cosio-Leon M, Ojeda-Carreno D, Vazquez-Briseno M, Nieto-Hipolito J (2015) Designing a gateway ieee1451-hl7 for e-health telemonitoring services. In: 2015 international conference on computing systems and telematics (ICCSAT), Oct 2015, pp 1–6
4. Tung HC, Tsang KF, Lam KL, Tung HY, Li BYS, Yeung LF, Ko KT, Lau WH, Rakocevic V (2014) A mobility enabled inpatient monitoring system using a zigbee medical sensor network. Sensors 14(2):2397

Design of Smart Embedded System for Auto Toll Billing System Using IoTs

Chandan Kumar and Vijay Nath

Abstract Road transportation in India is assumed to be the second largest network. To manage such a large network, Toll collection method is used. Toll is collected manually, which requires a toll attendant. Manual process is time-consuming, resulting in a huge rush at the toll booth. Toll collection is one of the important problems in India. For making the system smooth IoTs-based systems are required, that maintain the vehicle health as well as toll tax when passing through any toll booth. In one toll, approximately 24 people are working in shift in classical systems. But, in the proposed system no one is required on booth. Everything is monitored remotely. For this setup, main items required are server-based system with a collection of high-speed Internet Aadhar Card, ZigBee and GPS. Then toll charges will display on monitor and message will be forwarded to the owner of vehicles and payment will be deducted from the account of owner. No need to pay in every booth by driver. For this whole system, when vehicles enter in the city, booth officers will collect the relevant information (vehicle papers, owner Aadhar card, etc.) from driver if vehicles are old and they will paste the barcode in that vehicles. In newly purchased vehicles, this system database is created during delivering the vehicles to customers. When compared to the traditional method, this method is very easy and also reduces the user's wastage of time.

Keywords Aadhar card · Internet of Things (IOT) · ZigBee · Raspberry Pi CAN · GPS

C. Kumar (✉) · V. Nath
VLSI Design Group, Department of Electronics & Communication Engineering, Birla Institute of Technology Mesra, Ranchi 835215, Jharkhand, India
e-mail: chandanbcrec@gmail.com

V. Nath
e-mail: vijaynath@bitmesra.ac.in

V. Nath and J. K. Mandal (eds.), *Nanoelectronics, Circuits and Communication Systems*, Lecture Notes in Electrical Engineering 511,
https://doi.org/10.1007/978-981-13-0776-8_38

1 Introduction

Automation or automatic means replacement of human by machines. The task carried out by human will now be handled by a machine, once it is programmed. Use of machine reduces time, risk factor and increases efficiency. In the same manner, the toll collector manually can now be replaced by machine to handle the task such as collection of money and closing, opening the toll gate. Due to manual processing at toll booth, we are facing lots of problem during travelling like traffic, air pollution, wastage of time, fuel waste, etc. To overcome this, we transformed the manual toll collection into the automatic toll collection by means of coin machine and electronic toll collection (ETC). ETC is the latest method for collecting tolls which aim to eliminate the delay on toll roads. The ETC method allows the vehicle to pass through a toll booth without requiring any action or stopping by the driver. ETC improves the speed and efficiency of traffic flow and saves driver time. Taking into consideration the present development, the numbers of vehicles passing through a tollbooth is considerably high; hence there is a need for an alternative solution for the toll collection method on the national highway which should be cost-effective and efficient. The main aim of ETC is to eliminate delay on toll roads by the way of collecting tax mechanically without any user. The use of ETC technology provides the opportunity to reduce the rush at toll booth during festive seasons on the other hand it can also be used to detect the culprit in hit and run cases. The number of hit and run cases is increasing day by day. It is difficult to detect the culprit in hit and run case. It is observed that when the vehicle is stolen it is very difficult to track the vehicle. It is very necessary to control these above problems. So the system also has an additional feature of detecting the vehicles against which stolen and accident cases are registered. It is observed that overloading the vehicles may lead to accident and also damage the roads. This problem is also addressed in system by weighing goods carrying vehicle and charge them with extra toll if it is overloaded so that they will not overload again. System also makes payment system easy by making all transaction in online mode [1–3].

2 Methodology

To overcome the normal toll collection system, we introduce Smart Card for easily identifies the vehicle details and also at this time we suggest the technique of IOT for easily logging in the website in online mode and enter the vehicle details for authenticating or communicatinge with toll booth collection. Now, GPS is used to find the vehicle location and transmit the details to the toll booth with the help of ZigBee [4, 5]. In this methodology, we create two modes for collecting tolls automatically in toll collection.

1. User registration or log in mode
2. Toll collection transaction mode

2.1 User's Registration Mode

Here, Fig. 1 elaborates the basic module of user's registration while entering into a toll booth. First of all, users update their Aadhar card number into online with the help of IoT. Here, first, the user loads the IOT core into SD card and then insert this SD card into Raspberry Pi SD card Slot. Then automatically system will update and install the OS for IOT. After that users open the server and upload their Aadhar card number. Then, they display user's vehicle details and information recording collection of tolls. Here, IoT mainly used to communicate either the user's PC or mobile phone and also toll booth workers PC [6, 7].

In Fig. 2, we explain the concept of location identification and transmits the data to the toll booth. Here, GPS is connected with Raspberry Pi with the help of Serial Peripheral Interface (SPI). This GPS is mainly used to identify the locations easily during driving a vehicle. This GPS gives some road directions and give some information to drivers. Here, ZigBee is used to communicate the users and toll booth workers with a short distance. Here, CAN Protocol is used to communicate the Raspberry Pi board and the ZigBee. At that time ZigBee transmitter transmits data to the receiver.

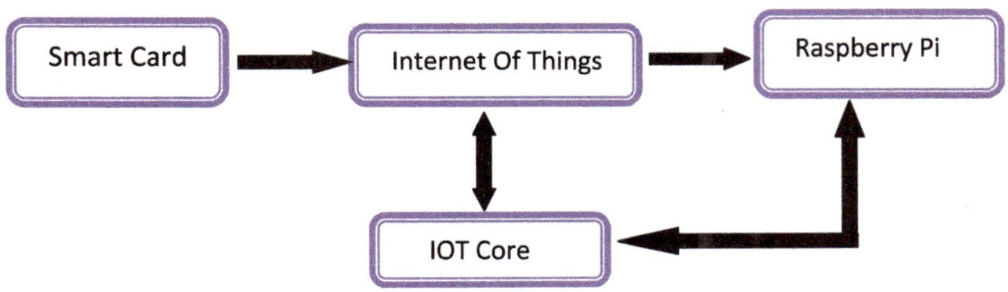

Fig. 1 User registration

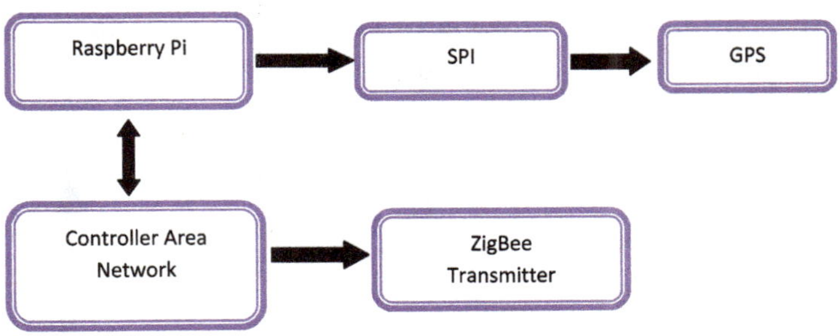

Fig. 2 Location identification and transmission

2.2 Toll Collection Transaction Mode

Figure 3 shows the basic module of toll collection transaction and data maintenance. The first transmitter sends data to the receiver. At that time, ZigBee receiver receives the data with short distance communication. The IOT server communicates with both the users and toll collection workers. Depending upon this process, they communicate easily and transmit amount easily without any waiting. Once the data like user's vehicle details, toll transactions are collected successfully means the Raspberry Pi communicates or interacts with stepper motor and opens the gate. Here, stepper motor rotates in clockwise direction and opens the gate. Here, the toll transaction is done online mode without any assistant [8, 9]. Fig. 4 demonstrate the process for active ZigBee communication. Because this process is performed when the user reaches the toll booth in previous 510 m distance.

Smart card is one type of a device that includes an embedded-based IC. It can be a secured microcontroller with internal memory. The smart card can be directly connected to a reader with direct physical contact. This smart card has a unique ability to store large amounts of data. They are performing with the help of both reader and terminals. Both support read and write functions into smart cards.

Internet of Things (IOT) is a way of the process; they perform communication between two devices without any collision and risk. This IOT is mainly performed

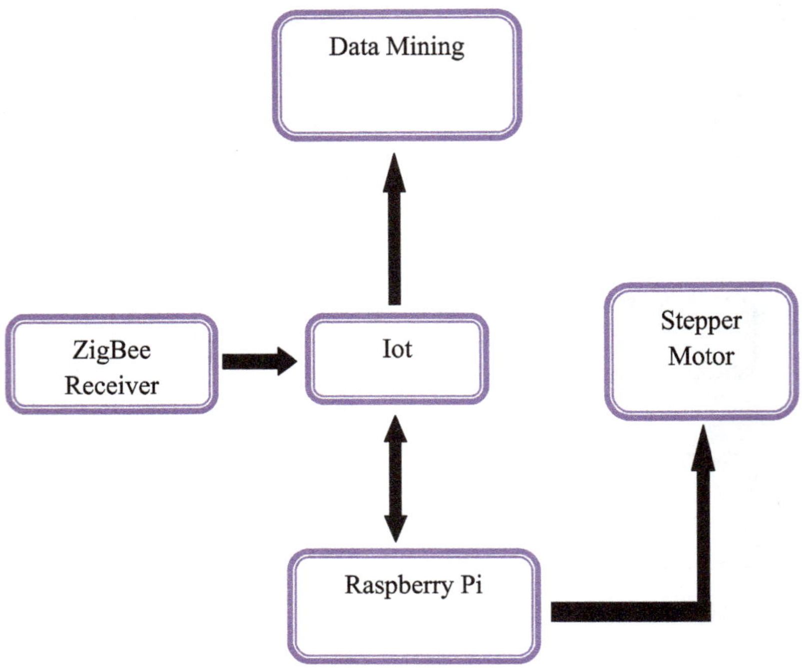

Fig. 3 Toll transaction and maintenance

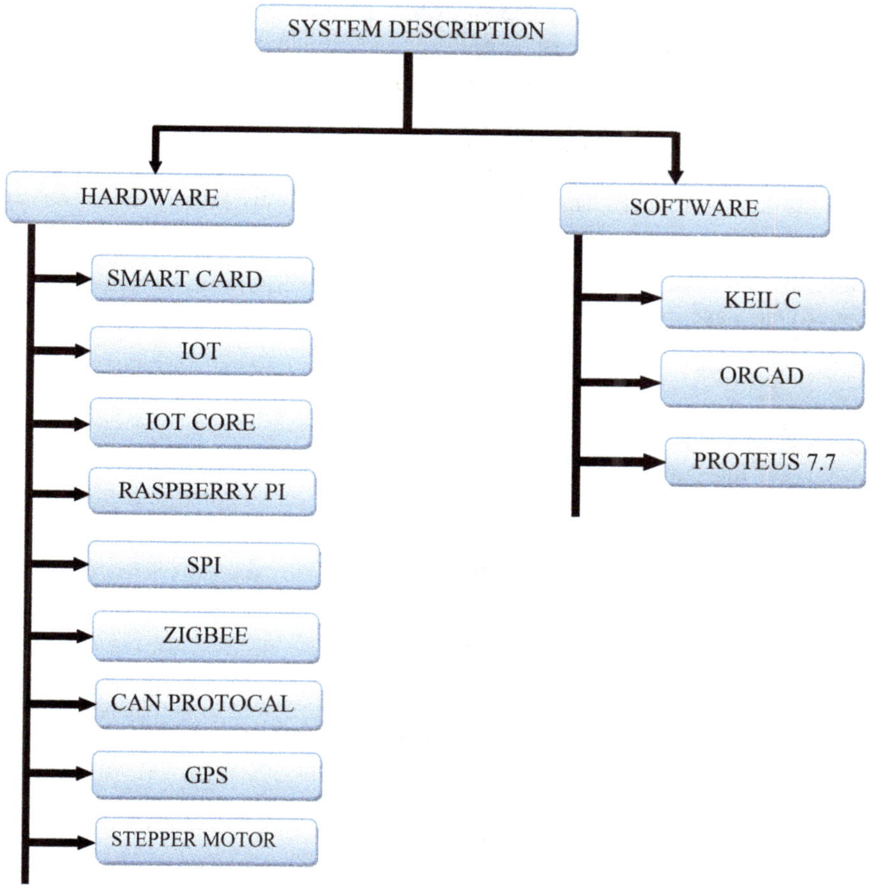

Fig. 4 ZigBee based Toll Billing System

with real-time system and they are performed with certain time task. They also mainly perform with risk management and time sharing. The IOTs is the clever connectivity of physical devices pouring huge gains in efficiency, business growth, and quality of life.

IOT core is one type of process or OS and this makes to run the IOT with depends upon Windows or Linux.

Function

Installer

Device driver

Raspberry Pi is one type of processor, which supports and operates the OS which depends upon SD slot. This Raspberry Pi supports communication buses like CAN, I2c and also supports UART, SPI, and General-Purpose Input Output (GPIO) Registers.

SPI is one type of Synchronous Interface which allows several microcontrollers or several peripherals to be interconnected. In each SPI, separate wires are required for both data and clock. This SPI supports and performs four types of various signals.

Master in Slave Out (MISO)
Master Out Slave in (MOSI)
Serial Clock (SCLK)
Slave Select (SS)

Features

1. Support full duplex and perform three wire synchronous processes
2. Perform both master and slave operations
3. Support 1.5 MHz for master bit frequency
4. Support 3 MHz for slave bit frequency
5. Four programmable master bit rates
6. Programmable clock polarity and phase
7. Master–Master mode Fault production

The Controller Area Network (CAN) is one type of serial communication protocol developed by BOSCH in 1980s. CAN was standardized as ISO-11898 and ISO-11519. This CAN protocol is mainly based on bus topology and only we need two wires for communication. This bus has a multimaster structure, where each device on the bus can send or receive data. At this time, only one device can send data at any time while all others listen. The CAN bus is mainly based on CSMA/CD mode while transmitting data. There are mainly two types of CAN protocols [10, 11].

CAN 2.0 A
CAN 2.0 B (Fig. 5)

Features

- CAN bus is a multimaster and this is a flexible one.
- CAN bus communication speed is not fixed
- All the devices on the bus can detect an error.

Advantages

- Low-cost network infrastructure which is often built into microcontrollers.
- Large market segment with broad availability of hardware, software and systems engineering tools.
- Lightweight, low latency, highly deterministic design specifically for real-time embedded applications.
- Reliable with strong error detection, fault-tolerant versions available.

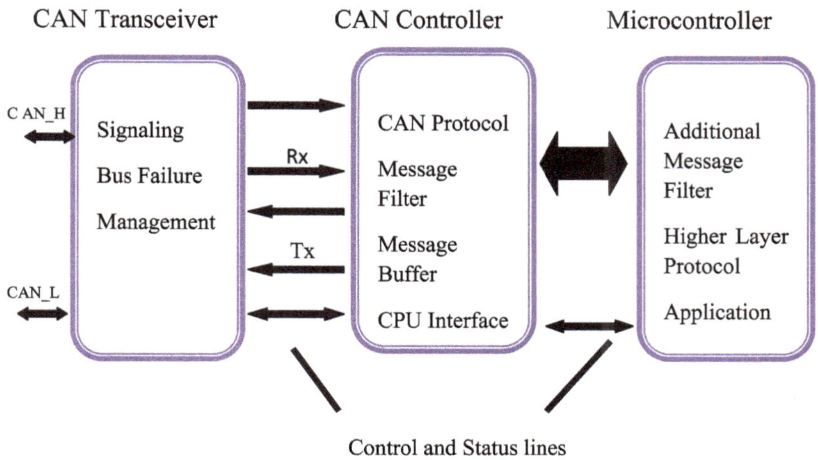

Fig. 5 Function of CAN controller

- Flexible and highly configurable with various higher level application protocols.
- Foundation for next-generation technology CANs.

ZigBee is one type of communication protocol; they perform both the transmission and reception operations. This ZigBee is performing short distance communication. They are approximately performed and communicate up to 5–50 m distance. This the ZigBee module design with low-power and low-cost wireless sensor networks. This module operates with ISM 2.4 GHZ frequency band.

GPS stands for global positioning system. The GPS comprises of three segments:

- The space segment (all functional satellites)
- The control segment (all ground stations involved in the monitoring of the system master control station, monitor stations, and ground control stations)
- The user segment (all civil and military GPS users).

During the development of the GPS system, particular emphasis was placed on the following three aspects:

- It had to provide users with the capability of determining position, speed and time, whether in motion at rest.
- It had to have a continuous, global, 3-dimensional positioning capability with a high degree of accuracy, irrespective of the weather.
- It had to offer the potential for civilian use (Fig. 6)

A stepper motor is one type of brushless synchronous electric motor that converts the digital pulses into mechanical shaft rotation. This stepper motor rotates step by step with depends upon shaft revolution. The stepper motor is connected to the microcontroller with the help of ULN2803A array [12, 13].

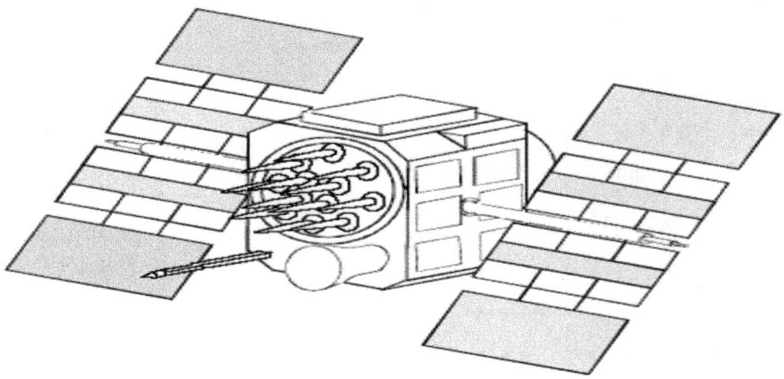

Fig. 6 GPS satellite

Principles of Operation

Consider a stator consisting of two pairs of pole pieces

- Rotor is a toothed structure with permanent magnets
- All teeth have the same magnetic pole
- Reverse the magnetization in horizontal pole pieces
- Rotor moves to a new equilibrium position, which is 30° clockwise from the old position.
- Now reverse the vertical poles.
- Rotor moves another 30° clockwise
- Direction of rotation of the rotor is opposite to that of stator magnetic field.

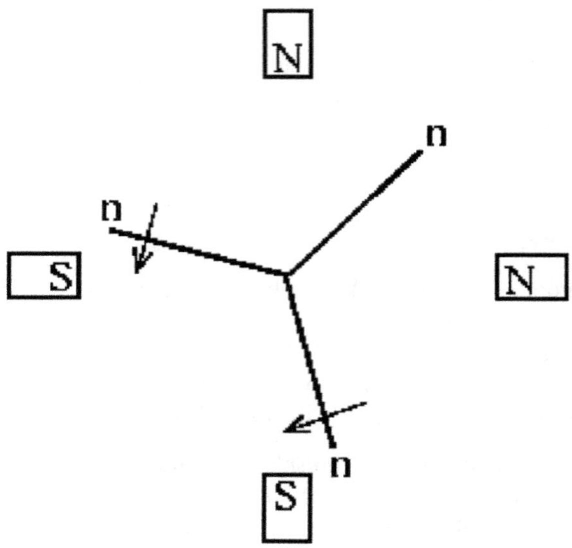

Data mining is mainly performed to easily identify the user details (like user vehicle details, amount transaction…) and to extract or minimizing the data from large databases. Here, data mining is mainly used to perform good relationship between users and toll booth workers with the usage of CRM.

Steps in Data Mining

i. Data cleaning: It is the process of removing noise and inconsistent data.
ii. Data integrating: It is the process of combining data from multiple sources.
iii. Data selection: It is the process of retrieving relevant data from the databases.
iv. Data transformation: In this process, data are transformed or consolidated into forms appropriate for mining by performing summary of aggregation operations.
v. Data mining: It is an essential process where intelligent methods are applied in order to extract data patterns.
vi. Pattern evaluation: The patterns obtained in the data mining stage are converted into knowledge based on some interestingness measures.
vii. Knowledge Presentation: Visualization and knowledge representation techniques are used to present the mined knowledge to the user.

1. KEIL C

KEIL C Version 4 is used to run the embedded coding with C platform.

2. ORCAD

ORCAD is one type of PCB design tool. This tool is mainly used to design and create the circuit with the hardware board.

3. Proteus 7.7

This Proteus tool is mainly used to design the circuit and to give the output in the form of simulation.

3 Conclusion

Nowadays, the numbers of vehicles is increased and also air pollution. To overcome this, governments introduced national highway scheme with toll collection system. But nowadays to maintain the toll booth is very hard for the manual toll collection methods. So, to overcome this, we introduce smart toll collection system with IOT technology. Here the users pay the tolls in online with the help of server communication (IOT) and ZigBee is used to transmit and receives the user's vehicle details. And also GPS is performed to find the correct location and give some information to users. So, this proposed scheme is reduced the user's long waiting time, collision, air pollution, etc.

Acknowledgments I would like to thank Dr. S. Pal HOD, Department of Electronics and Communication and M. K. Mishra Vice Chancellor, Birla Institute of Technology, Mesra, Ranchi for their support, encouragement, and sharing their experience to carry out my research work and to explore new ideas in my field of study.

References

1. Mehendale DK, Masurekar RS (2015) To study the implications of electronic toll collection system using RFID technology. Int Res J Multidiscip Stud I(V) (ISSN (Online): 2454-8499, Dec 2015)
2. Andurkar GK, Ramteke VR (2015) Smart highway electronic toll collection system. Int J Innov Res Comput Commun Eng 3(5) (ISSN (Online): 2320-9801, ISSN (Print): 23209798, May 2015)
3. Dhurat A, Magal P, Chheda M, Ingle D (2014) Gateless electronic toll collection using RFID. IOSR J Comput Eng (IOSR-JCE) 16(2):7380 (e-ISSN: 2278-0661, p-ISSN: 2278-8727, Mar–Apr 2014)
4. Piyare R (2013) Internet of things: ubiquitous home control and monitoring system using android based smart phone. Int J Internet Things, 5–11
5. Ning H, Zhang Y (2008) RFID and internet of things. Electronic Industry Press, Beijing, pp 18–26 (China)
6. Jerry L, Barbara C (2001) Shrouds of time: the history of RFID. AIM Publication, ver.1.0
7. Avoub Khan M, Manoj S, Prabhu RB (2009) A survey of RFID tags. Int J Recents Eng 1(4) (May 2009)
8. Tang WC, Ho TV (2007) Electronic toll collection system, US Patent, US 7233260 B2, June 2007
9. Radhika et al (2011) Electronic toll collection system. UNIASCIT 1(1):05–08
10. Nikitin PV, Ramamurthy S, Martinez R (2013) Simple low cost RFID UHF reader. In: IEEE international conference on RFID
11. Bhalekar S, Adesh Chanegiri G, Chauhan IP (2013) Automatic toll tax using RFID. IJCTEE, vol 3, Special Issue, March–April 2013, An ISO 9001:2008 Certified Journal
12. http://wordinfo.info/unit/3994/ip:1/il:R—Features of RFID
13. https://www.ingenuityworking.com/blog/b/weblog/archive/2010/08/04/the-unique-features-ofrfid.aspx—Features of RFID

Two Step Coupled Gap Resonator and Its Application as Bandpass Filter

Pratik Mondal and Susanta Kumar Parui

Abstract In this paper, a two step gap coupled resonator is modelled for designing a bandpass filter (BPF). The proposed structure is viewed simultaneously as edge coupled section and parallel coupled section. A detailed nodal analysis of the proposed coupled gap sections using different current and voltage equations are described, which is further derived to obtain different admittance parameter equations. Finally comparing those equations with normal two port network, even and odd mode impedance difference (Z_{0o}−Z_{0e}) equations are obtained in terms of parallel coupled admittance inverter (J_{g1}) and shunt susceptance (B_{g1}) at the edge coupled section. This equation is further used for designing the resonator with band accepted characteristics. A three pole wideband bandpass filter is demonstrated by series arrangement of four such coupled resonator unit. The proposed bandpass filter exhibits fractional bandwidth (FBW) of 53.5%. All the results are well verified by numerical simulation and validated through experiments.

Keywords Bandpass filter · Edge coupled · Open ended stub · Parallel coupled Stepped type coupled gap resonator

1 Introduction

Bandpass filter (BPF) is always one of the prime component for designing any trans-receiver system. It is a component which allows certain frequency band and rejects out of band one. Several bandpass filters were designed in planar forms on stripline, microstrip and Coplanar waveguide (CPW) [1, 2]. Edge coupled bandpass filter was the primary invention in the era of bandpass filter design techniques.

P. Mondal (✉) · S. K. Parui
Department of Electronics and Telecommunication Engineering, Indian Institute of Engineering Science and Technology, Shibpur, Howrah 711103, India
e-mail: pratik.mondal.1987@ieee.org

S. K. Parui
e-mail: arkapv@yahoo.com

V. Nath and J. K. Mandal (eds.), *Nanoelectronics, Circuits and Communication Systems*, Lecture Notes in Electrical Engineering 511,
https://doi.org/10.1007/978-981-13-0776-8_39

Later, Parallel-Coupled Wideband Bandpass filters were made of Image Parameter Method [3]. Several bandpass filters have been proposed by the use of transformation techniques also [1]. The major concerns for designing any bandpass filter include bandwidth, insertion loss and selectivity. In [4], the parallel coupled bandpass filters were designed and improvements of the design were also introduced. Moreover, in [5], a modified microstrip bandpass filter using resonating properties of the half wavelength coupled resonators was introduced, which reduced the size of a parallel coupled filter and improved the performance compared to the conventional one in terms of bandwidth and compactness. A planar bandpass filter was also realized earlier by CPW technology with open ended stub discontinuity [6]. Many research articles showed the design mechanism of compact and high selective bandpass filter using a combination of series open ended stub and shunt short ended stub to design CPW bandpass filter [7, 8]. Stepped-impedance resonators were synthesized with higher fractional bandwidth (FBW) but lower insertion loss [9, 10]. Band-pass filters based on ring resonators described in [11] have a high-Q factor and compact size. Several multimode resonator filters were developed for wide pass-band, good in-band and out-of-band performances [12, 13]. Wideband ring resonators with loading of open circuited stubs [14] as well as a wideband band-pass filter using transversal resonator and quadruple mode ring resonator [15] were proposed. A high selective bandpass filter with adequate isolation diplexer is proposed recently by mixed electromagnetic coupling [16].

In this paper, modelling of a two step coupled gap resonator is realized and used to design a bandpass filter. Basically, the proposed resonator is a combination of edge coupled and parallel coupled sections. Detailed analysis of the proposed resonator are derived to get different admittance parameter equations. These equations are then compared with normal two port network equations to synthesize the even and odd mode impedance difference ($Z_{0o}-Z_{0e}$) at the edge coupled section. These equations are very effective for designing bandpass filter. The operating frequency depends on the length of the resonator. Due to larger coupling area than the individual end coupled or parallel coupled resonating section, the proposed resonator resonates at a lower frequency. A three pole wideband bandpass filter is also demonstrated by series connection of four such proposed coupled resonator. This synthesis technique may be found very effective to analyze and design of different circuits.

2 Design Equations

The proposed two step coupled gap resonator filter is treated as a combination of both edge coupled and parallel coupled sections as shown in Fig. 1a. The resonating stubs are positioned such that adjacent resonators are in parallel to each other along the full of their length. To analyse such resonator, nodal analysis method or KCL is used. Thus, different current and voltage equations are obtained, which are further simplified and different admittance parameter equations are generated. Comparing those equations with normal two port network even and odd mode impedance difference (Z_{0o}−Z_{0e}) equation is obtained which is fruitful in designing of any bandpass filter having both the edge and parallel coupling.

Figure 1b, c shows the circuit modeling of the unit cell of the proposed resonator. The ABCD matrix of the coupled section of the proposed resonator can be represented as

$$\begin{bmatrix} A & B \\ C & D \end{bmatrix} = \begin{bmatrix} 0 & +\frac{1}{jJ_{g1}} \\ -jJ_{g1} & 0 \end{bmatrix} \tag{1}$$

where, 'J_{g1}' is the admittance parameter of J-inverter and 'B_{g1}' is the shunt susceptance. Now, from equivalent ABCD matrix of transmission line we get,

$$\begin{bmatrix} A & B \\ C & D \end{bmatrix} = \begin{bmatrix} \text{Cos}\theta_C & jZ_{01}\text{Sin}\theta_C \\ j\frac{\text{Sin}\theta_C}{Z_{01}} & \text{Cos}\theta_C \end{bmatrix} \tag{2}$$

Applying KCL at node 1 and node 2 and by solving enormous equations the admittance parameter values are obtained as

$$Y_{11} = \left[\frac{\left(\frac{B^2}{B_{g1}} - \frac{AB}{B_{g1}} - BjJ_{g1}\right)\left(DjJ_{g1} - \frac{1}{B_{g1}}\right) + \left(A - \frac{B}{B_{g1}} + \frac{B^2 jJ_{g1}}{B_{g1}}\right)\left(C - \frac{D}{B_{g1}}\right)}{\left(A - \frac{B}{B_{g1}}\right)^2 - \left(BjJ_{g1}\right)^2} + \frac{1}{B_{g1}}\right] \tag{3}$$

and

$$Y_{12} = \left[\frac{\left(\frac{-B}{B_{g1}} + A + \frac{B^2 jJ_{g1}}{B_{g1}}\right)\left(DjJ_{g1} - \frac{1}{B_{g1}}\right) + \left(\frac{B^2}{B_{g1}} - \frac{AB}{B_{g1}} - BjJ_{g1}\right)\left(C - \frac{D}{B_{g1}}\right)}{\left(A - \frac{B}{B_{g1}}\right)^2 - \left(BjJ_{g1}\right)^2} + \frac{D}{B_{g1}}\right] \tag{4}$$

Considering half wavelength resonator section ($\theta = 90°$), Eq. (3) may be simplified as

$$\therefore Y_{11} = \frac{\left(J_{g1}B_{g1}Z_0 + B_{g1} + Z_0^2 - jJ_{g1}B_{g1}Z_0^2\right)}{-Z_0^2 - J_{g1}^2 B_{g1}^2 Z_0^2} + \frac{1}{B_{g1}} \tag{5}$$

Similarly, considering $\theta = 90°$ in Eq. (4) may be simplified as

$$\therefore Y_{12} = -\frac{j\left(Z_0 + J_{g1}Z_0^2 - B_{g1}Z_0 + J_{g1}B_{g1}^2\right)}{Z_0^2\left(1 + J_{g1}^2 B_{g1}^2\right)} \tag{6}$$

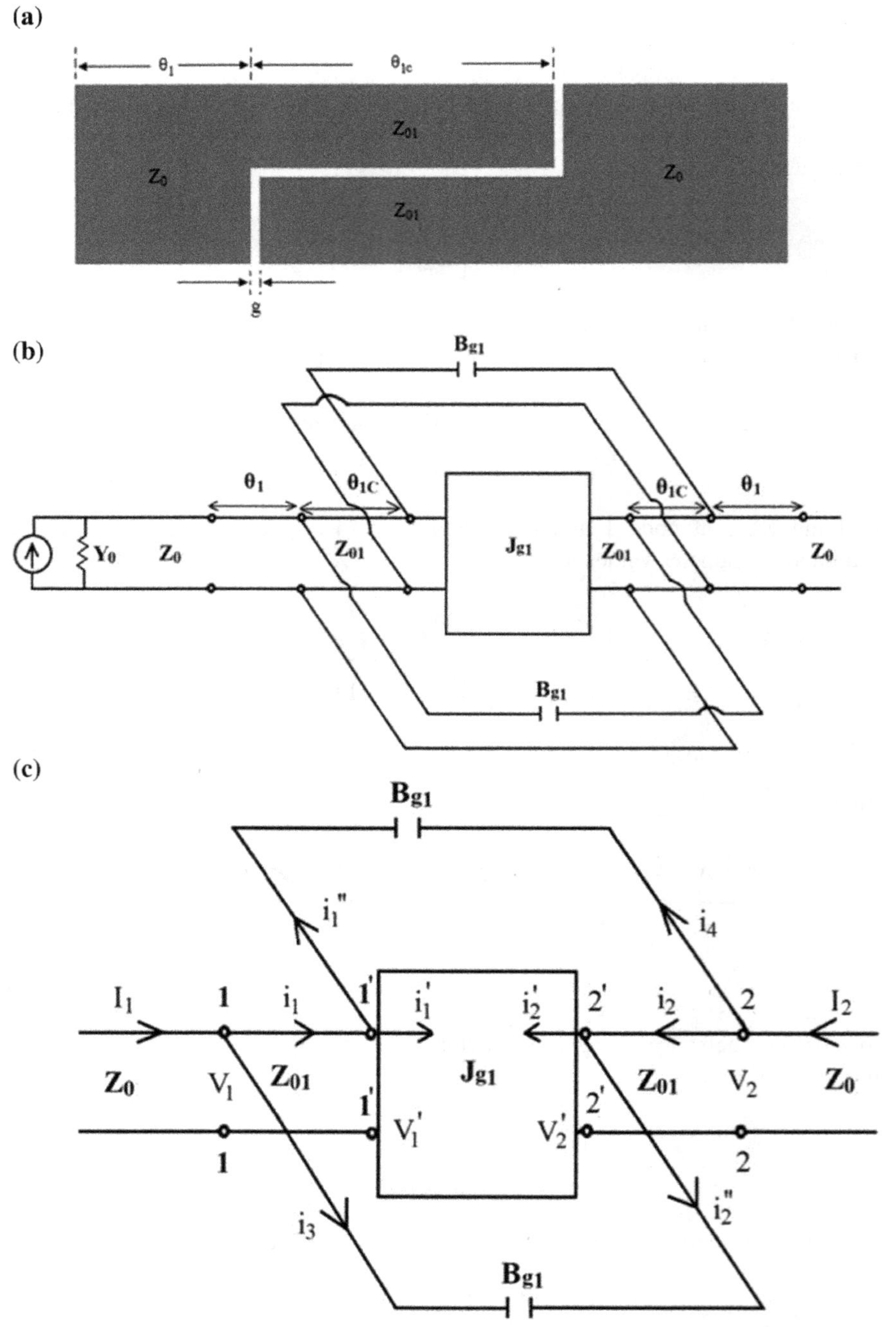

Fig. 1 **a** Schematic of proposed two step type gap coupled resonator; **b** Simplified circuit model; and **c** Circuit model including current and voltage

The Y-matrix may also be represented as

$$\begin{bmatrix} I_1 \\ I_2 \end{bmatrix} = \begin{bmatrix} V_1 \\ V_2 \end{bmatrix} \begin{bmatrix} 0 & -\frac{j(Z_0 + J_{g1} Z_0^2 - B_{g1} Z_0 + J_{g1} B_{g1}^2)}{Z_0^2 (1 + J_{g1}^2 B_{g1}^2)} \\ -\frac{j(Z_0 + J_{g1} Z_0^2 - B_{g1} Z_0 + J_{g1} B_{g1}^2)}{Z_0^2 (1 + J_{g1}^2 B_{g1}^2)} & 0 \end{bmatrix} \quad (7)$$

Comparing Eqs. (5) and (6), even and odd mode impedance difference is obtained as

$$(Z_{0e} - Z_{0o}) = \frac{Z_0^2 (1 + J_{g1}^2 B_{g1}^2)}{(Z_0 + J_{g1} Z_0^2 - B_{g1} Z_0 + J_{g1} B_{g1}^2)} \quad (8)$$

Using Eq. (8), the different dimensions of coupled lines and gaps are calculated. Figure 2 represents the three dimensional mesh view of even and odd mode impedance difference with respect to admittance of J-inverter (J_{g1}) and shunt susceptance (B_{g1}).

3 Design of a Two Step Coupled Gap Resonator

The design is implemented by two slots originating from the edge of the central strip on both sides such that the slots are connected to each other as shown in Fig. 3a. The resonator discontinuity formed is that of two quarter wavelength ($\lambda_g/4$) series open ended stubs. The discontinuity further transforms a short circuit at the starting terminal which forms a band acceptance response.

The proposed band accepted filter is designed by considering a Chebyshev lowpass prototype with 0.1 dB passband ripple. From (8) the even and odd mode impedance is obtained. Arlon AD320 material of permittivity of 3.2, thickness of 1.58 mm and loss tangent of 0.003 are considered for the proposed design. The width of 3.8 mm is chosen to match 50 Ω transmission line. J_{g1} values are calculated from normal parallel coupled bandpass filter design equation [1] and simultaneous differences of even and odd mode impedances are obtained from Eq. (8) for different FBW and tabulated in Table 1. B_{g1} values are calculated from existing coupled mode formulas [1].

From Eq. (8),

$$(Z_{0e} - Z_{0o}) = \frac{Z_0^2 (1 + J_{g1}^2 B_{g1}^2)}{(Z_0 + J_{g1} Z_0^2 - B_{g1} Z_0 + J_{g1} B_{g1}^2)}$$

$$= \frac{50^2 [1 + (-j1190)^2 \times (0.0351)^2}{[50 + 50^2 \times 0.02 - 50 \times (-j1190) + (-j1190)^2 \times 0.0351]} = 35.6628 + j44.93$$

$$|(Z_{0e} - Z_{0o})| = 57.4173 \cong 57.4\,\Omega$$

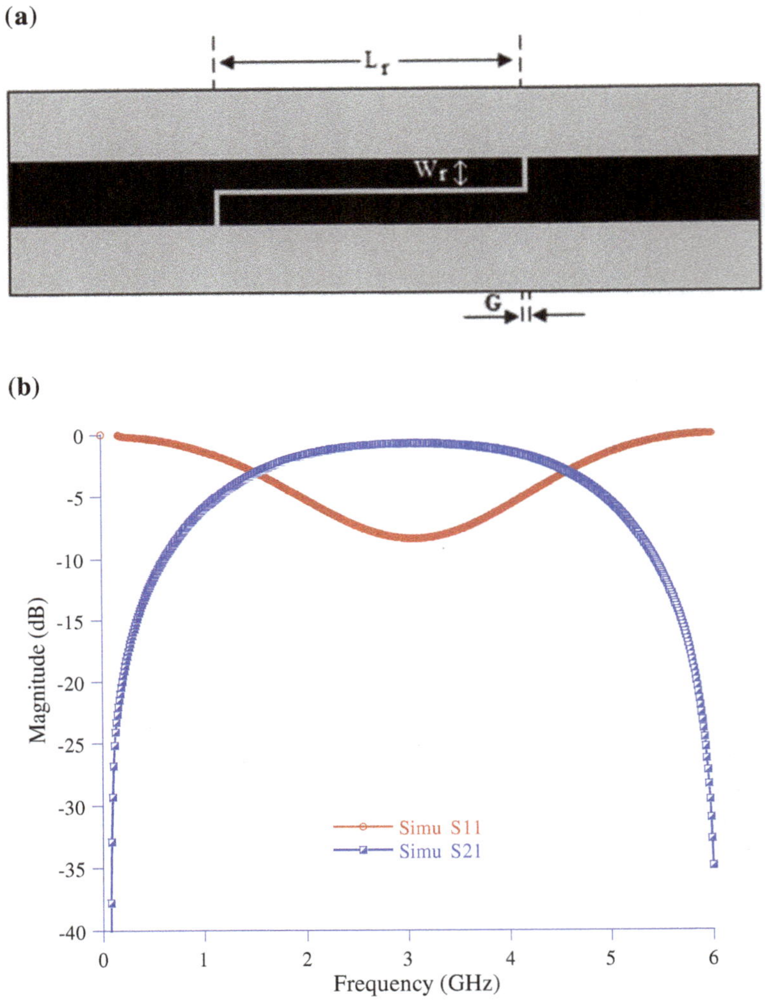

Fig. 2 **a** Schematic of unit cell of proposed step type coupled gap resonator; **b** Simulated S-parameter responses

The designed filter have a fractional bandwidth of (FBW) = 60% at the center frequency f_0 = 3 GHz. From Table 1, J_{g1} = 0.0351 considering the FBW and the elemental values. Simply by solving Eq. (8), the difference between even and odd mode characteristics equation is found to be 57.4 Ω which is depicted in Table 1. If the gap (g) becomes small, the even mode impedance goes high and lowers the odd mode impedance. In order to achieve the desired value of impedance difference, width of coupled line section or stub considered as W_r = 1.8 mm, gap, g = 0.2 mm and the length of the resonator required is L_r = 15 mm with a complete length of the coupled section is 18.4 mm [1, 3].

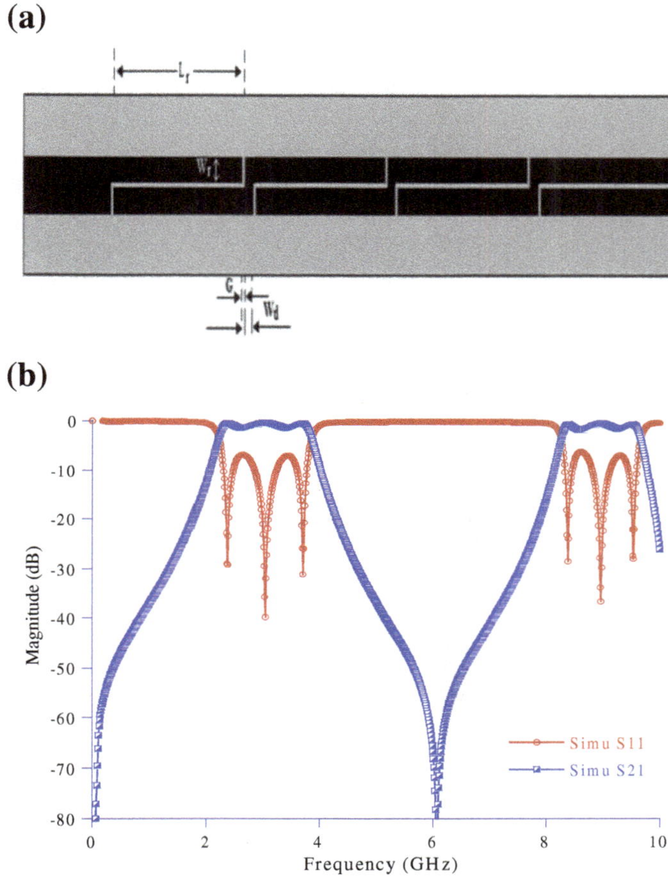

Fig. 3 **a** Schematic of proposed three pole series resonator based bandpass filter; **b** Simulated S-parameter responses

Table 1 Design parameters due to change of FBW

FBW (%)	Admittance "J_{g1}"	Shunt Susceptance "B_{g1}"	(Z_{0e}–Z_{0o})
20	0.0203	−j1190	22.4021
30	0.0249	−j1190	32.2905
40	0.0287	−j1190	41.2402
50	0.0321	−j1190	49.6941
60	0.0351	−j1190	57.4173
70	0.0380	−j1190	65.0603
80	0.0406	−j1190	72.0244
90	0.0430	−j1190	78.5232
100	0.0454	−j1190	85.1023

(a)

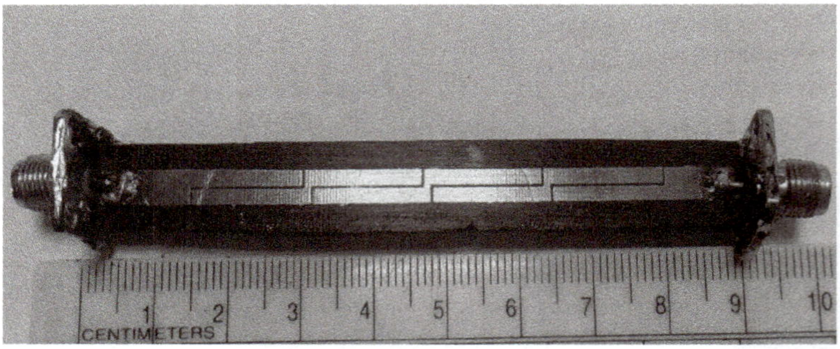

(b)

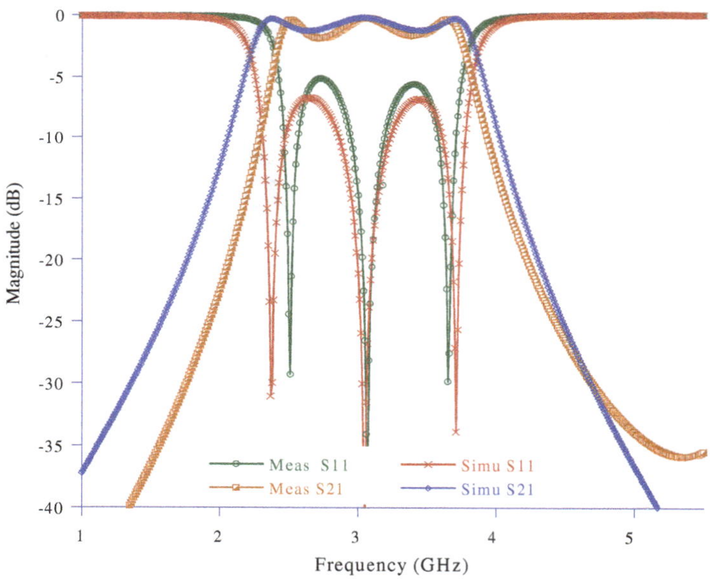

Fig. 4 **a** Fabricated prototype of proposed bandpass filter; **b** Comparative study between simulated and measured S-parameters

The length of the resonator is chosen to have center frequency near to 3 GHz and the gaps have been optimized to have a good response. This schematic diagram is then simulated using HFSS EM Simulator software and simulation result shows a band accepted response at center frequency 3.1 GHz.

4 Three Pole Bandpass Filter Design Using Series Arrangement of Proposed Resonators

Four such proposed two step gap coupled resonators are placed in series such that the discontinuities in the central transmission line is kept closer to each other which provide higher field confinement. Hence, a good possible compactness is obtained with adequate higher bandwidth. Numbers of pole increases as the number of discontinuities are increased in the design.

The filter has designed with a fractional bandwidth FBW = 60% at the center frequency f_0 = 3 GHz. Same substrate considered for the design. Assuming gap between two consecutive resonators W_d = 1 mm, a three pole bandpass filter has been obtained. The layout design of the filter is portrayed in Fig. 3a. The simulated result in Fig. 3b shows 3 dB bandwidth from 2.23 to 3.86 GHz with a center frequency of 3.045 GHz and maximum insertion loss of −1.27 dB, thus covering the entire S-Band. Also, the selectivity of 37.77 dB/GHz at the rising edge and 36.97 dB/GHz at the falling edge are observed which is quite high.

The prototype of proposed third order stepped type coupled gap resonator bandpass filter is fabricated as shown in Fig. 4a. The fabricated filter structure is then experimentally measured using Anritsu make vector network analyzer as depicted in Fig. 4b.

Measured results show a third order S-Band bandpass filter response with 3 dB bandwidth of 1.35 GHz, ranging from 2.41 to 3.78 GHz with maximum insertion loss of −1.67 dB. Measured selectivity has been found of 48.57 dB/GHz at the rising edge and 37.77 dB/GHz at the falling edge as depicted in Fig. 4b. The measured s-parameter responses thus found are in near agreement with the simulated results. The deviation in measurements is due to the fabrication tolerances and instrumental non-linearity.

5 Conclusion

Design of bandpass filter using two step coupled gap resonator is demonstrated. Detailed circuit analysis of the proposed resonator has been done. This equation forms are found very effective in designing such coupled circuits. Combination of four resonators in series provides a wideband bandpass filter with low insertion loss. Close arrangements of such resonators give rise to field confinement, which further improves the bandwidth of the filter around 53.5% (FBW), which is quite high. This methodology of designing BPF might be an appealing candidate for compact and wideband systems.

Acknowledgements This work is financially supported by U.G.C, Govt. of India.

References

1. Hong JS, Lancaster MJ (2001) Microwave filters for RF/microwave applications, 1st edn. Wiley, New York
2. Simons RN (2001) Coplanar wave guide circuits, components, and systems, 1st edn. Wiley, New York
3. Ye CS, Su YK, Weng MH, Yang RY (2010) Design of the compact parallel-coupled lines wideband bandpass filters using image parameter method. Prog Electromagn Res 100:153–173
4. Alaydrus M (2010) Designing microstrip bandpass filter at 3.2 GHz. Int J Electr Eng Inform 2(2):71–83
5. Mondal P, Moyra T, Parui SK, Das S (2012) New concept for designing of compact parallel coupled bandpass filter. In: Proceedings in international conference on communication, circuits and systems (iC3S-2012), Oct 2012, pp 19–21
6. Simons RN, Ponchak GE (1988) Modeling of some coplanar waveguide discontinuities. IEEE Trans Microw Theory Tech 36(12):1796–1803
7. Wang SN, Chen NW (2009) Compact ultra-broadband coplanar-waveguide bandpass filter with excellent stopband rejection. Prog Electromagn Res B 17:15–28
8. Mondal P, Sahoo M, Parui SK (2016) Improvement of stop-band performance of a CPW bandpass filter using DGS. Microw Opt Technol Lett 58(3):593–597
9. Makimoto M, Yamashita S (1980) Bandpass filters using parallel coupled stripline stepped impedance resonators. IEEE Trans Microw Theory Tech 28(12):1413–1417
10. Mirzaee M (2011) A novel small ultra-wideband bandpass filter including narrow notched band utilizing folded-t-shaped stepped impedance resonator (SIR). Prog Electromagn Res C 22:85–96
11. Kim CH, Chang K (2010) Ring resonator bandpass filter with switchable bandwidth using stepped-impedance stubs. IEEE Trans Microw Theory Tech 58(12):3936–3944
12. Zhu L, Sun S, Menzel W (2005) Ultra-wideband (UWB) bandpass filters using multiple-mode resonator. IEEE Microw Wirel Compon Lett 15(11):796–798
13. Li R, Zhu L (2006) Compact UWB bandpass filter using stub loaded multiple-mode resonator. IEEE Microw Wirel Compon Lett 16(8):440–442
14. Chiou YC, Kuo JT, Cheng E (2006) Broadband quasi-Chebyshev bandpass filters with multimode stepped-impedance resonators (SIRs). IEEE Trans Microw Theory Tech 54(8): 3352–3358
15. Li R, Zhu L (2007) Compact UWB bandpass filters using stub-loaded multiple-mode resonator. IEEE Microw Wirel Compon Lett 17(1):40–42
16. Sun S, Zhu L (2007) Wideband microstrip ring resonator bandpass filters under multiple resonances. IEEE Trans Microw Theory Tech 55(10):2176–2182

Performance Comparison of DCF and FBG as Dispersion Compensation Techniques at 100 Gbps Over 120 km Using SMF

Ashwani Sharma, Inder Singh, Suman Bhattacharya and Shalini Sharma

Abstract The fusion of EDFA and optical fiber causes an expansion in the transmission capacity over the large distances and thus making optical fiber a prevalent broadband communication technique. The practical execution of EDFA offers long transmission distances with less attenuation. However, in order to get high transmission range with high data rates using existing SMF, then techniques must be there to compensate the dispersion caused by fiber nonlinearity. In optical fiber communication, DCF and FBG are the trending dispersion compensation techniques. The use of DCF and FBG as a method of compensation of dispersion can notably heighten the overall performance of the system. In this paper, DCF and FBG as dispersion compensator are compared in terms of Q-factor and BER at 100 Gbps launched over an SMF of 120 km by using Optisystem 7.0 software. On the basis of results, it is suggested to use DCF as a best chromatic dispersion compensation technique in pre-compensation mode.

Keywords Dispersion · Dispersion compensation · Optisystem 7.0
Fiber Bragg grating (FBG) · Dispersion compensating fiber (DCF)
Bit error rate (BER) · Chromatic dispersion (CD) · Q-factor

A. Sharma (✉) · S. Sharma
School of Electrical and Computer Sciences, Shoolini University,
Solan 173229, Himachal Pradesh, India
e-mail: ashwanisharma@shooliniuniversity.com

A. Sharma · I. Singh
School of Computer Science Engineering, UPES, Dehradun 248007
Uttarakhand, India

S. Bhattacharya
TATA Consultancy Services, Mumbai, India

V. Nath and J. K. Mandal (eds.), *Nanoelectronics, Circuits and Communication Systems*, Lecture Notes in Electrical Engineering 511,
https://doi.org/10.1007/978-981-13-0776-8_40

1 Introduction

After T. H. Maimam invented laser in 1960, the optical communication developed swiftly. Today, the precise and brisk exchange of information has become a required need of the people. This demand of people encourages the use of optical fiber as a communication medium and to expand it as a large capacity, high transmission bit rate, high speed, and long distance transmission medium with low amount of noise and attenuation present in it. While considering all these requirements, it is found that information carrying capacity of these optical fibers is affected by dispersion, attenuation, and other fiber nonlinearity. Among these, attenuation and dispersion are the dominating elements to degrade the signal quality over optical fiber. Therefore, how to bring down and abolish these elements has become a major issue of concern. In past years with the disclosure of EDFA, attenuation is no longer a capacity limiting issue. But dispersion is still a major concerning factor which corrupts the signal quality. As a result, quality of communication declines which further restrains the information carrying capacity and transmission distance [1].

Two different perspectives are there to withstand chromatic dispersion. One is to outline fibers with diminished dispersion and other is to use dispersion compensation modules for the compensation of dispersion. Standard SMF having dispersion of 17 ps/nm/km at 1550 nm restrains the transmission distance to under 100 km for a system with a rate 10 Gbps or higher. There are different advancements in figuring out dispersion compensation, for example, dispersion compensation fibers (either fundamental mode or higher mode), fiber Bragg grating, etc. [2].

1.1 Transmission Fiber Types

Fibers that have been introduced or are being introduced in longer distances can be ordered into three categories—standard single mode fiber, dispersion-shifted fibers, and nonzero dispersion-shifted fibers. At 1550 nm, SMF has a high dispersion of around 17 ps/nm/km, which is a key confinement for high data rate systems. Dispersion-shifted fibers having zero dispersion at 1550 nm were presented to lower the dispersion. In spite of the fact that dispersion-shifted fiber is useful for high-speed transmission in single channel, it was soon found that it is not reasonable for WDM systems because of the fact that zero dispersion is a favorable condition for four-wave mixing (FWM) [3].

1.2 Dispersion

In communication systems, information to be transmitted is initially modulated and then sent in the form of pulses over large distances. Transmission of information is reliable as long as these pulses go through fiber without contortion. But normally while traveling through the fiber the pulses deprived of their original shape and overlap with their adjacent pulses which end up noticeably undefined at receiver. Dispersion is the general term associated to this and the impact is called intersymbol interference (ISI). Fiber dispersion acts as an essential part in the propagation of the short pulses as all the pulses having different spectral components travels at different speeds. All the optical pulses move inside the envelope with a group velocity where β_2 is a parameter representing the dispersion of group velocity, responsible for pulse widening. This concept is known as group velocity dispersion (GVD), and β_2 is GVD parameter. β_2 becomes negative for extensive wavelengths and disappears at zero dispersion wavelength (λ_D).

The relationship between dispersion parameter (D) and β_2 is given as follows (Eq. (1)):

$$D = \frac{-2\pi c}{\lambda^2}\beta_2 \tag{1}$$

For $\lambda < \lambda_D$, the fiber is said to have normal dispersion, i.e., β_2 is positive.

Chromatic Dispersion: A phenomenon in optical fiber caused due to the dependency of group index on wavelength results in the temporal widening of optical pulses as they pass through the fiber is called chromatic dispersion. Inside the SMF it can be classified into two parts—waveguide dispersion and material dispersion. In waveguide dispersion, distinct wavelengths travel at distinct velocities in optical fiber which is due to the physical structure of core and cladding. Material dispersion is a dispersion caused by the change in the refractive index of optical fiber with the change in the wavelengths. The pulse widening restricts the transmission distance and also the channel capacity. Most commonly used chromatic dispersion compensation techniques are Dispersion Compensating Fiber (DCF) and Fiber Bragg Grating (FBG) [4].

1.3 Dispersion Compensation Techniques

(a) Dispersion Compensating Fiber (DCF): For the data transmission at low fiber loss, 1550 nm window is the most favorable because EDFA operates around this wavelength and dispersion is about 17 ps/nm/km. For a data rate of 2.5 Gbps, the dispersion can be reduced by using distributed feedback laser (DFL). As the data rate increments to 10 Gbps, the transmission distance restricted to about 60 km [5].

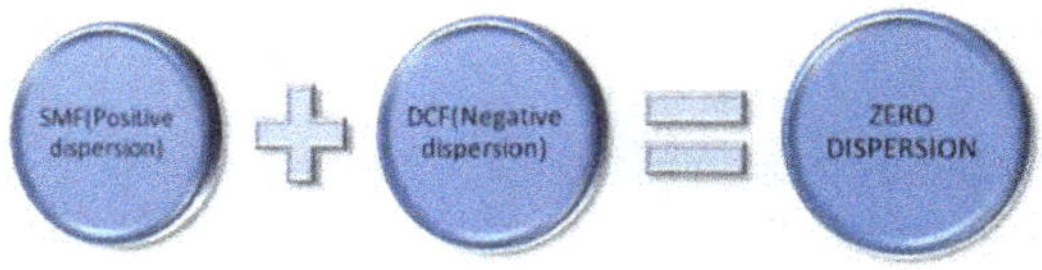

Fig. 1 Dispersion compensation fiber principle

DCF is a sort of optical fiber which has the characteristics of dispersion inverse to the characteristics of transmission link. The principle of DCF is shown in Fig. 1 as follows.

Today, most commonly used transmission link is common single mode optical fiber (CSMF). This fiber has a dispersion of about 18 ps/nm/km in 1550 nm window and a positive dispersion in the *C*-band window. The assembled positive dispersion becomes immense for a longer transmission distance which in turn reduces signal-to-noise ratio (SNR) and causes an increase in bit error rate. Hence, the overall performance of the system deteriorates. So the DCF design should be like that can compensate positive dispersion, i.e., DCF should have negative dispersion in the operating range of wavelength [6].

Relationship of dispersion with second derivative of propagation constant (β) is shown in Eq. (2)

$$D = \frac{-2\pi c}{\lambda^2}\frac{\mathrm{d}^2\beta}{\mathrm{d}\omega^2} = \frac{1}{c}\left(2\frac{\mathrm{d}n_\mathrm{e}}{\mathrm{d}\omega} + \omega\frac{\mathrm{d}^2 n_\mathrm{e}}{\mathrm{d}w^2}\right) \quad (2)$$

where

c velocity of light in vacuum,
λ wavelength,
ω frequency, and
n_e effective index.

The propagation constant can be defined using free space wave number and effective index as follows (Eq. (3)):

$$\beta = k_0 n_\mathrm{e} = \frac{\omega}{c} n_\mathrm{e} = \frac{2\pi}{\lambda} n_\mathrm{e} \quad (3)$$

Effective index can be further defined in Eq. (4)

$$n_\mathrm{e} = \Delta n_\mathrm{e} + n_0 \quad (4)$$

where

Δn_e effective index difference, and
n_0 refractive index of the cladding.

Thus, dispersion can be defined as follows (Eq. (5)):

$$D = \frac{-2\pi c}{\lambda^2}\frac{\mathrm{d}^2 k_0 \Delta n_e}{\mathrm{d}\omega^2} + \frac{-2\pi c}{\lambda^2}\frac{\mathrm{d}^2 k_0 n_0}{\mathrm{d}\omega^2} = D_{\mathrm{waveguide}} + D_{\mathrm{material}} \quad (5)$$

Thus to cancel the total dispersion, the principal in Eq. (6) is followed:

$$D_{\mathrm{TF}} L_{\mathrm{TF}} + D_{\mathrm{DCF}} L_{\mathrm{DCF}} = 0 \quad (6)$$

where

D_{TF} dispersion coefficient of transmission fiber,
L_{TF} length of transmission fiber,
D_{DCF} dispersion coefficient of DCF, and
L_{DCF} length of DCF.

(b) Fiber Bragg Grating (FBG): It is one of the most flexible and evolved techniques for dispersion compensation. For the most part, it can perform as a single and multichannel DCM and in addition give extra characteristic of dynamic compensation of dispersion, which are impractical with DCF and turned out to be essential parameter for WDM and OTDM systems [7]. Due to the better optical characteristics of chirped FBG, it is more convenient for WDM systems. The concept of Fiber Bragg grating is shown in Fig. (2) as follows.

The grating period of FBG is not a fixed value but alters linearly with the change in the length of the grating by keeping the shorter grating period at the starting of the grating. The grating period of FBG can be defined as the distance between two nearby most extremes estimations of refractive index. FBG mirrors the narrow spectrum of wavelengths that are focused at λ_B and allows all other wavelengths to pass through. For getting the reflected wavelength, following Bragg's condition in Eq. (7) should be satisfied:

$$\lambda_B = 2\Lambda n_g \quad (7)$$

where

λ_B is reflected wavelength (nm),
Λ is grating period (nm), and
n_g is effective refractive index.

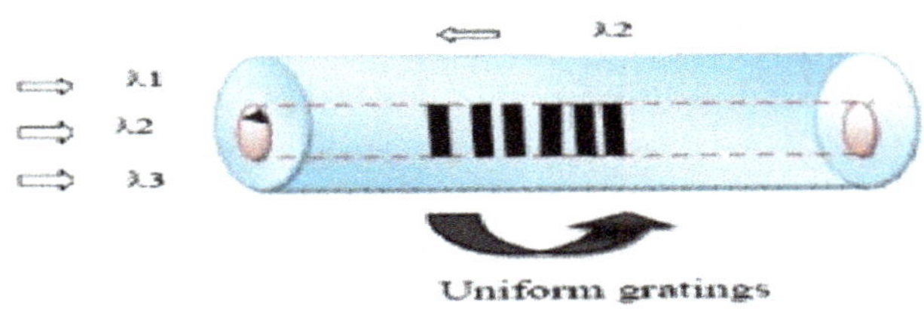

Fig. 2 Fiber Bragg grating principle

Table 1 Simulation parameters

Sr. no.	Research paper	Data rate (Gbps)	Distance (km)	Q-factor	BER
1	15	10	150	36.30	5.566e−287
2	16	20	150	21.58	1.14e−153
3	17	40	50	7.30	1.34e−13
4	18	40	80	10	1.2e−26
5	19	40	100	3.46	0.14e−3

At the point when dispersion influenced input pulse with width τ goes through the chirped FBG, the width of pulse is diminished by $\Delta\tau$ and its shape is re-established in the yield. At the starting point the grating period of chirped FBG are shorter but with the increase in the length of the grating, these periods also increases linearly. In this way, the wavelengths of shorter signals reflected sooner as compared to signals with longer wavelengths because these wavelengths propagate more into the fiber grating before their reflections [8]. FBG can be used as a dispersion compensator by limiting the bandwidth occurs due to the large oscillations present in grating dispersion. The main cause of these oscillations is Fresnel type reflections. These Fresnel reflections are accountable for leftover band, which are unenviable [9].

Fiber grating has a variety of merits over other dispersion compensations technologies used today. These include variations in physical parameters of grating at the time of inscription of grating. Changing these parameters allows us to make grating of required features like either narrow or wide bandwidth or other special characteristics [10]. Different authors have worked on different parameters to reduce the chromatic dispersion and achieved different results depending upon different parameters. Some of the results in terms of Q-factor and BER are shown in Table 1.

2 Simulation Setup

This section of the paper shows the implementation of dispersion compensation using Optisystem 7.0 software for single channel. Two most recently used dispersion compensation techniques are DCF and FBG. At the transmitter end, the simulation setups of both DCF and FBG consist of continuous wave laser (CW) used as a source of light, nonreturn to zero (NRZ) modulation format, and a Mach–Zehnder (MZ) modulator used for the modulation of the optical signal. All the optical pulses are sent at a transmission rate of 100 Gbps. Only a single frequency of 193.1 THz is used for the simulations. At the receiver terminals, pin detector photodiode is used to detect the optical pulses and then convert it to the electrical signals. At the time of simulations, the position of dispersion

Table 2 Simulation parameters

Sr. No.	Parameter	Value
1	Bit rate (Gbps)	100
2	Sample rate (THz)	6.4
3	Frequency (THz)	193.1
4	Power (mW)	1
5	Extinction ratio (dB)	30
6	Gain (dB)	20
7	Noise (dB)	2
8	Bandwidth (THz)	1
9	Dispersion FBG (ps/nm)	−2028

Table 3 DCF fiber parameters

Sr. No.	Parameter	Value
1	Length of fiber (km)	120
2	Reference wavelength (nm)	1550
3	Length of DCF (km)	24
4	Attenuation (db/km)	0.3
5	Differential slope (ps/nm^2/km)	0.21
6	Dispersion (ps/nm/km)	−80

Table 4 FBG fiber parameters

Sr. No.	Parameter	Value
1	Length of fiber (km)	120
2	Dispersion (ps/nm/km)	17
3	Attenuation (db/km)	0.2
4	Differential group delay (ps/km)	3
5	Differential slope (ps/nm^2/km)	0.008

compensating components (DCF and FBG) is relocated again and again with a motive to make clear how the position of dispersion compensating components affects the transmission distance and the bandwidth of the channel. This can be done in pre-compensation, post-compensation, and symmetrical compensation configuration. First, the simulations are carried out using DCF and then with FBG. EDFA is utilized for the compensation of the attenuation after each DCF and FBG components.

In this paper, six different types of combinations of dispersion compensation are simulated which are dispersion compensation by using DCF in pre, post, and symmetrical configurations and the same by using FBG. Finally, the outcomes of two compensation techniques are compared in terms of Q-factor and BER. Simulation parameters, DCF fiber parameters, and FBG fiber parameters are tabulated in Tables 2, 3, and 4, respectively. Simulation diagram for DCF in pre, post, and symmetrical configurations is shown in Figs. 3, 4, 5 and for IDCFBG in Figs. 6, 7, and 8, respectively.

Fig. 3 Pre-compensation using DCF simulation model

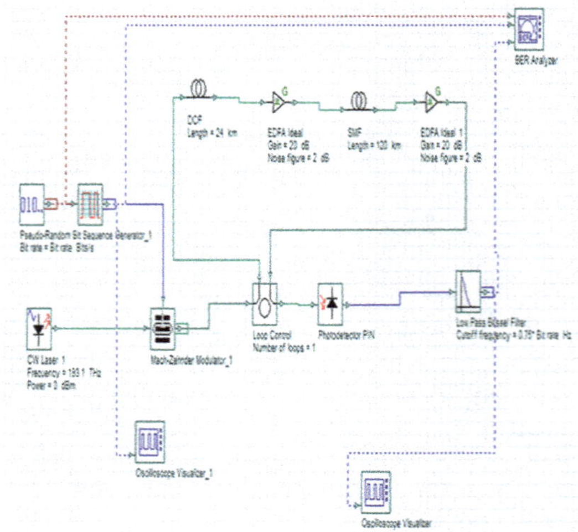

Fig. 4 Post-compensation using DCF simulation model

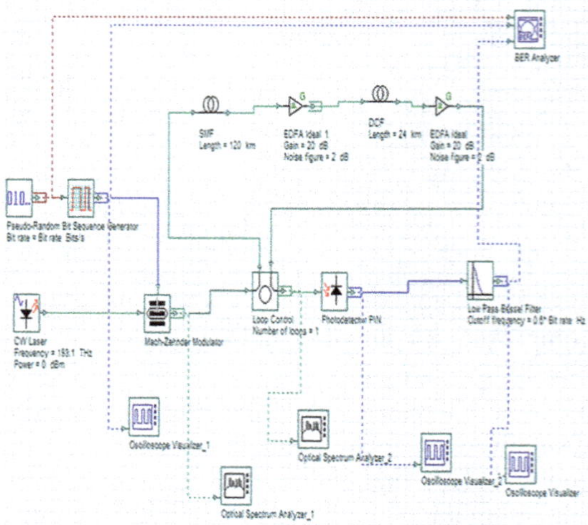

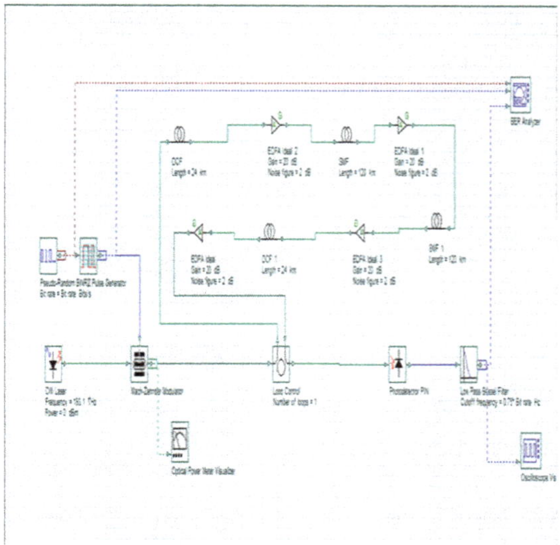

Fig. 5 Symmetrical compensation using DCF simulation model

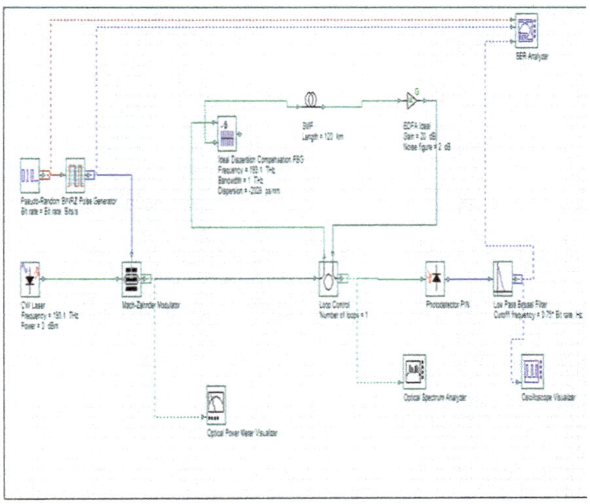

Fig. 6 Pre-compensation using IDCFBG simulation model

3 Simulation Results and Discussions

This section contains the analysis of the results obtained by using DCF and FBG in pre, post, and symmetrical configurations in terms of Q-factor and BER. The eye diagrams of DCF in pre, post, and symmetrical configurations are shown in Figs. 9,

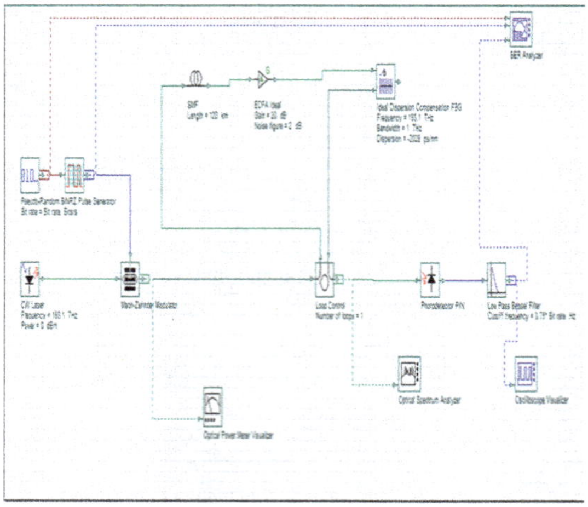

Fig. 7 Post-compensation using IDCFBG simulation model

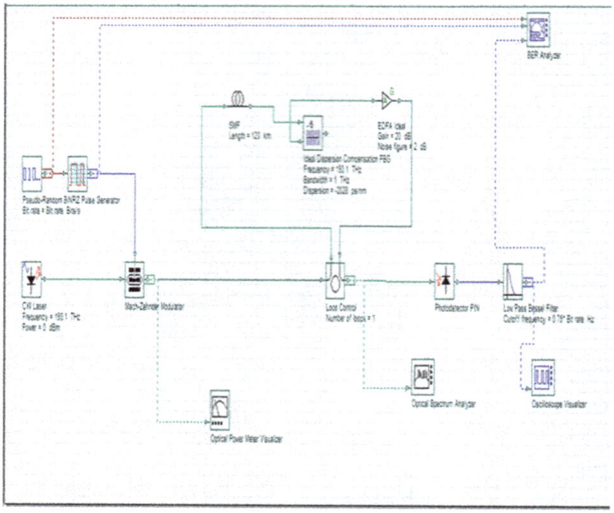

Fig. 8 Dispersion compensation using mix IDCFBG simulation model

10, and 11, respectively, whereas Figs. 12, 13, and 14 show the eye diagrams of pre, post, and mix compensation using IDCFBG. Table 5 contains the comparison results of DCF and IDCFBG, and it is found that pre-compensation of DCF is having greatest quality factor and least bit error rate in comparison to other schemes.

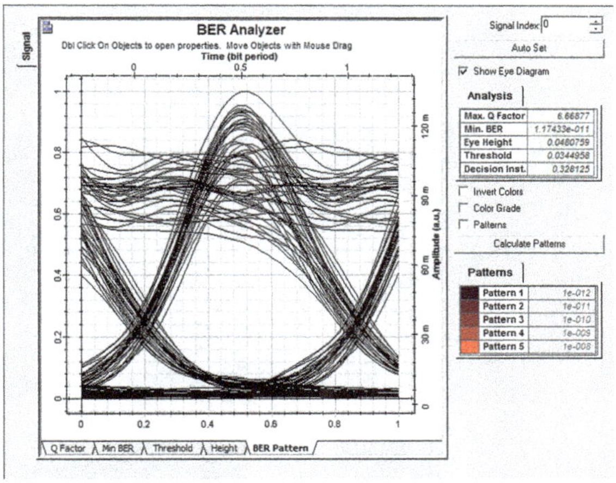

Fig. 9 DCF pre-compensation simulation result

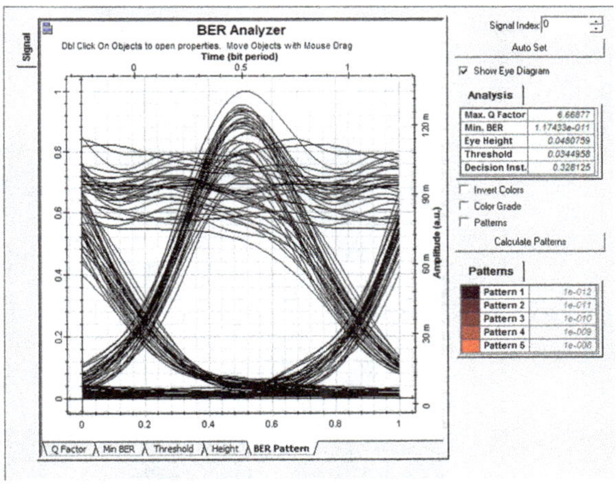

Fig. 10 DCF post-compensation simulation result

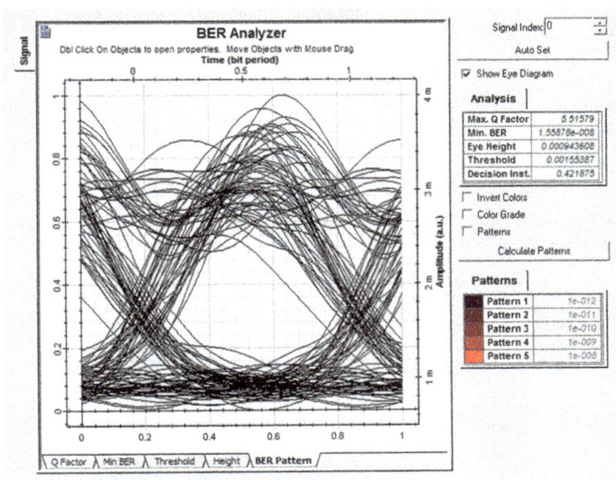

Fig. 11 DCF symmetrical compensation simulation result

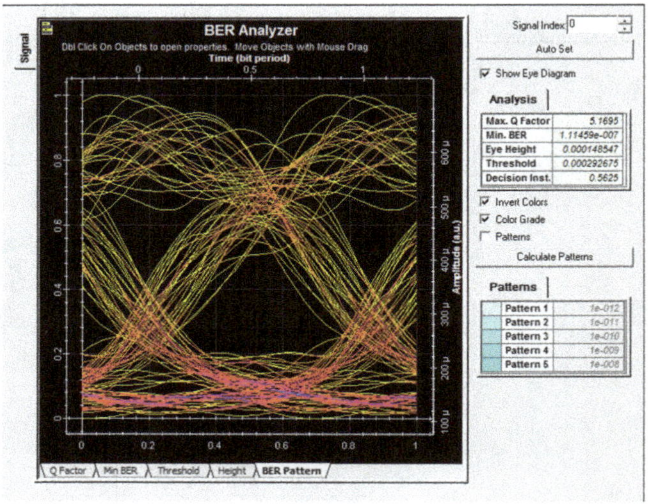

Fig. 12 FBG pre-compensation simulation results

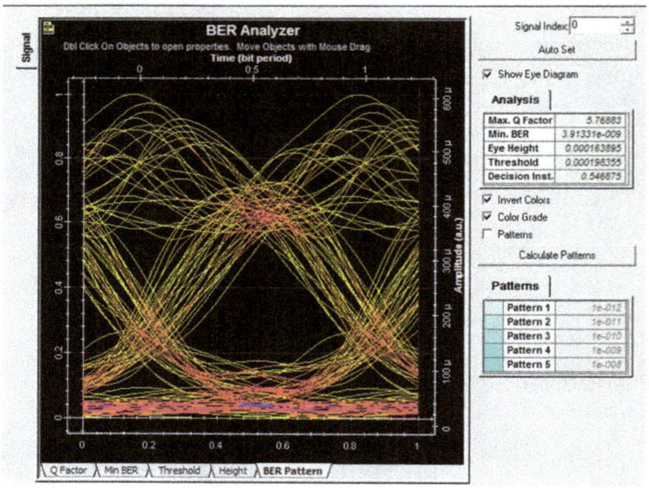

Fig. 13 FBG post-compensation simulation results

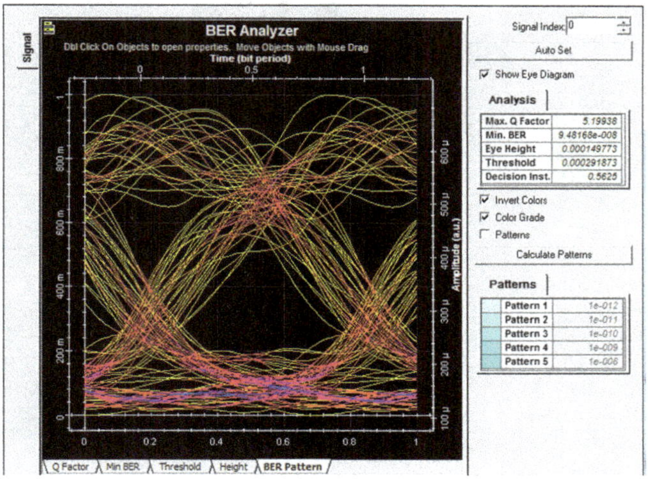

Fig. 14 FBG mix compensation simulation results

Table 5 Comparative results

DCF			FBG		
Model	*Q*-factor	BER	Model	*Q*-factor	BER
Pre	6.66	1.174e−011	Pre	5.16	1.114e−007
Post	6.45	5.018e−011	Post	5.76	3.913e−009
Symm.	5.51	1.558e−008	Mix	5.19	9.481e−008

Fig. 15 *Q*-factor comparison

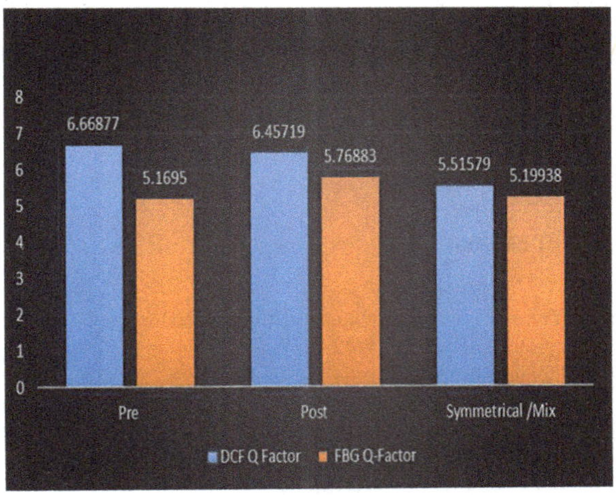

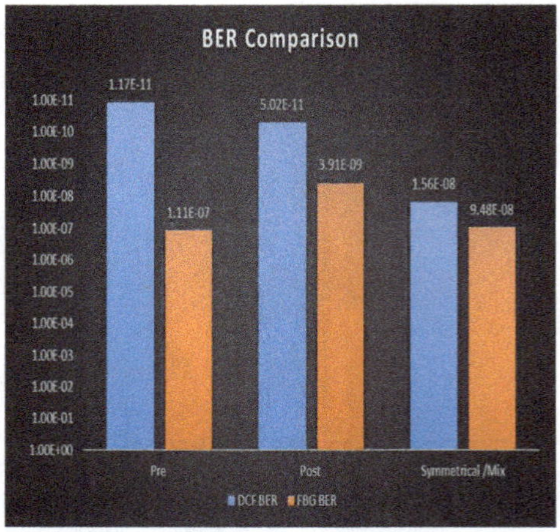

Fig. 16 Bit error rate comparison

4 Conclusions

The work in this paper is completely focused on implementing dispersion compensation techniques (DCF and FBG) in order to compensate for the dispersion occurs during transmission of signals over a distance of 120 km at 100 Gbps. DCF and FBG were used in three different configurations, i.e., pre, post, and symmetrical/mix. Results from the different configurations are compared in terms of *Q*-factor, eye diagram, and BER. Moreover, an analogy is made between DCF and FBG using graphs as shown in Figs. 15, and 16, respectively. After analyzing the graphs, it is clear that DCF with pre-compensation scheme has a high *Q*-factor when compared with the other configurations of DCF and FBG, altogether it has least BER among all the schemes. However, it is not possible to completely expel the dispersion during transmission at higher bit rates but using DCF in pre-compensation at 100 Gbps for an SMF of 120 km can lower down the dispersion.

References

1. Yuhu S (2011) Research on the dispersion problem in high speed optical communication systems. IEEE, pp 4742–4745
2. Dilendorfs V, Spolitis S, Bobrovs V (2016) Effectiveness evaluation of dispersion compensation methods for fiber optical transmission systems. In: Progress in electromagnetic research symposium (PIERS), 8–11 Aug 2016, pp 3759–3763
3. Li MJ (2001) Recent progress in fiber dispersion compensators. ECOC, pp 486–489
4. Ilavarasan T, Meenakshi M (2015) An overview of fiber dispersion and nonlinearity compensation techniques in optical orthogonal frequency division multiplexing systems. Springer, Heidelberg

5. Dar AB, Jha RK (2017) Chromatic dispersion compensation techniques and characterization of fiber Bragg grating for dispersion compensation. Springer, Heidelberg
6. Chan W, Li S, Lu P, Wang D, Luo W (2010) Dispersion compensation optical fiber modules for 40 Gbps WDM communication systems. Springer, Heidelberg
7. Sumetsky M, Eggleton BJ (2005) Fiber Bragg gratings for dispersion compensation in optical communication systems. Springer, Heidelberg, pp 256–278
8. Spolitis S, Ivanovs G (2011) Extending the reach of DWDM-PON access network using chromatic dispersion compensation. IEEE, pp 29–33
9. Eggleton BJ, Stephens T, Krug PA, Dhosi G, Brodzeli Z, Ouellette F (1996) Dispersion compensation using a fiber grating in transmission. Electron Lett 32(17):1610–1611
10. Brennan III JF (2005) Broadband fiber Bragg gratings for dispersion management. Springer, Heidelberg, pp 397–433
11. Bobrovs V, Spolitis S, Ivanovs G (2012) Comparison of chromatic dispersion compensation techniques for WDM-PON solution. IEEE, pp 64–67
12. Fews HS, Stephens MFC, Straw A, Forysiak W, Nayar BK, Gleeson LM (2006) Experimental comparison of fiber and grating based dispersion compensation schemes for 40 channel 10 Gbps WDM systems. IEEE
13. Patnaik B, Sahu PK (2013) Ultra high capacity 1.28 Tbps DWDM system design and simulation using optimized modulation formats. Elsevier, New York, pp 1567–1573
14. Park SG, Gnauck AH, Weisenfeld JM, Garrett LD (2000) 40 Gbps Transmission over multiple 120 km spans of conventional single mode fiber using highly dispersed pulses. IEEE 12(8):1085–1087
15. Kaur R (2016) Analysis on dispersion compensation with DCF on opti system—a review. IJES 17:390–396
16. Panda TK, Parakram K, Mishra R, Sinha A (2016) Performance comparison of dispersion compensation in a pre, post and symmetrical arrangement using DCF for long haul optical communication. AIJET 3(7):14–20
17. Yadhav M, Jaiswal AK, Agrawal N (2015) Design performance of high speed optical fiber WDM system with optimally placed DCF for dispersion compensation. IJCA 122(20)
18. Khatoon S, Jaiswal AK, Agrawal A (2017) Performance evaluation of post and symmetrical DCF technique with EDFA in 32 × 10, 32 × 20 and 32 × 40 Gbps WDM systems. IJCET 7(4)
19. Xie T (2014) Reparation of chromatic dispersion using dispersion compensation bank and BER analysis at various power level in 40 Gbps fiber optics system. IEEE, pp 1058–1062

Different Aspects of Smart Grid: An Overview

Neha Nidhi, Deepak Prasad and Vijay Nath

Abstract Smart grid involves the application of computer, transmission line, and networks into an inactive power distribution system. It is used to improve the maintenance, operations, and technologies of the smart grid, so that the component of the power grid can easily interchange data. However, concerns associated with communication and management must be addressed before full benefits of the smart grid can be accomplished. In this paper, all aspects that are relevant to smart grid communication and information technologies, and also its challenges, have been discussed.

Keywords Smart grid · Network · Transmission line · Topology Energy

1 Introduction

Power grid refers that distribute electricity from the power plant to the users. It contains power plant, transformer, transmission line, substations and distribution line, distribution transformer, etc. Power grid was made in the 1890s and upgraded day by day as technology enhanced each decade. At present, electric grid consists of beyond 9,200 power generating units which can generate over 1 million megawatts of power, which is linked to above 300,000 miles of transmission line. Further, a new kind of electric grid is required, which is made from the lowermost to manage the groundswell of computerized and digital equipment and technology that depend on it, also which can systematize and accomplish the rising difficulty and requirements of electricity in the twenty-first century. In other words, we can say to make a grid smart, a digital technology is used. This technology permits the two-way communication between the services and its customers. Smart grid

N. Nidhi · D. Prasad (✉) · V. Nath
VLSI Design Group, Department of Electronics and Communication Engineering, Birla Institute of Technology Mesra, Ranchi 835215, Jharkhand, India
e-mail: prasaddeepak007@gmail.com

V. Nath and J. K. Mandal (eds.), *Nanoelectronics, Circuits and Communication Systems*, Lecture Notes in Electrical Engineering 511,
https://doi.org/10.1007/978-981-13-0776-8_41

includes controls, computers, automation, new technologies, and equipment which all are working together. These technologies respond digitally to electrical grid. Through this reliability, efficiency, sustainability of the generation and economics, transmission, and supply of electrical power can be improved. The explanation of the smart grid is not essentially unique, as its visualization to the investors and the technological complications can be different [1]. The US Department of Energy (DOE) has suggested the definition of smart grid as "Smart Grid is an automated broadly distributed energy delivery network". It will be categorized by a bidirectional flow of electricity and information. Also, it will be capable of monitoring everything from power plants to customer preferences to individual appliances. The benefits of distributed computing and communications incorporate into the grid to deliver real-time information and allow the near-instantaneous balance of supply and demand at the device level [2].

Normally, smart grid is the combination of a traditional distribution network and a two-way communication network for sensing, monitoring, and spreading of information on energy consumptions. A smart grid communication architecture is shown in Fig. 1. A typical smart grid consists of several power generating entities and power consuming entities, all connected through a network. The generators feed the energy into the grid and further consumers draw energy from the grid.

The Smart Grid benefits include as follows:

(a) More effective transmission of electric power.
(b) Restoration of electricity will be faster after power disturbances.
(c) By reducing higher demand will also help lesser electricity rates.
(d) By this large-scale renewable energy systems will be increased.
(e) Customer-owner power generation systems will be improved.
(f) Costs of operations and management services will be reduced.
(g) Improved security.

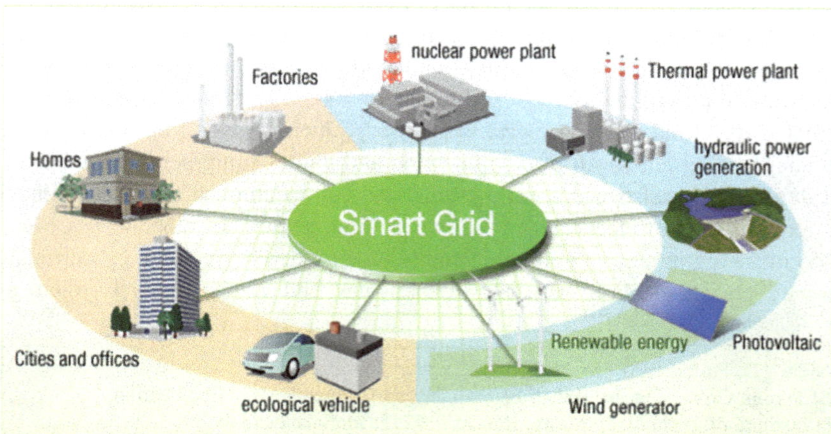

Fig. 1 A smart grid communication architecture

Smart Grid will enhance resiliency to our electric power system and prepared it for to address emergencies such as earthquakes, large solar flares, severe storms, and terrorist attacks. When instruments fail or outages occur, smart grid will permit for automatic rerouting because of its two-way communicating capacity. Through this outage, their effects will be minimized.

2 Smart Grid Features

The smart grid signifies the full set of current. It proposed responses to the challenges of electric power supply. There are many competing classifications and no agreement on a universal definition because of the different ranges of factors. However, possible classification is given here:

2.1 Reliability

By using technologies such as state approximation [3], the smart grid will improve the error detection and also allow the network to be self-curing without interference of operators. Through this reliable distribution of electric power will be ensured. Also, exposure to natural disasters will be reduced. By the economic effect of upgraded version of grid resilience, reliability becomes the subject of a number of studies. Also, by using a US DOE-funded methodology, it can be calculated by using only one calculation tool.

2.2 Efficiency

By the utilization of smart grid technology, overall enhancement of the efficiency of energy substructure can be predicted. Like in energy demand management, e.g., air conditioners will be turnoff, when short-term changes occur in the price of electricity, through VAR optimization technique (VVO) voltage on transmission lines are reduced and so is eliminating truck-rolls for meter reading. Redundancy of overall effect in transmission line and distribution line will be less. Also, using greater number of generators leads to lesser electricity prices.

2.3 Network Topology

Flexible transmission and distribution setup in next generation will capable to deal with probable bidirectional energy drifts. Also it permittes fair distribution of power generated from photovoltaic panels, from the fuel cells uses, wind turbines, from the electric cars batteries, pumped hydroelectric power, and from other sources. Traditional networks were operating on unidirectional flow of electric power. But it is consuming if local sub-net will generate more amount of power. And the inverse movement can increase reliability and security issues [4]. Aims of the smart grid are to manage these circumstances [5].

2.4 Load Matching/Load Balancing

Electric grid load can be changed expressively. The overall load is not only unstable but also varies solwly which makes total amount of load dependent upon the summation of several individual selections of the users. Traditionally, certain additional generators are added on a dissipative mode to give response to a quickly changed power consumption. To reduce the load temporarily or continuously, a smart grid can give the warning to all different television sets or another customer [6]. By the use of precise calculation techniques, it would be possible to calculate how many standby generators are required to be used and to obtain a certain failure rate. In customary grid at the cost of more standby generator, only the failure rate can be decreased. In smart grid, the problem will be reduced by the load reduction in small portion of the clients.

2.5 Sustainability

Smart Grid allows more saturation of extremely variable renewable energy resources such as wind power and solar power due to its enhanced flexibility, even though without adding of energy storage. For several distributed feed-in points, modern network setup is not made to allow it. And if some feed-in is allowed, transmission-level setup cannot accommodate it at the local (distribution) level. Because of gusty or cloudy weather, quick fluctuations in distributed generation occurred. This becomes challenging to electric engineers who have to make sure the steady power level by fluctuating the more controllable generators output, such as hydroelectric generators and gas turbine. For this reason, smart grid technology is an essential condition for a very large quantity of renewable electric power on the grid.

3 Challenges in Smart Grid

Smart grid involves substantial challenges because major changes occurred in it. From the DOE's National Energy Technology Laboratory report, modern grid has the following major barriers to attaining smart grids [7]:

- **Commercial Resources**

  The commercial instance for a self-healing grid is good, especially if it contains social benefits. Before authorizing major investments, regulators will require general proof based heavily on social benefits.

- **Government Sustenance**

  Without the help of government, the industry may not have the financial capability to fund new. Even though some utility industry has capital-intensive, with $800 billion in resources, but it has suffered from hard times in the marketplace. Also, some utilities have reduced its financial ratings.

- **Compatible Equipment**

  New smart grid technologies cannot fit with older equipment. So it must be replaced. Because of this, a problem exists for services and controllers since keeping equipment beyond its devalued life minimizes the cost to users. Early retirement of apparatus can become an issue.

- **Speed of Technology Improvement**

  It was predicted 50 years ago that the basement fuel cell, the solar shingle, and the chimney wind generator as an integral part of the home of the future. This retiring historical progress will need to increase the speed.

- **Regulation and Policy**

  A regular parochial view of new construction projects is taken by utility commission. The state financing on the project may not always be the one who benefits the most from it. Smart grid investment is encouraged so that the attractive return is held. To invest in a new technology utilities will remain reluctant.

- **Cooperation**

  To implement the smart grid concept, it is become challenging for 3,000 different utilities to cooperate while installing.

4 Conclusion

The smart grid is a completely automatic energy transmission network. It has the ability to guarantee two-way information and power flows between generation plants, final users, and applications among others. In this paper, some features of the smart grid have been discussed that include reliability, efficiency, network topology, load matching/load balancing, and sustainability. The development of smart grid will mean an extreme transformation in electric power administration and its uses. So, several technical challenges and issues associated with effective and safe communication and information processing must be resolved before realizing the vision of a smart power grid.

References

1. CEA Canadian Electricity Association (2010) The smart grid: a pragmatic approach, tech report, Sept 2010
2. US DOE US Department of Energy (2010) Smart grid: an introduction, tech report
3. Huang Y-F, Werner S, Huang, Kashyap N, Gupta (2012) State estimation in electric power grids: meeting new challenges presented by the requirements of the future grid. IEEE Signal Process Mag 29(5)
4. Tomoiaga B, Chindris M, Sumper A, Sudria-Andreu A, Villafafila-Robles R (2013) Pareto optimal reconfiguration of power distribution systems using a genetic algorithm based on NAGA-II. Energies 1439–1455, Mar 2013
5. Berger LT, Imewski K (2012) Smart grid applications, communications, and security. ISBN: 978-1-118-00439-5, April 2012
6. Smitsyn NA, Kundu S, Backhaus S (2012) Safe protocols for generating power pulses with heterogeneous populations of thermostatically controlled loads 67, Nov 2012
7. Bari A, Jiang J, Saad W, Jaekel A (2014) Challenges in the smart grid applications: an overview. Int J Distrib Sens Netw 10(2)

Robust Voiceprint Based Audio Watermarking Using Wavelet Transform

Sumit Srivastava, Mahesh Chandra and G. Sahoo

Abstract The audio system of the humans is more complicated and not easy to analyze or understand than the visual systems of the humans. Due to this, various effective watermarking algorithms have been implemented and introduced for digital text and digital images. Only fewer algorithms have been introduced or selected for the audio watermarking. This paper proposes a novel robust approach for the audio watermarking using voiceprint of a speech signal with wavelet transform. Haar and different orders of Daubechies wavelets have been used for the abovementioned task. The performance of voiceprint-based audio watermarking using wavelet transform has been compared based on percentage compression, retained signal energy, and signal-to-noise ratio (SNR). Higher order Daubechies wavelets have shown better performance for audio watermarking in terms of various performance parameters as mentioned.

Keywords Discrete wavelet transform · Haar wavelet · Daubechies
Audio watermarking · Human auditory system

S. Srivastava (✉) · G. Sahoo
Department of Computer Science and Engineering, Birla Institute of Technology Mesra, Ranchi 835215, India
e-mail: sumit.srivs88@gmail.com

G. Sahoo
e-mail: gsahoo@bitmesra.ac.in

M. Chandra
Department of Electronics and Communication Engineering, Birla Institute of Technology Mesra, Ranchi 835215, India
e-mail: shrotriya@bitmesra.ac.in

V. Nath and J. K. Mandal (eds.), *Nanoelectronics, Circuits and Communication Systems*, Lecture Notes in Electrical Engineering 511,
https://doi.org/10.1007/978-981-13-0776-8_42

1 Introduction

In this modern era of Internet where all the multimedia data are freely available in digital format, i.e., audio, video, and image, data security is a major concern. During the transmission of any type of data, there is always a possibility of attack (destruction, tampering, and modification of data) by any malicious user. That is why we need to provide security to our data or to the medium by which we are transmitting our data.

Digital watermarking is a way of identifying the ownership of the copyright or in simple words, it is the process of identifying whether the sender is genuine or not [1–3]. Watermark is nothing but a way of hiding information into a signal. We need watermarking for authentication of data and preserving the copyright [4]. Hiding of information is the core for watermarking. Watermarking can be used in voice conferencing to indicate the current speaking person of the party to other parties [5, 6]. The ability to detect/extract after the manipulation of common signals is known as the robustness [7]. The watermarking consists of two phenomena "embedding" and "extracting" [8–10]. The process of watermarking is shown in Fig. 1. The embedding process should be done in such a way that the quality of the input audio file is not degraded. The hiding of secret information in a signal using some algorithm is known as embedding [11–14]. It is essential to make our embedding mechanism robust to resist intentional or unintentional attacks. The extraction of the secret information using the algorithm is known as extracting.

Extraction of watermark is possible in case we need it. It only depends on the way of embedding and the nature of the watermark algorithm. There are several audio watermarking techniques. The one of the best techniques of audio watermarking is through using wavelet transform. A wavelet may be defined as the small wave that is related to energy. It has a move and swing back like properties and also has the ability to analyze both time and frequency, simultaneously. Wavelet transform, in other words, is the decomposition of the signal into various set of wavelets [13]. This technique is very useful due to its high compression ratio. It is a dominant mathematical tool in various areas of science and engineering. The wavelets provide stronger multi-resolution tool for the determination of the moving signals with better localization information. The watermarking of the audio file can be done by using the discrete wavelet transform. Discrete wavelet transform (DWT) breaks the speech signals into various groups or bandwidth lesser bandwidths and minimum sample rates. The DWT can provide a time–frequency representation of a given signal and break the signal into more than one coefficient.

Fig. 1 Watermarking process

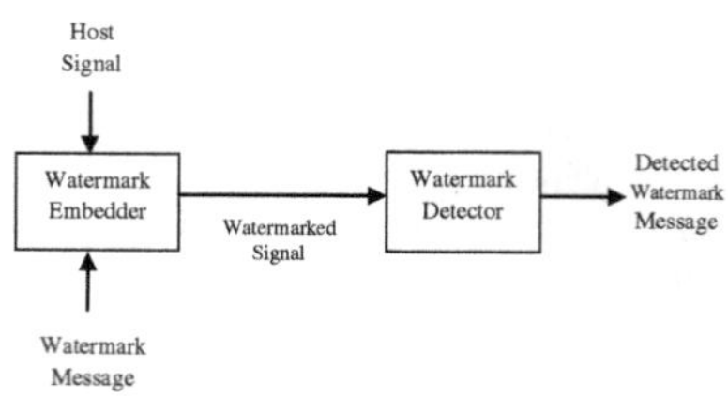

Digital watermarking is an approach, which shelves copyright information in digital media in order such that the inserted bit cannot be identified by the human senses [15]. It is very necessary to make the wrapped information so robust so that it can easily resist the common intentional or unintentional attacks [16]. A considerable approach focusing on video and image watermarking has been taken out, but only a handful number of algorithms have been reported on the watermarking of audio from human auditory system (HAS). HAS is more sensible compared to human's visual system [17].

Majority of the audio watermarking algorithms takes benefit from the drawback of human auditory system [18]. The watermark data in numerous parts of the speech signals is embedded in such a way that the distortion in the watermarked signal is not audible [18].

2 Wavelet Transformation (WT)

The wavelet transform is very popular in signal processing and image processing technique. The wavelet transform divides the signal into shifted and scaled versions of the mother wavelet. A wavelet is a waveform of a duration which has zero mean value. Wavelet transform of audio signals is used to transform the audio signal into time–frequency domain. For digital audio signals, discrete wavelet transforms (DWTs) are used.

A discrete wavelet transform is a transformation technique in which the wavelets are discretely sampled. The wavelet transform divides the signal into two components, approximation and detail. The approximation components have low-frequency mechanism of the signal, and the detail components are high-frequency mechanism of the signal.

A signal function fn(t) is analyzed when linear decomposition is performed as

$$\mathrm{fn}(t) = \sum_{l} a_l \Psi_l(t) \tag{1}$$

where

a_l is expansion coefficients, and
$\Psi_l(t)$ is expansion set (set of real-valued function of t).

$$a_l = (\mathrm{fn}(t), \Psi_l(t)) = \int \mathrm{fn}(t)\Psi_l(t)\mathrm{d}t \tag{2}$$

a_l is obtained by substituting (1) into (2)

$$\text{fn}(t) = \sum_{m}\sum_{l} a_{m,l}\Psi_{m,l}(t) \tag{3}$$

where

m, l are integer indices,
$\Psi_{m,l}(t)$ is wavelet expansion, and
$a_{m,l}$ is discrete wavelet transform of fn(t).

Equation (3) is its inverse.

The wavelet family is generated from a single scaling function by translation and scaling. The function $\Psi(t)$ also called mother wavelet is represented by

$$\Psi_{m,l}(t) = 2^{j/2}\Psi(2^m t - l) \quad m, l \in Z \tag{4}$$

Wavelet family satisfies multi-resolution condition.

It means set of scaling function can be calculated by

$$\varphi_l(t) = \varphi(t - l) \quad l \in Z,\ \varphi \in L2 \tag{5}$$

$\varphi(t - l)$ is the representation of a set of signals, and
$\varphi(2t - l)$ is the representation of a larger set of signals.

Due to spanning of pace of $\varphi(2t)$ by $\varphi(t)$ can be expressed as

$$\varphi(t) = \sum_{n} h(n)\sqrt{2}\varphi(2t - n), \quad n \in Z \tag{6}$$

$h(n)$ may be real or complex called scaling function coefficient.

But the important features of a signal can be described by $\Psi_{m,l}(t)$. These functions are wavelets and can be represented by

$$\Psi(t) = \sum_{n} h(n)\sqrt{2}\varphi(2t - n), \quad n \in Z \tag{7}$$

Equation (7) gives mother wavelet $\Psi(t)$ of the form given by Eq. (4)

$$\text{fn}(t) = \sum_{l=-\infty}^{\infty} c(l)\varphi l(t) + \sum_{m=0}^{\infty}\sum_{l=-\infty}^{\infty} (\text{d}(m,l))\Psi_{m,l}(t) \tag{8}$$

The coefficient in this wavelet expansion is called DWT of signal $f(t)$. For a large class of signals, m and l increase.

As a result, the DWT is efficient for audio and image processing.

2.1 Signal Decomposition

The first step in DWT is to decompose the signal into small groups or set. They are called the approximation coefficient (CA_1: data with low frequency) and detail coefficients (CD_1: data with high frequency). To get the coefficient vectors the original signal(s) is passed through the low pass filter (to obtain low-frequency data) and the high pass filter (to obtain high-frequency data). Then, a downsampling (reducing the signal rate) by factor 2 is applied to get the approximation coefficients.

Equation of approximation coefficient:

$$\mathrm{CA}_1 = \sum_k c_k S_{2i-k}$$

Equation of detail coefficient:

$$\mathrm{CD}_1 = \sum_k g_k S_{2i-k}$$

This equation gives the coefficients after which the forward discrete wavelet transform is performed. If the length of the signal S, and length of filters = $2N$, then CA_1 and CD_1 are calculated as

$$\text{floor}\,((S-1)/2)+N$$

This shows that the wavelet coefficient vector is a bit greater than the length of the signal S because of the filtering process. A wavelet decomposition tree can be formed while using these decompositions of signal with successive approximation. The input signal is broken down into numerous subsidiary resolution parts.

2.2 Signal Reconstruction

The signal can be rebuild using the inverse process of discrete wavelet transform, also known as inverse discrete wavelet transform (IDWT). Follow the above steps in reverse order to rebuild the signal.

In the reconstruction process first, the coefficients (CA_m and CD_m) are up-sampled and filtered with the reconstruction filters.

1. Selection of the Mother Wavelet

To design a high-quality coder selection of mother wavelet is of great importance. We can follow some criteria in order to choose the mother wavelet. Since the main purpose is to maximize the ratio of signal-to-noise (SNR), so it can be determined by calculating the energy concentrated by a wavelet function in level 1 approximation, which can be criteria to choose the mother wavelet.

2. Decomposition of Wavelet

The decomposition of a wavelet can reach up to $L = 2K$ levels, where K = discrete signal's length. Thus, the transform can be performed at any of these levels. Choice of level depends upon the type of signal being analyzed. During compression, it is very important to select the appropriate number of approximation coefficients and detail coefficients in order to represent the signal accurately.

3. Calculation of Threshold

After the application of DWT, the obtained coefficients give attention to energy in few neighbors. So, we can eliminate coefficients with low energy and keep the coefficients with high energy. We can calculate the threshold in two ways.

In the first method, we consider the wavelet expansion of signal and preserve the highest absolute value coefficients. This method is called "Global Thresholding". We manually select the global threshold, so only a single factor is selected. The values of coefficients below this threshold value are set to zero in order to achieve compression. The second method is called "By Level Thresholding". In this method, level-dependent threshold is applied.

4. Coefficient of Encoding

Compression of signal is attained by eliminating undersized valued coefficients and after that encoding them. In order to represent high-level coefficients, the coefficients are stored with the respective positions inside the coefficients in transformation of the wavelet vector. For a signal of size F, with DWT create a block of space and size T which is just greater than F. If the greater coefficients of L are restored, then the ratio C is given by

$$C = F/2L$$

Watermarking can be done in another way also. The reputative zero-valued coefficients are encrypted with the two bytes. The first byte is used to represent series of zeroes and second byte represents number of consecutive zeroes in the wavelet transform vector. We need to compress the transform vector based on threshold calculation or selecting a truncation percentage after zeroing the coefficients of wavelet with negligible values.

3 Experimental Setup and Results

The simulation operation for this wavelet-based watermarking is performed with the help of Matlab R2014 on a Hindi dataset. In this experiment a speech utterance "check" has been used as a watermark on a Hindi speech sample, to evaluate the system performance. Initially, optimum wavelets are chosen based on the power conservation properties and methods in the estimation part of the wavelet values and coefficients. In this work, Haar and different order Daubechies wavelets are taken to perform the experiment. The db1 wavelets are similar to the Haar wavelet. Wavelets operate by crumbling a signal in various resolutions or frequency bandwidths, and the above task is taken out by selecting wavelet function and evaluating the discrete wavelet transform (DWT). After crumbling of speech signals into different frequency bands, it was observed that 90% of wavelet coefficients have less 5% energy of the maximum value of energy. Global threshold includes taking the wavelet expansion properties of the signals and keeping the absolute value coefficients, which was the largest. Level threshold includes applying visually focused level and dependent thresholds to each crumbling level in the transformation of the wavelet. New variables are used for storing the compressed coefficients. Each variable represented each level detail coefficients. The locations in the original vector where we found the coefficient value other than zero were kept as it is in the corresponding new variable. But for zero-valued coefficients, we counted their numbers in between two nonzero coefficients. As, for example, say "X" be any vector and "X_1" be the compressed vector, then

$$X = [0.56, 0, 0, 0.34, 0.46, 0, 0, 0, 0, 0, 0, 1.23, 0, 0.9]$$
$$X_1 = [0.56, 2, 0.34, 0.46, 6, 1.23, 1, 0.9]$$

Decomposition as shown in Fig. 2 using wavelets is carried by breaking the signal into estimation and detail coefficients. The estimations are high scales and low-frequency components whereas the details are low scale, high-frequency constituents of the signals. Reconstruction of speech signal is done with inverse mechanism of discrete wavelet transform (IDWT). The process of synthesis starts with the estimation and detailed coefficients CA_m and CD_m, and after that again constructs CA_{m-1} with the help of upsampling and filtering inside the filters after their reconstruction as shown in Fig. 3. In this work, five-level decompositions are applied to speech signal for audio watermarking as shown in Fig. 4.

In audio watermarking process, implementation of supplementary information is possible in a soundtrack. The sound in digital watermark process has been

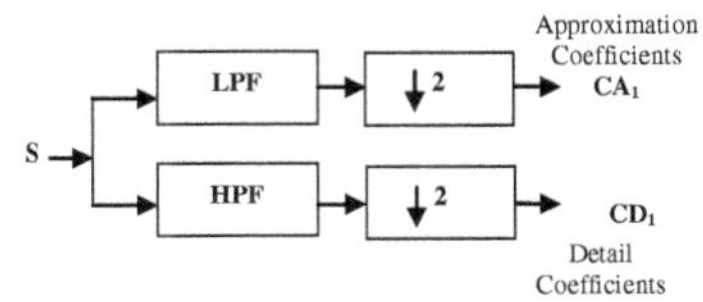

Fig. 2 Decomposition of signal into coefficients

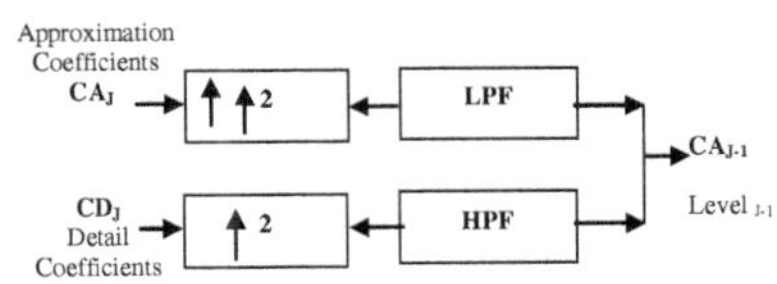

Fig. 3 Reconstruction of speech signal

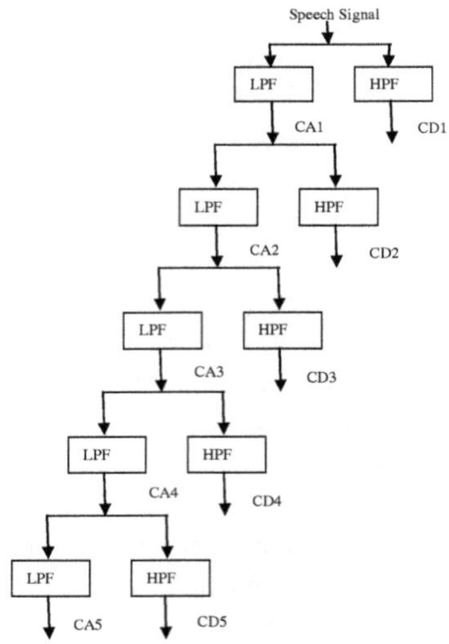

Fig. 4 Level of decomposition

originated to be one of the latest processes of sound reproduction which can implement again a specific data in the form of watermark to audio signals, which exceeds the data and verifies the user authenticity. No changes are done in the algorithms used in this process and in future we will evaluate two audio marking algorithms which will be existing. The data was recorded at three different sampling rates, 8, 16, and 22.05 kHz. Each block is iterated for two different wavelet families, Haar and Daubechies. Db2, Db4, Db6, Db8, Db10, Db12, and Db14 versions of Daubechies are used for performing the experiments. The selection of these two appropriate mother wavelets (Haar, Daubechies) is based on the quantity of the power; a wavelet field function can distillate in level-1 approximation coefficients. The signal is crumbled in 5-level estimation and detailed coefficients. A global threshold approximating data is used for the crumbling of signals. The performance measure of a data signal is the comparison among the parental signals and the watermarked signals. For performance evaluation of wavelet-based watermarking techniques, objective performances are evaluated. For objective measures, a variety of quantitative parameters are utilized to assess the efficiency of speech watermarking techniques. The quality of the signal was restrained depending on SNR, retained signal energy, compressing ratios, and percentage compression. Comparative results of these parameters at three different sampling frequencies are shown in Figs. 5, 6, and 7.

The improvement of the compressed signal is comparable for db10, db12, and db14. The performance in terms of % compression is almost same for db10, db12, and db14 but it shows a slight improvement on increment of order of wavelet. The performance in terms of % compression is best for 8 kHz signal. The comparative results in terms of various parameters are shown in the given Figs. 5, 6, and 7.

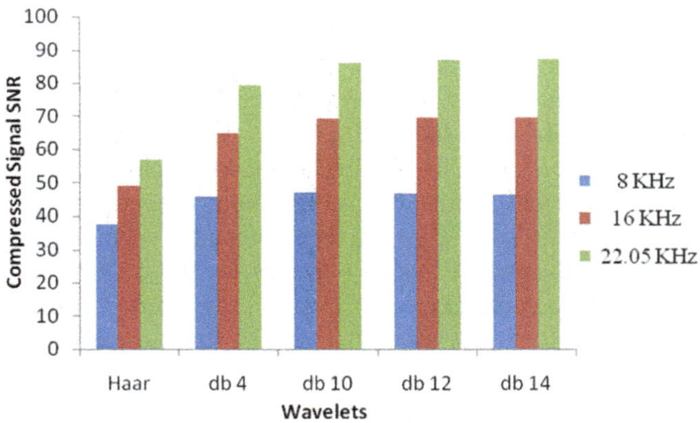

Fig. 5 Performance evaluation in terms of SNR

Fig. 6 Performance evaluation in terms of % compression

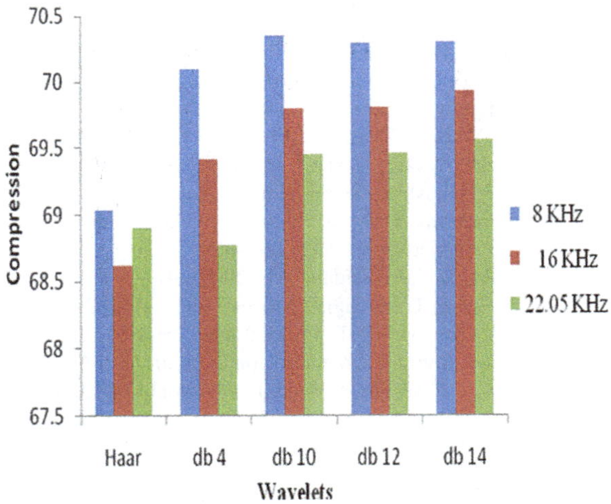

Fig. 7 Performance evaluation in terms of % energy retained

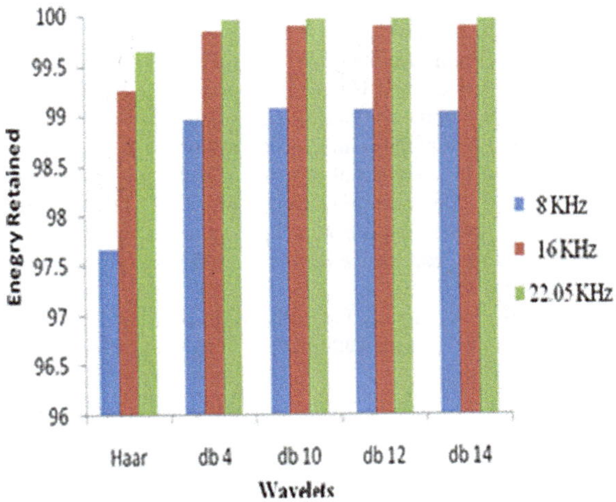

4 Conclusion

This paper proposes a novel robust audio watermarking algorithm which is based on wavelet transform. The Haar wavelet transform is the simple and fastest one to be used for signal processing and audio watermarking. However, due to its discontinuity, it is not optimal for simulation of continuous signal. Daubechies wavelet has shown its superiority over Haar wavelet for audio watermarking of speech signal. A significant advantage of using wavelets for the watermarking in speech signal data is that the compression ratio can be easily varied, while most of the others have fixed compression ratio. The results show the robustness and the audibility of the audio watermarking technique.

References

1. Neethu V, Kalaivani R (2016) Efficient and robust audio watermarking for content authentication and copyright protection. In: 2016 international conference on circuit, power and computing technologies (ICCPCT), pp 1–6
2. Guo Y, Yuan Q, Zhao X, Liu J, Yan Y (2013) Mixing-attack-proof randomized embedding audio watermarking system. JCP 8(12):3243–3250
3. Ghosh P, Ghosh R, Sinha S, Mukhopadhyay U, Kole DK, Chakroborty A (2012) A novel digital watermarking technique for video copyright protection. Computer Science and Information Technology, pp 601–609
4. Larbi SD, Jaïdane-Saïdane M (2005) Audio watermarking: a way to stationnarize audio signals. IEEE Trans Signal Process 53(2):816–823
5. Dhar PK, Kim JM (2011) Digital watermarking scheme based on fast Fourier transformation for audio copyright protection. Int J Secur Appl 5(2):33–48
6. Wang XY, Zhao H (2006) A novel synchronization invariant audio watermarking scheme based on DWT and DCT. IEEE Trans Signal Process 54(12):4835–4840
7. Elshazly AR, Fouad MM, Nasr ME (2012) Secure and robust high-quality DWT domain audio watermarking algorithm with binary image. In: 2012 seventh international conference on computer engineering & systems (ICCES), 27 Nov 2012, pp 207–212
8. Tianchi L, Guangming Y, Qi W (2011) A multiple audio watermarking algorithm based on shear resisting DWT and LSB. In: 7th international conference on networked computing (INC), 26 Sep 2011, pp 78–83
9. Das S, Bandyopadhyay B, Sanyal S (2011) Steganography and steganalysis: different approaches. Cornell University Library
10. Cvejic N, Seppanen T (2002) A wavelet domain LSB insertion algorithm for high capacity audio steganography. In: Proceedings of 2002 IEEE 10th digital signal processing workshop, 2002 and the 2nd signal processing education workshop, pp 53–55
11. Nehete S, Sawarkar SD, Sohani M (2011) Digital audio steganography using DWT with reduced embedding error and better extraction compared to DCT. In: Proceedings of the international conference & workshop on emerging trends in technology 2011 (ACM), 25 Feb 2011, pp 167–168
12. Ray AK, Padhihary S, Patra PK, Mohanty MN (2015) Development of a new algorithm based on SVD for image watermarking. In: Computational vision and robotics. Springer, India, pp 79–87
13. Anju R (2013) Modified algorithm for digital image watermarking using combined DCT and DWT. Int J Inf Technol 03:691–700

14. Vundela P, Sourirajan V (2013) A robust multiwavelet-based watermarking scheme for copyright protection of digital images using human visual system. Int Arab J. Inf Technol 10 (6):527–535
15. Shine KP, Kumar SK (2009) Extended bipolar echo kernel for audio watermarking. In: Proceedings of international conference on advances in recent technologies in communication and computing (ARTCom '09), IEEE Computer Society, Oct 2009, pp 487–489
16. Kalantari NK, Akhaee MA, Ahadi SM, Amindavar H (2009) Robust multiplicative patchwork method for audio watermarking. IEEE Trans Audio Speech Lang Process 17(6):1133–1141
17. Kirovski D, Malvar HS (2003) Spread-spectrum watermarking of audio signals. IEEE Trans Signal Process 51(4):1020–1033
18. Yeo I-K, Kim HJ (2003) Modified patchwork algorithm: a novel audio watermarking scheme. IEEE Trans Speech Audio Process 11(4):381–386

A High-Performance Energy-Efficient 75.17 dB Two-Stage Operational Amplifier

Neha Nidhi, Deepak Prasad and Vijay Nath

Abstract This paper discusses the design and analysis of two-stage CMOS operational amplifier. This design is operated at the supply of 1.5 V in 90 nm CMOS technology. In this design, 75.17 dB open-loop gain is achieved and having 7.73 MHz unity gain bandwidth and 148.8 m degree phase margin. This circuit has 10 pF capacitive load with 0.14 nW average power dissipation and slew rate is 0.25 V/μs. This proposed circuit is designed and simulated in cadence UMC 90 nm technology.

Keywords CMOS · Op-amp · Transconductance · Slew rate

1 Introduction

Operational amplifier is the most common and versatile used electronic devices in linear applications, also termed as op-amp. This classic op-amp which contains fully differential version is compatible with low-voltage applications; meanwhile, they do not need cascade output stages. Basic requirements of op-amp are as follows: high speed, high open-loop gain, high unity gain bandwidth, high input impedance, and low output impedance. In analog circuits, op-amp has wide-ranging applications because of its simplest structure and strong design methodology. Here, two stages refer to the number of gain stages in the op-amp. Here, the block diagram of two-stage op-amp is given in Fig. 1. Modeling, simulation, and layout design are demonstrated [1]. The first stage is differential transconductance stage. Through this input is given to the op-amp. Sometimes it also offers the differential to single-ended conversion. This differential-input stage provides a good overall gain, and also improves the performance of offset and noise. The second stage is usually an inverter. When the differential-input stage does not achieve the differential to single-ended conversion, then it is accomplished in the second stage

N. Nidhi (✉) · D. Prasad · V. Nath
VLSI Design Group, Department of Electronics and Communication Engineering, Birla Institute of Technology, Mesra, Ranchi, India
e-mail: nehanidhi199@gmail.com

V. Nath and J. K. Mandal (eds.), *Nanoelectronics, Circuits and Communication Systems*, Lecture Notes in Electrical Engineering 511,
https://doi.org/10.1007/978-981-13-0776-8_43

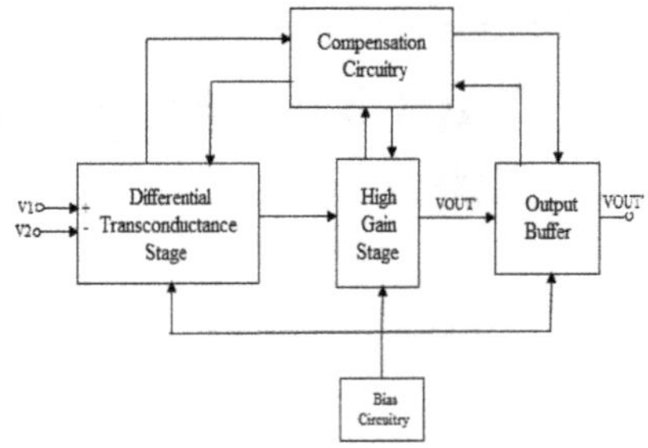

Fig. 1 Block diagram of a basic two-stage op-amp

inverter. Bias circuits are used to form the proper operating point for each transistor in quiescent state. To accomplish the stable closed-loop performance, compensation circuitry is required. To confirm the stability factor of the op-amp, a capacitor C_c is used in the op-amp when it is used with feedback. C_c is called as Miller capacitance because it presents in between the input and output of the second gain stage.

In this paper, a common two-stage CMOS op-amp design is presented by assuming the current buffer compensation technique. This proposed op-amp is achieved with enhanced UGB, high gain, and lower offset voltage having the power consumption of 0.14 nW.

2 Design Consideration

In the two-stage op-amp, the first stage consists of an n-channel current-mirror active load which is pair with a p-channel differential input [2–6]. The numbers shown beside the transistor indicate the width of transistor for a 1-μm process. Whereas the transistor length must be in between 1.5 and 2 times the minimum length of the transistor of that specific technology. Some op-amp specifications are as follows.

2.1 Op-amp Gain

The gain of the first stage in op-amp is

$$A_{v1} = \frac{g_{m1}}{g_{ds2} + g_{ds4}} = \frac{2g_{m1}}{I_5(I_2 + I_4)} \tag{1}$$

And, the gain of second stage [5] is

$$A_{v2} = \frac{g_{m6}}{g_{ds6} + g_{ds7}} = \frac{g_{m6}}{I_6(I_6 + I_7)} \tag{2}$$

$$\text{Gain Bandwidth, GB} = \frac{g_{m1}}{C_c} \tag{3}$$

2.2 Slew Rate

The maximum rate at which an output changes in volts per microsecond (V/μs) is known as slew rate (SR) [7].

$$\text{SR} = \left.\frac{\Delta v_o}{\Delta t}\right|_{\max} = \frac{(I_{cc})_{\max}}{C_c} = \frac{I_{D5}}{C_c} \tag{4}$$

2.3 Offset Voltage

If the inputs of the first stage are grounded and the matching is also perfect, then

$$V_{SD3} = V_{SG3} = V_{SD4} = V_{SG6} \tag{5}$$

Also, if no input offset voltage is existing, then essential form which guarantees it is

$$\frac{(W/L)_7}{(W/L)_4} = 2\frac{(W/L)_6}{(W/L)_5} \tag{6}$$

The ratio of differential gain magnitude to the common mode gain magnitude is termed as common mode rejection ratio (CMRR). The objective of this is to increase the difference between two different voltages irrespective of the common mode gain.

2.4 PSRR

It is defined as the ratio of the change in voltage supply to the change in output voltage of the op-amp which is produced by change in open-loop gain of the op-amp and power supply.

$$\text{PSRR} = \frac{\Delta V_{DD}}{\Delta V_{out}} A_v(S) \tag{7}$$

3 Design Implementation and Parameters

A two-stage CMOS op-amp schematic [7–9] is shown in Fig. 1. In the two-stage op-amp, the first stage is differential amplifier. In a given design, two input NMOS pairs are used for the current-mirror load. PMOS transistors have lower

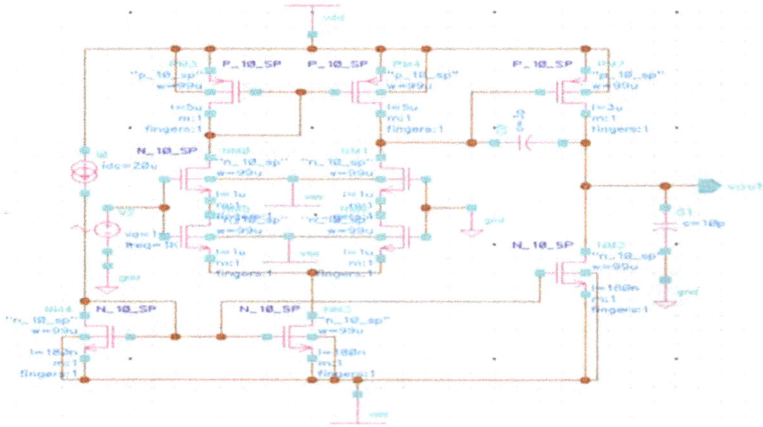

Fig. 2 Schematic diagram of two-stage operational amplifier

transconductance value than the NMOS transistors. Because of this for input couple, we are choosing NMOS transistors.

Through the understanding of unity gain bandwidth and compensation capacitance, transconductance of input transistors are predicted (Fig. 2).

3.1 Common Mode Rejection Ratio

To obtain the best performance, hold all the transistors in saturation mode. A common source amplifier is used to obtain a large amount of gain in output stage.

The *W*/*L* ratio [10–12] of all transistors and value of capacitances are shown in Table 1.

Table 1 Design parameters

Component	*W*/*L* (μm)
NM0, NM1	99/1
NM5, NM6	99/1
PM3, PM4	99/5
PM7	99/3
NM2, NM3, NM4	99/0.18
C_C	3 pF
C_L	10 pF

4 Simulation Results

In this section, the result of abovementioned design is discussed. Through this design, a maximum gain of 75.17 dB is obtained which is shown in Fig. 3 and also the phase margin obtained is shown in Fig. 4.

Here in Table 2, some specifications of op-amp are given, which are obtained after simulation of the given design.

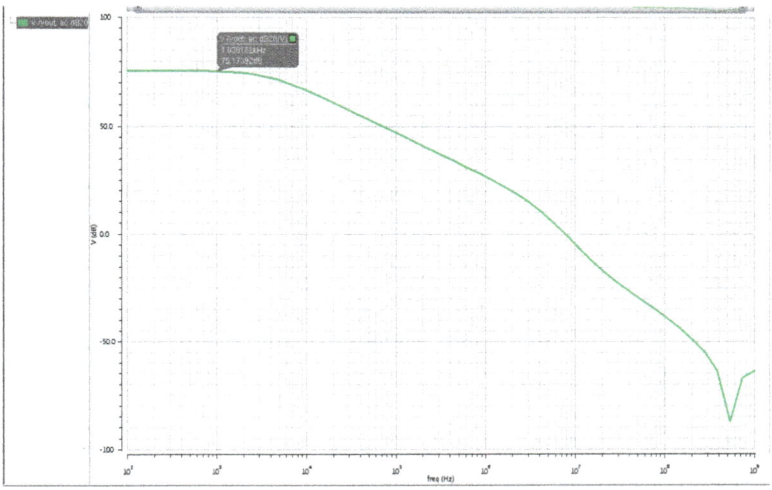

Fig. 3 Gain versus frequency plot of two-stage operational amplifier

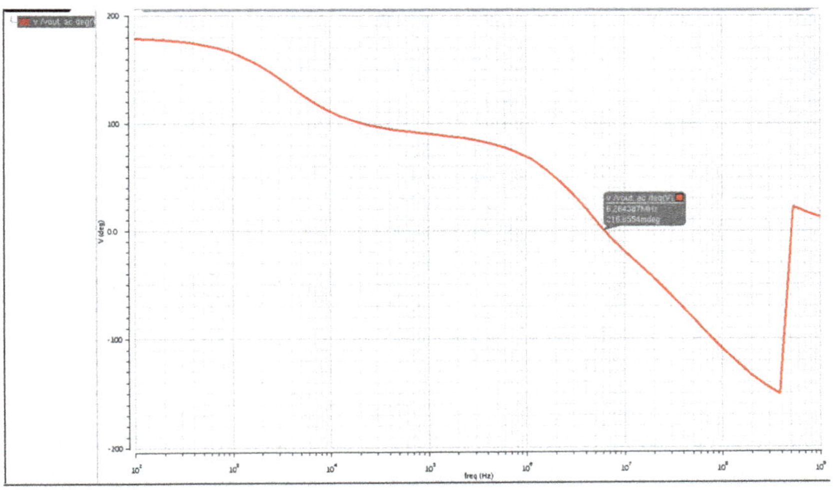

Fig. 4 Phase versus frequency of two-stage operational amplifier

Table 2 Process simulation results

Specifications	Results
DC gain (dB)	75.17 dB
Power dissipation (nW)	0.14 nW
UGB (MHz)	7.73 MHz
Phase margin	148.8 m degree
Slew rate	0.25 V/μs
Supply voltage	1.5 V

5 Conclusion

In this paper, low-voltage two-stage CMOS op-amp is designed. This design is operated at ±1.5 V power supply. Here, low power dissipation, high bandwidth, high output gain, and high slew rate are obtained. In designing the op-amp, many optimization problems occur in which optimization results of one parameter reduce the results of other parameters. Apart from this, by seeing the simulation results, it is confirmed that targeted specifications are achieve.

Acknowledgements The authors are heartily thankful to Dr. S. Pal, Head, Department of ECE and Dr. M. K. Mishra, Vice-Chancellor, BIT Mesra for their constant encouragement and providing the suitable laboratory to carry out this research.

References

1. Johns DA, Martin K (1997) Analog integrated circuit design, 2nd edn
2. Coughhlin RF, Driscoll F (2000) Operational amplifiers and liner integrated circuits, 6th edn, 2 Sept 2000
3. Sandhu M, Bala M (2012) Design of low voltage low power operational amplifier. ICACCT, 2012
4. Raut KJ, Kshirsagar RV, Bhgali AC (2014) A 180 nm low power CMOS operational amplifier, 28 Nov 2014
5. Jamuar SS, Kit TK (2004) Design of low voltage operational amplifier. IEEE, 2004
6. Allen PE, Holberg DR (2002) CMOS analog circuit design, 2nd edn
7. Patel PD, Shah KA (2013) Design of two stage operational amplifier. IJSR 2
8. Bandyopadhyay S, Mukherjee D, Chatterjee R (2014) Design of two stage CMOS operational amplifier in 180 nm technology with low power and high CMRR. IJRTET 11
9. Poonam MD, Saini H (2013) Design of two stage op-amp. IJATCSE 2
10. Priayadarshi S, Negi V, Nath V (2009) Design a low power stage CMOS operational amplifier in VLSI. In: International conference SCI-09, organized by Pentagram Research Centre, Pvt. Ltd., Hyderabad, (A.P.), India, 7–10 Jan 2009
11. Chakraborty S, Pandey A, Nath V (2014) A 141 dB gain of low power CMOS operational amplifier. In: 3rd International conference on computing, communication and sensor network organized by PIET Rourkela, 12–14 Dec 2014
12. Chakraborty S, Pandey A, Nath V (2015) A highly linear and high gain operational amplifier. In: IEEE International conference on electrical, electronics, signal, communication and optimization (EESCO-2015) organized by Vignan's Institute of Information Technology, 24–25 Jan 2015

Automated Toll Plaza Using Barcode-Laser Scanning Technology

The Way to an Efficient Lifestyle

Utkarsh Raj, Neha Nidhi and Vijay Nath

Abstract India is a developing country—a nation which always keeps on moving on wheels: stopping is not an option. Where the total number of vehicles (159,490,578 as of 2012) is more than the entire population of US, an efficient traffic system and smart toll booths are necessary for the efficient movement and proper tax collection; for the wheels should never stop.

Keywords Smart toll booths · Tax collection · Traffic system

1 Introduction (The Main Agenda)

Our research aims to improve tax collection method using a barcode-laser scanner technology which not only reduces the processing time per vehicle but also reduces the fuel consumption and hence reducing the cost of transportation improving the human lifestyle [1].

The main agenda of this paper is to create a system of toll booths where a vehicle does not have to stop for transaction; just the unique barcode which has been assigned to the vehicle will be scanned and the transaction will be behind the screens. This, as I mentioned, will not only maintain a smooth movement of vehicles which will create less jams but also reduce the fuel consumption that occurs during the long waits at the toll booth.

U. Raj (✉) · N. Nidhi · V. Nath
Department of Electronics and Communication Engineering,
Birla Institute of Technology Mesra, Ranchi 835215, India
e-mail: utkarsh77raj@gmail.com

V. Nath and J. K. Mandal (eds.), *Nanoelectronics, Circuits and Communication Systems*, Lecture Notes in Electrical Engineering 511,
https://doi.org/10.1007/978-981-13-0776-8_44

2 Current Toll Tax System

The tax collection system has been there for centuries, be it the Persian Silk Road or the Susa–Babylon highway travelers [2]. Tax system currently in use in India includes the manual toll tax collection and RFID technology [3]. Traditional Manual system is a slow, time-consuming method and can create long jams but RFID which is another new system which is prevalent is conditionally good i.e. it can be hacked easily and it becomes inefficient if there is a heavy fog. It also creates problem when there are a lot of vehicles in the line, and it gets confused as to which vehicle to scan first [3].

Apart from India, we have ETC system prevalent in US, Norway, and parts of European nations. This method is good if the price of software and hardware required per vehicles is not taken into account. Also, it has some speed limit issues as to what must be the optimum speed at which transaction can be successful.

3 Barcode—Laser Scanner Technology (Proposed Concept)

3.1 Elements of the New Toll System

A. *BARCODE: A barcode* (Fig. 1) *is an optical representation of data using parallel lines of different widths and spacing.*

The barcode has different parts which represent different information. The first part of the barcode represents the country where that product was manufactured. The second part tells about the manufacturer and the final part reveals about the product itself. This is where we store the user and vehicle database [4].

B. *BARCODE SCANNER: The scanner illuminates the barcode with a beam of red light and then captures the reflected light.* Figure 2 *shows a laser barcode scanner* [5].

It then converts it into a matching text and then generates a corresponding analog signal which is then sent to the decoder that reads the input signal and converts it into text. Barcode has an efficiency of 99.9999 that is it has an error of just 1 for every 10,000 [6].

C. *CENTRAL SERVER: To implement this project on a large scale, we need a database that can handle and process millions of records, has efficient searching algorithm, and provides quick response to the scanned input from the scanner.* Figure 3 *shows a general representation of server.*

It is also responsible to manage the taxes and maintaining records and checking if the dues of the user are cleared in proper time [6].

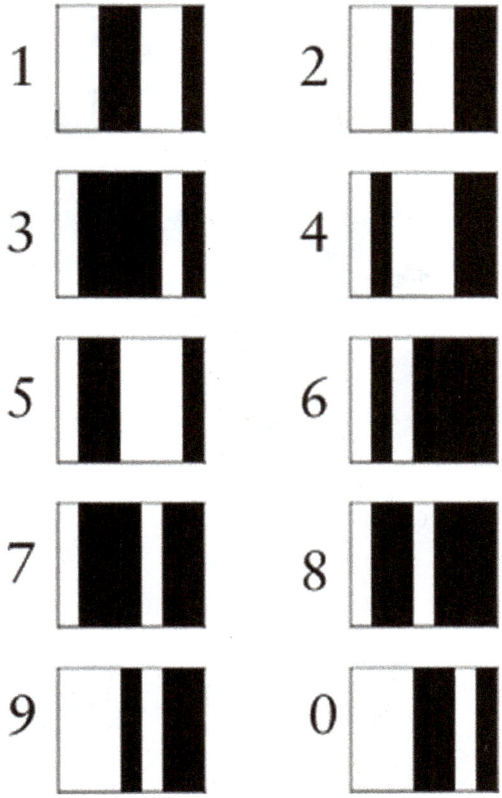

Fig. 1 Barcode representation

Fig. 2 Barcode scanner

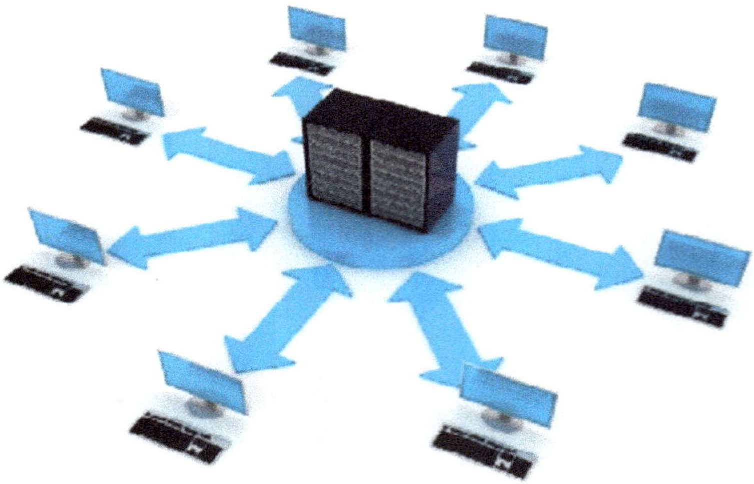

Fig. 3 Central server system

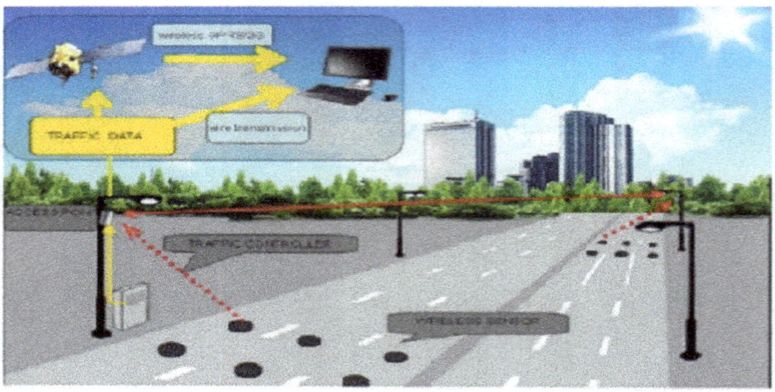

Fig. 4 IR detector system

D. *DETECTORS: The detectors will sense the vehicle from a given distance, and the scanner will start scanning the barcode. The vehicle will have a set speed limit to allow proper scanning. The detectors will signal the scanner that a vehicle is approaching and hence scanning starts.* Figure 4 *shows the vehicle detection method* [5].

This detector will use the infrared technology, and this detector can also be used in other ways by other organizations to check the vehicle density at a particular time of the day.

4 Working Mechanism

As soon as a vehicle comes to a toll booth, it is sensed by the detectors and the barcode embedded in the barrier reads the data with the help of laser scanning technology and send the information to the master server which searches the database for that user and vehicle information [4]. As soon as the user and the vehicle are identified, an amount of money depending on the vehicle type is deducted. The important thing is that the security of bank account is still intact as the user receives a message and OTP for granting permission to deduct that amount [3]. Now if the user has not cleared the dues for a toll tax and arrives at another toll booth, he/she will be stopped and will have to clear the dues manually along with the fine. This will not only get this system in habit of the users but also reduce the problems of hard cash transaction [7]. The records will be maintained by the server as to whether the vehicle is charged beforehand on a particular day so that when the vehicle arrives again it is allowed to pass smoothly without any manual verification.

Our model depicting this system will have an Arduino microcontroller which will work in coordination with the database [2]. The IR sensors will detect the presence of the vehicle, and then instructions will be sent to the scanner to scan the barcode. As soon as the scanning is done, the data is fed back to the database to complete the transaction [6].

5 Flowchart

This represents the direction of control flow and various possibilities that can arise in this system during data scanning and processing. At any time, vehicles can have at most one pending transaction which the database identifies by blacklisting them [7]. This tag is not removed until the dues are cleared along with the fine either manually or by online mode. Such blacklisted vehicles will be stopped at the barrier on the next toll booth and the tax will be charged.

Figure 5 shows the control flow diagram. It represents how this proposed method would respond when it detects a vehicle and what are the possibilities that can arise for a particular driver [8].

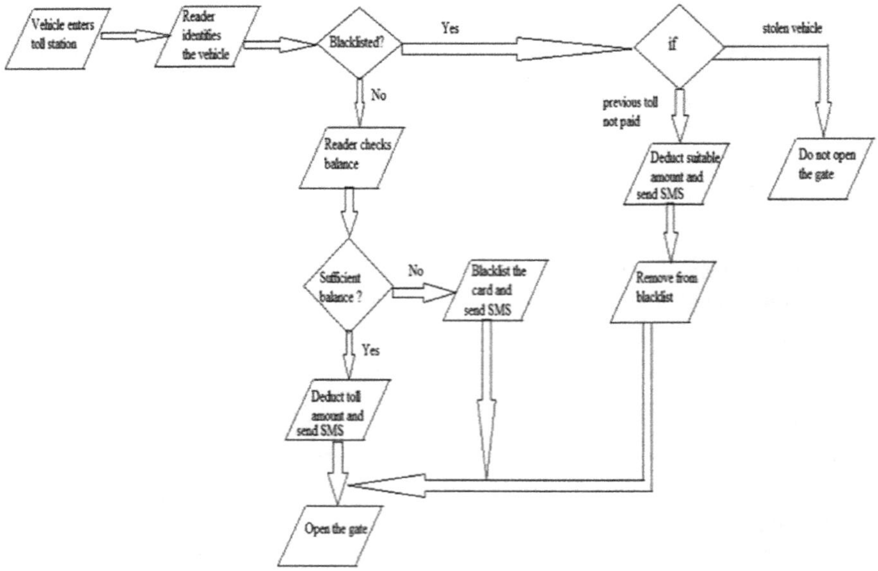

Fig. 5 Control flowchart showing the proposed concept

6 Pros and Cons

This method is a very efficient method for toll tax implementation. This system is not affected by weather conditions until and unless the weather is very bad that hinders the visibility. This is good for long run and offers the current method of transaction. This will lead to promotion of cashless mode of transaction [9].

Some of the problems we can face are due to the shelf life of the barcode sticker and barcode duplication which needs further encryption techniques and research work. But these cases are very rare and can detect if there are proper CCTV cameras installed on the toll booth [10].

7 Conclusion

The barcode-laser scanner technology-based toll booth system will be almost autonomous and make the traffic smooth running and efficient. It will also reduce transportation cost as it will reduce the fuel cost due to long jams and waiting at the toll booths. This in turn will reduce the emissions at the toll booths. It will increase reliability and also reduce manpower. Barcode can also be used to track the vehicle and increase security and reduce theft. Due to proper running of traffic, the number of accidents will reduce.

Acknowledgements I would like to thank DR. S. Pal for constantly supporting me through this project. His guidance was extremely helpful and I find myself fortunate to have this opportunity. I would also like to thank our Vice-Chancellor Prof. M. K. Mishra for providing such a great infrastructure.

Lastly, I would like to thank my parents who have always been by my side through thick and thin and motivated me toward my goal.

References

1. Copeland L (2008) Toll roads take cashless route. USA Today, 7/28/2008
2. Gilliet H (1990) Toll roads-the French experience. Transrouts International, Saint-Quentin-en-Yvelines
3. Staff (June 21, 2006) Road charging scheme: Europe - Italy, Rome. Road User Charging Worldwide. UK Commission for Integrated Transport. Archived from the original on 5 Oct 2006
4. John L, Mitra D, More S (2016) Automatic Toll Collection Using QR Code Scanning. International Conference on Recent Innovations in Engineering and Management, Mar. 2016
5. http://www.waspbarcode.com/buzz/laser-barcode-scanners/, accessed on 26/02/2016
6. http://www.explainthatstuff.com/barcodescanners.html, accessed on 10/01/2016
7. Kulkarni S, Kulkarni S, Kasture K (2016) Electronic toll tax collection and security system. Int J Eng Tech Res (IJETR) 4(1):2454–4698 (P). ISSN: 2321-0869 (O). Young M (1989) The technical writer's handbook. University Science, Mill Valley, CA
8. Satyasrikanth P, Penna M, Bolla DR (2016) Automatic toll collection system using RFID. IJCSMC 5(8):247–225
9. Vickrey W (June 1992) Principles of efficient congestion pricing. Columbia University. Retrieved 8 Mar 2012
10. Balasch J (October 2010) Privacy-preserving road charging. Katholieke Universiteit Leuven
11. http://tis.nhai.gov.in/, accessed on 15/02/2016

MIMO Wideband Antenna Technique in DGS for Enhanced Wireless Communication

Trisha Ghosh, Sneha Tiwari and J. Sahay

Abstract In this paper, MIMO antenna is designed, which consists of two planar symmetrical monopole antennas and T-slot adjoining to the common ground plane. The return loss is below −15 dB for a wider bandwidth. The overall antenna size is low as 35 mm × 22 mm but the effect of the mutual coupling among the antenna elements is reduced and is below −10 dB in lower range and below −20 dB at higher frequency range within wider frequency range of 4–10.5 GHz. The resonant frequency is taken as 4.4 GHz. The designed MIMO antenna has two ports whose reflection coefficient behaves as shadow of one another which is considered to be better in terms of efficiency. In this process, the maximum gain achieved is approximately 2 dB. The design has been verified and is ready for fabrication so that the MIMO antenna can be practically used. The design has been simulated using Ansoft HFSS software 15.0. The various parameters like S-parameters, VSWR, gain and radiation pattern are shown.

Keywords QoS—Quality of service · MIMO—Multiple input multiple output DGS—Defected ground structure

1 Introduction

The basic idea behind the MIMO technique is increasing the channel capacity and QoS of signal transmission. The increased number of antenna elements in both transmitter and receiver results in enhanced channel capacity. Due to number of

T. Ghosh (✉) · S. Tiwari · J. Sahay
Department of Electronics and Communication Engineering, BIT, Mesra, Ranchi, India
e-mail: putu.trisha@gmail.com

S. Tiwari
e-mail: snehasandilya14@gmail.com

J. Sahay
e-mail: jsahay@bitmesra.ac.in

V. Nath and J. K. Mandal (eds.), *Nanoelectronics, Circuits and Communication Systems*, Lecture Notes in Electrical Engineering 511,
https://doi.org/10.1007/978-981-13-0776-8_45

paths in between the transmitter and receiver, multipath propagation takes place. This multipath propagation causes signal fading which can be reduced by applying spatial diversity. MIMO is a major technique used for advanced wireless communication systems, for example, 4G, 5G, etc. It has become an integral part of the modern communication system. The range and robustness of the whole system are also increased but at the same time, it makes it complex.

In the implementation of MIMO antenna, any of the antennas such as vertically polarised multibeam antenna array, dual polarised spatially separated antennas, dual polarised multibeam antenna array or any combination of antennas with space, polarisation and pattern diversity may be used provided that the scattering environment is rich. It is observed that in a single beam array, there is an increase in capacity in spite of high interference and high correlation between multipath signals. In multibeam arrays, capacity decreases as compared to normal antenna arrays. It can be said that one of the biggest challenges in MIMO antenna technology is the issue of the coupling among the antenna elements with respect to smaller size of the antenna. This can be done by using various methods as desired like using slots, converting slots into slits, spacing between antenna elements, etc. MIMO antennas with smaller size have tremendous future scope when implemented in variable portable devices as per users [1–3].

2 Experiment

2.1 Antenna Design

In this section, we have designed an MIMO antenna consisting of two symmetrical planar monopole antennas that are separated by a distance of 8 mm, placed within the compact area of 35 × 22 mm^2. The substrate used for this design is Rogers with dielectric constant, ϵ_r of 3.5, loss tangent δ of 0.002 and thickness 1.6 mm (Table 1).

In Fig. 1, there are two symmetrical antennas where the radiator is square shaped that measures 8 mm. The feeding technique used for both the ports is microstrip feed. The impedance matching is simpler in case of microstrip feed as compared to other feeding techniques. Microstrip feed is reliable and easier to fabricate. The ground plane is slotted in a particular T-shaped. With all the dimensions mentioned in the geometry, the antenna is designed in Ansoft HFSS software [4–6] (Fig. 2).

Table 1 Dimensions of the MIMO antenna (in mm)

L_p	W_p	L_A	W_A	P
8	8	35	22	3.5
W_g	D_p	S_w	L_g	L_w
4	12	3	21	3

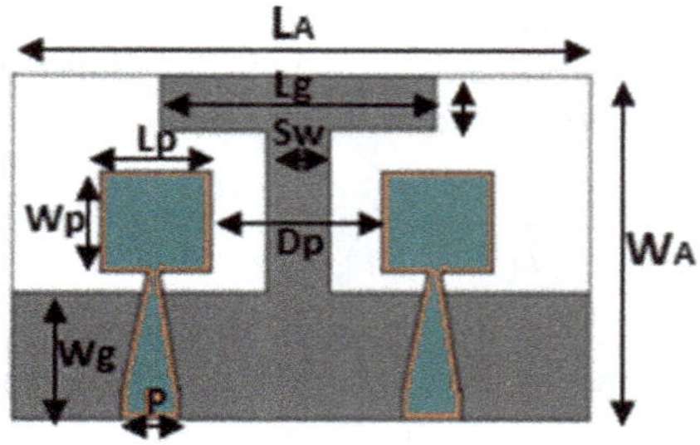

Fig. 1 Geometry of the MIMO antenna

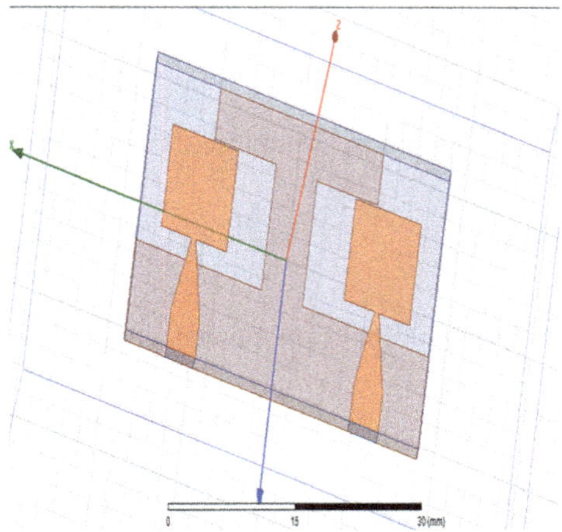

Fig. 2 Design of the MIMO antenna in software interface

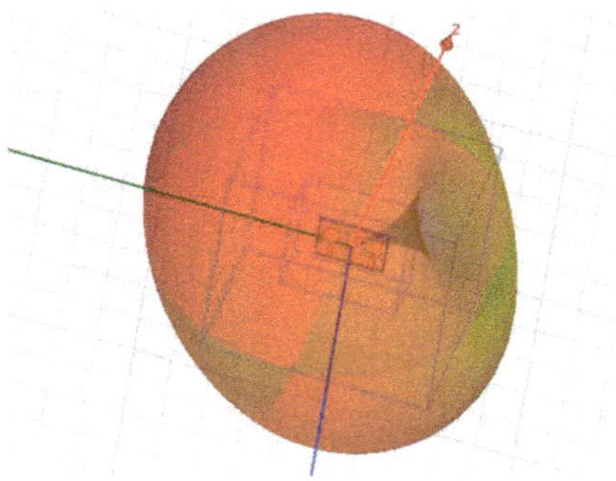

Fig. 3 Radiation far field applied in the design

In HFSS software, when radiation far field is applied to the air-box that encloses the antenna (in case of simulation design), we observe the direction of propagation of radiation that is evident in Fig. 3.

Derivation and Explanation:

All the design parameters are calculated using the following formulas.

We know that some waves travel in substrate and some in air; therefore, the concept of effective dielectric constant is introduced. Initial value of effective dielectric constant, ϵ_{reff} is given by

$$\epsilon_{reff} = \frac{\epsilon_r + 1}{2} + \frac{\epsilon_r - 1}{2}\left[1 + 12\frac{h}{W}\right]$$

where h is thickness of the antenna, and W is width of the patch.

Because of the fringing effects, electrically the patch of the microstrip antenna looks greater than its physical dimensions. Let the physical length of the patch be L, and length travelled by the field be $2\Delta L$. Thus, the effective length is given by

$$L_{eff} = L + 2\Delta L$$

The empirical values are derived and applied in the formula such as

$$\frac{\Delta L}{h} = 0.412\frac{(\epsilon_{reff} + 0.3)\left(\frac{W}{h} + 0.264\right)}{(\epsilon_{reff} - 0.258)\left(\frac{W}{h} + 0.8\right)}$$

For good radiation efficiency, the width of the radiator must be calculated by

$$W = \frac{c}{2f_r}\sqrt{\frac{2}{\epsilon_r + 1}}$$

The actual length of the patch can be determined as [7]

$$L = \frac{1}{2f_r\sqrt{\epsilon_{reff}}\sqrt{\mu_0\epsilon_0}} - 2\Delta L$$

Calculations:

The resonant frequency is taken as 4.4 GHz, and accordingly, the width is calculated at first. Once we calculate width W, we can easily get all other parameters by using formulas mentioned above.

$$W = \frac{3 \times 10^8}{2 \times 4.4}\sqrt{\frac{2}{3.5 + 1}} = 22.4 \text{ mm} \approx 22 \text{ mm}$$

Inference:

In this structure, the ground plane is slotted such that the mutual coupling between the antenna elements is reduced even in a compact size. The microstrip feed is used because it is simple and easy to fabricate. The cost of fabrication is also low. The design parameters are calculated and implemented in the structure.

Table 2 All the parameters used in simulation of the design

Parameters	Design metrics	Symbol
Dielectric constant	Rogers, 3.5	ϵ_r
Length of antenna	35 mm	L_A
Width of antenna	22 mm	W_A
Microstrip patch	8×8 mm^2, copper	$L_p * W_p$
Feeding technique	Microstrip feed	

3 Simulation Parameter

See Table 2.

4 Result and Discussion

The return loss for this MIMO antenna is shown.

In Fig. 4, the value of reflection coefficient, S_{11} for this MIMO antenna is below −10 dB for a wide range of frequencies for both port 1 and port 2. Patch 2 with port 2 shows better curve as compared to that of patch 1 with port 1, but overall if we observe the range of frequencies in which it is working, it can be said that this MIMO antenna works on wideband as in ultra wideband range almost, i.e. 4–10.5 GHz.

In Fig. 5, it is observed that S_{12} is below −10 dB for a wide range and maximum it has gone up to −50 dB. This indicates that the mutual coupling between the antenna elements is less and is in control.

In Fig. 6, it is seen that the VSWR for this particular MIMO antenna is less than 2, which is desired for an antenna, and it implies that the impedance matching in antenna is good (Figs. 7 and 8).

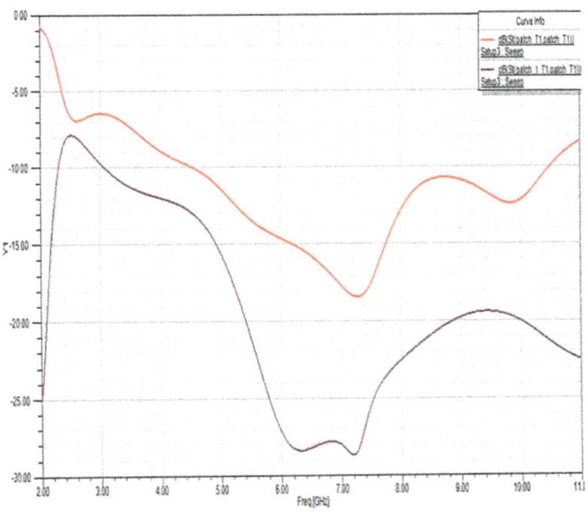

Fig. 4 Return loss S_{11} plot with resonant frequency 4.4 GHz

Fig. 5 The plot of S_{12} for the MIMO antenna

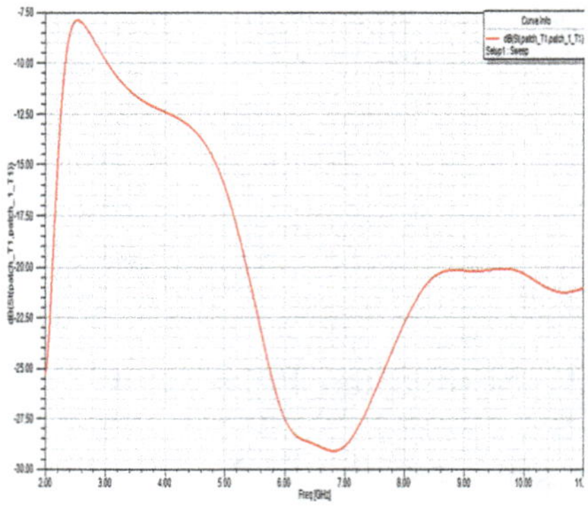

Fig. 6 VSWR for both the patches in MIMO antenna

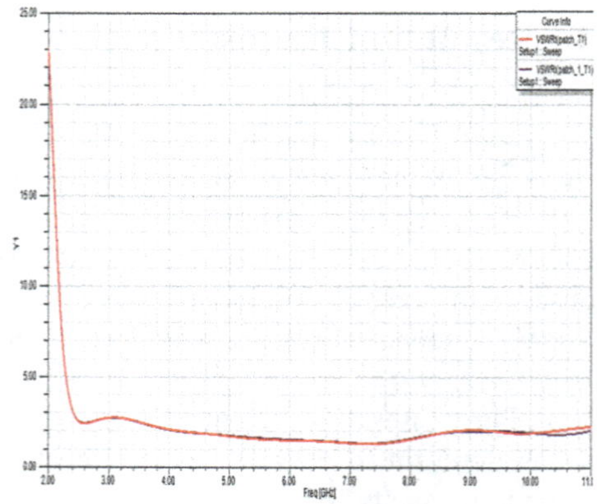

Fig. 7 Two-dimensional gain for the MIMO antenna

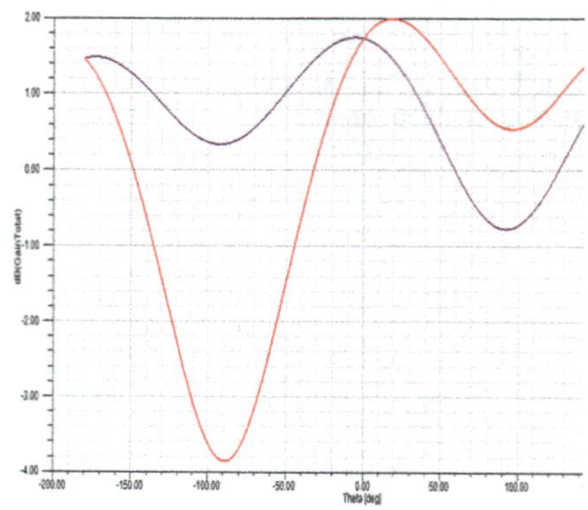

Fig. 8 Three-dimensional gain for the MIMO antenna

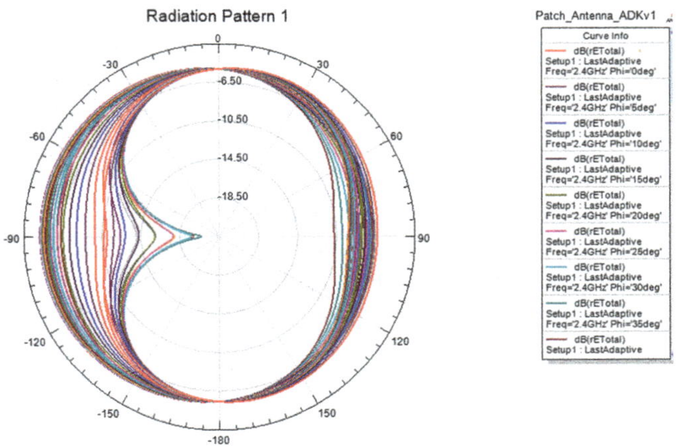

Fig. 9 Radiation pattern for designing MIMO antenna

The maximum gain achieved is observed approximately 2 dB that can be seen in two-dimensional plot, and the same has been represented in three-dimensional figure.

In Fig. 9, the radiation pattern in the far field is observed such that it is almost omnidirectional, i.e. directional in a particular direction and isotropic at all sides.

MIMO antenna is designed in such a way that the mutual coupling which is undesired has been reduced. The dielectric material used as substrate, i.e. Rogers, is easily available and hence easy to fabricate. The structure when simulated gives S_{11} below -10 dB for a wider bandwidth. We take -10 dB as a reference point because it indicates that 90% of the radiation coming from antenna is radiated and only 10% is reflected back. We calculated value of S_{11} and found nearly -50 dB which is very good as per literature survey. If this structure is fabricated, then with considerable fabrication error also, it can be expected that the S_{11} will have desirable values. The plot of S_{12} represents power from port 2 delivered to port 1. There must be good amount of isolation between the ports for the MIMO antenna to work well. The results show that it is below −10 dB and even lower than that for the wider range which states that the mutual coupling is low between the antenna elements. The VSWR represents how well the impedance matching is. When matching is good, more power is delivered to the antenna. The value of VSWR is generally greater than or equal to 1 but less than 2. In the graph shown above, the VSWR value is below 2, which implies that the antenna has good impedance matching. The gain does not hold much importance in case of portable devices since the distance of operation is not high but still, it is valuable. Here, we get maximum gain as 2 dB approximately. The radiation pattern at far field represents that it is almost omni-directional, which is desirable for the MIMO antenna. Since the impedance matching is good, the Smith chart reading shows that the path travels to the centre which means 50 Ω. Therefore, this MIMO antenna can be implemented for practical uses easily [8–12].

5 Applications of MIMO Wideband Antenna

- Can be used in LTE system which is in vogue nowadays.
- Wireless communication media such as WLAN, WiMax, etc.
- Can be used in medicinal and pathological departments.
- Numerous fields as per user-defined sphere since wideband [13–15].

6 Future Scope

In the ground plane, different slots or strips may be designed and the varied result can be observed. Without compromising the area of the radiator, and making changes on the ground plane using DGS, the mutual coupling should be made as low as possible with reduction in overall area of the antenna.

7 Conclusion

In today's world, wireless communication plays a key role. Antenna is considered to be the backbone of any wireless communication set-up. Wireless performance is completely dependent on high performance of antenna design and implementation. In this MIMO antenna, the design is done both in patches and ground plane. The main issue in MIMO antenna is reducing the mutual coupling when the size of the antenna is very small. This issue has been resolved by making the slotted ground plane. The reflection coefficient is also way below −10 dB throughout wider bandwidth, i.e. 4–10.5 GHz, which makes this antenna desirable for use as it would be usable for numerous applications. The antenna has been designed, and all the parameters have been successfully verified by simulating the structure. This structure is ready to be fabricated for practical usage.

References

1. Manteuffel D (2009) MIMO antenna design challenges. In: 2009 Lough-borough antennas & propagation conference, 2009
2. Jusoh M, Jamlos MFB, Kamarudin MR, Malek MFBA, A MIMO antenna design challenges for UWB application. Progress Electron Res 36:357–371. ISSN: 1937-6472
3. Honarvar MA, Hamidi N, Virdee BS (2015) Multiband antenna for portable device applications. Microwave Opt Technol Lett 57(4)
4. Liu L, Cheung SW, Yuk TI (2013) Compact MIMO antenna for portable devices in UWB applications. IEEE Trans Antennas Propag 61(8)
5. Ling XM, Li RL (2011) A novel dual-band MIMO antenna array with low mutual coupling for portable wireless devices. IEEE Antennas Wirel Propag Lett 10
6. Liu L, Cheung SW, Yuk TI (2015) Compact MIMO antenna for portable UWB applications with band-notched characteristic. IEEE Trans Antennas Propag 63(5)
7. Constantine AB, Antenna theory—analysis and design, 3rd edn. Wiley publication
8. Dong L, Choo H, Heath RW Jr, Ling H (2005) Simulation of MIMO channel capacity with antenna polarization diversity. IEEE Trans Wirel Commun 4(4)
9. Chiu C-Y, Murch RD (2008) Compact four-port antenna suitable for portable MIMO devices. IEEE Antennas Wirel Propag Lett 7
10. Lu Y, Huang Y, Chattha HT, Cao P (2011) Reducing ground-plane effects on UWB monopole antennas. IEEE Antennas Wirel Propag Lett 10:147–150
11. Yoon HK, Yoon YJ, Kim H, Lee CH (2011) Flexible ultra-wideband polarization diversity antenna with band-notch function. IET Microw Antennas Propag 5(12):1463–1470
12. Lee JM, Kim KB, Ryu HK, Woo JM (2012) A compact ultrawideband MIMO antenna with WLAN band-rejected operation for mobile devices. IEEE Antennas Wirel Propag Lett 11:990–993
13. Li JF, Chu QX, Li ZH, Xia XX (2013) Compact dual band-notched UWB MIMO antenna with high isolation. IEEE Trans Antennas Propag 61(9):4759–4766
14. Gao P et al (2014) Compact printed UWB diversity slot antenna with 5.5-GHz band-notched characteristics. IEEE Antennas Wirel Propag Lett 13:376–379
15. Chacko BP, Augustin G, Denidni TA (2013) Uniplanar polarization diversity antenna for wideband systems. IET Microw Antennas Propag 7(10):851–857

Designing of FIR Filter Using FPGA: A Review

Jyoti, Adesh Kumar and Anil Sangwan

Abstract The research article represents the literature review for the designing of the FIR filter. In this research review, various papers of different authors are reviewed. The FIR filter found a number of applications in large applications and designing of DSP and communication systems. As all the work nowadays revolves around the frequencies, so their detection become also an important factor. In the reviewed papers, various techniques, methods, and algorithms are used for the optimal designing of the FIR filter. In the work of the authors, digital filters (FIR) are mainly used as they have an enormous range of handling capability of the frequency and also find a wide range of applications. The design of FIR filter is focused on HDL implementation and FPGA synthesis. They all proposed FIR filter working and improvement in its various parameters to improve the circuit of various designs and reduce the hardware implementation.

Keywords FIR filter · DSP applications · FPGA synthesis

1 Introduction

In signal processing, the useful part of the signal is required to be extracted, and the unwanted part is to be removed. This is done mainly by the filtering action of the components which are used in the filter. In the filtering action, the desired signal is selected within the defined frequency range, and the random signal like noise is removed. The filters are mainly of two types:

Jyoti (✉) · A. Sangwan
Department of Electronics and Communication, UIET, Maharishi Dayanand University, Rohtak, India
e-mail: jrredhu95@gmail.com

A. Kumar
Department of Electronics, Instrumentation and Control Engineering, University of Petroleum and Energy Studies, Dehradun, India

V. Nath and J. K. Mandal (eds.), *Nanoelectronics, Circuits and Communication Systems*, Lecture Notes in Electrical Engineering 511,
https://doi.org/10.1007/978-981-13-0776-8_46

- Analog filter
- Digital filter

Analog filters are following analog or continuously varying signals using passive elements such as capacitors, inductors, and resistors. These filters are applicable to permit specific frequency component by rejecting other from analog or continuous signals. The analog filters are the main building block of DSP and used in many applications such as loudspeaker in which filter is used to separate audio signal before loudspeaker, telephone lines to separate conversions, radio receiver to select particular radio station, etc., which are the example of analog filters.

The digital filter removes the unnecessary signal, and numerical operations are performed by the digital filter on the sampled values of the signal. The digital filters are mainly of two types:

- FIR filter (finite impulse response filter)
- IIR filter (infinite impulse response filter)

FIR Filter

FIR filter is a digital filter. The filter is mainly used to take the discrete samples of a discrete sampled signal. The filter is used to retain only the selected component and discard the irrelevant part. The filtering action is done by Fourier transform. Impulse response can be defined as the response of a system when the input is an impulse. Finite impulse response is when the output of the system is finite in nature. Figure 1a, b presents an example of finite response and infinite response.

FIR represents a finite impulse response as the values of (n) are finite in nature. If values are not finite, the response will be infinite impulse response. For an example, $h(n) = \{1, 2, 3, 4\}$ present the finite values, and $h(n) = \{1, 2, 3, 4, \ldots\}$ are infinite sample value.

Figure 2a presents the finite response of rectangular window. It has nonzero component for finite duration. Figure 2b presents the infinite impulse response for $h(n) = a^n\, u(n)$.

Steps to design an FIR Filter

FIR filter output is a linear convolution over a window of "N" length or data samples with $N - 1$ tap coefficients. The output can be expressed mathematically. The direct form of FIR filter to support the Eq. (1) is shown in Fig. 3

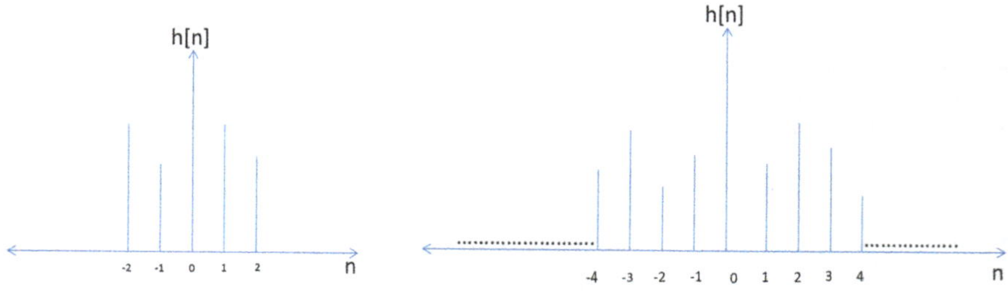

Fig. 1 **a** Finite impulse response, and **b** infinite sample sequences

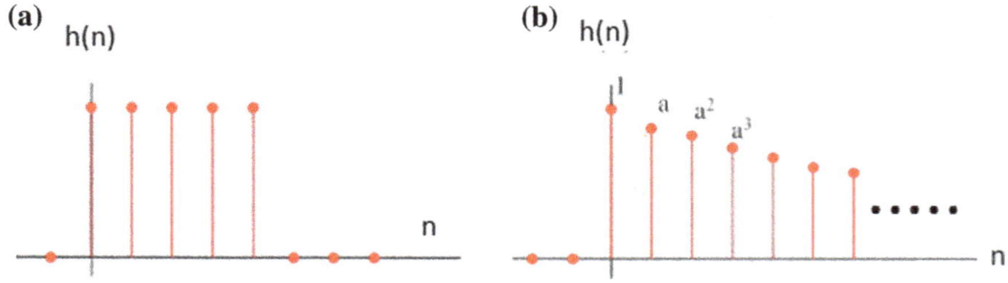

Fig. 2 **a** FIR response for rectangular window, and **b** infinite response for $a^n\ u(n)$

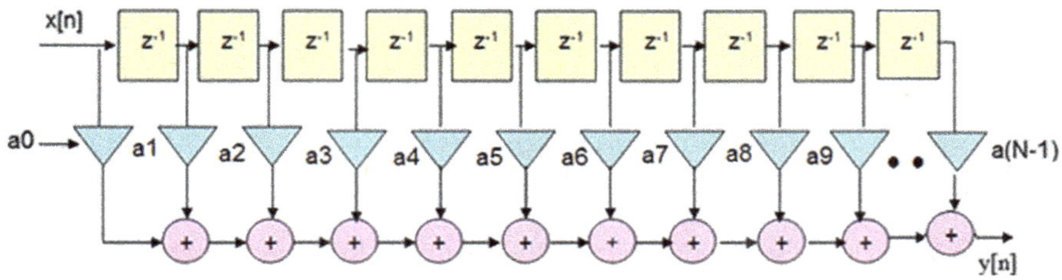

Fig. 3 FIR linear phase filter in direct form

$$y(n) = x(n) \otimes a(n) \tag{1}$$

$$y(n) = \sum_{k=0}^{N-1} a(k) \times (n-k) \tag{2}$$

The FIR can be designed by adding condition to the system we are designing. As we design the filter $H\ (e^{jw})$ is given to us.

$$H\left(e^{j\omega}\right)\overrightarrow{\text{IDFT}}\ h(n)\ \overrightarrow{\text{Algorithm}}\, h'(n) \tag{3}$$

where $h(n)$ = infinite impulse response and represented mathematically

$$h(n) = \frac{1}{2}\int H\left(e^{j\omega}\right)e^{j\omega n}\cdot d\omega \tag{4}$$

$h'(n)$ = finite response of filter.

The infinite response of the filter is changed to the finite impulse response by use of various algorithms.

Windowing technique in filter

In the designing of FIR filter, windowing plays an important role. As for the infinite responses of a system, windowing is used as tool to make the output finite. The infinite numbers of input are given to the system but only some outputs are coming out of the window. Mathematically, we can say

Table 1 Window type and corresponding weight function

Window type	Weight function
Rectangular window	$\begin{cases} W_R(n) = 1 & 0 \le n \le M-1 \\ \quad = 0 & \text{Otherwise} \end{cases}$
Hamming window	$\begin{cases} W_H(n) = 0.54 + 0.46\cos\left(\frac{2\pi n}{M-1}\right) & 0 \le n \le M-1 \\ \quad = 0 & \text{Otherwise} \end{cases}$
Hanning window	$\begin{cases} W_{HN}(n) = 0.5 - 0.5\cos\left(\frac{2\pi n}{M-1}\right) & 0 \le n \le M-1 \\ \quad = 0 & \text{Otherwise} \end{cases}$
Blackman window	$\begin{cases} W_B(n) = 0.42 + 0.5\cos\left(\frac{2\pi n}{M-1}\right) + 0.08\cos\left(\frac{4\pi}{M-1}\right) & 0 \le n \le M-1 \\ \quad = 0 & \text{Otherwise} \end{cases}$
Bartlett window	$\begin{cases} W_{BT}(n) = 1 - \frac{2n}{M-1} & \lvert n \rvert \le M-1 \\ \quad = 0 & \text{Otherwise} \end{cases}$

$$h'(n) = h(n) \cdot w(n) \tag{5}$$

where $w(n)$ = window function.

The infinite input is multiplied with the window function to obtain the finite output for the filter.

There are basically five major functions of windows. Table 1 lists the window technique and corresponding weight functions.

Windowing technique can be causal or anti-causal in nature. All the windowing techniques are causal in nature except the Bartlett windowing. Bartlett window is anti-causal in nature.

2 Related Work

The work on FIR filter is carried in different directions by many researchers. We have considered few papers to describe the work on FIR filter and its designing.

Yan et al. [1] proposed the design of FIR filter in the synthesis of the beamformer. With the increase in the frequency of the signal, the beam width of the signal also reduces and becomes narrow. The author proposed that the beam width of a signal become a function of the frequency and signal spectra and the beam is obtained in distorted form as with the frequency variations the beam is fed to the low-pass filter. To refocus the beam, FIB (frequency invariant beamformer) is used as a remedy method. Various approaches are provided for broadband beamformer by using FIR filters with frequency invariant method. Second-order cone programming method using frequency invariant patterns for the designing of the filter is used. The two which are briefly analyzed in the paper are as follows:

1. ***Step method***: In this method, by performing pattern synthesis of optimal array, the designing of array weights is done and using this approach, FIR filter is designed.
2. ***Joint optimization method***: In it, the frequency and spatial response are optimized jointly for designing the FIR filter.

Bhattacharjeea et al. [2] discussed that various FIR filter designs, and their main concern is on the speed and power consumption. The main concern is about the various applications of FIR filters. The FIR filter found the wide application because of its properties like linear phase response, absolute stability, etc. FIR found application in every field but the most extensively used in the communication for the pulse shaping, channel equalization, and matched filtering. For the implementation of the FIR filter, it makes use of the much-favored tool named FPGA, which is nowadays commonly used for the VLSI design. A major attention is given to the power consumption and the delays in both the pipelined and non-pipelined filters.

Chandra et al. [3] proposed the structure of the multipliers with using the SORIGA algorithm and discussed the complexity, which get introduced with the use of multipliers in the FIR filters, and this also increases the cost of the hardware. This hardware complexity can be removed or eliminated by making use of consecutive adders instead of multipliers, and then results are computed. Various algorithms are designed for removal of this bottleneck and in this paper, SORIGA a sophisticated algorithm is used as main effective tool. SORIGA is an intelligent optimization technique like some of the others which are genetic algorithm, differential algorithm, and many others. In this algorithm, an FIR filter with power-of-two is designed. A low-pass FIR filter with less multiplier is designed with SORIGA (self-organizing random immigrant genetic algorithm). Convergence speed is used to evaluate the effectiveness of the algorithm, and the frequency response of the FIR filter is also used to evaluate the effectiveness.

Thakur et al. [4] presented a brief review about the digital filters [4], how they are better over analog filters, its types (FIR and IIR), and how they are implemented using FPGA. Stress is given on the various advantages of FIR filters like the designing of the FIR filter is easy and its frequency response is better than IIR filter, and it can be implemented for any sort of frequency. And it also proposed that how a filter can make use of adders and shifters instead of multipliers to make the chip less complex that is to diminish the complexity. The main focus lies on the FPGA implementation of the designed FIR filter. With the use of Fpga, designed circuit are evaluated easily which is not possible on the developed DSP kits.

Alam et al. [5] proposed an FIR filter with finite word length and linear in phase, which is designed in the LNS domain [5]. The main concern of the author is to reduce the complexity of the filters, and improving the stability, by using logarithmic number system, the data and the coefficients effects the power consumption, switching capability, and complexity. Thus, the representation of the data and coefficients in the LNS reduces the complexity and improves the round off noise of the system and also improves the power dissipation (in the hearing aids, in video processing, etc.). The LNS generally simplifies the arithmetic operations such as it simply the power and root into division and multiplication, respectively. It reduces the power dissipation with the power saving up to 60%. Author proposed integer linear programming (ILP) for the designing of the finite word length and a linear phase FIR filter. And in it, the conversion to LNS is also shown by finding the largest minimum distance from the closest integer and branching them in the integer direction.

Llamocca et al. [6] proposed the main agreements of this paper which deals with the efficient implementation of 2D filter bank. Generally, static implementation describes the implementation of the filter bank. The static implementation has several disadvantages like hardware resources cannot be changed or adjusted in this implementation if the numbers of coefficient changes, i.e., get increased or decreased. Static implementations also have the disadvantage of consuming more power in the system when it is implemented. For such reasons, dynamic partial reconfiguration (DPR) technology is more suitable than static implementation and dynamic implementation works on the time—multiplexing of the FPGA resources. They also focused on the 2D real filter bank implementation. 2D filter is implemented by using two 1D FIR filters.

Grati et al. [7] proposed power consumption models for decimation FIR filters in multi-standard receivers. This paper presents the main concern about the power consumption of the decimation FIR filters. FIR filters found a wide application in the wireless communication that is why it is more necessary to know how much power is consumed by FIR filter as they have a substantial impact on the receivers in the communication. And what is the effect of the word length on the working of the receiver. The two main objectives of it are as follows:

1. Power consumption of the decimated FIR filters in the direct form.
2. The effect of the polyphase system on the power consumption of the FIR decimator filters.

Models are proposed in the paper for static and dynamic configuration (STM65 nm and STM95 nm, names of two models). By making the use of polyphase implementation, the power consumption is reduced up to 75% in some cases.

Qasim et al. [8] presented the design and FPGA implementation of sequential digital FIR filter using microprogrammed controller. This paper basically represents the structure of filter. In this, author proposed a novel approach for the designing of a digital filter using FPGA using the microprogrammed controller. A 4-tap filter is designed, synthesized using the VHDL coding. The filter design is simulated using the FPGA kit. They designed the filter using the delays, multipliers, and adders in the FIR filter basic structure. The working operation of the filter is controlled by the microprogrammed controller.

Lin et al. [9] designed techniques for high-speed multi-rate multistage FIR digital filters. The main need nowadays is high speed for processing the data and with the less complexity of the hardware. This paper represents the single rate and multi-rate FIR filters. It represents the working of the single rate filters and the pipelined structure of the single rate filter. The main bottleneck which arises mainly with the single rate filter is that the filter cannot handle multiple sample rate. This limitation is removed by using multi-rate filters. Decimator and interpolator are the basic blocks for the multi-rate filters. The multi-rate filter can deal with various samples of the signals of different sampling rates as multirate filter are able to handle multiple signal at a time, for the realization of the multi-rate filters, poly-phase structure is used. The multi-rate filter is designed using the IFIR filter design methodology and canonic signed digit (CSD) code representation.

Wei et al. [10] have designed a methodology for the implementation of FIR filter with smallest processor using FPGA. This paper is mainly concerned about the FIR implementation. In the methodology, the filter components that are delays, multipliers, and adders are substituted by the processor logical operations, and the multiplier is connected using the hardware circuitry. The basic purpose of this implementation is to reduce the complexity and make the hardware as small as possible, which itself reduces the cost relating to the filter as it decreases the chip size of the filter. The synthesis is done using the MATLAB.

Mirzaei et al. [11] represented the FPGA implementation for high-speed FIR filter using the add and shift method. For reducing the number of the adder, an elimination algorithm is used. This filter is compared with the MAC filter in which the embedded multipliers are used. This paper represented the designing of the filters without making the use of multipliers. In it, we use adder and shifts for its designing. In it, the basic blocks are configurable logic blocks in the designing of the FPGA. In it, X multiplication blocks are used, and X-1 addition blocks are used. In it, the AND and adder functions are replaced by LUTs.

Chen et al. [12] proposed the design of reconfigurable FIR filter. The FIR filter is made flexible using this implementation. The limitation which lies with the filter is that every system requires different numbers of taps and nonzero digits for the filter functioning. In this configuration, the filter is made flexible with implementing more hardware. The programmability and configurability are done in such a way that the same filter can be used for the various applications and with any change in the hardware. In this paper, a digit-reconfigurable FIR filter structure is shown. With the reconfigurability of the filter, the simplicity of the structure is somewhere reduced. In it, the number of nonzero digits and number of tap require an separate hardware for it to work. The improvement in the working of the filter is also done by using the thin chip. In the proposed filter type, not only the programs are considered but also the work is done on the chip structure and size.

Rani et al. [13] have proposed in this paper about the partial reconfigurable filters. The FIR filter found an enormous field of applications that is why it is becoming necessary that filter designed in such a way that one structure can be used for various applications. This paper is mainly concerned with the designing of the partial reconfigurable filters and its FPGA implementation and the problems which we face with the reconfigurable filters. In the reconfigurable filters, we make use of the adders and the lookup tables. With the increase in the number of coefficient which the filter has to handle the number of output also get increased in the lookup table which reduces the speed of the filter and speed is the most efficient feature for the working of the filter and other communication devices nowadays. So for the removal of this problem, partial reconfigurable FIR filter is designed. It generally makes use of the DA architecture which is systolic based, and it is dynamically reconfigurable. The use of this architecture reduces the . *BIT* file which is used for storing the LUTs file.

Rashidi et al. [14] represented the design of the filter with the reduced power consumption and reduced area. They make use of the serial adder, low-power serial multiplier, booth multiplier, folding transformation in linear phase technique, and multiplier based on adder or shifts [14]. In this research single pipeline clock is used in both the horizontal and vertical dataflow path. A common subexpression is used to reduce the adders time consumption. The filter designed for the high sample rates uses high power consumption. And if the power is reduced, the samples are taken at the moderate. In this paper, this technique is compared with the various other methods like sequential filter implementation, multipliers with more breadth in hardware than the depth, elimination algorithm, and DPDT technique. The result calculated shows less power consumption as compared to the other filters.

Singh et al. [15] proposed the sequential 7-tap FIR filter which is microprogram based. It is implemented and synthesized by using the SPARTN6. And the filter which is basically designed is synchronized by making the use of the clock pulse and reset signal. The filter designed is not significant for the small size filter but the filter can be significant for the large tap size.

Xiao et al. [16] have proposed the designing of the FIR filter using the DA algorithm. In it, all the possible outputs are stored in the LUTs which have 2^n combination already stored in it, which increases its speed but as the number of coefficients get increased, the speed get reduced. The distributed arithmetic algorithm works for improving the speed and reducing circuit complexity. The use of DA algorithm changes the multiplication into simple shifting and the adding process. The speed of the filter is also improved by using the parallel computing method. The limitation which arises in this model is that the size of the look-up table increases exponentially with the increase in number of the inputs, and this in turn reduces the speed of the filter and requires a large storage space.

Yagain et al. [17] retiming is a concept that can be used to increase the input clock frequency in the digital filter. Authors have proposed the same concept, and the transformation requires the shortest path and critical path at the several intermediate stages. They have introduced the high-level FPGA synthesis in the solution for computer-aided design (CAD) tool.

Sharma et al. [18], in their paper, proposed the concept of heterogeneous adder that supports the FIR filter implementation not using only decimal adders but using 4-bit RCA, 4-bit CSA, and 10-bit CLA adders. The size of the adder is of 18 bit. The proposed scheme in this paper is used to optimize the delay in FIR filter, and instead of using direct addition they have used three adders to form one 18-bit adder that replaces the use of 16-bit adder.

3 Various Design Methods for FIR Filter

3.1 *Microprogram-Based Designing of the FIR Filter*

In this designing methodology, a microprogram is used for designing a 7-tap FIR filter [15]. It mainly has two basic building blocks which are used in the implementation of the filter. The architecture consists of two building blocks control unit and the data path block as shown in Fig. 4.

The data path block consists of two multiplexers, multiplier, adder, and a register for the storage of data. The multiplexer used is a 4:1 mux of 8 bits, the adder is 16 bits, the multiplier can also handle 16 bits at a time, the register used for the storage purpose is a 8-bit register, and eight 8-bit coefficient register for input to the multiplexers is also used. The architecture to support the same functionality is shown in Fig. 5.

3.2 *2D Dynamically Reconfigurable Filter Bank Implementation*

This filter supports real- and complex-valued inputs. The reconfigurable filter is used for video processing applications. For implementation of 2D filter bank [6], we use 1D filter which can also work for real/complex coefficient. The 1D is implemented using distributed arithmetic. It can support:

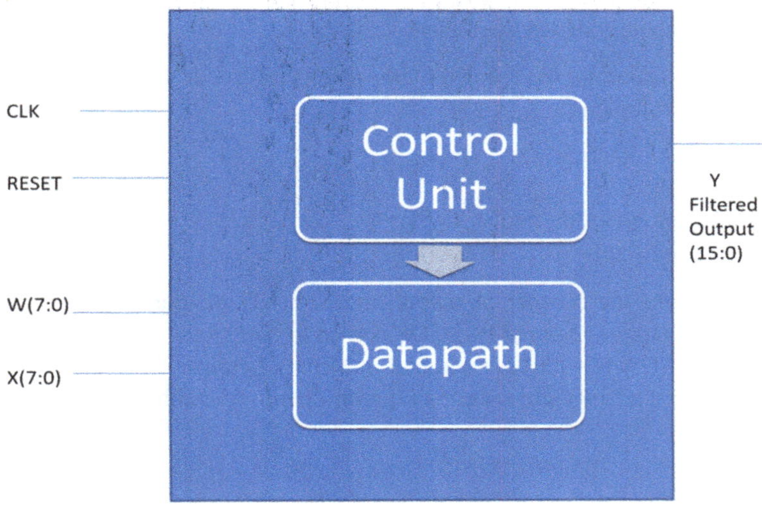

Fig. 4 FIR filter architecture

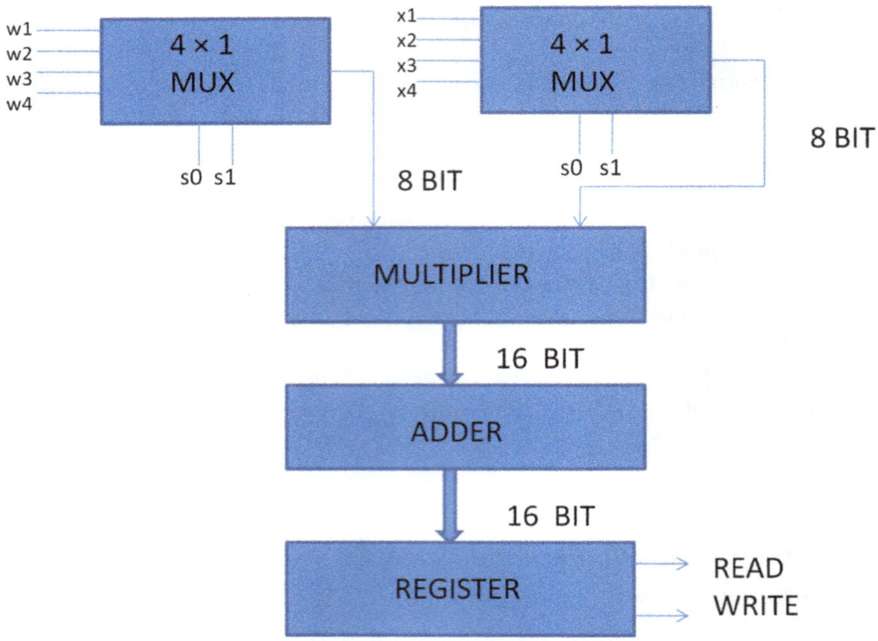

Fig. 5 Architecture of FIR microprogrammed filter

- Real input, complex coefficient,
- Complex input, real coefficient,
- Real input, real coefficient, and
- Complex input, complex coefficient.

These four modes are represented by the parameter "mode". The other parameter represents the following:

N Number of coefficient,
B I/P bitwidth,
NH Coefficient bitwidth, and
L LUT input size.

NO represents the number of output bits with NQ fractional bits. The output format is then expressed as [NONQ]. OP is used to control the saturation/truncation scheme, with NO bits with both real and imaginary outputs. The architecture with constant coefficient filter implementation with distributed arithmetic is depicted in Fig. 6.

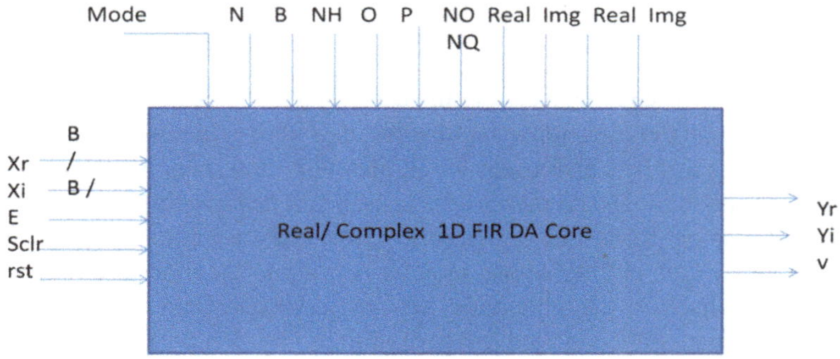

Fig. 6 Constant coefficient FIR filter with distributed arithmetic

3.3 FIR Filter Using Low-Power Multiplier and Adders

In this designing methodology, low-power multipliers and adders are used for low-power FIR filter. Figure 7 shows the FIR filter architecture to configure the FIR filter. The booth multiplier is used for low-power multiplication. The booth encoders accept the X and Y as the inputs of the multiplier, and output of the booth encoder is given to partial product generator and compressor which can have 2-bit, 3-bit, 4-bit adders, and so on with carry propagation.

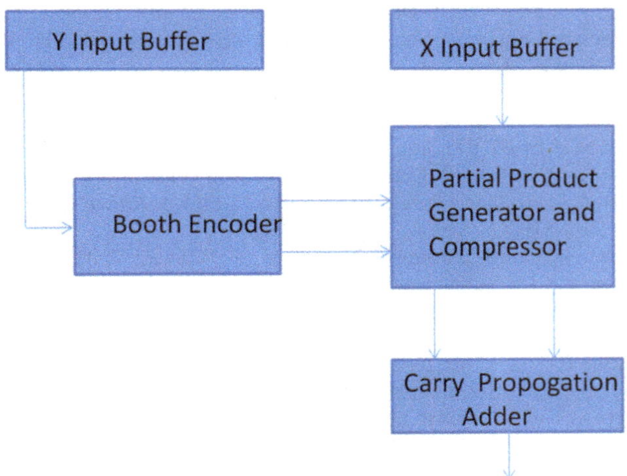

Fig. 7 Multiplier architecture for the implementation of low-power FIR filter

4 Conclusions

In this paper, the study is carried about the designing techniques of the FIR filter. In the signal processing applications, the function of a filter is to remove the unwanted portion of the signal. It is also used to eliminate the random noise and extract the useful part of the input signal such as the certain frequency component. Filters are classified as analog filter and digital filters. In DSP, a finite impulse response (FIR) filter is one type of filter whose impulse is of finite duration, because it settles to zero in finite time. On the other hand, infinite impulse response can have internal feedback and may be continued to give output infinite duration. FPGAs are high-performance data processing devices. In it, we studied the designing and synthesis of the FIR filter using different techniques.

References

1. Yan S (2006) Optimal design of FIR Beamformer with frequency invariant method. Appl Acoust 67:511–528
2. Bhattacharjeea S, Sil S, Chakrabartic A (2013) Evaluation of power efficient FIR filter for FPGA based DSP applications. Proc Technol 10:856–865
3. Chandra A, Chattopadhayay S, Ghosh B (2014) Design and implementation of SORIGA-optimized powers-of two FIR filters on FPGA. AASRI Proc 9:51–56
4. Thakur R, Khare J (2013) High speed FPGA implementation of FIR filter for DSP applications. Int J Model Optim 3(1):92–94
5. Alam SA, Gustafsson O (2014) Design of finite word length linear-phase FIR filters in the logarithmic number system domain. VLSI Des 2014:14 p, Article ID 217495. http://dx.doi.org/10.1155/2014/217495
6. Llamocca D, Pattichis M (2014) A self-reconfigurable platform for the implementation of 2D filterbanks with real and complex-valued inputs, outputs, and filter coefficients. VLSI Des 2014:24 p, Article ID 651943. http://dx.doi.org/10.1155/2014/651943
7. Grati K, Khouja N, Le Galn B, Ghazel A (2012) Power consumption models for decimation FIR filters in multi standard receivers. VLSI Des 2012:15 p, Article ID 870546. https://doi.org/10.1155/2012/870546
8. Qasim SM, BenSaleh MS, Bahaidarah M, AlObaisi H, AlSharif T, AlZahrani M, AlOnazi H (2012) Design and FPGA implementation of sequential digital FIR filter using microprogrammed controller. First workshop on emerging technologies and devices in signal processing, IEEE Xplorer, 2012
9. Lin M-C, Chen H-Y, Jou S-J (2006) Design techniques for high-speed multi-rate multistage FIR digital filters. Int J Electron 93(10):699–721
10. Wei C-H, Hsiao H-C, Tsai S-W (2005) FPGA implementation of FIR filter with smallest processor. IEEE, 2005, pp 337–340
11. Mirzaei S, Hosangadi A, Kastner R (2006) FPGA implementation of high speed FIR filters using add and shift method. IEEE, 2006, pp 1–6
12. Chen K-H, Chiueh T-D (2003) Design and implementation of a reconfigurable FIR filter. IEEE, 2003, pp 205–208
13. Rani JS, Phalghun CS (2014) FPGA based partial reconfigurable FIR filter design. In: IEEE international advance computing conference (IACC), 2014
14. Rashidi B, Rashidi B, Pourormazd M (2011) Design and implementation of low power digital FIR filter based on low power multipliers and adders on Xilinx FPGA. IEEE, 2011

15. Singh AK, Dangwal NK, Singh M, Kumar M, Singh G, Singh A (2014) Design and FPGA implementation of sequential digital 7-tap FIR filter using microprogrammed controller. Int J Sci Eng Res 5
16. Xiao S, Chen Y, Luo P (2010) The design of FIR filter base on improved DA algorithm and its FPGA implementation. In: 2nd IEEE international conference on computer and automation engineering (ICCAE), 2010
17. Yagain D, Krishna AV (2014) Design of synthesizable, retimed digital filters using FPGA based path solvers with MCM approach: comparison and CAD tool. VLSI Des 2014:1–18, Article ID 280701 (Hindawi Publishing Corporation)
18. Sharma P, Gupta M (2015) Area-efficient implementation of heterogeneous adder and its application in FIR filter. Int J Res Electron Comput Eng 3:1–5

Design of All-Terrain Rover Quadcopter for Military Engineering Services

Shantanu Chaudhary, Arka Prava, Neha Nidhi and Vijay Nath

Abstract In today's world, it is very much required to a country's military to be technically advanced. The most famous technology in military field is the use of drones or, in other words, QUADCOPTERS. Therefore, I intend to aid in improving this technology by creating an **all-terrain rover quadcopter** which can fly and can move also on any type of surfaces (including special case for water), which is required by military forces. It will work on normal principle of quadcopter, and the added feature to it is its wheel and streamlined and aerodynamic central body. It is so designed that in most of the cases, the propellers will not be harmed. Military can use it to pick up any small object from it (say up to dia. of 5 cm) with average payload capacity. It is of grass color so it can move in bushes creating a camouflage effect without being observed.

1 Introduction

A quadcopter is also known as quadrotor helicopter or quadrotor. It is a type of helicopter having man blades, i.e., raised and propelled by using four rotors. Quadcopters can be categorized as rotorcraft, as it is opposed to fixed-wing aircraft, because its lift is created by a pair of rotors.

Quadcopters normally use pairs of two rotor blades in which one pair move clockwise and other pair moves anticlockwise. When we change the speed of motor, thus rotor blades, it generates total thrust; to find the center of thrust both in laterally and longitudinally direction; and have required amount of total torque acting on the body or the turning force.

Quadcopters are different from normal helicopters in a way that it uses rotors which are able to vary the pitch of it dynamically when it moves a near the rotor hub. In the early days of flight, torque-induced effects lead to loss in control (as well

S. Chaudhary (✉) · A. Prava · N. Nidhi · V. Nath
Department of Electronics & Communication Engineering,
Birla Institute of Technology Mesra, Ranchi 835215, India
e-mail: shantanu10490@gmail.com

V. Nath and J. K. Mandal (eds.), *Nanoelectronics, Circuits and Communication Systems*, Lecture Notes in Electrical Engineering 511,
https://doi.org/10.1007/978-981-13-0776-8_47

as efficiency issues due to rotor) which is countered by using counterrotation and using short blades which are easy to build. A plenty of designs were found in the 1920s and 1930s. These are the first successful ever heavier-than-air, vertical takeoff and landing (VTOL) vehicles. However, the early ones were having the problems like poor performance, too much pilot workload.

In the late 2000s, technological advancement leads to the production of low cost and lightweight flight controllers, cameras accelerometers (IMU), and GPS. This resulted in the quadcopter's configuration turning popular for small self-flying vehicles. Having smaller body and planned movement, these quadcopters can fly in indoors and outdoors too.

Breguet-Richet Gyroplane (1907) was designed by Louis Breguet in early 1907. It has four wings blade copter. This was the first aircraft to lift off the ground, but only in particular flight at a height few feet from the ground. In 1908, it was reported that it was flying "several times", but details are very less in number.

Convertawings Model A Quadrotor (1956) was designed in order to have a military flying equipment. It was having two engines driving four rotors blades by a function of V belts. Without the use of tail rotor, it was controlled by varying the thrust acting on it. It successfully flew in the mid-1950s; this helicopter was approved and was also the first four-rotor helicopter to make a successful forward flight. As military was not interested, the project was terminated.

Curtiss-Wright VZ-7 (1958) was designed for the US Army. It was a VTOL aircraft. It was handled by varying the force of thrust acting on all four propellers accordingly.

2 Methodology

Quadcopter functionality is defined in Fig. 1. Quadcopters generally use pairs of two rotor blades in which one pair moves clockwise and other pair moves anti-clockwise. It is possible by varying motor speed.

It requires electronic stabilization. Its motors have different speeds. Its weight distributions were along the body. For performing this task, it requires ESC to control the speed. It also requires sensors and microcontroller unit for sensing and controlling the all activities in real domain. It has an excellent distribution of forces. The electronics working mechanism of quadcopter is shown in Fig. 2.

The quadcopter consists of main quadcopter frame body, propellers and four Motors, electronic speed controllers (ESC)*4, tank wheels, electronic stabilization (programmed board MCU), LIPO battery, and transmitter/receiver (Figs. 3, 4, 5, 6 and 7).

Various control schemes are used by open source projects which are shown in Fig. 8.

Lift force acts perpendicular to them moving direction, i.e., the rotor plane. The force of gravity is perpendicular to the ground. Quadcopter rolls when it turns—which causes a vertical and horizontal component of lift. Centripetal force is given

Fig. 1 A prototype of quadcopter

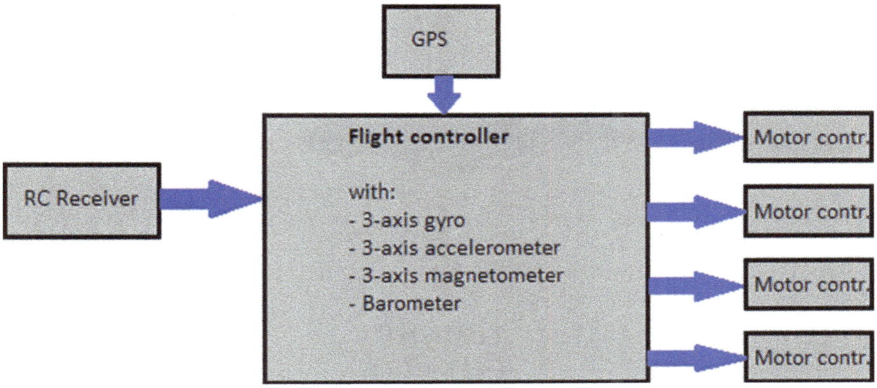

Fig. 2 Electronic working mechanism of quadcopter

by horizontal component of lift. The combined effect of weight and centripetal force cancels out the lift. This is known as g-force or load factor. *Z*-axis will be measured by the accelerometer, and the other ones are 0. Altitude can be determined by using different sensors like ultrasonic, barometric, etc. Local control helps in the stability of quadcopter, and fusion of sensors is necessary.

So this was how a basic quadcopter works. But in our case, it is different in a way that it has to move with respect to all terrain, i.e., an all-terrain vehicle. So we have to design a special type of programmed board (PCB) that is programmed to work simultaneously for propeller movement control and to control the movement of the continuous tread tank wheels. The programmed board will be serially communicating with a remote control. There will be different types of buttons designed on remote control panel to control extra features like picking up of any

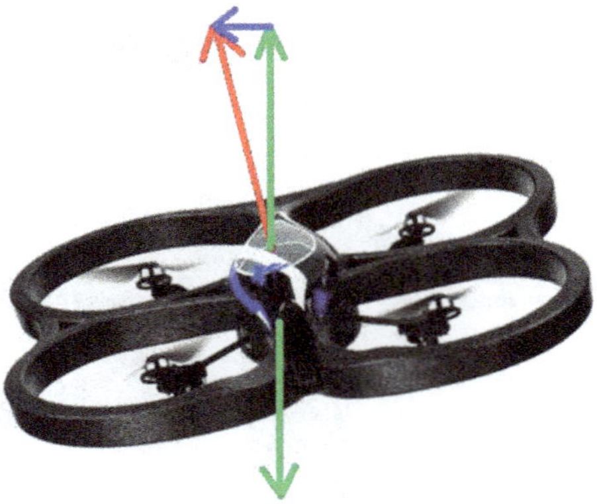

Fig. 3 Axial forces acting on a quadcopter

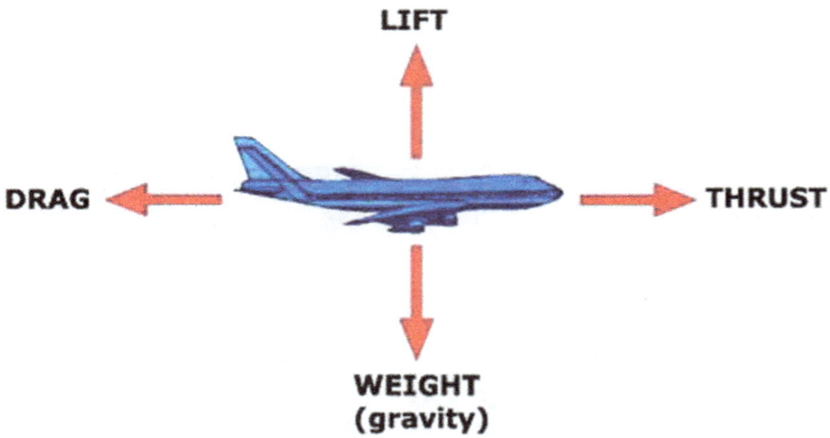

Fig. 4 Force diagram of flying vehicle

Fig. 5 Electronic stabilization using special programmed PCB

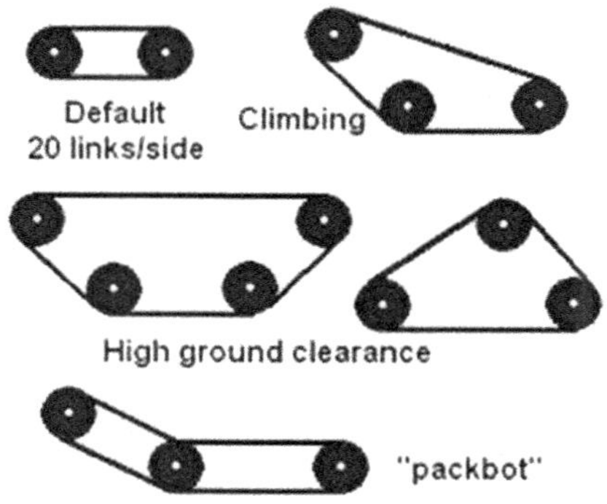

Fig. 6 Continuous tread tank wheels

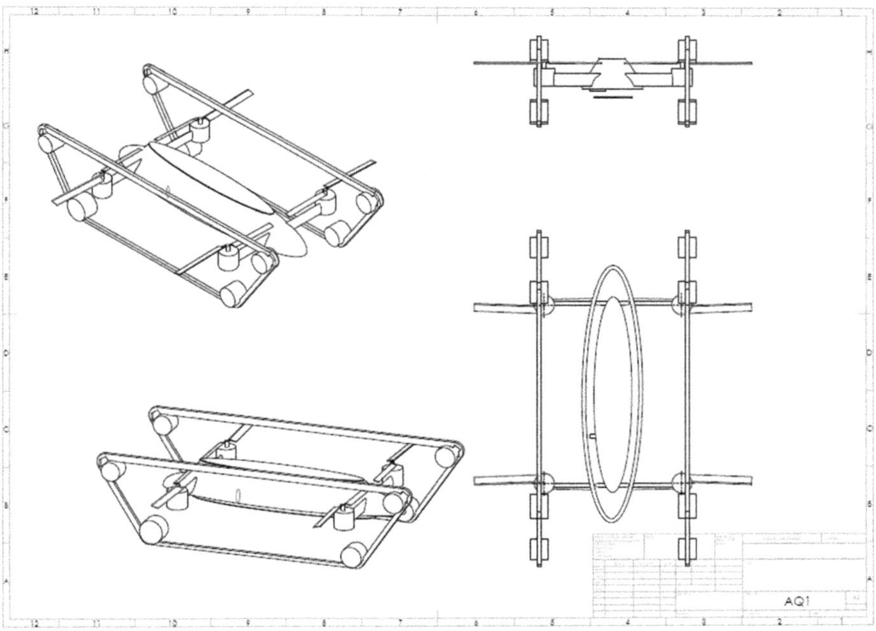

Fig. 7 Design of all-terrain rover quadcopter

object by a small robotic arm. It will move on grounds with help of wheels and will fly in air with the help of propellers. I will try to keep the size of MCU smaller so as to fit in my small aerodynamic central body. With the help of advanced accelerometer, gyroscopic sensors' accuracy and efficiency can be increased. We can also fix a camera on central body whose motion can be controlled by the remote. When the tank wheels move on grounds, it can cross or pass any type of obstacle on its way, which is approximated according to the size of quadcopter.

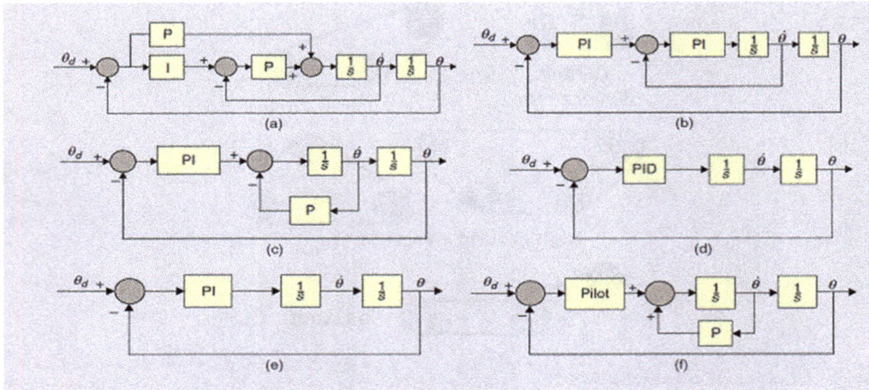

Fig. 8 Various PID control structures of the OSPs: **a** Arducopter, **b** Open Pilot, **c** Paporazzi, Multiwii, **d** Pixheawk, Aeroquad, **e** Mikrokopter, and **f** KK Multicopter

When it comes to its body manufacturing, it can use aluminum as body material. It is lightweight, corrosion resistance, and easy to manufacture making it easily available. It can also use carbon fiber as body material. It will be painted in green color to provide a camouflage effect when it moves in grass. IMPERX'S Bobcat 2.0 spying camera will be attached to its central body, which will enable the user to avoid obstacles while hovering around. IMPERX'S Bobcat 2.0 is an advanced, rugged, and extremely programmable camera series. GPS trackers which are used in current military equipment will be used for locating and tracking quadcopter. Also working on the idea that a claw type robotic hand will be installed that can be used to pick up any small object. The software which is used for designing the quadcopter is "Solid works Autocad".

3 Results

Finally, we designed a quadcopter which can work as an all-terrain type vehicle. It is very beneficial for military use. It will use different types of material for its body like aluminum or carbon fiber. It uses different types of sensors like accelerometer and gyroscopic sensor for its movement and control. Manual control is also possible using a remote controller. Here is used a pair of continuous tread tank wheels.

4 Conclusion and Discussion

So it has designed an all-terrain rover quadcopter which can move on all types of terrains and fly in air. The purpose of this is for betterment our military technology. It is smart and advances in many ways. It can also move on water with water level only below or up to the propellers.

Some more work is to be done such as designing of such type of programmed board (PCB) which can provide full electronic stability to quadcopter and specify the dimensions of body, propellers, etc.

Acknowledgements History of all great works says that neither of these works is possible without the support of people. I would like to show my gratitude toward all of the people who helped me to complete my research paper. I would like to thank Prof. M. K. Mishra, Vice-Chancellor BIT Mesra Ranchi and Head, Department of Electronics & Communication Engineering and Prof. S. Pal for providing infrastructure facility to carry out this research work.

References

1. Quadcopter introduction. https://en.wikipedia.org/wiki/Quadcopter
2. Gaponov I, Razinkova A (2012) Quadcopter design and implementation as a multidisciplinary engineering course, IEEE, 26 Nov 2012
3. Gheorghiţă D, Vîntu I, Mirea L (2015) Quadcopter control, IEEE, 09 Nov 2015
4. Hoffmann GM, Rajnarayan DG, Waslander SL, Dostal D, Jang JS, Tomlin CJ (2004) The stanford test bed of autonomous rotorcraft for multi agent control (STARMAC) (PDF). In: Proceedings of the 23rd digital avionics system conference, Salt Lake City, UT, pp 12.E.4/1–10
5. Stafford J (2014) How a Quadcopter works|Clay Allen. University of Alaska, Fairbanks. Retrieved 20 Jan 2015
6. Leishman JG (2000) Principles of helicopter aerodynamics. Cambridge University Press, New York, NY
7. Büchi R (2011) Fascination quadrocopter. ISBN 978-3-8423-6731-9
8. Pounds P, Mahony R, Corke P (2006) Modelling and control of a quad-rotor robot (PDF). In: Proceedings of the Australasian conference on robotics and automation, Auckland, New Zealand
9. Hoffman G, Huang H, Waslander SL, Tomlin CJ (2007) Quadrotor helicopter flight dynamics and control: theory and experiment (PDF). In: Conference of the American institute of aeronautics and astronautics, Hilton Head, South Carolina, 20–23 August 2007
10. How-to: quadrocopter based on Arduino. MAKE. Retrieved 29 Dec 2014
11. The seventh international conference on vibration problems ICOVP 2005. https://www.google.com. Retrieved 13 March 2015
12. Anderson SB (1997) Historical overview of V/STOL aircraft technology. NASA Technical Memorandum 81280
13. Flight archive—1956–1564. https://www.flightglobal.com. Retrieved 13 March 2015
14. US patent 3261407—Helicopter rotor system. https://www.google.com. Retrieved 13 March 2015
15. The seventh international conference on vibration problems ICOVP 2005. https://www.google.com. Retrieved 13 March 2015

A Systematic Approach Towards Development of Universal Software Fault Prediction Model Using Object-Oriented Design Measurement

Dipti Kumari and Kumar Rajnish

Abstract A new systematic approach towards development of a software fault prediction model for trustworthiness and excellence enhancement of software systems by predicting fault-proneness in different severity level of module after coding has been performed in this paper. The classification ability of data mining technique and important object-oriented metrics of object-oriented software system are used to develop this fault prediction model. This model predict the severity level of object-oriented software module based on the number of faults in nominal, low, mid, high and not fault prone (NFP). A different decision tree like Hoeffding tree (VFDT) is created for existing project data Eclipse 2.0 in order to gain information for the purpose of decision-making whether a particular module is faulty or not, and if faulty then in which category. The rules that are obtained from VFDT decision tree algorithm are transformed into fuzzy rules and join together with the fuzzy inference system (FIS) of soft computing approach to predict fault-proneness of object-oriented module in different categories of fault or also not fault free for target data that is other two version of existing project data Eclipse 2.1 and Eclipse 3.0. The main aim of developing this fault prediction model is that to give a direct help to software manager in identifying fault-prone module by including this model before testing phase and just after coding phase so that testing effort can be reduced. And indirectly, this model helps the common people in getting reliable and quality-based system.

Keywords Fuzzy profile · Object-oriented metrics · Decision tree
Fault prediction

D. Kumari (✉) · K. Rajnish
Department of Computer Science & Engineering, BIT Mesra,
Ranchi, Jharkhand 835215, India
e-mail: kumari_dipti0511@yahoo.co.in

V. Nath and J. K. Mandal (eds.), *Nanoelectronics, Circuits and Communication Systems*, Lecture Notes in Electrical Engineering 511,
https://doi.org/10.1007/978-981-13-0776-8_48

1 Introduction

In every field, the modern systems are dependent on software-based system. And these dependencies are increasing with time. That is why reliability of the software as well as quality of the software has become the most important issue throughout the development of software system. But it cannot be possible at first try to develop quality software. The reasons behind that are the complexity of problem, human behaviours complexity and the unavailability of resources. A fault can be defined as a kind of defect in source code that causes failures after execution [1]. Software failure causes system failures, which is a common problem that ends with undesirable outcomes like unreliable and unsafe software system. Different related software modules build a software system. As we know, the majority of faults are determined in some of its modules. Or in other words, we can say that 55% of faults exist inside 20% of source code of the module. Therefore, a lot of interest has been found to find out fault-prone software modules at early stage of a project [2]. Including the fault prediction model before testing phase and after design phase inside the software life cycle (SDLC) can be used to reduce software delivery time, and also reduce the testing effort and achieve the goal of developing reliable as well as quality-based software system. So, it will be better decision to classify the software module as faulty module with different categories based on the severity levels of fault or fault free just after the coding of the software in SDLC. Just after the classification of modules as fault prone (i.e. nominal, low, mid or high) or fault free, locating the total number of error in that class will become easy. So, severity level will become the decision factor of testing efforts that will be put on the different fault-prone module to produce quality software.

The main aim of this paper is the development of a model which is capable for prediction of fault-prone software module by the combination of both strategies data mining [3] and one of the soft computing approach, i.e. fuzzy logic [4, 5]. The reason for using data mining technique before fuzzy logic is that the extracted rule from decision tree can be used as input for FIS. From the literature, it has been discovered that at design time, the OO-metrics are of fuzzy in nature so crisp assignment is not right decision or in other words, it is impractical.

Identification of faulty software modules through prediction model is an interesting field of research of among the software quality researchers, software testing and debugging group. This interest has shown numerous efforts towards the development of software fault prediction model using methods like neural networks, Bayesian methods, decision trees, support vector machines, Dempster–Shafer belief networks, naive Bayes, fuzzy logic, case-based reasoning, genetic programming and logistic regression [6–18].

The remaining paper is structured as follows: the subsequent section presents experimental design associated with the problem. Section 3 describes the result and discussions. Section 4 presents results and conclusion.

2 Experimental Design

This section deals with all the steps that we have followed for achieving our goal:

a. Data collection (i.e. software modules and their OO-metrics and respective fault data),
b. Fuzzify independent (i.e. OO-metrics) and dependent (i.e. fault) data,
c. Building decision tree,
d. Extract rules from decision rule for development of FIS,
e. Proposed algorithm and
f. Evaluation of the model.

2.1 Data Collection

For this work, data collection is done in three steps:

i. Selection of object-oriented software system,
ii. Selection of OO-metrics containing different OO-properties and
iii. Arranging each class oriented OO-metrics value with the associated number of pre-release error data.

The object-oriented software system coded in Java programming language, i.e. Eclipse 2.0, has been used as data source. There are numerous varieties of OO-metrics that is why it is very difficult job to select the software metrics. So, to simplify the selection job, we have used three selection criteria:

a. All aspects of OO-design are taken into account in metrics.
b. Automated tool can able to estimate the selected metrics values.
c. Threshold value of different OO-metrics can able to distinguish between faulty and fault-free classes.

For measurement of selected OO-metrics from Eclipse 2.0 source code [19], an automated tool jHawk [20] is used. From Promise 2.0, which is publicly available has pre-release fault data of Eclipse 2.0 [21]. Pre-release faults which are reported in the 6 months before release for each Class in Eclipse 2.0 are tabulated.

We have used five OO-metrics covering different five OO-metrics properties listed in Table 1.

Table 1 Object-oriented metrics description

OO-metrics	Description	OO-property
NOS	Total number of statements	Size
TCC	Total cyclomatic complexity	Complexity
DIT	Depth of inheritance	Inheritance
COH	Cohesion	Cohesion
RFC	Response for class	Coupling

2.2 Fuzzification of OO-Metrics

Object-oriented properties are measured in form of different object-oriented metrics of software system. These software metrics at design phase have fuzziness in nature and specific crisp value assignment appears to impractical assignment [22]. Fuzzy representation of software metrics value as well as fault value will be helpful in form of good indicator of software quality. For fuzzy representation of any value, first normalization of the data has been done. After normalization, a particular membership function is used to give membership value to every element of the set. The widely used membership function is triangular membership functions (TMFs); the reasons of using this are that it is simple as well as easily understandable [11]. That is why we are using triangular membership function, which is linear in nature and is more suitable when the precise membership functions of a fuzzy set are not known [23]. Therefore, in this model, triangular membership are considered for representing the linguistic states of OO-metrics and fault data. We have used a technique for construction of a fuzzy profile of both object-oriented design metrics and fault data [24]. Let SM denote a set of N training patterns containing n OO-metrics values (O_1, O_2 … O_j, O_n). The OO-metric O_j has n values ov_{1j}, ov_{2j}, ov_{nj}. O_{jmin} and O_{jmax} denote the minimum and maximum value of OO-metric O_j. Each OO-metrics value is converted into linguistic variable like nominal, low, mid and high by assigning a range of value to each linguistic variable. The OO-metric is quantitative in nature, and then steps followed for membership function generation are as follows (Fig. 1):

Step 1. Perform the sorting operation on the values of OO-metrics O_j in ascending order as well as remove redundancy.
Step 2. Apply K-means clustering algorithm on the quantitative values of the OO-metrics O_j into k clusters (cy_1, cy_2, …., cy_i, …., cy_k) where cy_{imin} and cy_{imax} represent minimum and maximum value of ith cluster (cy_i).
Step 3. Compute the cluster centres (cc_1, cc_2, …, cc_i,…, cc_k) of k clusters (cy_1, cy_2, …, cy_i, …, cy_k).

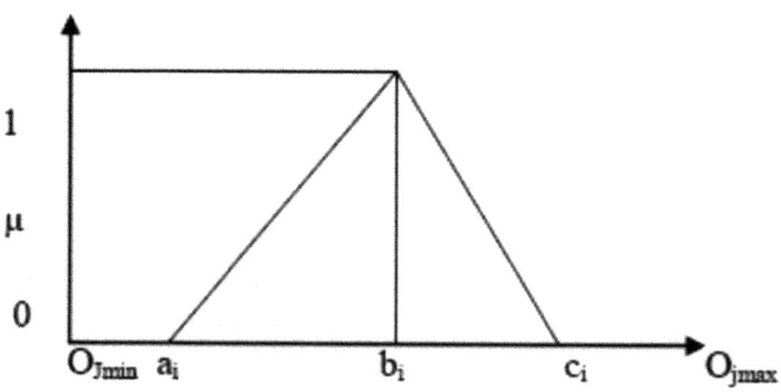

Fig. 1 A triangular membership function

Step 4. Calculate two boundary points with their membership value of every cluster.
Substep 4.1. Determine the difference of adjacent data. For each pair ov_i and ov_{i+1} ($i = 1, 2, \ldots, n-1$)
$ovdiff_i = ov_{i+1} - ov_i$.
Substep 4.2. Calculate the adjacent data similarity value of object-oriented metics O_j. The similarity value is obtained from the following formula [25]:

$$Os_m = \begin{cases} 1 - \frac{ovdiff_i}{C \times \sigma_s} & \text{for } ovdiff_i \leq C \times \sigma_s \\ 0 & \text{otherwise} \end{cases} \tag{1}$$

where

s_m denotes the adjacent data similarity value,
σ_s std. deviation of $ovdiff_i$ and
C is control parameter for the shape of membership functions.

Substep 4.3. The Os_m value of ith cluster which has minimum value is taken as the membership value of two boundary points of that cluster cy_{imin} and cy_{imax}.
Step 5. Determine by interpolation the left vertex point (a_i, 0).

$$a'_i = cc_i - \frac{cc_i - cy_{imin}}{1 - \mu(cy_{imin})} \tag{2}$$

$$a_i = \begin{cases} 0 & \text{for } a'_i \leq 0 \\ cc_{i-1} & \text{for } a'_i \leq cc_{i-1},\ i > 1 \\ a'_i & \text{for } a'_i > cc_{i-1} \end{cases} \tag{3}$$

Step 6. Determine again by interpolation the Right vertex point (c_i, 0).

$$c'_i = cc_i + \frac{cy_{imax} - cc_i}{1 - \mu(cy_{imax})} \tag{4}$$

$$c_i = \begin{cases} c'_i & \text{for } c'_i \leq cc_{i+1} \\ cc_{i+1} & \text{for } c'_i > cc_{i+1} \end{cases} \tag{5}$$

For last cluster if $c_n = c'_n > cy_{nmax}$, then $cc_n = cy_{nmax}$ where n shows the subscript of last cluster.

Step 7. Compute the value of membership of each OO-metrics O_j.

Table 2 Construction of membership function of all selected metrics

Metric	Cluster1 (nominal)	Cluster2 (low)	Cluster3 (mid)	Cluster4 (high)
NOS	$a_1 = 0$	$a_2 = 39$	$a_3 = 682$	$a_4 = 1941$
	$b_1 = 39$	$b_2 = 306$	$b_3 = 976$	$b_4 = 2539$
	$c_1 = 306$	$c_2 = 976$	$c_3 = 1787$	$c_4 = 3582$
TCC	$a_1 = 0$	$a_2 = 14$	$a_3 = 204.7148$	$a_4 = 613$
	$b_1 = 14$	$b_2 = 113$	$b_3 = 323$	$b_4 = 824$
	$c_1 = 113$	$c_2 = 323$	$c_3 = 663.9348$	$c_4 = 1222$
DIT	$a_1 = 1$	$a_2 = 2$	$a_3 = 5$	$a_4 = 7$
	$b_1 = 1$	$b_2 = 3$	$b_3 = 5$	$b_4 = 7$
	$c_1 = 2$	$c_2 = 4$	$c_3 = 6$	$c_4 = 8$
COH	$a_1 = 0$	$a_2 = 0.0205$	$a_3 = 0.2297$	$a_4 = 0.7$
	$b_1 = .0205$	$b_2 = .2297$	$b_3 = .4991$	$b_4 = .8830$
	$c_1 = 0.2297$	$c_2 = 0.4991$	$c_3 = 0.883$	$c_4 = 1$
RFC	$a_1 = 0$	$a_2 = 13$	$a_3 = 64$	$a_4 = 280$
	$b_1 = 13$	$b_2 = 64$	$b_3 = 155$	$b_4 = 390$
	$c_1 = 64$	$c_2 = 155$	$c_3 = 390$	$c_4 = 596$

$$\mu(\mathrm{ov_i}) = \begin{cases} \frac{\mathrm{cc}_i - \mathrm{ov}_i}{\mathrm{cc}_i - \mathrm{a}_i} & \text{for } \mathrm{ov}_i < \mathrm{cc}_i \\ \frac{\mathrm{c}_i - \mathrm{ov}_i}{\mathrm{c}_i - \mathrm{cc}_i} & \text{for } \mathrm{cc}_i \leq \mathrm{ov}_i < \mathrm{c}_i \\ 0 & \text{otherwise} \end{cases} \tag{6}$$

Using this method, we fuzzify the selected metrics results, which are shown in Table 2, and then carry the research forward.

2.3 *Decision Tree*

To analyse the research data which is huge in amount, the data mining [3] technique has emerged. Classification technique of data mining is used to classify the data into predefined fault class. The classification technique results in a model which is capable to classify future versions of software or any new developed software's modules as faulty or fault free in an automated way.

Decision tree is easily understandable, simple and also an efficient classification technique. There are many algorithms available in literature which has been proposed in past by researcher for decision trees formation. Among that a most popular decision tree is Hoeffding tree (VFDT) [26]. A decision tree has a flowchart type of structure where a test on an attribute is shown by the internal node, outcome of the test is shown by the branches and the target class is shown by leaf node.

VFDT algorithm As we know from literature, 55% of faults exist in 20% of source code. So, VFDT is best option for developing software fault prediction model. A model is trained by a complete set of training data. Each training data record has many attributes. A training record will take the form (O, f), where O is a vector of (O_1, O_2 …, O_n) n attributes of a software module and f is a class label where classification algorithm identifies the fault-prone modules (i.e. nominal, low, mid and high) and NFP. The VFDT algorithm for tree initializing and tree growing are as follows:

HT(Training record, δ)
Step 1. Initialize tree HT with a single node (i.e. root).
Step 2. Initialize counts n_{ijk} at root.
Step 3. **for** each record (O,) in training data.
Step 4. **do** HTGROW((X, f), HT, δ).
HTGROW ((O, f), HT, δ)
Step 1. Sort (O, f) to leaf l using HT.
Step 2. Update counts n_{ijk} at leaf l.
Step 3. **if** attributes at l are not all of the same class **then** Compute G for each attribute.
Step 4. **if** G(Best Attr.)-G(2nd best Attr.) $>\sqrt{\frac{R^2\ln(1/\delta)}{2N}}$, **then** split leaf on best attribute.
Step 5. **for** each branch.
Step 6. **do** Start new leaf and initialize counts.

The knowledge obtained by decision tree can be converted into the 'IF-THEN' rules. 'IF-THEN' rules are the classification rules which are taken out from the decision tree by going through the all paths which are started at root to a leaf node.

2.4 *Fuzzy Inference System*

A fuzzy set theory is used by fuzzy inference system (FIS) to map inputs to outputs.

Figure 2 is showing the architecture of the fuzzy inference system, where A is the fuzzifier, B is the Inference engine, C is the defuzzifier and D is the fuzzy knowledge base. By the help of membership function, the fuzzifier converts the crisp input value into fuzzy or linguistic input. Inference engine converts the fuzzy input to fuzzy output. Again with the help of membership function, the defuzzifier converts the fuzzy output to crisp value if required through membership fuction. Fuzzy knowledge base is the combination of database and the rule base.

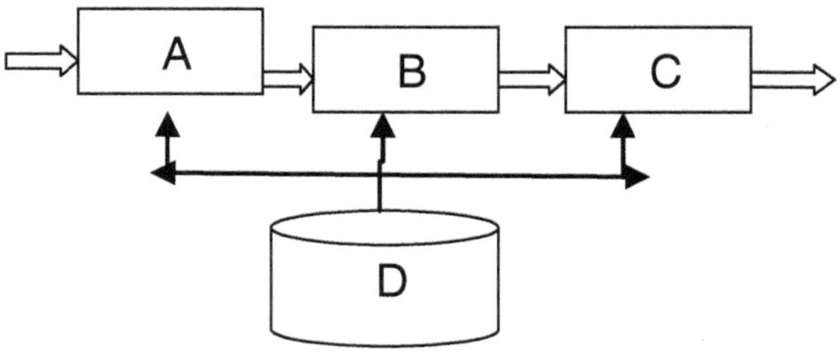

Fig. 2 Architecture of fuzzy inference systems

2.5 *Proposed Algorithm*

Step 1. Identify the actual fault-proneness of class C {FP (i.e. nominal, low, mid and high), Nfp}, for a required project data.
Step 2. Compute the OO-metric value of each class.
Step 3. Associate OO-metric value and actual fault value of respective classes.
Step 4. Devlope the fuzzy membership of each metrics and fault data.
Step 5. Apply decision tree algorithm on the processed data.
Step 6. Returns (Decision Tree).
Step 7. Take out classification rules form the decision tree.
Step 8. Use these rules as fuzzy rules into FIS at step 9.
Step 9. Develop a fuzzy inference system (inputs are object-oriented software metrics, and outputs are FP (i.e. nominal, low, mid and high) or Nfp modules).
Step 10. Classify the project data into target class (FP (i.e. nominal, low, mid and high) or Nfp) using developed FIS.

2.6 *Model Evaluation*

Accuracy of the model can be obtained by either comparing the proposed model's results with the traditional models or through confusion matrix [27] as shown in Table 3. Information of actual and predicted classifications is represented in confusion matrix in row and column, respectively.

Table 3 Confusion matrix

		Actual value		Total
		p	n	
Prediction outcome	p'	TP	FP	P'
	n'	FN	TN	N'
Total		P	N	

Accuracy is the number of correctly classified classes in this case faulty classes or fault-free classes divided by the total number of classes that exist in the software, where TP (true positive) and FP (false positive) are correct and incorrect predictions of positive, i.e. faulty class, respectively. And TN (true negative) and FN (false negative) are correct and incorrect predictions of fault-free class, respectively.

$$\text{Accuracy} = \frac{\text{TP} + \text{TN}}{\text{TP} + \text{TN} + \text{FP} + \text{FN}}$$

3 Results and Discussion

The decision tree learning algorithm VFDT has been examined and implemented using MATLAB programming. From Eclipse 2.0 project dataset, many different decision trees are built by taking 20, 40, 60 and 80% of that.

After generating decision trees, we can derive different classifiers, and accuracy of each classifier is listed in Table 4.

The 'Train/Test Percent age' of the data in 10 iterations has been performed as experiment. In Table 4, the accuracy of every experiment which is done in iterations is listed. From Table 4, it is obvious that with the increase of training data, the accuracy of the model is also increasing. The accuracy of the fault prediction model is calculated by the average accuracy results of the experiment that is done in 10 iterations. Next, we show the dependence of prediction accuracy on the training data. For this, we have developed six different FIS, namely, FP5_95, FP10_90, FP20_80, FP40_60, FP60_40 and FP20_80, using Eclipse 2.0 dataset. From the FIS name, it is easily identified that at the start of the experiment, data size is extremely small (i.e. 5%), prediction correctness is found to be 74.56% and it suddenly raises to 82.83 when training data increases to 10%. From the result, it is obvious that as size of training data increases, there is also increase in the prediction accuracy up to 87.80% as listed in Table 5.

Table 4 Prediction accuracy of the model on Eclipse 2.0 dataset

Exp.	Training (%)	Test (%)	Accuracy	Average
1	20	20	79.23	80.533
2		40	80.67	
3		60	81.78	
4		80	80.45	
5	40	20	82.34	84.226
6		40	84.56	
7		60	85.78	
8	60	20	86.45	86.115
9		40	87.78	
10	80	20	87.80	87.80

Table 5 Training effect on prediction accuracy

In (%)	FP5_95	FP10_90	FP20_80	FP40_60	FP60_40	FP80_20
Training data	5	10	20	40	60	80
Testing data	95	90	80	60	40	20
Accuracy value	74.56	82.83	83.34	84.67	85.78	87.80

Table 6 Performance result of the model on Eclipse 2.0 dataset

Model	Class prediction	Rank prediction	Accuracy
Catal et al. [28]	Yes	No	80.6
Pandey et al. [29]	Yes	Yes	82.67
Proposed model	Yes	Yes	84.66

The experiment is done in ten different iterations, and the model accuracy is calculated by taking the average of all those experiment accuracy value. The comparative study has been done with the developed model result and earlier models [28, 29] results, which are given in Table 6. Results are showing that developed model has better performance than earlier ones.

4 Conclusions

A software fault prediction model is presented in this paper for identifying fault-prone module with severity level of a large software system. For developing the model, classification technique of VFDT algorithm is used with fuzzy inference system. The reason for choosing VFDT among a number of decision trees is that it can very smoothly handle large dataset as well as it does not require the whole dataset during learning process. It also reduces the time required to build the tree. So, we can say that VFDT is efficient in terms of memory and time requirements. VFDT also maintains its accuracy for large dataset whereas others collapse. When we extract the rules from VFDT and convert it into fuzzy rules and after that integrate into FIS, then the fault severity level is also possible to predict through the model. This model is mainly useful for project managers, for reducing the time duration of testing phase. They can apply precedence on the effort that is applied during testing of modules to accomplish object-oriented software system with more reliability as well as quality. The outcome are good, which indicate future work for analysing testing effort, with the fault severity level and find out the testing effort saved by the proposed model.

References

1. Musa JD, Iannino A, Okumoto K (1987) Software reliability: measurement, prediction, and application, McGraw-Hill Publication
2. Benlarbi S, Emam Khaled El, Geol N (1999) Issues in validating object-oriented metrics for early risk prediction. In: 10th international symposium on software reliability engineering, ISSRE'99, Boca, pp 17–18
3. Han J, Kamber M (2001) Data mining: concepts and techniques. Morgan Kaufmann Publishers, USA
4. Ross TJ (2010) Fuzzy logic with engineering applications, 3rd edn. Willy, India
5. Zadeh LA (1965) Fuzzy sets. Inf Control 8:338–353
6. Khoshgoftaar TM, Seliya N (2002) Software quality classification modeling using the SPRINT decision tree algorithm. In: Proceedings of the 4th IEEE international conference on tools with artificial intelligence, Washington, DC, pp 365–374
7. Thwin MM, Quah T (2003) Application of neural networks for software quality prediction using object-oriented metrics. In: Proceedings of the 19th international conference on software maintenance, Amsterdam, The Netherlands, pp 113–122
8. Elish KO, Elish MO (2008) Predicting defect-prone software modules using support vector machines. J Syst Softw 81:649–660
9. Pai GJ, Dugan JB (2007) Empirical analysis of software fault content and fault proneness using bayesian methods. IEEE Trans Softw Eng 33:675–686
10. Menzies T, Greenwald J, Frank A (2007) Data mining static code attributes to learn defect predictors. IEEE Trans Softw Eng 33:2–13
11. Yadav OP, Singh N, Chinnam RB, Goel PS (2003) A fuzzy logic based approach to reliability improvement estimation during product development. Reliab Eng Syst Saf 80:63–74
12. Pizzi NJ (2007) Software quality prediction using fuzzy integration: a case study. Springer-Verlag, pp 67–76
13. Guo L, Cukic B, Singh H (2003) Predicting fault prone modules by the Dempster-Shafer belief networks. In: Proceedings of the 18th IEEE international conference on automated software engineering, IEEE Computer Society, Montreal, Canada, pp 249–252
14. Evett M, Khoshgoftaar T, Chien P, Allen E (1998) GP-based software quality prediction. In: Proceedings of the 3rd annual genetic programming conference, San Francisco, CA, pp 60–65
15. Khoshgoftaar TM, Seliya N, Sundaresh N (2006) An empirical study of predicting software faults with case-based reasoning. Softw Qual J 14:85–111
16. El Emam K, Benlarbi S, Goel N, Rai S (2001) Comparing case-based reasoning classifiers for predicting high risk software components. J Syst Softw 55:301–320
17. Olague HM, Gholston S, Quattlebaum S (2007) Empirical validation of three software metrics suites to predict fault-proneness of object oriented classes developed using highly iterative or agile software development processes. IEEE Trans Softw Eng 33:402–419
18. Schneidewind NF (2001) Investigation of logistic regression as a discriminant of software quality. In: Proceedings of 7th international software metrics symposium, London, UK, pp 328–337
19. Eclipse source code (for archived releases). http://archive.eclipse.org/eclipse/downloads/. Accessed 3 Dec 2013
20. JHAWK metrics reference. http://www.virtualmachinery.com/jhawkreferences.html. Accessed March 2014
21. Eclipse bug data (for archived releases). http://www.st.cs.uni-sb.de/softevo/bug-data/eclipse. Accessed 20 Nov 2013
22. Pandey AK, Goyal NK (2009) a fuzzy model for early software fault prediction using process maturity and software metrics. Int J Electron Eng 1:239–245

23. Yadav HB, Yadav DK (2014) A multistage model for defect prediction of software development life cycle using fuzzy logic. In: Proceedings of the third international conference on soft computing for problem solving (SOCPROS-2013), 2013, IIT Roorkee, India, Advances in Intelligent Systems and Computing, vol 259, pp 661–671, Springer India Publication
24. Hong TP, Lee CY (1996) Induction of fuzzy rules and membership functions from training examples. Fuzzy Sets Syst 84:33–47
25. Hattori K, Tor Y (1993) Effective algorithm for the nearest neighbour method in the clustering problem. Pattern Recogn 26:741–746
26. Pfahringer B, Holmes G, Kirkby R, Orgun MA, Thornton J (2007) New options for Hoeffding trees. AI, pp 90–99
27. Witten I, Frank E (2005) Data mining: practical machine learning tools and techniques. Morgan Kaufmann, San Francisco
28. Catal C, Diri B (2008) A fault prediction model with limited fault data to improve test process. In: Proceedings of the 9th international conference on product focused software process improvement, LNCS 5089, pp 244–257
29. Pandey A, Goyal N (2010) Predicting fault prone software module using data mining and fuzzy logic, In the Special Issue of IJCCT 2010, vol 2 Issue 2, 3, 4

6T SRAM Cell Design and Investigation for Ultra-Low-Power Application

Priyanka Parihar, Neha Gupta, Vaibhav Neema and Praveen Singh

Abstract In this paper, multi-threshold voltage CMOS technique (MTCMOS) is investigated on static random access memory (SRAM) cell. Cell or/and pass transistors with high threshold and various aspect ratios (β) of transistor are presented. This work is focused on stability, power consumption, delay and write-trip point of SRAM cell, which are very important parameters to design any memory circuit. The read, write and hold stability of data is improved by 2.13×, 1.06× and 1.25×, respectively, with new MTCMOS SRAM cells. The read and write power is suppressed by 1.08× and 2.83×, respectively, and the read delay is suppressed by up to 1.96×, and write-trip point is enhanced by 1.03× with proposed MTCMOS SRAM cells compared with conventional 6T SRAM circuit in 70 nm with MTCMOS technique.

Keywords SRAM · Static noise margin · Write-trip point · MTCMOS

1 Introduction

SRAM is an essential building block for any VLSI system design. According to the recent trends, in VLSI system design at nanometre and beyond technology node, leakage power dissipation and noise margin are the challenging cell parameters for

P. Parihar (✉) · N. Gupta · V. Neema · P. Singh
Department of Electronics & Telecommunication Engineering,
IET—Devi Ahilya University, Indore 452017, India
e-mail: priyankapariharpriyanka@gmail.com

N. Gupta
e-mail: nehagupta121192@gmail.com

V. Neema
e-mail: vaibhav.neema@gmail.com

P. Singh
e-mail: psinghdavv@yahoo.com

V. Nath and J. K. Mandal (eds.), *Nanoelectronics, Circuits and Communication Systems*, Lecture Notes in Electrical Engineering 511,
https://doi.org/10.1007/978-981-13-0776-8_49

memory design engineers. Traditionally, 6T SRAM cell [1] is used for cache memory designing for VLSI systems.

In this paper, MTCMOS technology [2, 3] is used to reduce power consumption and delays and enhancing write, read and hold static noise margin (SNM) and write-trip point of the 6T SRAM circuit. The paper is organized in the following pattern. 6T SRAM cell with different threshold voltages is presented in Sect. 2. Simulation results based on stability, write-trip point, delay and power consumption are explored in Sect. 3. The conclusion of paper is shown in Sect. 4.

2 Conventional 6T SRAM Circuits with Different Threshold Voltages

The conventional 6T SRAM (Fig. 1) [4] cell is widely used in modern microprocessors. It has two access transistors, connected with bit lines, which are complementary to each other. It has two cross-coupled CMOS inverters. The pass transistors, word lines (WL) and bit lines (bl and blbar) are used for reading and writing data to and from the cell [5, 6] (Fig. 2).

2.1 Write Operation

For write operation in the cell, the data is given to the bit line, and the WL is high, so that the pass transistors get turned on. The bit line is kept at Vdd or 0 V that depends if a '1' or a '0' is written in the cell. After writing a '1', the output q goes high and qbar goes low. When writing '0' in the cell, q goes low and qbar goes high.

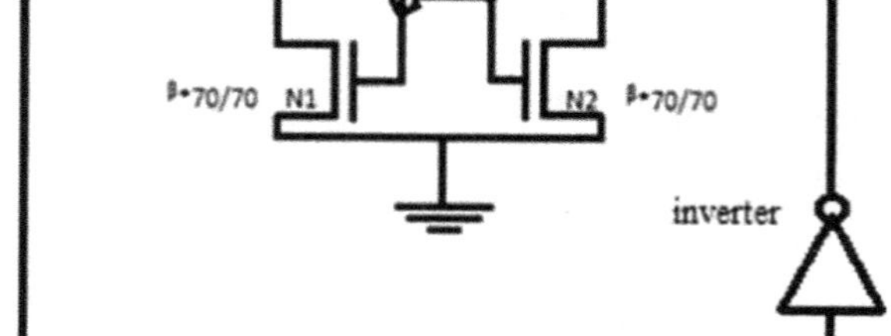

Fig. 1 6T SRAM cell. Transistor sizes are in nanometres using 70 nm CMOS technology. *wl* word line, *bl* bit line

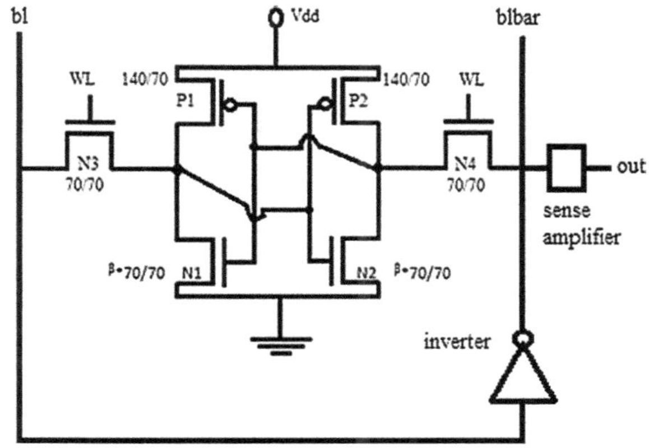

Fig. 2 Conventional 6T SRAM cell circuit for read operation. *Out* output port of sense amplifier

2.2 *Read Operation*

To store the data in memory, the WL is low, which makes the transistors off. After the data is stored, it is read by again turning ON the pass transistors. In read operation, bl is '0' so that blbar becomes high which is connected to bl through an inverter. The data stored in qbar is read through the sense amplifier output. blbar is recharged to Vdd to again read the data.

The blbar voltage gets transferred to the sense amplifier. The schematic of the sense amplifier has been shown in Fig. 3.

The conventional 6T SRAM circuits with different values of threshold voltages have been shown in Fig. 4. In this paper, low threshold voltage (LVT) SRAM circuits (Fig. 4a) are explored for applications of high performance. High threshold voltage (HVT) SRAM cells (Fig. 4b) are explored for applications of low power. Dual threshold voltage (DVT) SRAM circuits (Fig. 4c) are characterized in this paper.

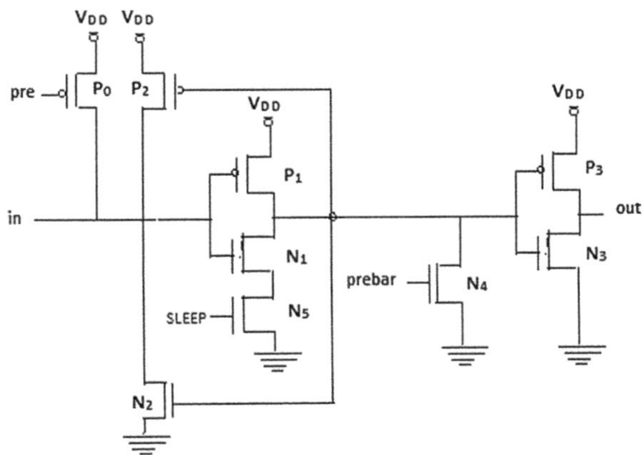

Fig. 3 Single-ended sense amplifier [5]

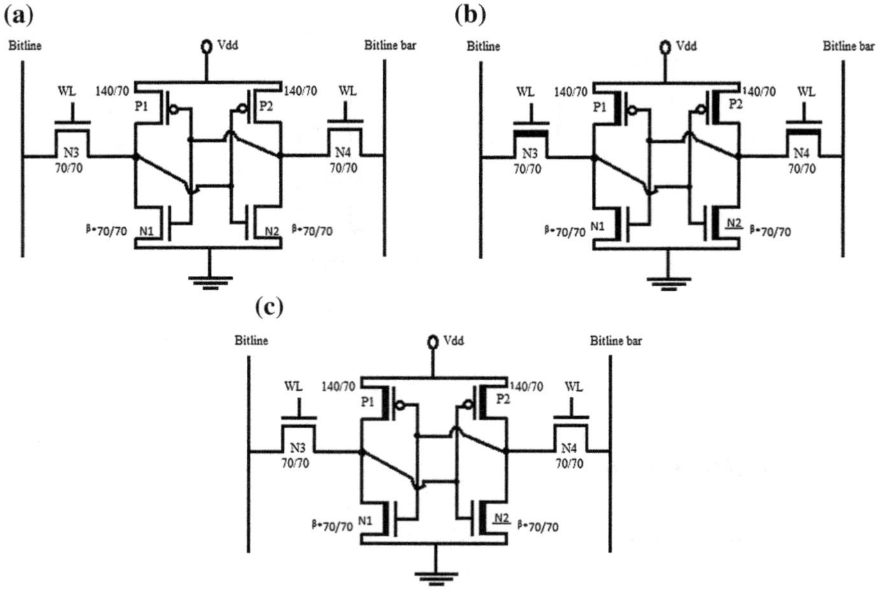

Fig. 4 6T SRAM circuit with different values of threshold voltage

6T SRAM [4] cells with larger W/L ratio ($\beta > 1$) are characterized for higher stability compared with low size SRAM cells. In these circuits, different parameters are evaluated with different β values ($\beta = 1, 2, 3, 5$) and with different threshold voltage techniques. Simulations are done in TSPICE using 70 nm technology node.

3 Simulation and Analysis

Four important parameters are taken for analysis: power dissipation, delay, write-trip point and stability. These four parameters on three different threshold voltages [4] are optimized.

3.1 Data Stability

Stability of data is the most important parameter in any memory system. The read, hold and write SNM [7, 8] of SRAM are explored here. The SNMs are calculated in Origin software by using butterfly curve method. The DC transfer characteristics of CMOS inverters are used to construct a butterfly diagram which is shown in Fig. 5.

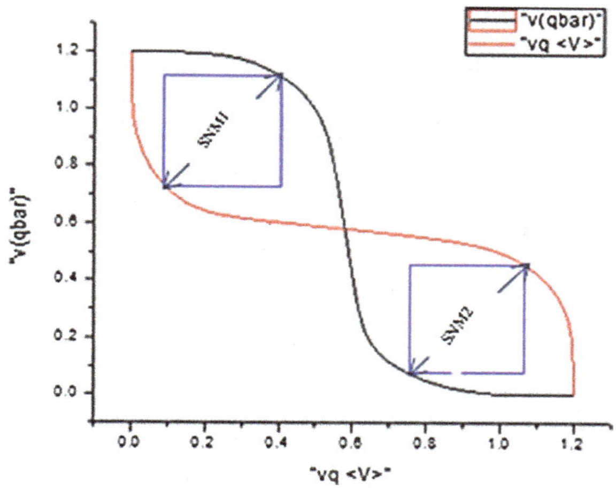

Fig. 5 Butterfly diagram for SRAM cell stability [5]

$$\text{SNM} = \min(\text{SNM1}, \text{SNM2})$$

(i) ***Read SNM***

Read SNM [9] is maximum DC noise voltage that is tolerated by SRAM during read operation without data loss. While reading the data, it is accessed from the bit lines. The stability is therefore reduced in 6T SRAM cells. So, the data stability is enhanced by using MTCMOS technology. From Fig. 6, 6T SRAM circuits with dual threshold and high threshold voltage have high read stability.

(ii) ***Write voltage margin***

The write voltage margin is deviation of bit line voltage from Vdd or 0 V by which transfer of '1' or '0' is done into the cell, respectively. From Fig. 6, 6T SRAM cells with dual threshold and high threshold voltage have high write stability.

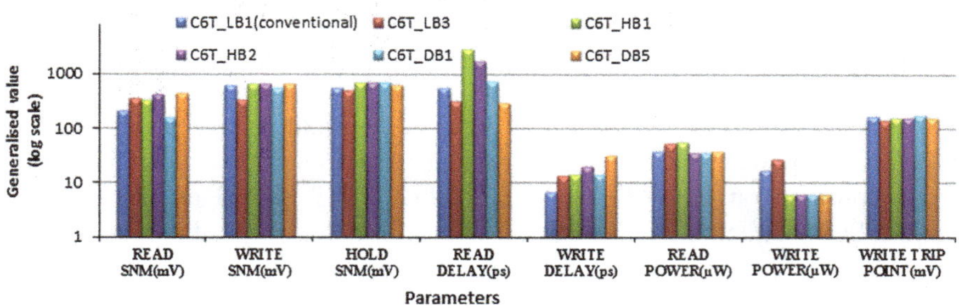

Fig. 6 Performance comparison of different SRAM cells on logarithmic scale. Conventional 6T SRAM cell parameters are with dark blue colour in the above figure

(iii) ***Hold SNM***

Hold SNM is that maximum DC noise voltage that an SRAM circuit can tolerate in idle state without data loss. From Fig. 6, 6T SRAM circuits with dual threshold and high threshold voltage have high hold stability.

3.2 Delay

(i) ***Read delay***

The read delay is the time between 50% word line (WL) value when it goes from low to high and 50% sense amplifier output when it goes from high to low and low to high transitions. From Fig. 6, 6T SRAM cell with dual threshold (β = 5) and low threshold (β = 3) voltage has reduced read delay.

(ii) ***Write delay***

The write delay is the time between 50% 'q' node low to high transition and 50% 'qbar' node high to low transition. From Fig. 6, conventional 6T SRAM cell with dual threshold (β = 1), high threshold (β = 1) and low threshold (β = 1) voltage has reduced write delay.

3.3 Power Consumption

The write and read power consumption is described in this section. Power consumption is very important which should be kept in mind while designing a memory cell [10, 11].

(i) ***Write power***

The write power consumption with different threshold voltage values has been shown in Fig. 6. It is the power which an SRAM circuit consumes during its write operation. From Fig. 6, 6T SRAM circuits with dual threshold (β = 1 and β = 5), high threshold (β = 1 and β = 2) voltage have reduced write power.

(ii) ***Read power***

The read power of SRAM cell with different threshold voltage values is shown in Fig. 6. The power which an SRAM circuit consumes in its read operation is read power consumption of SRAM cell. From Fig. 6, 6T SRAM circuit with dual threshold (β = 1) and high threshold (β = 2) voltage has reduced read power.

3.4 *Write-Trip Point*

The write-trip point [12] is an important parameter of SRAM. It is the DC voltage value in bit line which when transits from low to high, the voltage at 'qbar' node trip its value going from high to low transition. The write-trip point is important to consider during data write operation in the SRAM cell. It should be high.

From Fig. 6, conventional 6T SRAM cells with dual threshold ($\beta = 1$), high threshold ($\beta = 1$) and low threshold ($\beta = 1$) voltage have high write-trip point values.

4 Conclusion

The MTCMOS technique is used in this paper. The delay and power consumed by the SRAM get reduced in MTCMOS technique. In dual threshold voltage SRAM both write and read delay are comparatively less than the novel 6T SRAM cell. The write and read power consumption are suppressed by 2.83× and 1.08×, respectively, with

Table 1 Simulation results for conventional 6T and MTCMOS SRAM cells on different values of threshold voltage and β ratio

Parameters	6T_LB1 (conventional) [3]	6T_LB3 (proposed)	6T_HB1 (proposed)	6T_HB2 (proposed)	6T_DB1 (proposed)	6T_DB5 (proposed)
Read SNM (mV)	212.1	353	332.3	424.3	155.54	452.5
Write SNM (mV)	608	325	650	650.44	551.46	650.44
Hold SNM (mV)	565	494.9	692.8	678.7	707	636.3
Read delay (pS)	570	306	2950	1691	726	290
Write delay (pS)	6.94	13.7	14.5	19.9	14.2	32.7
Read power (μW)	38.8	52.9	56.1	35.6	35.5	38.1
Write power (nW)	17.3	27.5	6.11	6.12	6.11	6.14
write-trip point (mV)	168.28	142.9	160.54	155.48	173.66	153.83

MTCMOS SRAM cell with dual and high threshold voltage values. The read, write and hold data stability are increased by 2.13×, 1.06× and 1.25×, respectively, with MTCMOS technology (Table 1).

Acknowledgements We would like to express our regards to M.P. Council of Science & Technology, Bhopal, India, for the finical support under R&D project scheme (No. 1950/CST/R&D/Phy&EnggSc/2015: 27th Aug 2015.0).

References

1. Giraud B, Amara A, Thomas O (2010) An innovative 6T hybrid SRAM cell in sub-32 nm double-gate MOS technology. In: Fifth IEEE international symposium on electronic design, test and application
2. Kursan V, Friedman EG (2006) Multi-voltage CMOS circuit design. Wiley Ltd., ISBN 0-470-01023-1
3. Liu Z (2008) Multi-voltage nanoscale CMOS circuit techniques. Ph. D. thesis, University of Wisconsin, Madison
4. Jiao H et al (2016) Low power and robust memory circuits with asymmetrical ground gating. Microelectron J 48:109–119
5. Azam T (2010) Robust asymmetric 6T-SRAM cell for low-power operation in nano-CMOS technologies. Electron Lett 46(4):273–274
6. Harnzaoglu F, Wang Y, Kolar P, Wei L, Ng Y, Bhattacharya U, Zhang K (2011) Bit cell optimizations and circuit techniques for nanoscale SRAM design. IEEE Des Test Comput 28 (1):22–31
7. Seevinck E, List FJ, Lohstroh J (1987) Static-noise margin analysis of MOS SRAM cells. IEEE J Solid-State Circuits SC-22(5)
8. Oniciuc L, Andrei P (2008) Random doping fluctuation effects on static noise margins of 6TSRAM cells. In: Solid-state and integrated circuit technology, pp 488–491
9. Majumdar B (2011) Low power single bitline 6T SRAM cell with high read stability. In: 2011 International conference on recent trends in information systems, pp 169–174
10. Liu D, Svenson C (1994) Power consumption estimation in CMOS VLSI Chips. IEEE J Solid State Circuits 29(6):663–670
11. Tawfik SA (2009) Digital circuit methodologies for low power and robust nanoscale integration. Ph. D. thesis, University of Wisconsin, Madison
12. Hobson RF (2007)A new single-ended SRAM cell with write-assist. IEEE Trans Very Large Scale Integr (VLSI) Syst 15(2)

Design of Robot Monitoring System for Aviation

Rohit Mohan, Akash Keneth Suraj, Sakshi Agarawal, Sananaya Majumdar and Vijay Nath

Abstract A significant share of the cost of owning an aircraft is comprised of the cost incurred on its maintenance. The paper purports to provide an autonomous framework for the maintenance cost reduction. The suggested framework is built upon Robot Operating System (ROS), implementing navigation, semantic mapping and manipulator motion planning to facilitate daily check tasks. In this framework, a robot mounted with a robotic arm having end effector suitable for performing daily check tasks is capable of doing visual inspection of the aircraft, performing maintenance tasks and can change its end effector depending on the tasks needed to be performed. Gazebo was used for simulation purpose.

1 Introduction

It is widely accepted, when an aircraft is owned either by an individual or commercial airline company, maintenance of aircraft plays a crucial part. In case of individual owners even if the aircraft is rarely put to use, its maintenance can increase its lifespan. On the other hand, commercial aircraft have daily flights with passengers, and maintenance then becomes an utmost important task to be undertaken to avoid any mishaps. Consequently, every airline company has to invest in

R. Mohan (✉) · A. K. Suraj · S. Agarawal · S. Majumdar · V. Nath
Department of Electronics & Communication Engineering, Birla Institute of Technology, Mesra, Ranchi 835215, Jharkhand, India
e-mail: rohitmohan1914@gmail.com

A. K. Suraj
e-mail: keneth.akash@gmail.com

S. Agarawal
e-mail: sakshi.agrawal1221@gmail.com

S. Majumdar
e-mail: shana.intouch@gmail.com

V. Nath
e-mail: vijaynath@bitmesra.ac.in

V. Nath and J. K. Mandal (eds.), *Nanoelectronics, Circuits and Communication Systems*, Lecture Notes in Electrical Engineering 511,
https://doi.org/10.1007/978-981-13-0776-8_50

different tiers of maintenance checks laid down by the Federal Aviation Administration. To cut the cost incurred, one aspect that has been vastly explored is the optimization of scheduling of maintenance. The authors of [1–5] take different approaches to resolve this particular problem. Our paper proposes how we can bring down the cost further once optimized scheduling of maintenance and assignment of personnel is employed. This paper focuses on only 'Type A Check', one of many checks mandated by FAA as listed out in [1]. Type A check is conducted roughly every 65 flight-hours or about once a week. It involves inspection of all major systems such as landing gear, engines and control surfaces. Building upon the idea hinted in [6], automation of an operation reduces its cost; we propose an automation framework for Type A Check. We selected this type because of its frequent employment. This paper is organized into five parts: 1. Introduction, 2. Methodology, 3. Result and Discussion, 4. Acknowledgement and 5. References. This paper assumes the environment in which our autonomous robot navigates to carry out maintenance tasks comprises social systems or paths which give access to all parts of aircraft in need of inspection and repair. It also assumes that the tools used to carry out the inspection and repairs are robot operable.

2 Methodology

The proposed framework consists of an autonomous robot capable of navigating through dynamic environments, building both metric and topological map and object detection. It consists of a 7-DOF arm with variable end effector capable of autonomous manipulation. The end effector is the tools that are needed for inspection and repair works. Robot operating system is used to implement all the autonomous functions. This section is divided into different aspects that make the autonomous framework to give in-depth details for each.

A. Motion Planning

Given a map of the surroundings in which the robot functions, motion planning algorithms plan out trajectories from the initial position of the robot to its desired final position, avoiding all static obstacles. There are II types of planners used that are global planner and local planner. The global plan lays out the course from initial to final while avoiding all static obstacles marked on the map. In this paper, Dijkstra's algorithm [7] has been used as global planner. The local planner, on the other hand, access only a small area of the map centred around the robot (local to the robot), updates the map with current sensor information, plans out trajectory which is close to the global planned course at the same time avoids any dynamic and static obstacles. It generates the velocity and angular velocity for the robot to travel along. This paper uses [8] for local planner implementation along with [9, 10] for efficient computation. Time elastic band [8] takes into account both time and space constraints for trajectory planning, proposing the fastest route for the robot to travel along.

B. Mapping

Mapping is a process to build illustrations of an environment. This paper utilizes two types of map: metric map and topological map (Figs. 1 and 2).

(1) Metric Map: A metric map can be 2D or 3D representation of an environment. The most common 2D metric map used is occupancy grid map which can be defined as a pair $M = (L, S)$, where $L = fl_1, \ldots, l_ng$ is a list of cell locations $l_i = (x, y)$, and $S = fs_1, \ldots, s_ng$ is a list of values s_i that represent if a cell is occupied, free or unknown. While mapping, several issues have to be dealt with such as accumulation of error in odometry data, loop closing, etc. as mentioned in [11]. Simultaneous Localization and Mapping (SLAM) is a technique for creating maps. This paper uses gmapping SLAM implementation in ROS which uses Rao-Blackwellized particle filters [12] (Fig. 3).

(2) Topological Map: A topological map represents the given environment discretely. It is typically defined by a pair $T = (N, E)$ where $N = fn_1, \ldots, n_ng$ is a set of nodes which describe discrete places, and $E = fe_1, \ldots, e_mg$ Ł N N is a set of edges, where each edge connects two nodes. There are many ways of creating such maps, one of them being by using Voronoi diagram. In this paper, we build topological map on top of a 2D metric map. The topological map gives semantic information and is used for global planning whereas 2D metric map gives local information and is used for local planning. This paper uses semantic map [13] package in ROS (Fig. 4).

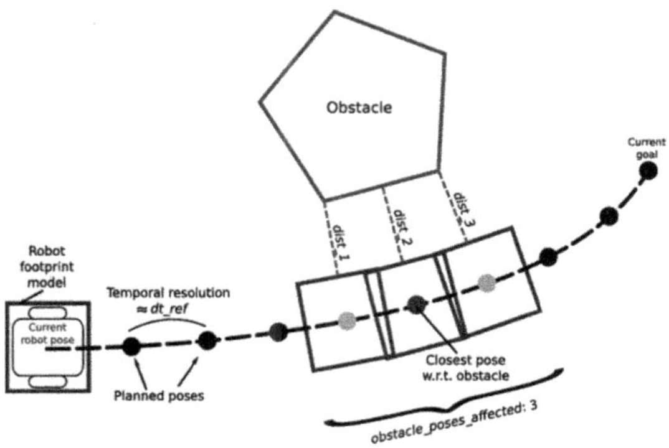

Fig. 1 Obstacle avoidance (local planner)

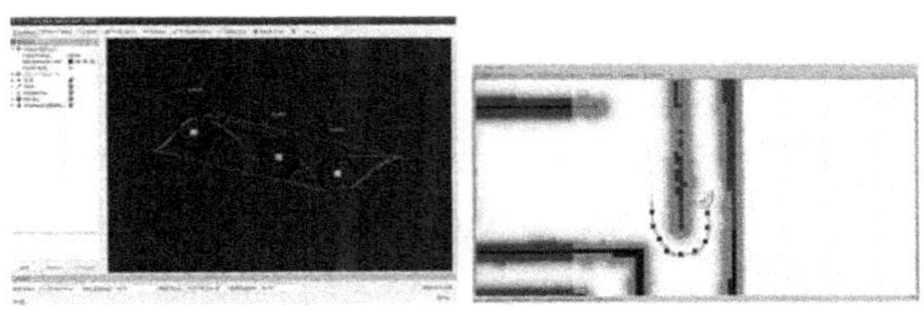

Fig. 2 Left: multiple trajectory, and right: differential drive motion planning

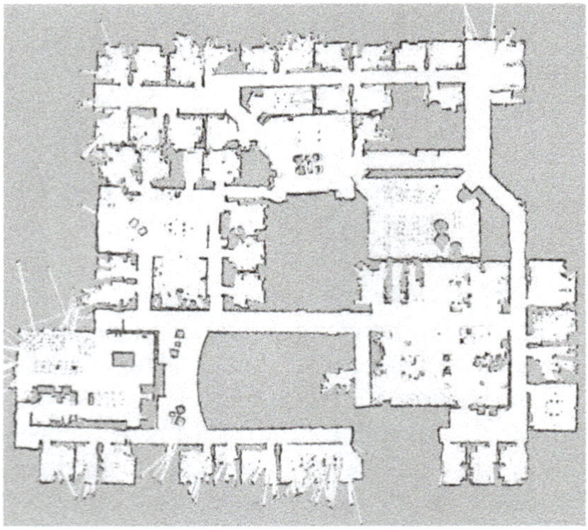

Fig. 3 Occupancy grid map of Willow Garage world. White pixel represents free space, black pixel represents occupied space (obstacles) and grey pixel represents unknown

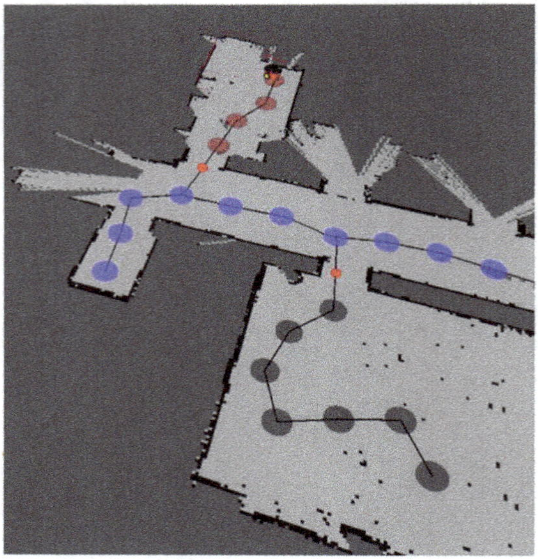

Fig. 4 Topological map built on top of a 2D metric map

C. Object Detection

Object detection consists of solution to two problems: recognition and localization. Given an image of an environment, an object of interest has to be recognized, i.e. classify whether the image contains the desired object. If the object is present, then a bounding box is drawn around the object, localizing the object in the given image. A number of algorithms exist for object detection, and we discuss three of them briefly.

(1) Histogram of Oriented Gradients (HOG): Histogram of Oriented Gradients (HOG) is feature descriptor. They are based on edge orientation and local intensity gradients. They are invariant to translation and rotation. To acquire these descriptors, a picture is partitioned into little districts (cells) and the edge orientation is collected. The HOG descriptor is made out of all the cell histograms after a standardization procedure [14].
(2) Speeded Up Robust Features (SURF): Speeded Up Robust Features (SURF) [15] is an interest point detector and descriptor. It is invariant to image scaling and rotation. It depends upon Hessian matrix determinant for its real-time performance and high accuracy.
(3) RoboEarth: RoboEarth [16] is a ROS stack that provides algorithms for object detection and object modelling. It is divided into three main parts: object scanning, object detection and the RoboEarth database (Figs. 5, 6, 7 and 8).

Fig. 5 HOG feature representation of a given input image

Fig. 6 SURF features

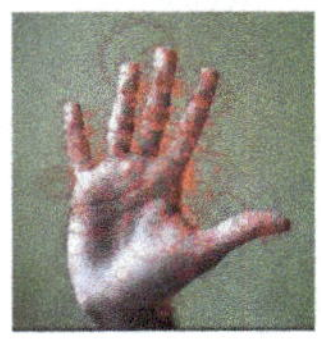

Fig. 7 Box in the marker template

Fig. 8 3D model

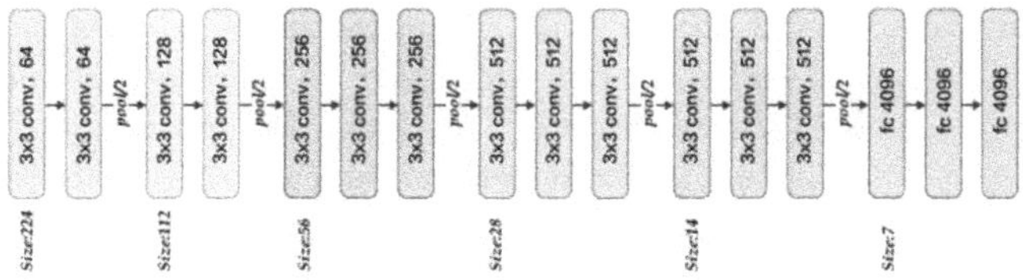

Fig. 9 VGGNet-16 architecture

There are various advantages and disadvantages of each object detection method mentioned above. In this paper instead of taking traditional object detection methods, we departed for the deep learning approach. We train a deep convolutional network (DCN) on aircraft mechanical parts dataset [17]. Since training a DCN from scratch is not an easy feat as it requires large dataset, high GPU configuration and significant quantity of time for training, we opted for a more subtle method: transfer learning. Transfer learning or inductive transfer is a research problem in machine learning that concentrates on putting away learning picked up while taking care of one issue and applying it to an alternate yet related issue. For example, knowledge gained while learning to recognize cars could apply when attempting to recognize trucks. VGGNet-16 [18] is a DCN trained on ImageNet dataset. Using [19, 20] as our guide for fine tuning VGGNet-16, we obtained an aircraft parts detector. This detector was used for building semantic maps.

D. Robot Operating System (ROS)

Robot Operating System (ROS) [21] will be a mechanical technology middleware (i.e. gathering from claiming product frameworks to robot product development). As we aware that ROS is not an operating system, but it gives administrations planned for heterogeneous PC group, for example, fittings abstraction, low-level gadget control, execution from claiming regularly utilized functionality, message-passing the middle of courses which is more economy bundle oversaw. Running sets from claiming ROS-based forms are known as a chart structural engineering. It prepare the place that put in hubs that might receive, post what's more multiplex sensor, control, state, planning and actuator, and also different messages. The suggested framework has been fabricated utilizing ROS because of its effortlessness for usage and more proficient building design (Fig. 9).

E. Tensorflow

TensorFlow [22] is an open source programming library for numerical calculation utilizing information stream diagrams. Hubs in the diagram speak to scientific operations, while the chart edges speak to the multidimensional information clusters (tensors) imparted between them. The adaptable engineering allows you to send calculation to at least one CPUs or GPUs in a desktop, server or cell phone with a solitary API. TensorFlow is utilized for machine learning over a scope of errands. It is an emblematic math library, and furthermore utilized as a framework for building and preparing neural systems to identify and translate examples and connections, similar to human learning and thinking. It was initially created by analysts and specialists dealing with the Google Brain Team inside Google Machine Intelligence explore association for the reasons for directing machine learning and profound neural systems look into, however, the framework is sufficiently general to be appropriate in a wide assortment of different areas too. Tensorflow system was utilized to tweak pre-prepared model (VGGNet-16) on the dataset of [17] to create airplane mechanical parts locator.

F. Robot Manipulators

Robot controllers are made from an arrangement of connection and joint blends. The connections are the unbending individuals associating the joints, or tomahawks. The tomahawks are the portable parts of the mechanical controller that reason relative movement between bordering joins. The mechanical joints used to develop the automated arm controller comprise five main sorts. Two of the joints are direct, in which the relative movement between neighbouring connections is non-rotational, and three are revolving sorts, in which the relative movement includes revolution between joins. The ROBOTIS-MANIPULATOR which is a moderate multi-reason controller intended for research and computerization was utilized as a part of this paper with variable end effector. It depends on the h-arrangement dynamixel genius.

G. Gazebo

Gazebo [23] is a free robot simulation platform for testing new algorithms, new robot's architecture, simulating real-world scenarios which might be quite expensive to realize in real world or very dangerous. This tool allows accurate and efficient simulation of the robot's interaction with the environment in which it is supposed to function. The environment can be both indoor or outdoor. This all is possible due to its robust physics engine, a good graphical user interface, excellent quality graphics (Figs. 10, 11 and 12).

3 Results and Discussion

Algorithm [13] for semantic mapping was implemented. For localization of the robot in indoor environments as this paper assumed that the robot will carry out its function of maintenance of aircraft, in a hangar, we employ robot localization in indoor environments [24] using FastSLAM [25] (called gmapping in ROS) and Particle filters [26]. The framework of our robot is closely related to KUKA Youbot, compatible with ROS. The KUKA arm was swapped for the ROBOTIS-MANIPULATOR whose manipulation capability were implementated based on [27] . A Microsoft Kinect (RGBD sensor) [28] is mounted on the robotic arm along with a laser rangefinder on robot's base [29]. This covers the robot's part of simulation. For the environment

Fig. 10 ROBOTIS-MANIPULATOR, robotic manipulator used in this paper

Fig. 11 Different types of end effectors

Fig. 12 Youbot specifications

kinematics	serial chain, 5 revolute axes
height	655 mm
work envelope	$0.513\ m^3$
weight	6.3 kg
payload	0.5 kg
structure	magnesium cast
communication	EtherCAT
power supply	24 V DC
power limitation	80 W

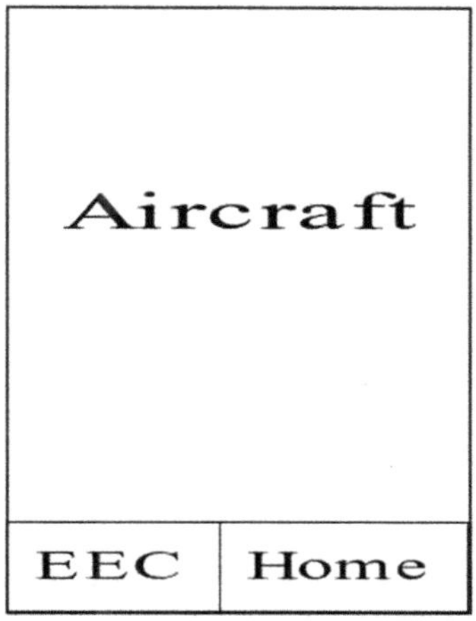

Fig. 13 World area 1: Aircraft area, where all the maintenance tasks are carried out; 2: EEC area (end effector change), where the robot can change its end effectors depending on the task it is going to perform; 3: home, where the robot will reside when not in use

aspect, different tasks as listed by FAA were simulated. Both the simulations were combined to get a simulation world where the robot had to navigate and carry out the different tasks. The world had three main areas: aircraft, home and end effector change area. The explanation of each area is straightforward and is mentioned in Fig. 13. First, we discuss conceptual mapping and reasoning, necessary for semantic mapping, and then we will discuss our robot's operation flow followed by the experimentations performed by us.

A. Conceptual Mapping and Interpretation

The following mapping and interpretation part consolidates derived data from the robot's surrounding with natural human knowledge into a probabilistic chain diagram. Its point is to incorporate the spatial information portrayal in a probabilistic structure. Each area found by the topological mapping calculation is expressed by a random variable for area classes. They are associated with each other after the topological structure of the robot's surrounding. The learning about the topology of the robot's surrounding and the identified objects are obtained from the saved documents. The potential capacities express how certain classes of areas are commonly associated; for instance, that a restroom will probably be associated with a passageway than to another lavatory or that workplaces are generally associated with hallways. The nearness of articles is additionally expressed by random factors specifically associated with the area where they have been identified. We dole out to these variable factors a likelihood of 0.9 that the variable has been recognized accurately and 0.1 that it is a false positive discovery. POI portrays information about the likelihood of finding certain items in a specific area (e.g. will probably discover an end effector in the EEC area than in the home area or that crash survival memory unit are generally be in the aircraft area).

B. Experiment

Object detection model training was carried out as mentioned previously, i.e. using VGGnet-16 and transfer learning concept [17]. The dataset consists of images of 20 different parts of an aircraft. With the help of data augmentation [30], 50,000 balanced image dataset was prepared; each aircraft part has 2,500 images. Training was done for a million iteration using Adam optimizer (lr = 0.00001).

Object detection approach may also be used for detecting different types of end effectors in EEC area but we chose to follow a more simple approach. Another arm was installed at the EEC base station which was responsible for changing of end effectors of our Robot's arm. Each end effector was given a proper position defined by 3D coordinates with respect to the additional arm or the fixed arm. The robot needs to park itself in a specific zone (in EEC area) with its robotic arm in a particular configuration, again with respect to the fixed arm. The manipulation trajectory from each end effector position to constant robot's arm configuration is recorded and hard coding is done for changing end effectors.

In this experiment, it is assumed that all parts to be maintained are on the same plane for simplicity. This can be easily extended to a non-planar configuration with suitable mediums for the particular parts access by the robot. All simulations were done using Gazebo simulator.

Once the plane is parked in the hangar, the robot is teleoperated through the hanger; it builds the map of the area with semantics wherever needed (aircraft parts). It is also assumed that the plane is always parked in the same state (slight changes can be easily accounted). Whenever maintenance procedures need to be carried out, a human personnel signals the robot remotely. The robot navigates from home area to aircraft area first. There, it analyses different parts of the aircraft through image analysis [17] and decides which parts need to be replaced or whether lubrication has to be done. In case of part replacement, it sends a report to human personnel, he verifies it and gives a green signal if the robot's assessment is correct. All the new parts of aircrafts are assumed to be placed in EEC area, and manipulation requirement for each is stored in robot's database. The robot then plans out the order in which each faulty part will be attended to. It also tabulates the requirements of end effectors and parts for replacement and carries out all the needed steps. Once all the tasks are performed, the robot moves back to its home area (Figs. 14 and 15, Table 1).

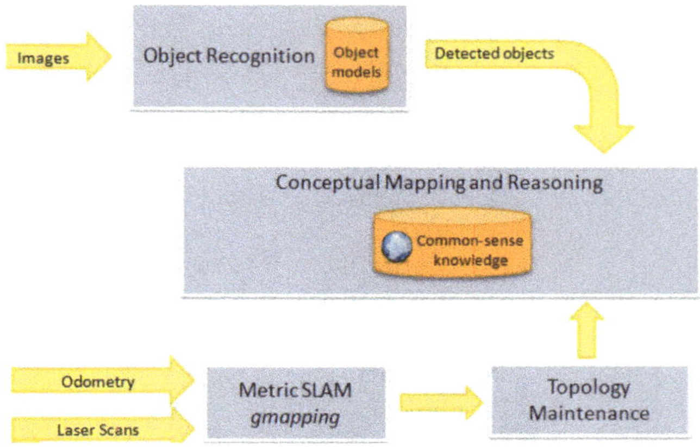

Fig. 14 Robot operation flow

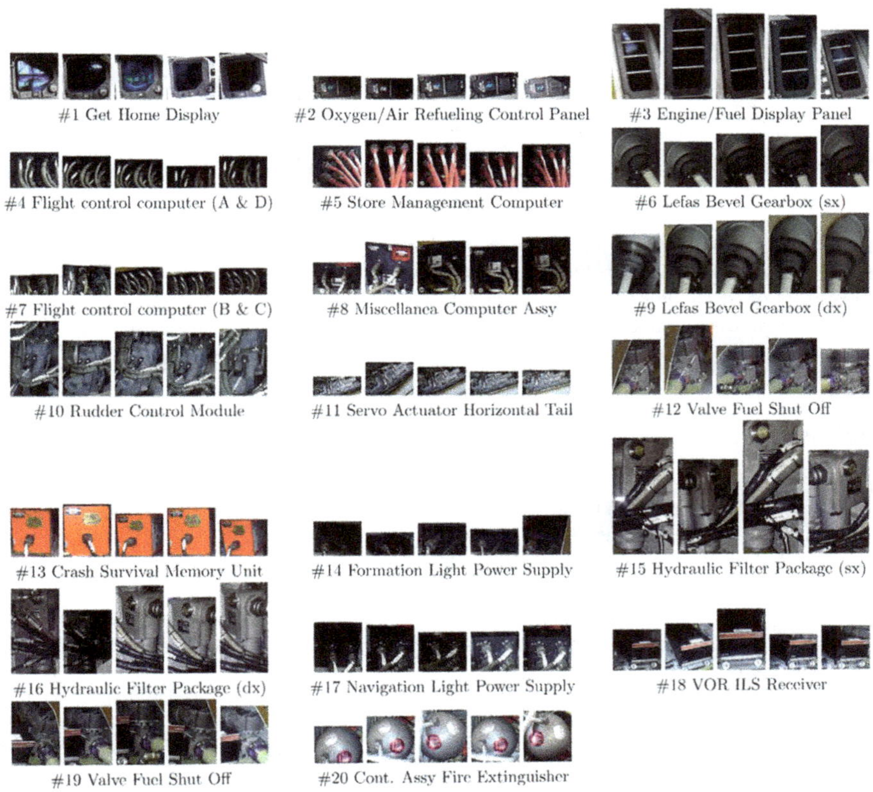

Fig. 15 Images of different aircraft parts (taken from [17])

Table 1 Accuracy of object detector

Part's name	True positive	False positive
Get home display	100	0
Oxygen/air refuelling CP	99.8	0.2
Engine/fuel DP	97	3
Flight control	100	0
Store management	100	0
Lefa bevel	98	2
Flight control (II)	98.4	1.6
Miscellanea computer	100	0
Flight control	99.2	0.8
Rudder control	97.5	2.5
Servo actuator	100	0
Valve fuel	100	0
Crash survival MU	100	0
Formation light	100	0
Hydraulic filter (sx)	99	1
Hydraulic filter (dx)	98.7	1.3
Navigation light	100	0
VOR	100	0
Valve fuel (II)	100	0
Assy fire	100	0

Acknowledgements We would wish to express our gratitude to Vice-Chancellor of our college, Birla Institute of Technology, Mesra, Dr. M.K Mishra. We would also like to thank Head of Department of Electronics and Communication, BIT Mesra, Dr. Srikanta Pal for his patience and valuable help. We would also like to thank Indigo Airlines personnel for their useful insights.

References

1. Sriram C, Haghani A (2003) An optimization model for aircraft maintenance scheduling and re-assignment. Transp Res Part A: Policy Prac 37(1):29–48
2. Wu H, Liu Y, Ding Y, Liu J (2004) Methods to reduce direct maintenance costs for commercial aircraft. Aircr Eng Aerosp Technol 76(1):15–18
3. Samaranayake P, Lewis GS, Woxvold ERA, Toncich D (2002) Development of engineering structures for scheduling and control of aircraft maintenance. Int J Oper Prod Manage 22 (8):843–867
4. Badere M, Bilge U (2014) Operational aircraft mainte-nance routing problem with remaining time consideration. Eur J Oper Res 235(1):315–328
5. Liang Z, Feng Y, Zhang X, Wu T, Chaovalitwongse WA (2015) Robust weekly aircraft maintenance routing problem and the extension to the tail assignment problem. Transp Res Part B: Methodol 78(null):238–259
6. Garg A, Deshmukh SG (2006) Maintenance management: literature review and directions. J Qual Maint Eng 12(3):205–238

7. Skiena S (1990) "Dijkstras algorithm." Implementing discrete mathematics: combinatorics and graph theory with mathematica. Addison-Wesley, Reading, MA, pp 225–227
8. Rsmann C et al (2012) Trajectory modification considering dynamic constraints of autonomous robots. In: Robotics; Proceedings of ROBOTIK 2012; 7th German conference on. VDE
9. Rosmann C et al (2013) Efficient trajectory optimization using a sparse model. In: Mobile robots (ECMR), 2013 European conference on. IEEE
10. Rsmann C, Frank H, Bertram T (2015) Planning of multiple robot trajectories in distinctive topologies. Mobile robots (ECMR), 2015 European conference on. IEEE
11. Kohlbrecher S et al (2011) A flexible and scalable slam system with full 3d motion estimation. In: Safety, security, and rescue robotics (SSRR), 2011 IEEE international symposium on. IEEE
12. Grisetti G, Stachniss C, Burgard W (2007) Improved techniques for grid mapping with rao blackwellized particle filters. IEEE Trans Rob 23(1):34–46
13. Pronobis A, Jensfelt P (2012) Large-scale semantic mapping and reasoning with heterogeneous modalities. In: Robotics and automation (ICRA), 2012 IEEE international conference on. IEEE
14. Dalal N, Triggs B (2005) Histograms of oriented gradients for human detection. In: Computer vision and pattern recognition, 2005. CVPR 2005. IEEE computer society conference on. vol 1. IEEE
15. Bay H, Tuytelaars T, Van Gool L (2006) Surf: Speeded up robust features. In: Computer vision ECCV 2006, pp 404–417
16. Waibel M et al (2011) Roboearth. IEEE Robot Autom Mag 18(2):69–82
17. Cusano C, Napoletano P (2017) Visual recognition of aircraft mechanical parts for smart maintenance. Comput Ind 86(null):26–33
18. Simonyan K, Andrew Z (2014) Very deep convolution networks for large-scale image recognition. arXiv preprint arXiv:1409.1556
19. Razavian AS et al (2014) CNN features off-the-shelf: an astound-ing baseline for recognition. In: Proceedings of the IEEE conference on computer vision and pattern recognition workshops
20. Azizpour H et al (2015) From generic to specific deep representa-tions for visual recognition. In: Proceedings of the IEEE Conference on computer vision and pattern recognition workshops
21. Quigley M et al (2009) ROS: an open-source robot operating system. In: ICRA workshop on open source software. vol 3. No. 3.2
22. Abadi M et al (2016) Tensorflow: large-scale machine learn-ing on heterogeneous distributed systems. arXiv preprint arXiv:1603.04467
23. Koenig N, Howard A (2004) Design and use paradigms for gazebo, an open-source multi-robot simulator. In: Intelligent robots and systems, 2004. (IROS 2004). Proceedings. 2004 IEEE/RSJ international conference on. vol 3. IEEE
24. Fox D (2001) Particle filters for mobile robot localization. Sequential Monte Carlo methods in practice. Springer, New York, pp 401–428
25. Montemerlo M, Thrun S (2007) FastSLAM: A scalable method for the simultaneous localization and mapping problem in robotics, vol. 27. Springer
26. Jensfelt P (2001) Approaches to mobile robot localization in indoor environments. Dissertation KTH
27. Bagnell JA et al (2012) An integrated system for autonomous robotics manipulation. In: Intelligent robots and systems (IROS), 2012 IEEE/RSJ international conference on. IEEE
28. Zhang Z (2012) Microsoft kinect sensor and its effect. IEEE Multimedia 19(2):4–10
29. Shyy YH (1993) Laser range finder." U.S. Patent No. 5,262,837. 16 Nov 1993
30. Frhwirth-Schnatter S (1994) Data augmentation and dynamic linear models. J Time Ser Anal 15(2):183–202

Electronic Toll Collection System Using Barcode Technology

E. V. V. Hari Charan, Indrajit Pal, Akash Sinha, Raj Kamal Roye Baro and Vijay Nath

Abstract This paper emphasizes the barcode technology for automatic Electronic Toll Collection (ETC) systems, in order to avoid the ever-increasing stream of traffic and the long queues at the tollbooths of the highways. The proposed techniques make use of digital image processing techniques to scan the barcode and to match it with the existing database, by utilizing the decoded data. The implemented hardware setup has been illustrated and discussed in detail. The proposed technique for barcode detection significantly improves the speed, efficiency and lowers the cost of implementation. The method has been implemented by using Python and OpenCV.

Keywords Barcode · QR code · RFID tag · ETC · Tollbooth

1 Introduction

The Electronic Toll Collection (ETC) system was developed in order to overcome the problems associated with the manual collection of the toll. The manual toll deduction system has been replaced by the automated system, in order to avoid problems of long queues and congested traffic. The automated ETC system allows

E. V. V. Hari Charan (✉) · I. Pal · A. Sinha · R. K. R. Baro · V. Nath
Department of Electronics & Communication Engineering,
Birla Institute of Technology, Mesra, Ranchi 835215, Jharkhand, India
e-mail: hc17760@gmail.com

I. Pal
e-mail: pal.indrajit99@gmail.com

A. Sinha
e-mail: akashshubham95@gmail.com

R. K. R. Baro
e-mail: rkrb3112@gmail.com

V. Nath
e-mail: vijaynath@bitmesra.ac.in

V. Nath and J. K. Mandal (eds.), *Nanoelectronics, Circuits and Communication Systems*, Lecture Notes in Electrical Engineering 511,
https://doi.org/10.1007/978-981-13-0776-8_51

Fig. 1 Barcode

the user to avoid the wait for payment of cash or for collection of token in order to cross the toll. The automatic system collects toll electronically by using RFID, barcode technology, laser technology, etc. More efficient than the RIFD and the laser, the barcode technology is discussed in detail. A parallel black bars and white spaces series of varying widths constitutes a barcode. The elements of a barcode are the spaces and bars. Different characters are represented by the different combinations of spaces and bars, such as numbers or letters. The automated system works faster and more efficiently. It is capable of determining the registration status of a car and informing authorities of violation in payment of toll and informing owner via email or SMS of the location and the deduced toll for a vehicle. The objective of this technology is to prevent tollbooths from becoming congestion by a vehicle queue. It helps the government to maintain a proper and an updated record of the collection of toll. The front number plate of a vehicle can also have a tag installed on it.

The paper attempts to address certain developments required in order to take into account the drawbacks of the existing ETC system. The paper proposes a new developed embedded system, which can effectively detect the barcode, and also recognize and transfer the necessary encoded data efficiently (Fig. 1).

The paper organization is as follows. A survey of the existing work is provided in Sect. 2. Section 3 explains the simulation setup for the embedded system, the algorithm utilized, and offers a discussion of the results obtained. Finally, Sect. 4 concludes the paper with a brief overview of the findings of the paper.

2 Survey of Existing Work

Authors in [1] have emphasized a system, where a vehicle approaches a toll plaza, a laser embedded in the barrier reads the barcode, and information is retrieved and then deduces a toll amount from the account of the vehicle's owner. If the owner's account does not have sufficient balance, even so the barrier would be lifted, but an email or SMS would be sent to the owner notifying the due balance. Otherwise, the amount which was deducted is sent to the owner via email or SMS along with the location. Before the barrier opened, the server checks the vehicle information whether the vehicle is valid/registeredor not. If found registered then gate is opened and vehicle pass the barrier without delay and toll taxt messagedelivered to owners.

Barcode is embedded in the vehicles of the registered owners, and if the vehicle is stolenthen it is sent to the police via the database of the toll server, making the vehicle invalid. Hence, valid vehicles do not have any complaint lodged against them. If a vehicle is registered and valid, then the barrier is lifted; otherwise, police alert is aroused by an alarm and the police will contact. The barrier has a barcode reader placed on it and is connected to a backend central server. Barcode reader has a software implementation and a central server. The tags are stored in the backend database from where the data comes. The flow of information is provided by the software between a reader and backend central server database. There is a connection to a reader, which is placed in the barrier. LAN connection is also present in the central server. The entire computer system of the LAN is connected via a WAN to a central database. The record and data for the toll reduction shall be stored at the central server and records of the monthly transaction from the central server will be sent to the owner.

Authors in [2] have made use of a system architecture, which involves a login and a registration system. The client registered for user, vehicle, tollgate, respectively, and then only the vehicle is allowed to enter ETC system. The prepaid process of recharge will be implemented. A vehicle, which is already registered, having sufficient balance in account can move as tracking of vehicle is automatically done to ensure the registration of the vehicle. A theft complaint is issued for the lost vehicle and a search is conducted for that vehicle. Further, RFID card tags have also been used in order to automate the toll collection process. The authors in [3] have proposed an automated RFID-based automatic toll collection system. An RFID tag is a transponder that comprises reception apparatus and chip. The chip has a unique serial variety or different information based totally on the tag's kind of memory, which may be read-best, write once examine-many (bug) or examine-write.

Authors in [4] have also explored the domains of digital image processing and have implemented the same for the ETC system. To capture the number plate of vehicle, a camera is utilized. The database will be checked by the system; for a match, the toll amount will be deduced from the user's account and the gate will be opened. Via a message, notification shall be sent to the user. Image processing algorithms like plate localization, character segmentation and optical character recognition are used for the appropriate image processing. Quick Response (QR) codes, as shown in Fig. 2, are also used for toll collection. Authors in [5, 6] have proposed a novel ETC system, which is based on the QR detection technique. The QR has numerous advantages like high capacity, readable from all directions, structured appending, low storage space, etc. QR code is detected from a camera/imaging device by utilizing various QR detection algorithms, which are based on the pattern recognition techniques. Solomon error correction technique is also utilized for the detection of the authenticity of the code.

A comparative study of the different techniques of implementing the ETC system was discussed by authors in [7]. Authors have compared between the RFID, QR and barcode techniques, and have concluded that RFID is a more faster and flexible technology.

Fig. 2 Quick response (QR) code

3 Proposed Technique for the Barcode Detection

This paper primarily focuses on the barcode detection algorithm utilized to obtain the optimum working efficiency. Concepts of machine learning along with an IDE environment of Python and OpenCV have been used in order to obtain the working model. An imaging device like a camera was used for the appropriate detection of the barcode. The barcode to be scanned is shown in Fig. 3. It stores the bank account details, mobile number and Aadhaar card details of the vehicle owner.

The Schaar gradients were calculated along both the axes, and the x gradient was subtracted from the y-gradient in order to get regions in image having high horizontal gradients and low vertical gradients. The segmented image is a shown in Fig. 4. The filter noise from the background was removed. Despite that, the image was blurred, and the bars of the graph were not visible. As shown in Fig. 5, the morphological operations were performed to obtain a refined version of the thresholded image.

Fig. 3 The barcode detected and it is to be scanned

Fig. 4 The scanned barcode, by applying Schaar gradient

Fig. 5 Final refined image for barcode detection

The various other blobs detected interfere with contour detection and hence a series of erosions and dilations were performed. Finally, the contour with the largest area was retained. After detection, the barcode is decoded to extract the useful information.

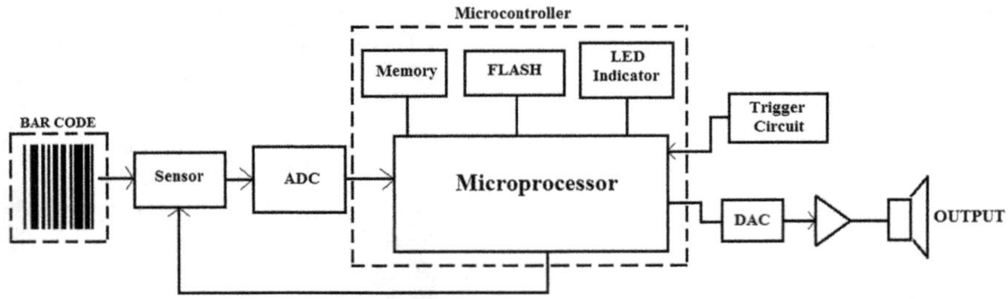

Fig. 6 The hardware setup required in order to implement the barcode scanner

4 Working of the Proposed Technique

The hardware required for the implementation of this system is shown with the help of block diagram shown in Fig. 6. The trigger circuit comprises infrared sensor interfaced with the microcontroller. The IR sensor detects the presence of the vehicle which is present close to the tollbooth and sends a strobe signal to the microcontroller to initiate the process of barcode detection. After successful detection of barcode and deduction of toll from the account of the vehicle owner, a message is sent to the registered mobile number of the vehicle owner notifying the amount deducted and due balance, and an alarm beeps indicating the successful completion of the process.

Figure 7 shows the flowchart of the working and the architecture of the proposed system. The logical decision to be made by the system is also shown in the flowchart. The components of the system as depicted in Fig. 7 govern the working of the proposed technique.

Further, authors in [8–20] have also proposed techniques aiming to improve the process of barcode detection. The respective research works in [8–20] have also proposed flowcharts to illustrate the operation of the technique and the architecture of the proposed methodology.

5 Conclusion

In the research article barcode detection for ETC systems has been proposed. The present work emphasizes the implementation of the proposed barcode detection technique in order to obtain better efficiency. The implementation of the technique illustrated in this paper shows consisted results. The hardware implementation depicts a significant improvement in the process of barcode detection, over the traditional methodology.

Simultaneously, the simplicity in the hardware design and the vast improvement in performance makes the proposed technique for ETC implementation very

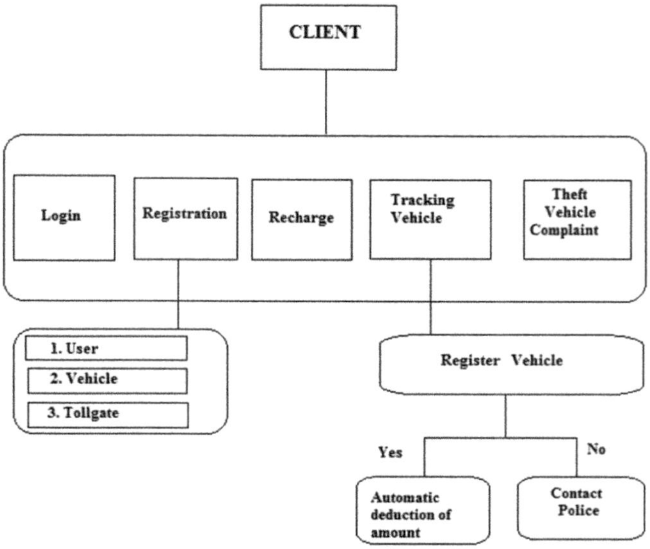

Fig. 7 Flowchart of the working and the incorporated system architecture

attractive. Figure 7 shows the overall operating flowchart of the proposed technique, along with Fig. 6, depicting the hardware implementation of the system.

References

1. Agarwal S, Gupta S, Sharma N (2014) Electronic toll collection system using barcode laser technology. Int J Emerg Trends Technol Comput Sci (IJETTCS) 3(2) (Mar–Apr 2014, ISSN 2278-6856)
2. Croon JA, Sansen W, Maes HE (2015) Electronic toll collection using barcode reader. Int J Adv Res Comput Sci Softw Eng 5(2) (Feb 2015, ISSN: 2277 128X)
3. Sunanda M, Kapilya (2017) Automatic detection of vehicles in toll plaza using sensors. Int J Pharm Technol 8(4) (ISSN: 0975-766X)
4. Cardoso AJ, de Carli LG, Galup-Montoro C, Schneider MC (2016) Vehicle counting and automated toll collection system using image processing. Int J Comput Sci Netw 5(2) (ISSN: 2277-5420)
5. Gaikwad SY, Audhutwar CS (2017) Automatic toll collection by using QR code capturing. Int J Comput Sci Inf Technol 5(5) (ISSN: 0975-9646)
6. Shah K, Josh P, Garg D (2015) Automatic toll collection by using QR code. Int Res J Eng Technol 3(10) (ISSN: 2395-0056)
7. Shigyo N (2017) A comparative study of different technologies for electronic toll collection system. Int J Innov Res Comput Commun Eng 4(2) (ISSN: 2320-9801)
8. Blythe P (1999) RFID for road tolling, road-use pricing and vehicle access control. In: IEE colloquium on RFID technology (Ref. No. 1999/123), pp 8/1-816
9. Chawla V, Ha DS (2007) An overview of passive RFID. Communications Magazine, IEEE 45:11–17

10. Gabriel N, Mitraszewska I, Tomasz K (2009) The Polish pilot project of automatic toll collection system. In: Proceedings of the 6th international scientific conference TRANSBALTICA
11. Cheng Y, Wang M (2011) An RFID electronic tag based automatic vehicle identification system for traffic iota applications. In: Control and decision conference (CCDC), 2011 Chinese, 2011, pp 4192–4197
12. Kamal R (2009) Embedded systems architecture, programming and design, 2nd edn
13. Liu T, Tan T, Chu Y (2007) 2D barcode and augmented reality supported english learning system. In: Proceeding of the 6th IEEE/ACIS international conference on computer and information science
14. Alapetite A (2010) Dynamic 2d-barcodes for multi-device web session migration including mobile phones. Personal and Ubiquitous Computing
15. Gao J, Kulkarni V, Ranavat H, Chang L, Mei H (2009) A 2d barcode-based mobile payment system. In: MUE, pp 320–329
16. Gao JZ, Prakash L, Jagatesan R (2007) Understanding 2d-barcode technology and applications in m-commerce—design and implementation of a 2d barcode processing solution. COMPSAC 2:49–56
17. Gao JZ, Veeraragavathatham H, Savanur S, Xia J (2009) A 2-d barcode based mobile advertising solution. In: SEKE, pp 466–472
18. Lisa S, Piersantelli G (2008) Use of 2d barcode to access multimedia content and the web from a mobile handset. In: GLOBECOM, pp 5594–5596
19. Mhatre D, Kamble R, Pimple S, Sankhe A (2015) Electronic toll collection using barcode reader. Int J SS & SE (ISSN: 2277–128X)
20. Kamarulazizi K, Ismail W (2010) Electronic toll collection system using passive RFID technology. J Theor Appl Inf Technol, pp:70–76 (ISSN: 2005–2010)

Design of Smartphone Controlled Robot Using Bluetooth

Vidushi Goel, Riya, Pinki Kumari, Prachi Shikha, Tanushree, Deepak Prasad and Vijay Nath

Abstract The paper is designed to develop a robot for stocking shelves, using Arduino UNO, Android, and RFID technology. This robot can grab items off a shelf and put them back again using information obtained from RFID cards and RFID card readers. According to the gesture commands received from Android, the movements of the robot are controlled. Gestures are captured through the Android smartphone. The Bluetooth modules are controlled by the microcontroller.

Keywords Arduino UNO · RFID sensor · Bluetooth module · Smartphones DC motor

1 Introduction

Robots are microcontroller-programmable machines that are designed to execute one or more tasks automatically with speed and precision. Due to its capability to work continuously, production shall be increased. Smartphones have become powerful devices with faster processors, smarter sensors, and more communication methods. Today, gestures can be used to interact with machines. A gesture is a form of nonverbal communication in which movements of body parts communicate particular messages. Modern smartphones are embedded with accelerometer sensor, Bluetooth module and are powered by Android OS [1, 2]. Robots can be controlled by hand gestures. Gestures are captured with the help of inbuilt accelerometer of smartphones. Android applications can be developed to convert gestures captured by accelerometers to information and send these information using Bluetooth module of the smartphone [3, 4]. This data is transmitted to a robot through Bluetooth module of smartphone using an Android app. After that, it is processed by Arduino chip embedded on the robot for its desirable motions. Here in this

V. Goel (✉) · Riya · P. Kumari · P. Shikha · Tanushree · D. Prasad · V. Nath
Department of Electronics and Communication Engineering, Birla Institute of Technology, Mesra, Ranchi 835215, Jharkhand, India
e-mail: vidushigoggle@gmail.com

V. Nath and J. K. Mandal (eds.), *Nanoelectronics, Circuits and Communication Systems*, Lecture Notes in Electrical Engineering 511,
https://doi.org/10.1007/978-981-13-0776-8_52

process, RFID device has been recommended over bar code or magnetic strip which is generally used in ATM card or credit card. RFID does the same job as bar code or magnetic strip [5–7].

2 Methodology

2.1 Direct Current (DC) Motor

It is an electric motor which can be used for mechanical movements by converting conventional energy. Here, electrical energy gets converted into mechanical energy. It works on the principle that a torque is experienced when a current carrying conductor comes under magnetic field. Above phenomenon is generally referred as motoring action. They are used for clockwise as well as anticlockwise operations. By varying the supply voltage or changing the strength of current, the speed of the motor can be varied. The direction of rotation will depend on the polarity of current.

2.2 L293D (Motor Driver)

This is a motor driver IC that can drive two motors simultaneously. L293D has 16 pins, and it allows the motors to move in either direction. L293D and L293NE are the most commonly used ICs of this series. Controlling of two DC motors simultaneously can be done using these ICs. Basically, it adopts two H-bridge which is used to control the current of motor. The motor drivers will be referred as L293D only. With the help of these motor drivers, two DC motors can be attached to a single IC and both of them are moved in two different directions. It has 16 digital pins that can be used as input and output [8]. This driver can drive both high and low rpm motors. To decide the direction of motor's rotation, the voltage may be varied.

2.3 Radio-Frequency Identification Device (RFID)

It is a device that records the existence of an object by using radio signals. This technology has not only used to identify an object but also person, a package, or an item. It can also be used for tracking the packages in the airports and the railway stations. This technology can be used to provide additional functionality and benefits for product authentication [6]. It cannot be a replacement for magnetic tapes used in ATM cards, but is a complement for distant reading of codes. It can be utilized in the packaged items for detection of their presence in the shopping malls. The main information is stored in the tags. Among active tags and passive tags, passive tags get energy from nearby radio waves [5, 6]. On the other hand, active tags draw energy from the provided local power source, i.e., battery and its wide range of operation.

2.4 Arduino UNO

Arduino UNO is basically a microcontroller adopting ATmega328P with 14 digital input–output pins; 6 pins act as analog input pins, and 6 pins are used for PWM output pins and a 16 MHz crystal oscillator. The power to the board is provided by using USB power jack. Arduino is used to connect digital devices and various sensors. Serial communication is interfaced to load programs from system using USB cable, HC-05 Bluetooth Module [1, 3, 4]. Here, HC-05 is adopted. It is a qualified Bluetooth and can be used simultaneously as both master and slave.

2.5 Arduino IDE

It has multifunctional property, i.e., it is used as text editor for writing code, a text console, a message area, and a toolbar with buttons for common functions. Its main function is to interface and program uploading between Arduino board and hardware. The program here written and to be uploaded using Arduino Software (IDE) are called sketches [2, 9]. Text editor is used to write sketches which are protected with the file extension.ino, and they consist of two functions which are compiled:

- setup(): this always runs at the starting of every program and is generally referred to initialize the program.
- Loop(): this term is called again and again until the power of the microcontroller is cut off.

3 Design and Operation

The design of robot constitutes of Arduino UNO microcontroller, a motor driver L293D, 2 DC motors, an HC-05 (Bluetooth module), and an RFID sensor. The process starts with the transfer of data from Android phone to HC-05, a Bluetooth module which further transfers the data to Arduino UNO board. Here, the data is manipulated and performs the function which is a controller that controls the signals and performs the given functions, and it forwards the signals telling the motors which direction to rotate signals. The user can operate the robot by using smartphone-based application. Using the applications not only movement but other functions of robot will be decided.

Figure 1 depicts the basic model of the smartphone application-based proposed system. The buttons decide the movement of robot. The input will be given using the buttons, and the input signal will be sent further by using Bluetooth module [10]. The motor driver has a lot of pins and these pins are for ground power supply. Each of the four dc motors has their own respective pins that on getting high signal allow the dc motor to move accordingly, like two pins are assigned for left motor and two for right motor. The RFID module is connected to the robot that is directly connected with the microcontroller which senses the RFID code of the tags present on the blocks that are present in the warehouses in order to compare the different codes of the blocks and then segregate them according to the RFID code of each one in the warehouse [5–7].

Application Instructions

The application is generated through Processing 3 software:

- The first step is to enable Bluetooth module.
- Mainly five buttons play the vital role. These buttons are LEFT, RIGHT, FORWARD, and BACKWARD and STOP which is shown in Fig. 1. On pressing one of the five buttons, the robot can be moved in the specified direction. The pressing of stop button instantaneously stops the motor [2, 4] (Figs. 2 and 3).

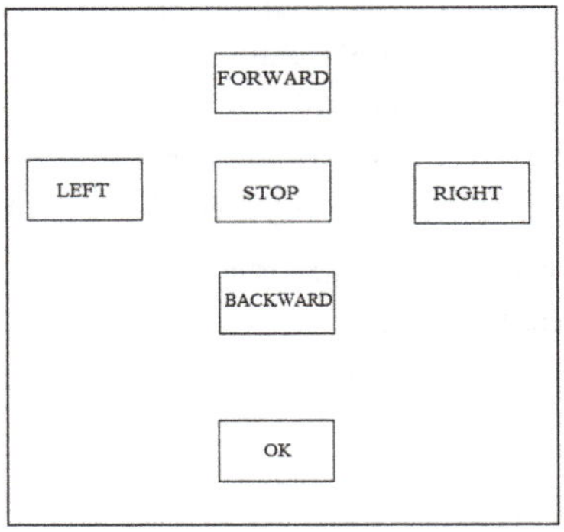

Fig. 1 Application instruction block diagram

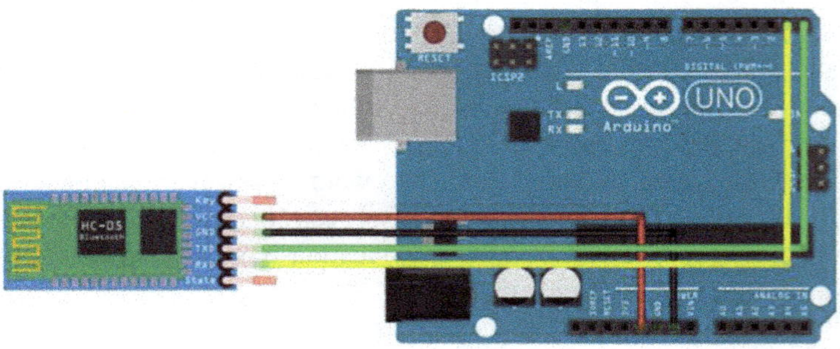

Fig. 2 Connection of Bluetooth module with Arduino

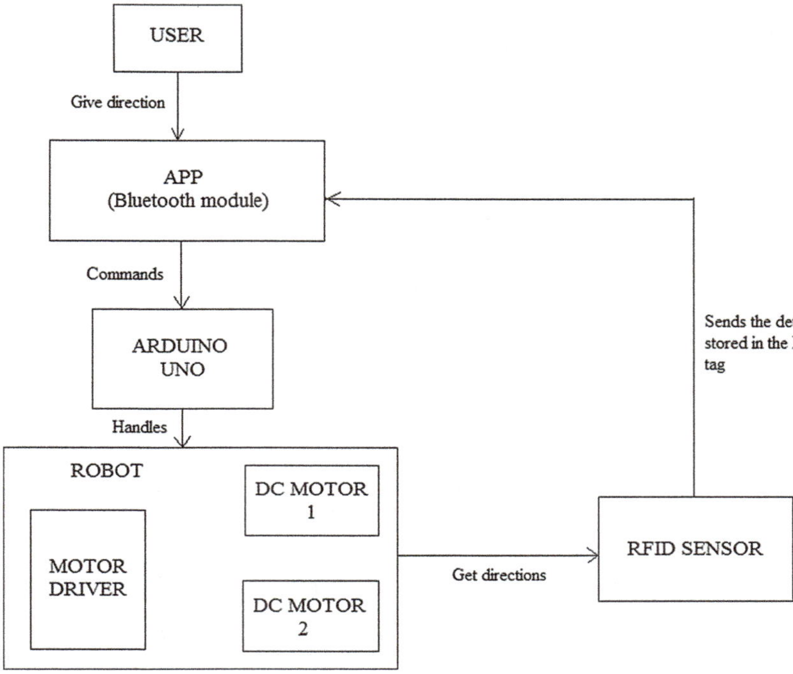

Fig. 3 Block diagram

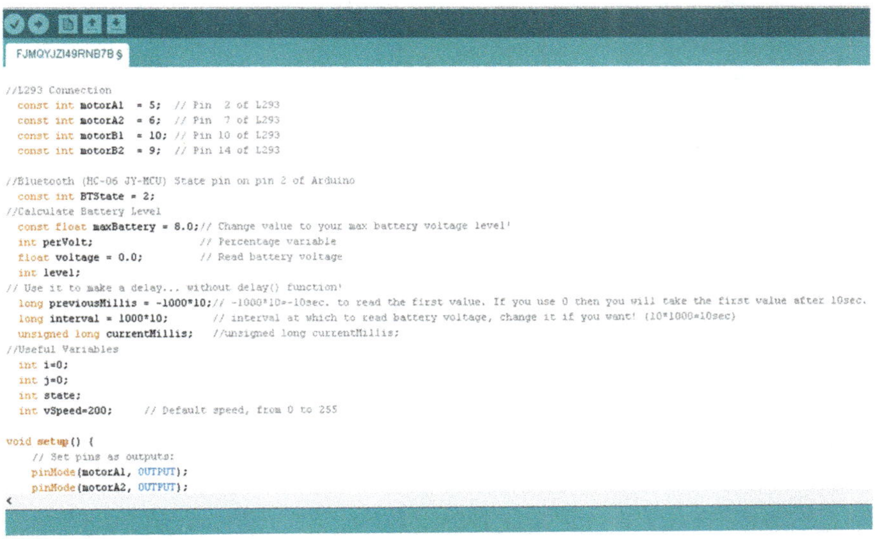


```
//L293 Connection
  const int motorA1  = 5;  // Pin  2 of L293
  const int motorA2  = 6;  // Pin  7 of L293
  const int motorB1  = 10; // Pin 10 of L293
  const int motorB2  = 9;  // Pin 14 of L293

//Bluetooth (HC-06 JY-MCU) State pin on pin 2 of Arduino
  const int BTState = 2;
//Calculate Battery Level
  const float maxBattery = 8.0;// Change value to your max battery voltage level!
  int perVolt;                 // Percentage variable
  float voltage = 0.0;         // Read battery voltage
  int level;
// Use it to make a delay... without delay() function!
  long previousMillis = -1000*10;// -1000*10=-10sec. to read the first value. If you use 0 then you will take the first value after 10sec.
  long interval = 1000*10;       // interval at which to read battery voltage, change it if you want! (10*1000=10sec)
  unsigned long currentMillis;   //unsigned long currentMillis;
//Useful Variables
  int i=0;
  int j=0;
  int state;
  int vSpeed=200;     // Default speed, from 0 to 255

void setup() {
    // Set pins as outputs:
    pinMode(motorA1, OUTPUT);
    pinMode(motorA2, OUTPUT);
```

```
void loop() {
  //Stop car when connection lost or bluetooth disconnected
     if(digitalRead(BTState)==LOW) { state='S'; }

  //Save income data to variable 'state'
    if(Serial.available() > 0){
      state = Serial.read();
    }

  //Change speed if state is equal from 0 to 4. Values must be from 0 to 255 (PWM)
    if (state == '0'){
      vSpeed=0;}
    else if (state == '1'){
      vSpeed=100;}
    else if (state == '2'){
      vSpeed=180;}
    else if (state == '3'){
      vSpeed=200;}
    else if (state == '4'){
      vSpeed=255;}

  /***********************Forward****************************/
  //If state is equal with letter 'F', car will go forward!
    if (state == 'F') {
        analogWrite(motorA1, vSpeed); analogWrite(motorA2, 0);
        analogWrite(motorB1, 0);      analogWrite(motorB2, 0);
    }
  /***********************Forward Left************************/
```

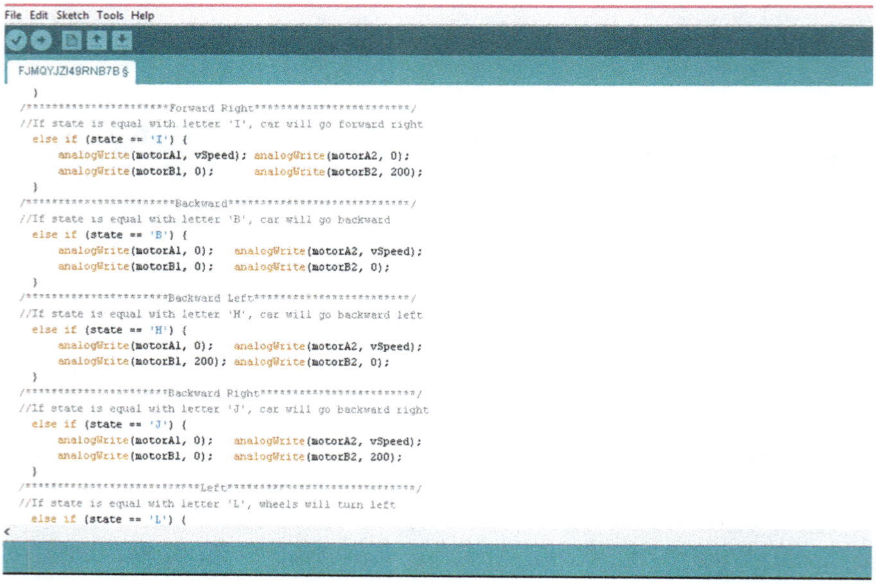


```
    }
  /**********************Forward Right***********************/
  //If state is equal with letter 'I', car will go forward right
    else if (state == 'I') {
        analogWrite(motorA1, vSpeed); analogWrite(motorA2, 0);
        analogWrite(motorB1, 0);      analogWrite(motorB2, 200);
    }
  /***********************Backward****************************/
  //If state is equal with letter 'B', car will go backward
    else if (state == 'B') {
        analogWrite(motorA1, 0);   analogWrite(motorA2, vSpeed);
        analogWrite(motorB1, 0);   analogWrite(motorB2, 0);
    }
  /**********************Backward Left************************/
  //If state is equal with letter 'H', car will go backward left
    else if (state == 'H') {
        analogWrite(motorA1, 0);   analogWrite(motorA2, vSpeed);
        analogWrite(motorB1, 200); analogWrite(motorB2, 0);
    }
  /**********************Backward Right***********************/
  //If state is equal with letter 'J', car will go backward right
    else if (state == 'J') {
        analogWrite(motorA1, 0);   analogWrite(motorA2, vSpeed);
        analogWrite(motorB1, 0);   analogWrite(motorB2, 200);
    }
  /*************************Left*****************************/
  //If state is equal with letter 'L', wheels will turn left
    else if (state == 'L') {
```

4 Result and Discussion

A robot is prepared using which items can be managed in a warehouse, and it would help to keep an account of details of the items available and its quantity. It would help in better management and utmost utilization of the technologies available for the better management of the items being added or removed from the warehouse. It would provide safety to the workers as they are not directly involved in the displacement of item as it is being carried out by smartphones which are manhandled. Semi-automation in the workspace would provide high performance in less time.

Acknowledgements We would wish to express our gratitude to Vice Chancellor of Birla Institute of Technology, Mesra, Dr. M. K. Mishra. We would also like to thank Head of Department of Electronics and Communication, BIT Mesra, Dr. Srikanta Pal for his patience and valuable help.

References

1. Osorio SP, Aristizábal LM, Zuluaga CA (2016) Development of a command interface based on handheld devices for remotely operated vehicles. In: IEEE Colombian conference on robotics and automation (CCRA), pp 1–5
2. Harding RJ, Whitty MA (2015) Employing android devices for autonomous control of a UGV. In: European conference on mobile robots (ECMR), pp 1–6
3. Subankar R, Tashi RW, Bhatt R (2016) Arduino based bluetooth controlled robot. Int J Eng Trends Technol (IJETT) 32(5):216–219
4. Scherhäufl M, Pichler M, Stelzer A (2015) UHF RFID localization based on phase evaluation of passive tag arrays. IEEE Trans Instrum Meas 64:913–922
5. Yang P, Wu W (2014) Efficient particle filter localization algorithm in dense passive RFID tag environment. IEEE Trans Ind Electron 61:5641–5651
6. Elara MR, Rojas N, Chua A (2014) Design principles for robot inclusive spaces: a case study with Roomba. In: IEEE international conference on robotics and automation (ICRA), pp 5593–5599
7. Piyare R, Tazil M (2011) Bluetooth based home automation system using Android phones. In: IEEE 15th international symposium on consumer electronics (ISCE), 14–17 June 2011, Singapore
8. Sharma A, Verma R, Gupta S, Bhatia SK (2014) Android phone controlled robot using bluetooth. Int J Electron Electr Eng 7(5):443–448
9. Franquiz JJP, Cruz MG, Yuriy SS (2015) Self-localization over RFID tag grid excess channels using extended filtering techniques. Sel Top Signal Process IEEE J 9:229–238
10. Goud RK, Kumar BS (2014) Android based robot implementation for pick and retain of objects. Int J Eng Trends Technol (IJETT), 16(3) (Oct 2014)

Design of Earthquake Indicator System Using ATmega328p and ADXL335 for Disaster Management

Paritosh Kumar Sinha, Satyam Saraiyan, Momojit Ghosh and Vijay Nath

Abstract Disaster management is the field dealing with or organizing proper management and use of resources, technology, and information to deal with any disastrous event or emergency. Disaster management is the proper planning to coordinate natural and any man-made disastrous event and is measured how efficiently and effectively is one's preparedness to respond to calamity and recovery in order to lessen the impact of such events. The purpose of this paper is to design a system to detect an earthquake and give a warning message for home and industry. The design uses Arduino (ATmega328p), accelerometer (ADXL335), LEDs, and buzzer.

Keywords Earthquake · ADXL335 · ATmega328p

1 Introduction

An earthquake is a natural unpredictable disaster which often causes huge damage to both lives and property. Earthquake is an unavoidable event which is caused by sudden movement of tectonic plates, which in turn leads to abrupt movement of earth surface. The main problem lies in the prediction of earthquake and information such as magnitude and location [5]. There have been numerous attempts to

P. K. Sinha (✉) · S. Saraiyan · M. Ghosh · V. Nath
Department of Electronics and Communication Engineering,
Birla Institute of Technology, Mesra, Ranchi 835215, Jharkhand, India
e-mail: paritoshks10@gmail.com

S. Saraiyan
e-mail: satyamsaraiyan@gmail.com

M. Ghosh
e-mail: momojit.ghosh@gmail.com

V. Nath
e-mail: vijaynath@bitmesra.ac.in

V. Nath and J. K. Mandal (eds.), *Nanoelectronics, Circuits and Communication Systems*, Lecture Notes in Electrical Engineering 511,
https://doi.org/10.1007/978-981-13-0776-8_53

make an early warning system for earthquake. A few of them are compared in the following table below.

Willy H. K. Lee et al. (2003) explained that mainly there can be two ways to achieve an early earthquake warning system, either using only one station method or an array method. In one station method, seismic waves are handled sectionally and then compared with the predefined threshold value which gives the output accordingly. In array method, the main station handles signals from the array of stations and then checks whether the incoming signal is greater than the threshold value or not [1].

Sanjib Kalita et al. (2012) demonstrated the idea that when the intensity of vibration exceeds a prefixed value, an alarm is set off. The intensity of vibration is measured by using an accelerometer [2].

USGS (2012) has been working on the method to detect the earthquake just after it begins. In this method, first, P-wave energy is calculated, which provides the location and intensity of the earthquake [3].

According to C. Buzduga et al. (2015), if there is any survivor in the collapsed building, then the electrostatic sensor transmits the information via radio transmitter which is received by radio receiver, and the microcontroller attached to it provides the exact location for the survivor [4].

Hima MS et al. (2016) in their paper have shown an alarming or indicating system for earthquake event. The paper worked on earthquake indicator using Arduino and an accelerometer that can precisely detect even small tremors and displays the vibration intensity on LCD [5].

Pei Yanglin et al. have worked on development of an onsite earthquake warning system (EEWS) which relies on the fact that P-wave travels faster than S-wave. The proposed system receives signal from onsite sensors and provides a warning message. The warning system can be aided with disaster control system that parks the elevator and switches off electrical power and gas supply [6].

Dan Wang et al. implemented an earthquake alarm system using wireless sensors. This implementation is faster than the other methods used. The key is to position the sensor in place which makes the overall system more efficient [7].

Alphonsa A et al. demonstrated a method in which P-waves are sensed by the sensors which are transferred by a Zigbee transmitter to a Zigbee receiver which is an IOT, which transfers the warning message to the smartphones [8].

Erika Yamasaki has worked on detection of an earthquake based upon the data collected from ground movement using dense seismic network. The analysis of this information reveals whether an earthquake warning is required or not [9].

Pratiksha P Kamble's design consists of a sensor unit and base unit. The data is given in the sensor unit which is analyzed in the loop window. If this data exceeds the preset threshold window size, the data is given to the base node, which is connected to the flame sensor. The base unit provides audible alert and prompts warning message to the server and the user [10].

2 Methodology

This earthquake indicator or alarm can be designed with the help of an Arduino microcontroller (ATmega328p) and a highly sensitive accelerometer (ADXL335). Such a system works on the principle that during an earthquake due to the movement of earth surfaces when vibration crosses or is increased beyond a threshold value, then ADXL335 accelerometer senses these vibrations and gives indications either by glowing LEDs or by creating a buzzer sound. Even if the system provides with warning 60 s before the earthquake, the time can be utilized to shut the factory, stop traffic and provide cover to population [13]. The warning can be generated based on the fact that the speed of P-wave and S-wave differs from each other. And on detection of P-wave, the warning is issued. These alarming system can not only help one to be prepared for such an emergency at household level but such a system can be used at industrial level by wiring it with a PLC to shut off the moving machines, industrial appliances, and control furnace whenever vibrations are indicated by the system to avoid and control damages to these machines. Threshold adjustment buttons are provided for this purpose.

Here, in this paper, we have tried to present a working model of an earthquake detection system which can help people to save lives and damage caused by earthquake and to tackle such calamity in much more efficient and effective manner. If the ADXL335 accelerometer senses acceleration or vibrations which are greater than the preset value of the system then blows the buzzer.

This system can be used in houses, industries, or in high rise building as shown in Fig. 1.

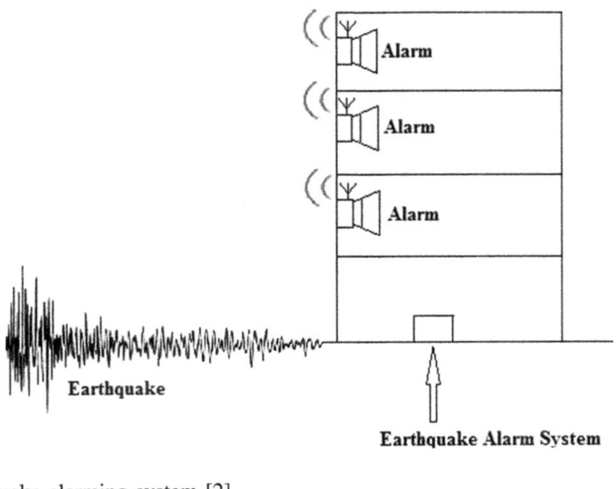

Fig. 1 Earthquake alarming system [2]

2.1 Framework of the System

The block diagram of earthquake alarm system comprises one accelerometer block, the next block is an Arduino, buzzer and indication comes next, and the last block processes plot of earthquake over computer.

As shown in Fig. 2, the first part of the block diagram includes the ADXL335 (MEMS) accelerometer which can detect the vibrations produced during the earthquake. This block also consists of a microcontroller to work on the values received from accelerometer (ADXL335), and then it generates a signal only when the accelerations during the earthquake are more than the already set value. The information generated by the microcontroller is then transferred to the system containing buzzer and led indicator.

ADXL335: As stated earlier, ADXL335 is an accelerometer. MEMS accelerometer is a device which can recognize vibrations, shocks, and gravity. It has been found that microelectromechanical system (MEMS) accelerometer has several

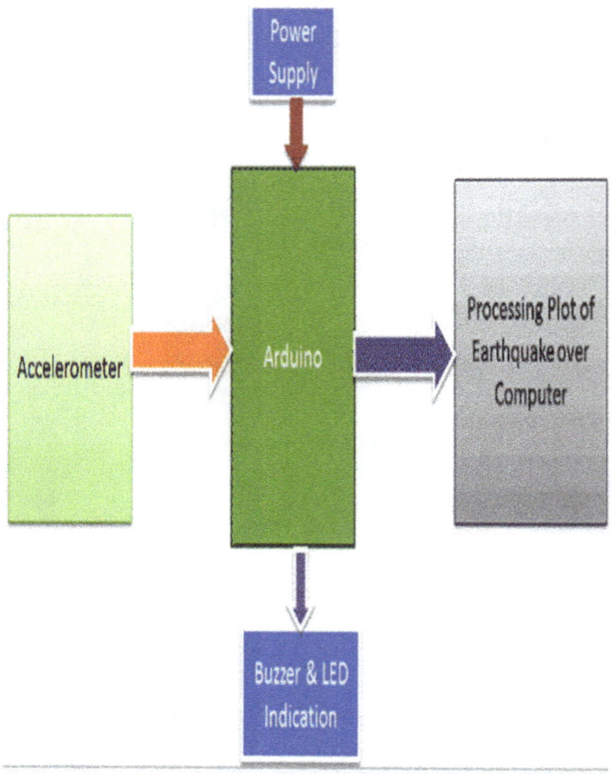

Fig. 2 Block diagram of earthquake indicator using Arduino and accelerometer

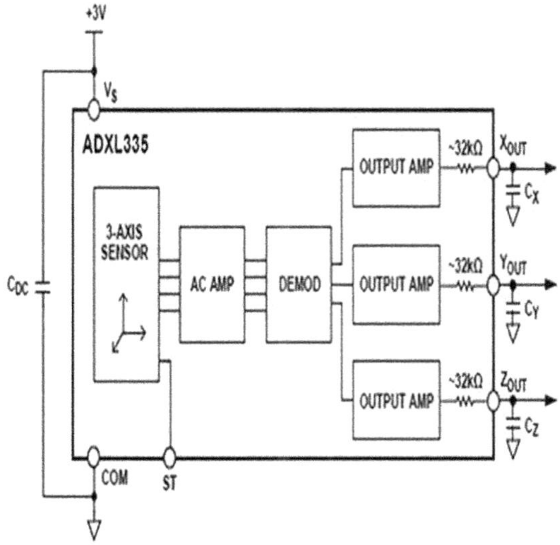

Fig. 3 Functional block diagram of ADXL335 [2]

applications such as for image rotation and stabilization in digital camera, automotive airbags, etc. [15].

ADXL335 is a light weighted, three-axis accelerometer with a minimum full-scale range of ±3g. The power consumption of this accelerometer is also quite low. Figure 3 shows the functional block diagram of ADXL335.

ADXL335 is connected to analog-to-digital converter pins of Atmega328p. It sends voltage elevation or the vibrations caused during the earthquake to the microcontroller.

Atmega328p: Atmega328p is a high-performance AVR 8-bit microcontroller with low power consumption. It has 23 programmable pins, and the voltage at which it operates is low (1.8–5.5 V). The main concern of this module is to process the signals received from accelerometer and to compare it with the preset values. Atmega328p has 10-bit successive approximation analog-to-digital converter. The ADC is attached to an eight-channel Analog Mux. The microcontroller adjusts the voltage levels received from ADXL335 and calculates the peak ground acceleration. As stated earlier, if the intended value is greater than preset value, then it produces an alarm, and this information is sent to the buzzer or led indicator, which is also the last block in the diagram.

2.2 *System Flowchart*

A system flowchart is the step-by-step portrayal of the functioning of the system. Each function is being represented by boxes of different shapes and sizes for better understanding of the model.

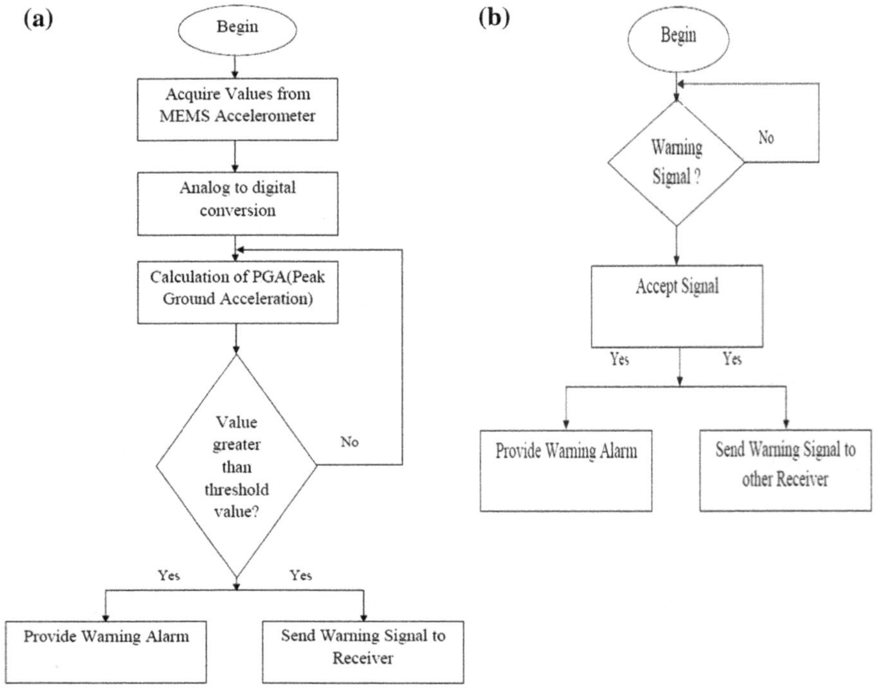

Fig. 4 **a** Flowchart showing the general working of earthquake indicator [2]. **b** Flowchart showing the general working of earthquake indicator [2]

Figure 4 shows the general flowchart of the earthquake indicator.

When the system begins its working, it acquires a value from MEMS accelerometer. The data is then converted from analog to digital form. From the digital data, the peak ground acceleration is calculated and compared with the threshold value. Depending upon the comparison, the warning signal is generated.

The microprocessor runs a loop for checking of peak ground acceleration; if the peak ground acceleration exceeds the threshold value, it comes out of the loop and provides warning alarm.

3 Result and Discussion

Reference	Author	Contribution
[1]	Willy H. K. Lee, Juan Manuel Espinosa-Aranda	Earthquake early warning systems: Current status and perspective
[2]	Sanjib Kalita, J. N. Borole, Dr. K. P. Rane	Wireless earthquake alarm system

(continued)

(continued)

Reference	Author	Contribution
[3]	USGS	Shake Alert, Early Earthquake warning
[4]	C. Buzduga, A. Graur, C. Ciufudean, V. Vlad	System for the detection earthquake victims—construction and principle of operation
[5]	Hima M. S., Josphin Subha, Kalyani Devi G., Karthika Sankaran, Navin S.	Earthquake indicator using Arduino
[6]	PeiYangLin, Shieh-Kung Huang, Hung-Wei Chiang, Zhe-Ping Shen	Development of the on-site earthquake early warning system in Taiwan
[7]	Dan Wang, Yiking Ni	Wireless sensor networks for earthquake early warning systems of railway lines
[8]	Alphonsa A., Ravi G.	Earthquake early warning system by IOT using Wireless sensor networks
[9]	Erika Yamasaki	What we can learn from Japan's earthquake early warning system
[10]	Pratiksha P. Kamble	Wireless sensor networks for earthquake detection and damage mitigation system
This work	Paritosh S., Momojit G., Satyam S., Vijay N.	Design of earthquake indicator system using ATmega328p and ADXL335

This work is based on the idea that P-waves travel faster than S-wave and can be detected to give an early warning. A value is prefixed in the microcontroller, and the intensity of vibrations is measured by accelerometer; when the intensity of vibration exceeds the value, an alarm is set off.

4 Conclusion

With the help of this paper, we have tried to present an earthquake indicator system. The basic components of this system are microcontroller, accelerometer, buzzer, and LED. It relies on the principle that if the vibrations exceed a minimum value, it gives an alarm. The earthquake detection system can be used in multistoried building and industries. The advantages of the system include efficiency in cost and size apart from low power consumption.

Acknowledgements We are grateful to Dr. M. K. Mishra, Vice-Chancellor and Dr. S. Pal, HOD, Department of Electronics & Communication Engineering B.I.T. Mesra, Ranchi (Jharkhand) for encouraging us to prepare this paper and for the valuable suggestions and feedback for completing this paper.

References

1. Lee WHK, Espinosa-Aranda JM, Earthquake early warning system: current status and perspectives. http://usgsprojects.org/lee/Lee_Espinosa_2003_EWSbook.pdf
2. Kalita S, Borole JN, Rane KP, Wireless earthquake alarm system. Int J Eng Trends Technol (IJETT) 12(3), June 2014. http://www.ijettjournal.org
3. USGS, ShakeAlert, early earthquake warning. https://www.shakealert.org
4. Buzduga C, Graur A, Ciufudean C, Vlad V, System for the detection earthquake victims—construction and principle of operation. www.inase.org/library/2015/vienna/bypaper/csscc/csscc-15.pdf
5. Hima MS, Josphin Subha, Kalyani Devi G, Karthika S, Navin S (2016) Earthquake indicator using Arduino. Int J Res Advent Technol 4(8), August 2016
6. Lin PY, Huang S-K, Chiang H-W, Shen Z-P, Development of an on-site earthquake early warning system in Taiwan. www.iitk.ac.in/nicee/wcee/article/WCEE2012_5064.pdf
7. Wang D, Ni Y, Wireless sensor networks for earthquake early warning systems of railway lines. https://link.springer.com/chapter/10.1007/978-3-642-27963-8_38
8. Alphonsa A, Ravi G, Earthquake early warning system by IOT using wireless sensor networks. http://ieeexplore.ieee.org/stamp/stamp.jsp?arnumber=7566327
9. Yamasaki E, What we can learn from Japan's early earthquake warning system. http://repository.upenn.edu/momentumvol1/iss1/2
10. Kamble PP, Wireless sensor networks for earthquake detection and damage mitigation system. https://www.ijircce.com/upload/2016/march/272_Wireless.pdf
11. Earthquake indicator using Arduino by Ronie Adhiraaj Ghosh. http://electronicsforu.com
12. https://www.ieee.org
13. Jordan TH (2014) The prediction problems of earthquake system science. Seismol Res Lett 85 (4):767–769
14. Kalita S, Borole JN (2014) Needs of early earthquake warning system in north-eastern region of India. Int J Eng Res Technol (IJERT) 3(3):501–504, ISSN: 2278-0181. Mar 2014
15. Kalita S, Borole JN (2014) Application of MEMS accelerometer to consumer electronics. In: International conference on modeling and simulation in engineering and technology, ICMSET-2014, pp 188–190, Feb 15–16

Study and Design of Smart Embedded System for Aviation System: A Review

Deril Raju, Lalitha Eleswarapu, Rohith Saiv and Vijay Nath

Abstract It is because of the technological advancements that we can now reach from one place to the other one within few hours. Aviation industry is concerned as one of the busiest industries in the market. Aviation industry not only deals with the design and development but also plays a crucial role in the operations and production of the aircrafts that are to be operational. Aviation deals with the Civil as well as general aviation. However, advanced avionics do not enable either the engineers or the pilot to break the laws of physics. It was designed for keeping utility and safety in mind. Safety field is done by increasing its awareness. In this paper, we discussed the methods adopted by aviation technology and gave a brief description about the principles followed. The various methodologies and technological advancements involved in the aviation system are also discussed which include various improvisation techniques as well. These include the ways to improvise engines for more efficiency, managing fuel system and improving its efficiency. Causes of noise in aircrafts and ways to reduce noise are also discussed. The very important aspect of aviation system is the reduction of noise in its navigation and its engine during flight. Managing air traffic is another major hurdle in aviation system. The methods to control air traffic by improving navigation system are put in a nutshell. There should be a proper use of checklists and systematic training for decreasing the count of casualties during an accident. Aviation is a flourishing industry with many upcoming technologies to enhance the avionics system. Researches have been going on across all countries in improving avionics system. This also helps them to improve the armed forces of their nation. Military

D. Raju (✉) · L. Eleswarapu · R. Saiv · V. Nath
Department of Electronics and Communication Engineering,
Birla Institute of Technology Mesra, Ranchi 835215, Jharkhand, India
e-mail: derilraju@gmail.com

L. Eleswarapu
e-mail: lalithaeleswarapu98@gmail.com

R. Saiv
e-mail: rohithsaiv98@gmail.com

V. Nath
e-mail: profvnath@gmail.com

V. Nath and J. K. Mandal (eds.), *Nanoelectronics, Circuits and Communication Systems*, Lecture Notes in Electrical Engineering 511,
https://doi.org/10.1007/978-981-13-0776-8_54

aircrafts are imbibed with latest aviation technologies everyday which improve their target sensor and also other functions like multi-targeting, etc. Various researches on how to reduce the weight of the aircrafts by using better heat-resistant alloys and how to make the aircraft robust and resistant to drastic weather conditions and turbulence are going on widely.

Keywords Aviation · Fuel improvements · Air traffic · Military aviation

1 Introduction

Aviation technology has been advancing with leaps and bounds. Every year over two billion people travel and there are over 23,000 aircrafts in commercial use. Aviation industry is a global necessity, and it connects the world and provides lifeline to many communities. We can now transport people and goods across the globe covering large distances in very less time which was once thought impracticable with small amounts of energy. Aviation industries are still striving for greater fuel efficiency.

Aviation can be divided into two broad categories, namely commercial and military aviation. Commercial aviation is related to transport of people and other goods to different places via air. It includes airlines, choppers, etc, while military aviation deals with the defense of the nation which includes fighter planes, helicopters, etc. [1, 2, 3, 4].

Smart aviation concept has arrived in the very recent times. The main aim of the topic was to develop efficient and much safer aircrafts for the society. This includes engine improvements which mainly focuses on a safer and much powerful engine, weight reduction [1, 5, 2, 6], for a much efficient aircraft and less fuel consumption [5, 7], better design of the aircraft to reduce the net drag which automatically improves the overall efficiency and many improvements. Smart aviation clearly has a wide scope in the coming future than the other aircrafts flying today. The main aim of this concept is to improve the safety of the whole aircraft [5, 8], which ultimately leads to lesser accidents and ultimately improves the economy of this sector.

1.1 How Does It Fly

Aerodynamic lift is caused as a result of positive angle of attack where the wing's shape is designed with precision and the oriented accurately. It is designed such that the front part of the wing is tilted slightly higher than the back part. According to the Bernoulli's principle, this produces low pressure comparatively on the top part of the wing. The excess force per unit is that the pressure on the lower side of the wing pushes the wing upwards. This net resultant upward force is called aerodynamic lift [1].

The wings are equipped with devices which can be extended or retracted depending on the flight speed and conditions changing the shape and size of the wing. These devices help the aircraft not to lose balance even while flying with very high speed or with low speed while taking off or landing. They also help in producing the same amount of lift throughout the time of flight over wide range of speeds.

To change the direction of the flight, the amount of air flowing between left and right wings is controlled by ailerons, whereas the amount of lift is controlled by slats and flaps, for example, while landing the speed of the aircraft drops suddenly. Hence, to maintain the lift, slats and flaps extend to increase the surface area.

When the aircraft flies, it experiences resistance [1] from air which is called 'drag'. In olden days, propellers were thought to be the only solution. But aircraft design has been advanced a lot that the jet engines can overcome this drag by using the force of thrust.

1.2 Current Problems

The main reason for which the concept came up was the problems faced by the aviation sector in recent times. These problems lead to lowering of the overall economy [8] of the industry. The problems may include human errors and failures causing accidents.

In today's aviation industry, human fingers have an important role in it. However, this leads to many errors which cause many faults and accidents. Humans do not have the potential to work for a longer time which if they do leads to multiple discrepancies and thus leading to accidents. These accidents lower the economy of the industry which is becoming a major problem for the sector.

In aviation system, there are corresponding procedures to manage all the situations. We need to choose well between various response outputs in times of a problem. Aviation psychologists [6] say that the decision errors and knowledge-based errors have been of great hazard over the last several decades.

Sometimes, crew members face such situations that none of the responses or the formal procedures can help them and solutions are nowhere to be found. In these cases, novel solution is required in which the individuals should go through effortful reasoning procedure [6] (Fig. 1).

Smart aviation mainly aims for more computerized approach which will lead to lesser errors and thus ultimately improves the economy of the industry with being the safest mode of transport.

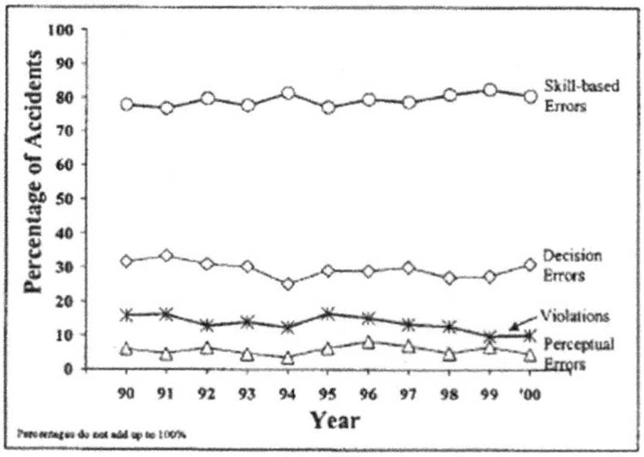

Fig. 1 Percentage of accidents by error category by year. *Source* Federal Aviation Administration govt.

2 Methodology

2.1 Enhancement in Aerodynamics

Over the past centuries, there has been continuous development in aerodynamics but the changes appear very rarely. The main factor which affects the improvement in aerodynamic is the overall fuel consumption. Much meagre, smooth and easy to move wings, smooth stratified flow and an activated flow control, are the factors which improve the dynamics of the vehicle.

Laminar flow is the best method to reduce the drag but they are tough to manufacture and require high tolerances; this leads to increase the overall cost and requires more time [9, 1]. Seeing to this researchers predict that the laminar flow concept might come up in 2030s. Much slender and flexible wings require more control techniques; however, they reduce the net drag and weight efficiently. The main focus for the aerodynamics improvement is the high-speed cruise; [5] another one is the lift of aircraft at smaller speed and another can be proper use of tractable surfaces with can reduce the noise and drag from flaps and slats. Another advantage of active flow control is that it can improve noise reduction and can decrease the landing and takeoff times.

Researchers predict 20–40 years later that the conventional aircraft wings will no longer be effective and cannot deliver the same expected output. Two major problems [10, 5] can be regarded the mounting of engine and to keep the noise low so that it cannot leave the boundaries of airport.

Many researches are being conducted regarding the placement of large engine radius over the wing or on the tail—and also a design of an airframe which can restrict the noise to go to surroundings. For cleaners, wing and reduced drag aft mounted engines will come in handy. For longer span and reduced drag, truss-braced wing can be adopted.

Turbofans and electric propulsors are embedded in the tail to reduce the drag and repower the aircraft. Blended or hybrid wing body (BWB/HWB) [1] is a wing capable of flying with improved aerodynamic efficiency. There is still confusion about this design, but the HWB is a most anticipated airlifter configuration. The wide airframe can provide significant shielding for turbofans, open rotors and distributed propulsors.

2.2 Engine Improvements

Several advancements have been made in the field of engine improvements. Integrated High-Performance Turbine Engine Technology (IHPTET) program is the best evidence for advanced research on readymade technology. The IHPTET program [5, 2] share was 25%–25%–50% amongst NASA—Department of Defense—industry program. Readiness level (TRL) being 6, IHPTET produced useful research results and were changed into useful products. High-level goals like thrust-to-weight ratio and component-level goals such as efficiency and cooling were achieved by each phase of IHPTET. Depending upon on the accepted systems approach, it allowed groups to participate both in high-level and low-level goals which were the main reason for its success. The program's flexible nature was the main reasons for its success and it did not try to win in the technical sense.

Advanced Affordable Turbine Engine (AATE) [1] provides advanced propulsion versions to army aviation. To support modernization requirements, AATE fixes its mind on creating a new engine which is advanced in its turboshaft.

AMRDEC's Aviation Development Directorate (ADD) led the Science and Technology (S&T) reasons to create a new advanced engine, which focuses on three objectives:

- Significantly reducing the fuel consumption (SFC).
- Significantly improving the inverse of weight to shaft horsepower ratio.
- Significantly reducing the cost of life cycle (LCC).

The AATE had two main engine approaches: one was implemented with general electric and the other one was implemented with Advanced Turbine Engine Company (ATEC). The AATE program reached its maximum point in 2014 by accomplishing the target of originally planned, demonstrating complete engine systems including performance, durability and sand tests for the two competing architectures. The ITEP [6, 2] managed by PEO Aviation has completed its analysis of alternatives (AoA) phase and is ready to enter into a matured phase of technology.

The AATE concept was prototyped to fit into both the Black Hawk and Apache installing processes. The AATE provides 50% more power output when compared to conventional engines. The use of particle separation technologies using advanced inlets, coatings resistant to erosion and shock-resistant blading are some of the significant efforts to ensure vigorous operation in severe or constraint environments.

Fuel efficiency improvements, better maintenance and long-lasting engines are few of the reasons for a reduced aviation overall cost. Range improvements, increased capacity and increased capability are the few high-level technologies of the Black Hawk platforms [1] and can be applied for future vertical lift (FVL) rotorcraft.

2.3 Fuel Improvements

This topic mainly focuses on the advancement in fuel industry and their effects on engines. Alternative fuels have quite some advantages over the conventional fuels, and this property brings the synthetic fuel closer to the concept of DoD [2, 1]. Coal, shale, etc. which are domestic fuels have three times the potential than the fuel of Middle East. The main challenge in this field is to develop this concept with the private industries. Barriers such as economic and environmental ones should be overcome for these fuels to hit the markets. A proper and systematic certification is required. Fuel blending, storage, transportation and handling are the few main topics to be dealt with. All the services of these alternate fuels allow them to be made usable because of their unique properties. These fuels can cause the decreased dependencies on the foreign fuel.

The commercialization [6] of these fuels is not possible unless the standards of these fuels allowed to modify them. Many initiatives have been taken up for instating these synthetic fuels.

The fuel properties of interest are as follows:

- Conductivity and gravity,
- Composition of different elements present,
- Stability toward temperature changes,
- Reactivity toward lower temperatures,
- Swelling of elastomers,
- Fuel lubrication properties, and
- Heating characteristics.

Cost is one of the prime reasons for these fuels to not come in market. Due to increased demand and pricing of commercial fuels, alternative fuels [2] are finding a path to establish strong firm ground in the market.

New challenges include establishing fine firms with the private industries. Price, availability, and demand are a few obstacles which have to be dealt with. CO_2 production is one of the main environmental obstacles which gets produced during the production of synthetic fuels. Transportation and job security are the few barriers of commercial sector.

2.4 Weight Reduction

One of the major drawbacks in the efficiency of aircraft is the additional weight present in it. Main factor of the additional weight is the additional fuel carried for safety reasons and this is called tankering. Tankering [2] is done so as the flight does not fail in the worst situations of low fuel conditions and is done for commercial reasons. It is estimated that because of tankering, 0.5% of total fuel is burnt. All commercial flights are mandated to carry additional fuel for safety of the civilians and the onboard passengers. Factors [2] such as mass of the aircraft, the route, weather forecast and others are taken into account for the calculation of how much fuel should be taken onboard. Usually, this estimation is carried out on a well setup computerized database. The pilot is provided with the overall flight plan and the calculated amount of fuel; however, the captain finally decided how much fuel should be taken onboard. This factor of safety is one of the main reasons for the increase in additional weight of the aircraft.

A weight reduction like this can save up to 1–2% of the total fuel. The total reduction of fuel burnt is less than 1%.

Other factors of weight reduction may include reduction of weight from the passenger seats.

For making the seat-back profile [6] thinner, companies have adopted the method of using polyester netting material rather than conventional foam padding [6]. Reducing seat pitch from 32 in. to a less by 2 in. that is 30 in. was obtained for the pocket to position above tray table from the knee position which ultimately led to achieve 4 cm more legroom space. Weight was then reduced by 30% on each seat.

Improving manufacturing for more consistent and accurate parts for better construction and tolerances, using materials which have better strength to weight ratios and reducing the number of parts to reduce the joints which needs proper strengthening are few methods applied by sector to cut down the weight.

Few airlines have adopted the method of using composite oxygen bottles. This did not only remove 16 kg per aircraft but are cheaper and have lower maintenance costs.

Even the lavatories [2] can be improved to reduce the overall weight. This is done keeping in mind the comfort of passengers so that they need not find the space too clumsy. Weights can be removed from removing unnecessary stuff from the lavatories. The main of the company is to reduce the weight up to at least 26%. Thin dimension walls, better non-conducting materials, less mass motors and new techniques—such as the glue type usage rather than screws made of metals—are among the new innovations which can be adopted. A reduction of 20% has been found in the weight of galleys.

2.5 Noise Reduction

How noise is monitored? For monitoring the noise produced by an aircraft, first it flies over a monitor. Usually, the noise level reaches its peak value within 10 s and then slowly gets reduced to the normal background value. This is called as an event and it lasts for over 20–40 s depending upon the height of the flight, type of aircraft and loudness. A threshold noise [5] level is set in the monitor. When the aircraft noise is above this level, the event begins and ends when the noise drops below it. The noise level is noted for every second by the monitor and is compared with the peak voltage. The data is averaged to avoid extreme results and then stored and matched with the radar data.

How can noise be reduced? The major noise reduction has been achieved by reducing the jet noise. Nowadays, the aircrafts in use have a higher bypass ratio (BPR up to 10–12) [1]. This led to the dramatic reduction in the jet noise. The fan size plays a major role in noise. A larger fan means a stronger noise source.

Rotor boundary [8] layer interface noise is caused when the rotor fan blade interacts with the boundary layer of the inner duct. Eddies produced in the rotor boundary layer with the rotor trailing edge produce rotor boundary self-noise, whereas with the vane trailing edge produces OGV self-noise. OGV interaction noise is caused when the rotor wake interacts with the downstream outlet guide vanes.

Nacelles [7] are considered as a major support to the turbo-machinery as it reduces noise through acoustic liners but it adds a lot of weight and spurious drag. Hence, the ultrahigh bypass ratio nacelles should be reduced in length and volume. In order to lighten the nacelle, both the upstream and downstream are shortened. Internal noises are absorbed by shortened ducts. Noise absorbers (liners) should be optimized. These liners behave as Helmholtz resonators and allow noise of a certain frequency range to be reduced.

However, while considering the noises, we should also consider the noises that are produced by the open engines, core machinery noise, such as compressor noise [7, 2], turbine noise or even combustion noise. Noise can also be changed by tuning the parameters such as changing the shape and length of blades, adjusting the gap between the propellers or even their rotation speed and direction. Latest noise reduction technologies involving 3D-optimized designing [2] of blades are emerging. By the upcoming technologies to reduce the drag and the noise, fuel consumption can also be reduced. The future configurations would strongly reduce both fuel consumption—through a dramatic reduction of the drag—and noise, with masking effects.

2.6 Smart Navigation

Non-directional beacons (NDBs) [9, 6] are used with a digital display. This digital display directs the beacon from the flight. Cross-cut is made by intersection of two beacon lines to locate the aircraft.

NDBs [1] can also lead to unexpected readings. These are caused because very high range of wavelengths is used, which are affected by the ground and the atmosphere. With the help of few navigation aids, NDBs still continue to be the main subject for navigation. Decent navigation aids are provided by VOR, and it uses the main system of navigation maintained for aircrafts those fly under IFR. Beacon generates sine waves which being out of phase with each other. Distance measuring equipment (DME) [1, 5] allows a receiver which is suitable for calculating the exact displacement from the station. Using this feature, the distance can be calculated with a single beacon itself. Transmission of data with the local weather stations can be a great aid for the pilot to manoeuver. VOR and DME correlated come under the category of TACAN [7]. On inert continental routes, airliners used inertial navigation system. GPS was made available for the civilians after the incident of shooting down of Korean Air Lines Flight 007 took place. After all the steps, ATC [7] can finally convey any information to the pilot which can be the position of aircraft. GNSS in aircraft is getting increasingly common in today's world. In terms of aircraft position, altitude, heading, etc., GNSS provides the least accuracy. GNSS instrument approach is day by day increasing its reputation.

Smart tracking ensures that the aircraft flies within a specific route which is defined by smart navigation. Smart tracking [5], therefore, delivers safety of the aircraft through precise and smart navigation. This increases the predictability of the aircraft by traffic controllers which helps in reducing air traffic as well. This sophisticated navigation system [1] reduces the workload and reduces the physical work of air traffic controllers and pilots. This system alerts the crew members if there is any deviation from the precisely planned path by the aircraft. Smart tracking [6] conserves hundreds of tonnes of fuel every year. Smart tracking reduces the efforts in case of bad weather conditions. This also means very few arrivals and departure delays for the passengers, and aircrafts need not engage themselves in air waiting for the clearance of runway or get diverted to other airports (Fig. 2).

2.7 Smarter Air Traffic Systems

The growing body mass of craft, vehicular increment in traffic volume, is the current scenario that keeps us thought-provoking. The government has been exploring many measures to make sure air transportation safety and to forestall accidental incidents [8]. They constitute improving aviation safety facilities, variedly modernizing aviation safety services, ensuring improvements in systems for safety of aircraft and improving ATM data systems.

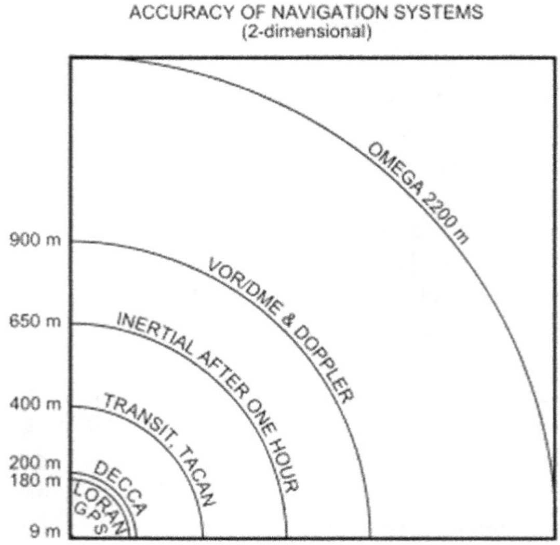

Fig. 2 Accuracy of navigation systems. *Source* Air Services Australia

I. **New RNAV (Air Navigation System)**:

(a) **Improving Advanced control Systems**:
The government can look forward to the development of modern engine control systems designed to boost control proficiency and refuge, utilizing numerous management support functions together with the function of evaluating up and down airplanes, and to scale back the assignment of the controllers on hand by introducing a well-developed integrated interface for preventing human errors, done by linking the aviation control information processing systems.

(b) **Improving the Air Traffic Management (ATM) System**:
To aid for wide operation of ATS, ATFM and ASM [6], the government can reform the surveillance reviews and controlling the potency of traffic for endorsing the progress of ATM systems. The government also will bring up a solution counterfeit function for outlining air routes in airspace, and intensify Collaborative Decision Making (CDM) among the air transportation corporations and management depots to bring progress in the systems.

(c) **Terrain light and telegraphy facilities**:
The government can install secondary surveillance radio detection and ranging mode S (SSR Mode S) to aid for strong aircraft surveillance in high-traffic airspace.

(d) **Improving flight inspections**:
The government is initiating the idea of area navigation (RNAV) and also the unusual air navigation systems [6] by promoting a replacement and an improved flight scrutiny system.

II. **Improving Airspace Utilization**:
To upgrade the air traffic control, the government can reorganize air zones and air courses effectively, hence exploiting the ATM and RNAV function operations, and wisely employing the quantity of airfield for continual flight routes. The government can perform a reorganization of the air routes (sky highway project) by developing RNAV routes nationwide, due to traffic by its significant establishment in addition to existing air routes, which parted the gradient of the RNAV tracks and the existing routes operationally.

Utilizing the functions of ATM centres:
The government can attempt for secure and dynamic use of airspace, in alliance with suitable body agencies and air transportation corporations. Hence, this can only be achieved by desiring and executing multiple traffic conditions across entire agrarian in a unified manner at the ATM centres. Furthermore, action should be taken by airliners to cut back flight delays by an alteration for personal aircrafts to fly in training air zones unused by the autonomic armed forces.
Advanced traffic data systems based on ADS-B:
Determining the position and digitally exchange data with aerial and field-based facilities are most likely coming up in unborn traffic control systems. All the participating aircrafts can continually broadcast their own position in track, gradient, airspeed, trajectory line and identification to the air traffic control faculties, using the automatic-dependent surveillance—broadcast (ADS-B) [7] systems. ADS-B aircraft endlessly receive constant data of routes from like equipped aircraft within the space (line of sight) that grants for onboard displays for the encompassing traffic. ADS-B has been implemented with loads of success in Alaskan trials and needs some groundwork to be usable. ADS-B instrumentation is exhibiting promise for improved traffic partitioning on transoceanic stream routes of the scale of field-based systems. ADS-B beacons are relayed on the 978 MHz channel. The data gathered will then be broadcasted back to every TIS airplane, from all the incoming aircrafts [5], which hence provides for an in-depth image of the traffic scenario and present considerations, even in the absence of these ADS-B onboard systems in the aircraft.

2.8 *Smarter Airports*

A. **Improving major airports in large cities**:

Increased demand for domestic and international aerial services are arising facts that many developments can be taken as an initiative by government to establish main airports in large cities. For confirming traveller safety, the government will be looking up towards creating efficient air passenger terminals facilities ‘barrier-free’ by eliminating static steps and planning a better outlook for establishing better use by the senior and disabled individuals from point of view of achieving general environmental improvements by promoting universal-movable designs point of view.

B. **Improving airfield safety technology**:

Safety measures are highly intended for which it is vital that the amenities like runways are constructed strictly by laws, and suitable standards. Additionally, all these newborn facilities should be well-controlled and managed to make sure their prompt operation in any respective times. In this viewpoint, they will guarantee to enhance this technology concerning airfield safety [8] that directly causes consequences in traffic safety. This includes certain measures to be taken throughout construction work in airports, upgrading multiple way airfield signal facilities. Furthermore, the government can look upon comprehensive learning to investigations and promote active study for the introduction of a lot of modern airfield safety technology.

Airspace around airports:

The government can work to retort to the satellite RNAV (GNSS) to broaden the RNAV (Area Navigation) geological routes by recommending new approach, return routes and advent modern effective techniques. Therefore, the volume of traffic is predicted to extend further significantly, as the traffic congestion within the airspace of Kanto space as well as Tokyo International airfield and Narita International airfield is turning into pronounced, and therefore, the government can perform a reorganization of the Kanto airspace with the re-expansion project of Haneda field, by introducing RNAV and new systems [8]. This caused reorganization is anticipated to modify a rise of airspace capacity, improvement in operational potency, environmental measures and reduction of the work of controllers and pilots by efficiently utilizing the restricted airspace of Kanto space.

2.9 *Advanced Safety Measurements*

An altitude is set by pilot called 'pointing' [5] which is the minimum altitude below which the plane should not drop as it may lead to mid-air collisions. The co-pilots should verify the condition of speed brake before the flight. Speed brake failure might lead to missing of landing on the runway. Flaps are extensions on the aircraft which helps the flight to keep its balance and stay aloft when the flight is taking off at very slow speed. But travelling at high speed might damage their motor. It is the duty of the co-pilots to remember these speed limits. The flight needs to decrease its speed rapidly while maintaining its balance during landing. Spoilers are wing surfaces which reduce the lift of the plane when the speed is decreasing suddenly to land carefully (Fig. 3).

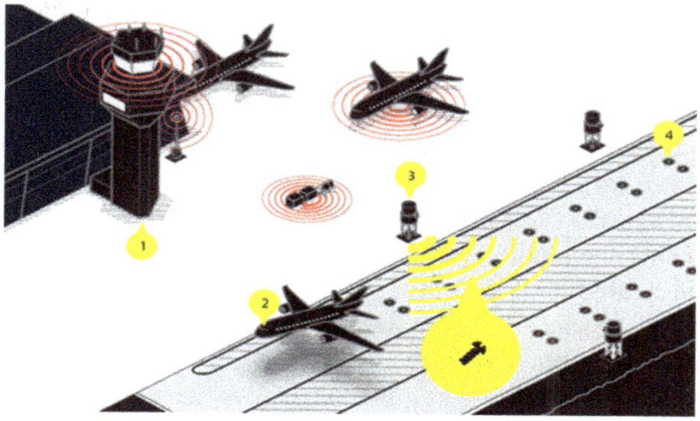

Fig. 3 Various components of an airport. *Source* Civil Aviation Office

2.10 Building a Safe Airframe

To make the aircraft resistant to temperature changes and various other physical stresses, ball bearings made of new ceramic formulas are used [8].

Any failures in the plane are detected by ultrasound. Materials embedded with resin filled with nanostructures are used to patch the cracks that are formed.

While generating lift [5] to the flight while taking off, high pressure is applied to the wings. The area where wing connects to the fuselage, the stress on the metal is high. Any hole or deformation on the skin of the metal makes it susceptible to cracking. Hence, high level of maintenance is required using ultrasound equipment for weaker signals.

To handle the changes that are caused by cabin pressurization, aluminium fuselages are used.

2.11 Building a Safer Airport

The digital map is updated instantaneously by integrating the data from the GPS of the main plane and from the different other aircrafts and also the vehicles on the ground. In order to avoid the regular updating problem of paper maps, digital maps are used. These are regularly updated with obstacles and routes. Electronic flight bags (EFBs) [5] are used that login with the cockpit navigation system. To spot small obstacles, detectors with radars having wavelength as small as millimetre are used. Runway lights help pilots to land in the night-time. Red light indicates that the runway is busy and green indicates it is ready for takeoff.

2.12 Improvements in Military Aviation

Military aviation is the comprehensive use of armed aircrafts and alternative flying machines for the needs of implementing (or) sanctioning aerial warfare, as well as national airlift (air cargo) [1] capacity to conduct logistic supply to forces stationed in a theatre or along a front in the warfield. Aerial power includes the national means of imposing such military warfare, as well as the intersection of transport and warcraft. Military armoured aircrafts embody bombers, fighters, transports, trainer aircraft and reconnaissance mission aircrafts [1, 2].

How to detect Stealth Aircraft?

Modern aircraft employ a sort of concealment technologies that scale back reflection of and emission of radar and radiations like infrared, visible light, radio frequency spectrum and audio, hence having an ability to avoid detection. Besides, technology always merely delays in detection and pursuit, since stealth is not a cloak of invisibleness, thus providing tactical advantages in aerial battles and enemy defence mechanisms.

Heat, sound and other the different emissions as well as the physical properties of the aircraft should be taken strategically taken into consideration to locate and attack enemy armed forces [3, 4].

I. If a stealth fighter is carrying external fuel tanks—as is commonly the case in reconnaissance missions—it is not technically in stealth mode. Also, they are easily detectable on microwave radar by fixing a device on the ventral aspect of aircraft which can also enhance its cross section.

II. Technically speaking, the aircraft should be well optimized to surpass higher frequency bands [1]. Once the frequency wavelength value becomes greater than the particular threshold value, resonance might occur in the lower part of the aircraft which is capable of being seen. This situation arises when a property on the aircraft, like a fish fin like a tail, is less than 8 times the dimensions of a specific wavelength frequency [2]. The surface of the stealth aircraft can further facilitate concealing the aircraft by microwave radar absorbent coating materials containing carbon black particles or small iron spheres. Civilian air traffic control [4] radars operate at a lower frequency, and hence can nearly detect military fighter-sized aircraft.

III. Additionally, a hearth control system would be required within the case of an ATC microwave radar. In order to avoid approximate approach by the radar, many nations are working to develop their military aviation by using longer wavelength UHF and VHF [1, 2, 4] early warning radars. Still, VHF and UHF band microwave radars accompany massive radar resolution cells which means that contacts do not seem to be tracked accurately and with the desired preciseness to guide a weapon not off course. In fact, missions do not require cloaking devices if the threat can see it but cannot do anything about it (Fig. 4).

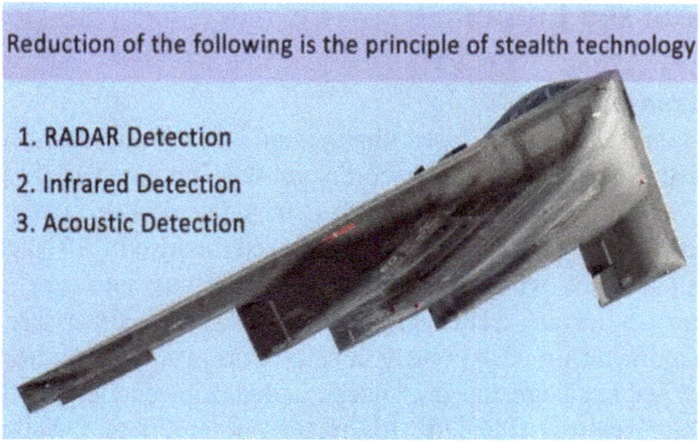

Fig. 4 Principle of stealth technology. *Source* Defence Aviation

Design Enhancement:

Basically, the design properties of the stealth aircraft tend at aiming towards reducing microwave radar and thermal detection. The design's prime priority ought to satisfy an inventory of conditions that ultimately decide the potential of the aircraft. These include the following [5, 2, 10]:

(a) Reducing the thermal emission from thrust.
(b) Reducing microwave radar detection by neutering physical characteristics of the overall configuration.
(c) Decreasing microwave radar detection once the weapon bay opens.
(d) Lessening the degree of infrared and microwave radar detection throughout adverse atmospheric condition.

It is unpredictable to operate the B-2 aircraft, in its fully operational gradient height which results in imposing a suitable time of flight for armed weapons leading to miss-targeting. And aircrafts like the F-22 and F-35 [5] easily can revert themselves back to stealth aircrafts in few seconds after their very actions of opening their hood bays and deploying ordnances. Whatsoever continually, there is always a time-off between stealth or range and payload.

Industry consultants [2] recommend a mix of mega-pace infolinks and less-frequency array-phased radars to spot the presence of stealth aircraft and generate suitable ordnance quality track. This can be overcome by typically operating the stealth aircrafts alongside electronic attack platforms similar to the Boeing EA-18G Growler [10]. Since detection is all concerning the signal/noise ratio, the low noticeable signal is reduced, whereas electronic attack will increase the noise. The U.S. Navy and Lockheed are already making the requirement to develop even additional refined sensors [2] that cue radars regarding the invisible blackbirds that cast our skies and are intensively operating in these areas of dynamic stealth technology.

3 A Look at the Future

Supersonic Transport

For the dream of travelling faster than sound, supersonic transport (SST) was introduced for the civilians. Only Concorde and the Tupolev Tu-144 [5] have been to regular service in terms of SST till today. In today's world, there is no SST in commercial service. Few companies have proposed the idea of SST again to bring it back for the business jets which might increase its commercialization again. Current ongoing design studies are related to supersonic airliners. Excessive noise generation, high manufacturing costs, costly construction materials, higher mass and an increased cost per head are the drawbacks and design challenges. Despite these challenges, Concorde was operated with quite some profits in a shallow market for over 27 years.

The Japanese Space Agency JAXA [5] developed the next-generation supersonic transport which is a supersonic transport (SST). It has the capability to carry 3 times the passengers, fly at almost the same speed and have twice the overall range. The main goal was to achieve a net ticket price almost as comparable to that of subsonic business class. An 11.5 m prototype has already been tested by the agency. The strength of the sonic boom SST generated is one of the most crucial factors in the viability of the transport. The boom created was so powerful that it could fly over land, thus not disturbing other passenger flights. Till today, many researches have been done regarding this effect.

ELON Musk's BFR:

A tall beast of 106 m slightly shorter than Saturn V and 9 m in diameter [5] is the latest plan according to Tesla CEO, Elon Musk. Having a capacity of 150 tonnes of cargo or 100 passengers, this beast is powered by 31 raptor engines, which has an improved thrust-to-weight ratio reason being the usage of mixture of methane and oxygen. It can be reused. The rocket goes into Earth's orbit that splits into two with upper second stage remaining in space holding an amount of fuel whilst lower returns to passengers to pick up passengers. The lower stage is refuelled in earth when required and blasts off with people in it. The seats will be partitioned into cabins and other areas with a great luxury. Musk foretold the attendees that the very first Raptor engines [5] have already been designed and built and test fired for 1200 s, so far. He confessed that the final designed rocket could be assembled within a year. By 2022, a couple of test flights would be sent to scout out suitable landing sites on Mars, accompanied by two cargo flights bringing supplies, before a manned mission in 2024.

The major bulk of that cargo would be the equipment required to manufacture fuel to make the trip home. Eventually, musk desires to move thousands of individuals to Mars and let mankind escape the surly bonds of Earth. Travelling at 18,000 miles per hour, the rockets promise quick connections between key economic hubs.

In the future, Musk believes that passengers can board rockets in New York, traverse 7000 miles and land in Beijing 39 min later.

80–200 individuals will blast from Hong Kong to Singapore in 22 min.; Los Angeles to Toronto in 24 min.; and Dubai will be reachable from London within half an hour. Undoubtedly, the price tag for a rocket ride would be appropriately astronomic, although musk claimed in an Instagram coda that the price would be no beyond a full-fare economy price ticket.

The earth-bound logistic potential for such speeds is important.

VTOL: Vertical takeoff [5] and landing vehicles made its mark owing to experiments administered during the years 1950–1970 and the majority came out to be problems which became failures. Generally, VTOL employs the concept of taking a short run before takeoff, and thus they are also known as to as STOL, short run takes off and lands the vehicle. The flight's overall stability and control of VTOL/STOL are extremely troublesome are is of major importance of analysis in the current field. VTOL constitutes primarily three configurations since the current developments in this field, they being configuration which is like the wing-type, chopper-type configuration and the configuration which is like the ducted type. Wing has mounted thrust enabled engine, and ducted kind has rotor equipped with duct type that enables elevation of chopper-type mounting on top of it to give the lift. The wing-type configuration was of initially developed in VTOL, and then military and primary purposes were physically human force operated; however, later its use was understood and many more pro-designs of it came into being. Design in direction of remote-controlled VTOL was the first major step taken by the Hover eye platform from Bertin Technologies [5]. The eye hover platform brought about a major success in the VTOL ducted-type conformation. Latest trends in the field of aerospace for unmanned aerial vehicles and application of pictorial are well known. Numerous chopper-type conformations implemented for UAV are explicated along with their benefits when compared to any other conformation here. The most recent current field research is hoverbike that could be a hybridized machine. It uses ducted rotors to realize its required elevations in order to achieve its goals. It may be either run or not run by a human force. Its conception came from the hovercraft that has hybridized capability and used fan type with ducted rotors to hover and manoeuvres. Design with one engine has issues of stabilization and therefore the design with engines in multiple numbers effectively. However, engines with open rotors prove to be hazardous, so it ought to be very well opted. The main aim is to design a rather cheap, less mass, heavy power to weight and more precise control for VTOL and stabilize its overall performance. Such VTOL may be used for nation's defence against terrorism for complete operations under observation.

4 Results and Conclusions

By improving the engines, using materials of lighter weight and improving the aerodynamics of the plane, we can improve the efficiency of the engine. Hence, less fuel is burnt and less CO_2 is emitted. Every year because of advancements in the aviation technology, carbon dioxide emission is being brought down. It is estimated that over 80% of carbon dioxide levels have been reduced since the first aircraft. The major objective of aviation industry is to improve the fuel efficiency during combustion. This is achieved by preparing alloys which are resistant to heat. To obtain more thrust by consuming less fuel, gear systems should be improved. Advanced technology developed materials that weigh less than the metals are used in the making of the aircraft. Carbon fibre is one such example. 3-D designing and printing also make the manufacturing process easier and effective. Emission of aerosols by aircrafts into the air causes air pollution, resulting in depletion of ozone and also lead to global warming. Hence, usage of alternative fuels instead of crude oil reduces the emissions into the air over more than half. The drag caused by the weight of the plane during its flight can be reduced by improving the winglets design and engines. This can cut down the fuel consumption and reduces the emission by 6%. By using smart navigation technology, extreme weather conditions can be overcome which reduces the fuel consumption by over 1.4 tonnes.

By improving airport structures and design with the help of electric vehicles in the runways, solar panels and also reduction of time of taxing are the key factors to improve an airport.

Acknowledgements We would like to express our gratitude to Vice Chancellor of our college, Birla Institute of Technology, Mesra, Dr. M. K. Mishra. We would also like to thank Head of Department of Electronics and Communication, BIT Mesra, Dr. Srikanta Pal for his patience and valuable help.

References

1. Avidyne Corporation, Cirrus Design, Inc. Garmin Ltd., Rockwell Collins Inc., S & Tec Corporation (2009) Advanced avionics handbook. U.S Department of Transportation, Washington, DC
2. The National Academics Press (2001) Improving the efficiency of engines for large nonfighter engines, Ch. 9, 500 Fifth St., Washington, DC
3. FLC, Army aviation and missile research development, Dept. of Defense, Redstone, Arsenal (AL)
4. Dsouza L (2016) Defense aviation, May 3, 2016
5. Warwick G (2016) Aviation week & space technology, July 29, 2016
6. Wise J (2009) Anatomy of plane crash. Hearst Communications, Dec 3, 2009
7. Bowditch N (2010) The American practical navigator. National Imagery and Mapping Agency, Bethesda, Maryland, p 815, Retrieved 14 Dec 2010
8. The National Academics Press (2001) Securing the future of U.S, air transportation: a system in Peril, Ch. 4, 500 Fifth St., Washington, DC
9. Grierson M (2008) Aviation history-demise of the flight navigator. FrancoFlyers.org website, Oct14, 2008
10. IPCC Special report aviation and the global atmosphere. Cambridge University Press, Cambridge, UK, 373 p, 1999

Study and Design of Smart Industry: A Review

Divyesh Kumar Maurya, Ankush Kumar, Suraj Kaunoujiya, Deepak Prasad and Vijay Nath

Abstract Now, the time industry is moving to a smart technique. Industry is facing a new worldwide revolution: integrating Internet-based information and communication technology (ICT) into their complete value chains. This development is often referred to as the Internet of Things (IoT), smart industry. Beyond leading to more efficient and adaptive production and supply chain processes, it also will lead to a complete new generation of always-connected products.

Keywords ICT · IoT · Automation

1 Introduction

Smart industry is a broad concept that covers a number of technological developments that will change the way companies function and help them improve their competitive position. Many entrepreneurs inside and outside the Netherlands are stressing the importance of standardisation to increase interoperability and innovation and boost the competitiveness of industry. The Dutch employers' organisation FME, TNO (Netherlands Organisation for Applied Scientific Research), the Ministry of Economic Affairs, VNO-NCW (Confederation of Netherlands Industry

D. K. Maurya · A. Kumar · S. Kaunoujiya · D. Prasad (✉) · V. Nath
Department of Electronics and Communication Engineering, Birla Institute of Technology, Ranchi 835215, Jharkhand, India
e-mail: prasaddeepak007@gmail.com

D. K. Maurya
e-mail: divyeshald121@gmail.com

A. Kumar
e-mail: chauhanankush631@gmail.com

S. Kaunoujiya
e-mail: Suraj12111998@gmail.com

V. Nath
e-mail: profvnath@gmail.com

V. Nath and J. K. Mandal (eds.), *Nanoelectronics, Circuits and Communication Systems*, Lecture Notes in Electrical Engineering 511,
https://doi.org/10.1007/978-981-13-0776-8_55

and Employers), the Chamber of Commerce and the Dutch ICT trade association Nederland ICT have developed the Smart Industry action agenda. This is currently being put into practice within so-called field labs, for example, field labs are practical environments within which companies and knowledge institutions are developing, testing and implementing smart industry solutions in a targeted manner, as well as environments within people [1].

Smart industry means that any industry where the technology is important for production and development. Industry performance relies on the advancement of planning and structure ability. It contains the idea that how to design manufacture, adapt and use modern instrument, IoT equipment and also admit new machinery which is need for industry such as mining and agricultural. Industries and technology have greater potential of how to apply products in the more developing areas. There is an extreme difference between developing and developed areas. Smart industries perform suitable work in developed areas, whereas in developing it needs to perform work in larger extent [2]. For example, in agriculture sector, it is necessary to design the control equipment that can be used for adoption and alternations.

Digitalisation is leading to dramatic changes within the Dutch manufacturing industry. Possibilities are emerging for companies to generate new business by taking advantage of new technologies such as big data processing, the Internet of things, new adaptive robots, 3D printing, nanotechnology, miniaturisation and new sensor technology. These developments are opening up all manner of opportunities to manufacture products more cheaply and with higher quality, as well as to align them much more closely with customers' needs, as they can increasingly be manufactured on a custom basis. Now, digital marketing is the great requirements in companies. It is a great process for the connection between customers and companies. Nowadays, it is a great challenge for companies to identify their customers. Because every company wants to know about his customers, ICT plays a most important role. ICT development is most important nowadays.

Standards are a prerequisite if data is to be exchanged effectively, and the necessary level of cybersecurity and the required quality of the ICT infrastructure are to be achieved. These standards may relate to technical protocols, but may also define regulations on questions of quality and security, for example. In some areas, standards are already available, while in others they are yet to be developed. Some of this development work is taking place in the Netherlands, but in many cases it is being carried out on an international level. In the production field, automotive is also so important. Maximum automobile companies are using automotive system. But now, we require other companies like agriculture industries, construction industries, etc. In these companies, pick up things and push things are required for automotive system as it is in many places required automotive system. India is the world largest automobile industry Indian government aims to develop India as overall construction as well as R&D in many cases. Smart industry standardisation is being carried out by formal, global standard organisations (ISO, IEC, ITU, etc.), consortia and associations [3]. An important role has also been reserved for companies. As sectors in which SMEs are strongly represented, such as the machinery and automotive sectors, have a greater focus on ICT implementation, SMEs—with their strong networks—are in an excellent position to transfer their strengths to other sectors, such as the ICT and service sectors. This process could contribute significantly to the development of standards. In addition, active steps are being taken in this area by various companies, such as Google, which announced in 2014 that it was developing an open innovation and research programme linked to the Internet of things (IoT), with the aim of further developing open standards 10. Companies such as SAP, Deutsche Telekom, Siemens, Bosch, Infineon, Volkswagen, ABB and IBM have also come together within Germany's Industries 4.0 platform to reach agreements on smart industry standards. After all, any party that determines the standards used to allow machines and products to communicate with each other is also in the best position to become a leading player. Internationally, the ambition is to implement smart industry with a limited number of new standards. This is in keeping with the principle acknowledged by the EU that no new standards should be created at European level if there is an existing international standard that can be used, which means that the EU is actively seeking to align itself with international standards.

In April 2016, the European Commission presented a priority plan for ICT standardisation as part of a technology package for the digital single market. The European approach consists of an integrated plan including five top priorities for standardisation (cloud computing, internet of things, 5G, cybersecurity and data) and a plan to promote key digital standards on an international level in cooperation with industry and standards organisations. Various technology platforms, such as the Alliance for Internet of Things Innovation (AIOTI) and the 5G public–private partnership, are currently guiding relevant roadmaps. In addition, the European Commission offers funding to the standard organisations ETSI, CEN and CENELEC, and, according to its communication for ICT standardisation, also plans to provide financial support via research and innovation programmes (Horizon 2020 and CEF). The Commission will also be offering financial support to large-scale pilot projects within these priority areas to validate and improve the adoption of standards.

The United States, China, Japan and South Korea 32 are also focusing greater attention on smart industry standards than countries of a similar size to the Netherlands. A number of examples of activities in this area are described here. One major initiative launched in the United States is the Smart Manufacturing Operations Planning and Control Program of NIST (National Institute of Standards and Technology) 33, which is focusing on improving innovation in the US and the competitiveness of industry by means of intelligent manufacturing systems, an area in which standards have an important role to play. This programme will enable smart manufacturing based on performance metrics, process control, prognostics, health management (including diagnostics and maintenance), integrated wireless platforms, industrial control security and efficient systems analysis during operations [4]. NIST has drawn up a report entitled 'Current Standards Landscape for Smart Manufacturing Systems' and has developed a reference architecture model.

The ‘Made in China 2025’ strategy is a Chinese initiative that aims to improve Chinese industry in the area of Advanced Manufacturing, with priority given to standards in the area of intelligent manufacturing. This goal will be achieved by accelerating the integration of domestic standards and creating a standardisation system for intelligent manufacturing.

In Japan, the ‘Industrial Value Chain Initiative’ has been set up. This was launched in mid-2015 by 30 companies, in response to the German standardisation initiatives in the area of industry 4.036. This initiative aims to create standards for linking factories and to internationalise Japanese industrial standards. The participants include Mitsubishi Electric, Fujitsu, Nissan Motor and Panasonic.

The Smart Factory Standard Research Council in South Korea is made up of private parties and is established to respond to international standardisation activities and promote locally developed standards in the area of smart industry. The South Korean government also has plans to set up a Smart Factory Standardization Forum, to enable it to respond to international smart industry standards and ensure compatibility with existing smart industry standards.

2 Structure of the System

Here, a block diagram for a smart industry a normal structure is given. Here, most important things for companies like Automation, improving quality, manufacturing technologies, etc. are mentioned (Fig. 1).

3 Methodology

The overarching smart industry action plan defines three key technological developments that constitute the basis for change within the manufacturing industry (see the inner circle of Figs. 2):

1. Manufacturing technologies: the emergence of new manufacturing technologies, such as industrial Robotics, 3D printing and printed electronics. These technologies make it possible to manufacture with fewer defects and (in many cases) at a lower cost. They also allow many more products to be manufactured on a custom basis. IOT is also a great use for manufacturing (Fig. 3).

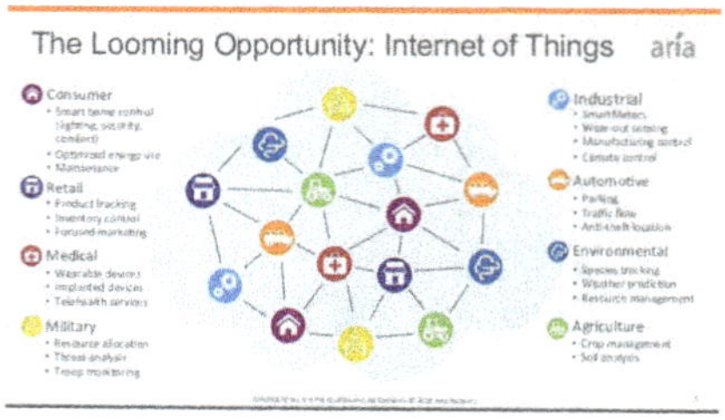

Fig. 1 Network model of Internet of things

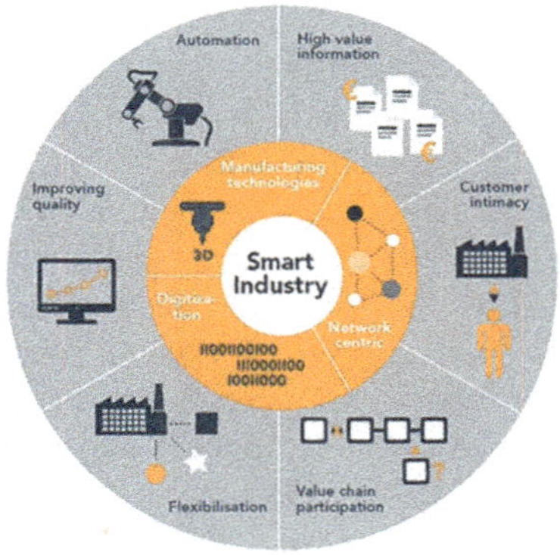

Fig. 2 Smart industry wheel. *Source* Smart Industry Action Agenda

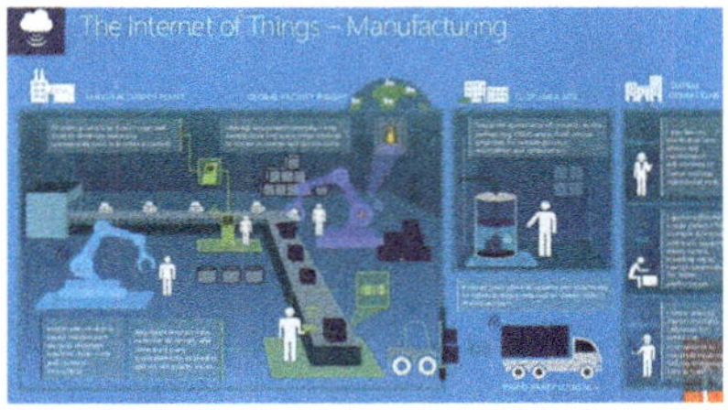

Fig. 3 Referential architecture model (Smart Manufacturing Ecosystem) NIST (National Institute of Standards and Technology)

2. Digitalisation: the far-reaching digitalisation of industry by means of sensors and high-quality ICT networks with good exchange protocols and the application of sensor technology to record and control production. Internet of things technology is resulting in more and more data sources: all kinds of devices and sensors are being connected to the Internet. Thanks to big data technology, new insights can then be gained from analyzing this data: companies have access to more information that can be put to use in the design and production process.
3. Network-centric: a network approach that involves connecting production equipment and people both within and beyond the value chain. New digital technologies allow data to be shared in a controlled way. This means that companies can increasingly specialise in a particular area of production and customers and suppliers can work together on a product more easily.
4. High-value information: Manufacturers can improve their products and processes by analyzing the constant flow of usage data from their systems. This may be the data from a production process, but also data received from smart sensors on a product that has already been delivered, for example. Manufacturers will therefore able to monitor the quality of their products more effectively during production, as well as over the life of the product, and in this way respond better to their customers' needs.

5. Customer intimacy: Customer intimacy is increasing, as it is possible to offer greater customization at a lower cost price and customers are able to participate in the product design process. The product can therefore be tailored entirely to the customer's needs.
6. Value chain participation: Digitalisation is changing production chains and leading to closer cooperation between various parties in the chain. As a result, production chains need to be organised in a different way. New parties may become part of a chain and existing parties may specialise in a particular production process. Parties are also starting to cooperate more closely: customers and suppliers are working together on designs to ensure the product meets the customer's needs and can be manufactured efficiently.
7. Flexibilisation: New manufacturing technologies are making it possible to increase the flexibility of the production process. Using robots, for example, makes it easier to switch from one product to another, allowing custom products to be supplied more easily at a lower cost price.
8. Improving quality: The availability of data on the product and process and new manufacturing technologies are improving knowledge of the production process and how it is controlled, resulting in better quality (e.g. 'zero defects').
9. Automation: Lastly, the far-reaching automation and robotization of production is resulting in consistently high quality and lower costs.
10. Smart products: Using the Internet of things makes it possible to monitor products after they have been manufactured and have left the factory. This creates opportunities in terms of monitoring the performance and maintenance of these products, for example. Standardisation is needed to collect the data from these products and sensors and to record and use it securely and responsibly.
11. Various reference architecture models have been developed for industry, with the US and German models being the most well known. In the United States, this is *the NIST* model (Smart Manufacturing Ecosystem), while in Germany the model in question is the RAMI (Reference Architecture Mod*el for Industry 4.0*) *model*, which has been developed by industry associations. The NIST model is used as a basis for categorisation and evaluation with a view to applying existing standards to intelligent manufacturing systems, presenting a detailed overview of emerging standardisation activities in the area of smart manufacturing and identifying areas in which new standards are required to implement intelligent manufacturing systems (see Fig. 4).
12. Internet of Things: this working group is working on the identification of use cases, network-level technologies and the areas within Internet of things in which standards are lacking. It also wants to offer other standardisation working groups, the opportunity to join in discussions on the development of global standards.
13. Transport and logistics: The main issue in this sector is the development of all manner of autonomous devices, such as self-driving carts in a distribution centre. Safety standards are important here, as are standards for controlling the equipment and exchanging data.

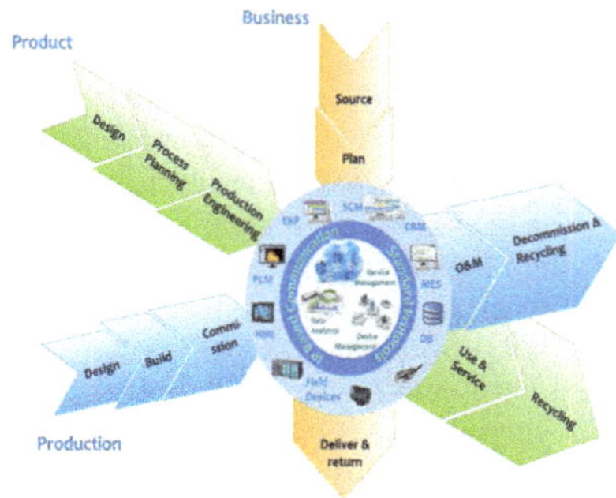

Fig. 4 The Internet of things manufacturing

Develop the human interaction standards and associated safety standards for (industrial) production and associated safety standards linked to intrinsic electrical safety and mechanical impact. Raise awareness amongst users.

Develop complex control standards and performance criteria for robots relating to accuracy, speed, output, margins for error, capacities, etc.

Develop robot modularity and standardise complex robot processes; standards for modules within robotics are important for the development of safety systems, while the standardisation of complex processes will create more flexible programming options.

Promote cooperation between companies, start-ups and knowledge/research and educational institutions in relation to the development and coordination of standards for industrial production using robots and the development of new robot products.

4 Conclusion

It is the most required technical ability advancement across all the fields, and also the necessary development of equipment. For smart industries, more necessary a smart management works on IoT, ICT, etc. in all fields of industry. Referent architecture model (Smart Manufacturing Ecosystem), NIST (National Institute of Standards and Technology) model, is the best model for smart industry.

Acknowledgements We would wish to express our gratitude to Vice Chancellor of our college Birla Institute of Technology Mesra Ranchi. We would also like to thank Head of Department of Electronics and Communication at BIT-Mesra Dr. Srikanta Pal for his patience and valuable help.

References

1. Smart Industry by Simon Robert, Corporate strategy and Industrial Development School of Economic and Businesses Sciences University of the Witwatersrand

2. IBEF research report, https://www.ibef.org accessed on 06/02/2016
3. Amsden A (2001) The rise of the rest. Oxford
4. Best M (2001) The new competitive advantage. Oxford University Press, Oxford
5. A STEP drawing is a 3D design for a product. Incorporating additional characteristics into these drawings is advisable, to make it easier for a manufacturer to interpret them. See: https://en.m.wikipedia.org/wiki/ISO_10303
6. http://www.smartindustry.nl/fieldlabs
7. Further information see http://www.rvo.nl/subsidies- regelingen/intelligente-netten/praktijkverhalen-0
8. http://www.smartindustry.nl/wpcontent/uploads/2015/09/15-0820-De-verdienkracht-van-Normalisatie-Smart-industry.pdf
9. http://library.fes.de/pdf-files/wiso/11479.pdf
10. http://postscapes.com/google-offering-research-grants-for-an-open-web-of-things
11. http://ec.europa.eu/growth/tools-databases/newsroom/cf/itemdetail.cfm?item_id=8609&lang=en
12. https://ec.europa.eu/digital-single-market/en/news/advancing-manufacturing-advancing-europe-report-task-force-advanced-manufacturing-clean
13. http://eur-lex.europa.eu/legal-content/EN/TXT/?uri=celex%3A52014DC0500
14. http://ec.europa.eu/newsroom/dae/itemdetail.cfm?item_id=24190&newsletter_id=0&lang=enwww.floricode.com/portals/0/downloads

Design of Smart Security Systems for Home Automation

S. Sanjay Kumar, Ayushman Khalkho, Sparsh Agarwal, Suraj Prakash, Deepak Prasad and Vijay Nath

Abstract With continuously increasing the use of energy and growth in population, conservation of energy is very much needed. Remote access to electronic appliances can solve the problem. Instructions can be given to these systems through a web or an android application. Some of the technologies which are used can be GSM, ZigBee, Wi-Fi, and Bluetooth, along with different controlling devices. This paper gives a survey of all those systems which have already been used in various different applications.

Keywords Bluetooth · ZigBee · GSM · Microcontroller · Linux

1 Introduction

Home automation allows one to remotely access fans, television sets, air conditioners, lighting, and heating systems more conveniently and efficiently. It can be as simple as remote or automatic control of a few lights, or it can be a whole system that controls all major parts of your home. A remotely accessible environment is defined where each appliance can be controlled using a user-friendly software which includes Android as well as a Web application. These systems have a large demand in the market. In this paper, we conduct a survey based on their advantages and disadvantages.

Home automation is one of the most widely known technologies, which is used by both public and private sectors in this era. Home automation is the new emerging technology which not only eases the living hood but also terminates unexpected causality. This leading technology is not only limited to home but also for industry, medical, building, etc. It also helps in regular health checkup without visiting hospitals. This technology mainly consists of different types of sensors, control

S. Sanjay Kumar (✉) · A. Khalkho · S. Agarwal · S. Prakash · D. Prasad · V. Nath
Department of Electronics and Communication Engineering, Birla Institute of Technology, Mesra, Ranchi 835215, Jharkhand, India
e-mail: sanjaysurishetty@gmail.com

V. Nath and J. K. Mandal (eds.), *Nanoelectronics, Circuits and Communication Systems*, Lecture Notes in Electrical Engineering 511,
https://doi.org/10.1007/978-981-13-0776-8_56

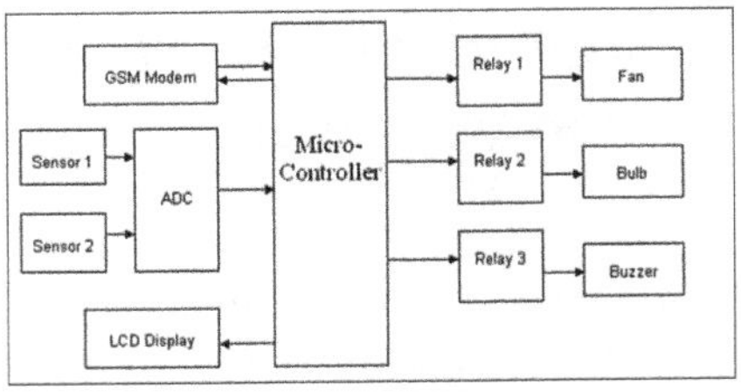

Fig. 1 Home automated system using GSM and ZigBee

units and module which assembled together to give their utility. Since many researches have been done in last few years, still improvement is required in terms of efficiency, accuracy, cost, and robustness. Here, in the present paper, an automated home system is designed by integrating GSM, ZigBee, and other modules together (Fig. 1).

2 Methodology

2.1 Home Automation System Using GSM

This technology is very useful today to provide home security. We can control our home appliances through this technology. In this technology, we use GSM mobile phone to give commands to our home appliances through SMS. We can check the status of our home appliances through this technology to know whether a particular appliance is in "on" or "off" state. To implement this technology, we need the following requirements:

Sensors:
We need various types of sensors for different appliances for its controlling, for example, for detecting smoke, we need sensor to detect it.
Microcontroller:
When our sensors detect anything, it passes the information or signal to the microcontroller. When the microcontroller receives the signal, according to the program which has been done on the microcontroller, it performs actions on the home Appliances.

2.2 Bluetooth-Based Home Automation

This system makes use of Bluetooth technology for home automation system. For using, we need microcontroller, Bluetooth module, and Bluetooth controller which can be installed from Google Play Store. Bluetooth devices can scan other devices which are located near it within the range of 15 m. For controlling the function of the appliances through Bluetooth technology, we need to program or code for different appliances. Then, after programming we need to "on" our mobile Bluetooth and open the Bluetooth controller. When we "on" the Bluetooth, then it will scan the nearby devices. We will connect our Bluetooth to the Bluetooth module which will receive the data sent by the mobile through Bluetooth controller. The Bluetooth module will send the received data to the microcontroller for the processing. The microcontroller reads the data and after reading the data it controls the appliances accordingly.

2.3 Phone-Based Home Automation

Home automation can be used to control the home appliances like bulbs, fans, television, etc. through a mobile from anywhere. There is no need for a remote, we can control the appliances through mobile, for example, if we forgot to switch off the appliances while going out, then we can use the mobile.

The cell phone can be used as a remote control. In the system, the mobile phone is connected to the headphone. It receives the "Dual Tone Multiple Frequency (DTMF)" by the headphone and passes it to the circuit. The system contains three components. The system starts with the first stage which consists of DTMF receiver and ring detector followed by an IO interface unit, and last stage is PC which completes online process. It detects the ringing of line and authenticates the user. Now the microcontroller sends the command to the IC which controls the relay.

2.4 ZigBee-Based Home Automation

Home automation can be done using ZigBee wireless communication technology. It requires a PIC microcontroller and voice recognition. Voice can be taken through a mike which is then compared with the already stored voice. The voice command is then transmitted from the PIC microcontroller to the receiver via ZigBee. Another PIC microcontroller is present in the receiver unit which processes the command and use relays to control the Appliances. The main drawback of the system is that ZigBee has a low range communication so we cannot use it from far away places. The second drawback is that the noise recognition can be unwieldy. Also, the ZigBee communication system comes with a smoke detector which sends a message to the mobile when smoke is sensed.

2.5 Wireless Technologies

Many systems use wireless communication and its main applications can be seen at home, office, workplace, or some other suitable place. Wi-Fi and Bluetooth can be used for this system. The Open Surface Gateway Interface (OSGi) is used. Different appliances are networked together through different ways. At the user end web browsers, pocket PC application and the main console are used. Speech recognition technology can also be used for the operation. Modern features such as device discovery and device connection are present or used here. The entire system uses LINUX OS. Many ways are used by the universal plug and play system for interoperability or the ability of computer systems or software to exchange and make use of information which is its main advantage which also contains the ability for sharing of service.

This system makes the use of Arduino's Wi-Fi. A smartphone is used to communicate with a home PC or a laptop which acts as a server. The Arduino board is connected to the PC and is controlled by it. MATLAB is used for GUI designing. Communication between the mobile phone and the PC is done with the help of Internet which can be accessed through Wi-Fi. The advantage of this technology is that it can be used from distant places through the smartphone. Disadvantage being that the PC needs to be left on all the time which consumes power and electricity. Also that Wi-Fi which is used for Internet access is not very reliable source and cannot be used widely and is short ranged.

2.6 Mixed Type

Our purpose here is to achieve home automation with the help of GSM, Bluetooth, and ZigBee. This can be done through an Android application which is user-friendly. In this application, the user's voice as an input and converts it to a textual command. These textual commands are being sent to some other phone present in the home through SMS. After this, the commands are sent to PIC controller through Bluetooth technology. The PIC controller sends these commands to ZigBee transceiver. The ZigBee transceiver sends these commands to the main controller where the commands are decoded and operation has been performed accordingly. In this way, this technology can provide remote access. However, the system can become expensive by using multiple controllers and technologies. Also, the SMS system which is used here is very slow and costly also.

Another way to use GSM and ZigBee is that it uses SMS to send commands to the home mobile from remote mobile. In case of any unnecessary activity, the system uses GSM technology to send SMS alerts to the user to warn him of his belongings measures. The user then takes appropriate actions.

ZigBee module, Atmega MCU (microcontroller unit), and a mobile set are the most important hardware to be used. Both the ZigBee and GSM network technologies are used by the microcontroller. Sensors present then communicate with ZigBee through control unit of microprocessor. Whole system is implemented using C language or any other user-friendly language. The relay controls the home appliances. This system is user-friendly and is less expensive. Home security is an additional feature. Disadvantage is that it cannot happen or operate in real time.

3 Discussion

This paper has been made to introduce all the technologies which can make our home smart. The technologies which have been used that have certain common features. All the technology can be implemented to make our home smart and provide security to our home. All the technology which is to be used has certain advantages and disadvantages. The good feature of the home automation system is that there is a very good interaction between users and the appliances through the technologies which are applied in this. Using these technologies, user can know the status of each appliance. Android smartphones are generally a handy tool for controlling the home appliances in the home automation systems like, for example, GSM and Bluetooth technology. Another technology called IOT (internet of things) [1] makes our work easier as it is the fastest technology among others which have been studied. But the biggest issue with this technology is the privacy because if it has been hacked by any hacker then it can be misused. The Bluetooth device in general can work within a very small range between (10–15) m.

4 Conclusion

According to the pros and cons of different technologies and methods, this paper provides a most suitable method for home automation with the help of GSM (data channel for internet access). But this requires 24 h Internet to ensure security and remote access. This will make use of the home appliances using the Internet protocol. This is made sure with the help of a mobile and web, so that everyone can access it easily. It should be easily available and a cost-free; only then can automated homes become commercially viable. A new device can be added to the system which is an additional feature and advantage too.

Coming future will surely see more advanced and yet simple methods for home automation which will conserve more power and electricity and making it reliable, cost-free, and efficient. It can use various sensors for the same system for detecting movement from a distance and could also give details about the change in the surroundings.

Acknowledgements We would like to express our gratitude toward our Vice Chancellor, Birla Institute of Technology Mesra, Dr. M. K. Mishra. We would also like to thank Head of Department of Electronics and Communication, BIT Mesra, Dr. Srikanta Pal for his patience and valuable help.

References

1. Sagar V, Kusum S (2015) Home automation using internet of things. Int Res J Eng Technol (IRJET) 02(03)
2. Satish P, Hariharan N, Kesh TN, Vidhyalakshmi S, Angel Deborah S (2015) Home automation systems—a study. Int J Comput Appl 116(11)
3. Ohsung D, Ha I (2015) A digital door lock system for the internet of things with improved security and usability. Adv Sci Technol Lett 109:33–38
4. Van Der Werff M, Gui X, Xu WL (2005) A mobile based home automation system, applications and systems. In: 2nd international conference on mobile technology, 15–17 Nov 2005, IEEE, Guangzhou, China
5. Jivani MN (2014) GSM based home automation system using a-inventor for android mobile phone. Int J Adv Res Electr Electron Instrum Eng 3(9)
6. Baig F, Baig S, Khan MF (2012) Controlling home appliance remotely through voice command. Int J Comput Appl 48(17)
7. Bharanialankar SR, Manikanda BC (2014) Intelligent home appliance status intimation control and system using GSM. Int J Adv Res Comput Sci Softw Eng 4(4)
8. Teymourzadef R, Addin SA, Chan KW, Hoong MV (2013) Smart GSM based home automation system. In: IEEE conference on systems, process & control, 2013, Kuala Lumpur, Malaysia
9. Piyare R, Tazil M (2011) Bluetooth based home automation system using cell phone. In: 15th international symposium on consumer electronics, IEEE, 14–17 June, 2011, Singapore
10. Lee KY, Choi JW (2003) Home automation system via bluetooth home network. In: SICE annual conference, Fukui 4–6 Aug 2003
11. Stauffer HB (1991) Smart enabling system for home automation. IEEE Trans Consum Electron 37(2), May 1991
12. Wong EMC (1994) A phone based remote controller for home and office automation. IEEE Trans Consum Electron 40(1), Feb 1994
13. Koyuncu B (1995) PC remote control of appliances by using telephone lines. IEEE Trans Consum Electron 41(1), Feb 1995
14. Narayanan VS, Gayathri S (2013) Design of wireless home automation and security system using PIC microcontroller. Int J Comput Appl Eng Sci 3, Aug 2013
15. Yuksekkaya B, Kayalar AAM, Tosun B, Ozcan MK, Alkar AZ (2006) A GSM, internet and speech controlled wireless interactive home automation system. IEEE Trans Consum Electron 52(3), Aug 2006
16. Al-thobaiti BMO, Abosolaiman IIM, Alzaharani MHM, Almalki SH, Soliman MS (2014) Design and implementation of a reliable wireless real-time home automation system based on Arduino Uno single-board microcontroller. Int J Control Autom Syst 3(3)

Auto-Train Track Fault Detection System

Vidushi Goel, Shubham Kumar, Aditya Muralidharan, Naveen Markham, Deepak Prasad and Vijay Nath

Abstract In this project, we introduce the railway track crack monitoring our system. In this system, sensors such as ultrasonic sensors are used to detect cracks in the railway track and measure distances, while infrared sensors are used to detect obstacles. Whenever a crack is detected in the railway track, a message is sent to the authorized user via GSM module. The GPS module advises the exact position with the help of SMS. This system can be used for daytime and nighttime crack detection process.

Keywords Microcontroller · GSM module · IR sensor · Ultrasonic sensor GPS

1 Introduction

No doubt, railway is the most frequently used transportation in India. In other words, it may be said that it is a lifeline of India and reason behind its high usability is its pocket-friendly fare. Also, one of the bitter truths regarding this is that very frequently we come across many accidents related to trains news. Since a single accident takes the lives of numerous people away, effort to make the ratio of accidents per year less should be implanted [1, 2]. Therefore, efforts in terms of improving safety should be done, as train accidents cause a lot of damage in terms of lives lost/affected as well as damage to property. Despite all precautionary measures, train accidents happen with an unfailing regularity, which is frightening. Our aim is to help railway administration detect faults in their track system. Overlapping of railway tracks and crossings are where most of the mis-happenings occur. In the proposed technology, the railway wagon carrying the control equipment is provided with ultrasonic sensor [3–5] to detect the crack, fire sensor to

V. Goel (✉) · S. Kumar · A. Muralidharan · N. Markham · D. Prasad · V. Nath
Department of Electronics and Communication Engineering, Birla Institute of Technology, Mesra, Ranchi 835215, Jharkhand, India
e-mail: vidushigoggle@gmail.com

V. Nath and J. K. Mandal (eds.), *Nanoelectronics, Circuits and Communication Systems*, Lecture Notes in Electrical Engineering 511,
https://doi.org/10.1007/978-981-13-0776-8_57

Fig. 1 Derailing of train

detect the fire and IR sensor to detect the obstacle in track, i.e., obstacle avoider (Fig. 1).

This system enables monitoring the condition of the railway tracks or more specifically, identifies the faulty location of track, and intimates to the train operator through LCD [6, 7]. Our paper is designed in consideration with cost efficiency and also the exact location of defect identification.

2 Literature Survey

The conventional system of crack detection on railway lines use trackmen/gangmen who apart from laying tracks are also involved in its surveillance, warning of any visible cracks or gaps, helping in its prompt repair, thus avoiding accidents. However, as with any human system, this too is prone to human failures/errors resulting in loss of life and property [8–11]. Therefore, a need is being felt world over, for finding a solution through automation which is foolproof and cost-effective. New technology in use in this direction as ultrasonic sensors, infrared systems, LED-LDR systems, etc. These new methods also have certain shortcomings, viz., their ability to detect only cracks on the surface, consumption of too much time and power, low efficiency due to external factors, etc. Of all the new systems, use of sensors is found to be closer to being accurate (Fig. 2).

The components used in this type of sensor system are as follows.

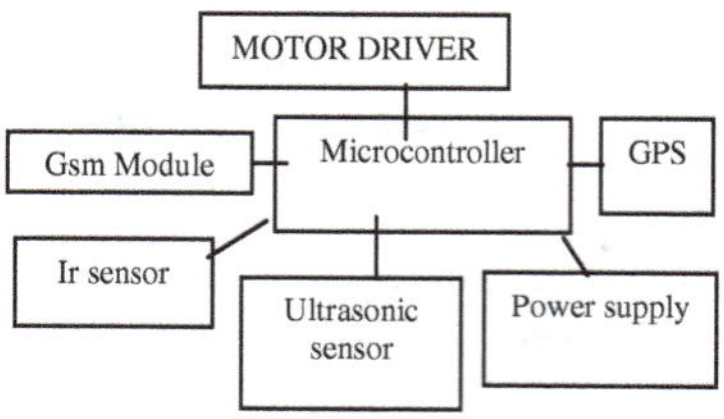

Fig. 2 Block diagram of proposed circuit

Microcontroller

The microcontroller used is a high-performance CMOS low powered, 8-bit microcomputer with flash memory of 8 Kb. Nonvolatile memory is the technology for manufacturing this Atmel device. It is especially compatible with MCS 51 [12]. The flash memory helps in reprogramming the program memory in the system with the help of microprocessor. It can also be reprogrammed by a conventional non-volatile memory programmer.

GSM Module

A GSM module or a GPRS module consists of a chip or a circuit that will send messages between the device at the site and the computer in the control unit. This system can send both voice message and text messages/data. It is majorly used for transmitting and receiving the messages. GSM wireless data module is for wireless remote applications, machine-to-machine or user-to-machine communications.

Power Supply

Power supply is required for any electrical or electronic circuit to run. In this case, +5 V and +12 V supply are used. +5 V is for the microcontroller board, ultrasonic sensor, etc. 1.5 mA current is required for the operation of the GSM module. +12 V is required for the operation of the motor driver.

Motor Driver

Most microprocessors operate at low voltage supply and require only a small amount of current for its operation. The motor driver requires a relatively higher voltage and current which cannot be supplied by the microprocessor. Hence, there is a primary need for the motor driver IC [13].

A motor driver is something which amplifies current; here, high current signal has been received from low current control signal, which can then run a motor. Motor driver acts as an interface between microprocessors and the motors in the robot. Some examples include the L293 series like L293D, L293NE, etc. Its applications include autonomous robotics.

GPS (Global Positioning System)

It is the global navigation satellite system or GPS which provides a geographic location and time information to a GPS receiver anywhere on or near the Earth to four or more GPS satellites. The GPS receives the location of the problem area using longitudes and latitudes and maps it so as to enable identify the exact position. GPS also has an interface for communication link. GPS also includes features such as video creation/sharing, tracking mode, vehicle history, and network help available.

Sensors

Sensor is a device which converts any physical input like temperature, pressure, or humidity and converts it into electrical signals as output. It can be combined with other tools to create a simple device or a complex one. Its applications can include speed synchronization of multiple motors in industries using a speed sensor or automatic closing and opening of doors using a PIR sensor (Passive Infrared sensor). The sensitivity of sensor explains the change in the rate of change of output with respect to input quantity. A sensor's sensitivity indicates how much the sensor's output changes when the input quantity being measured changes. For example, the IR sensor detects any obstacles in its path. The ultrasonic sensor detects the input signals and converts into electrical signals by using ultrasonic waves.

3 Methodology

The railway carriage carrying the control equipment is provided with a microcontroller, a motor driver, GSM module, GPS, and the ultrasonic and infrared sensors. A SIM card is inserted in the GSM module to send messages as commands. When there are fissures/gaps in rail lines, the ultrasonic sensor detects it and the device stops moving. When a message is sent, the SIM card in the GSM module sends it and passes it to the microcontroller. The microcontroller performs the action by sending it to the motor driver, so that it stops moving. This action is communicated to the control center by the GSM module in the shape of an SMS. By using the GPS, the position where the crack occurred can be triangulated and its coordinates are sent to the user. Since the GSM module is user protected, unauthorized users cannot access the channel (Fig. 3).

4 Result

From the result, we infer that the sensors detect the crack and send the information to the microcontroller, which in turn activates a system to bring the train to a halt, thus averting an accident. Time parameters are important and can be set/reset with the help of microcontrollers. Thus, this device would help reduce train collisions.

Fig. 3 Implemented system

5 Conclusion

Various types of sensors are used in this system. The ultrasonic sensor measures the crack and tracks distances and infrared sensors having its high-efficiency infrared waves which are not visible to the human eye are used for detecting obstacles. Therefore, accidents related to train derailment are avoided. Hence, putting in place such a system would reduce unnecessary loss of human life and property, by assessing the danger and averting it in time.

Possible Problems

1. Since the sections of the tracks are joined together, the ultrasonic sensors might not differentiate between the crack and a joint and stop the train unnecessarily.
2. If the infrared sensor used is not at the level of the obstacle, then the sensor might not detect the obstacle and will lead to a collision.

Acknowledgements The authors are ever so grateful to Birla Institute of Technology Mesra for providing the necessary support and encouragement in making this paper possible. We give our special thanks to our Vice Chancellor Prof. M. K. Mishra and our HOD Dr. S. Pal for giving us guidance and support for the research work.

References

1. Polivka AL, Matheson WL (2014) Automatic train control system and method. U.S. Patent no. 5828979, 27 Oct 2014
2. Ambegoda ALATD, Silva WTSD, Hemachandra KT, Samarasinghe TN, Samarasinghe ATLK (2013) Centralized traffic controlling system for Sri Lanka railways. In: 4th international conference on information and automation for sustainability (ICIAFS08), Sri Lanka, 12–14 Dec 2013
3. Khan M (2014) Automatic railway track switching system. Int J Adv Technol 54
4. Boylestad FRL, Nashelsky L (2012) Railway crack detection using gpa technology, 9th edn. Prentice Hall, USA, pp 196–199
5. Dipoppa G, D' Alessandro G, Semprini R, Tronci E (2011) Integrating automatic verification of safety requirements in railway interlocking system design. In: The 6th IEEE international symposium on high assurance systems engineering (HASE'01), Washington, USA, 2–24 Oct 2011
6. Tarnai G (2012) Safety verification for train traffic control communications. IEEE J Sel Areas Commun 4(1):1118–1120
7. Muralidharan V, Dinesh V, Manikandan P (2015) An enhanced crack detection system for railway track. Int J Eng Trends Technol 21(6)
8. Saravana Moorthy V, Murugananthan GN (2015) Automatic identification of obstacles & crack sensing scheme in rail tracking system. Int J Adv Res Electron Commun Eng 4(1)
9. Selvamraju S, Vigneshwar M, Gourav S (2012) Robust railway crack detection scheme (RRCDS) using LED LDR assembly. In: International conference on breakthrough in engineering, science & technology, Dr. V. Vaidehi, Madras Institute of Technology, Anna University, Chennai, Mar 2012
10. Avinash V, Arun DK (2013) Automatic broken track detection using LED-LDR assembly. Int J Eng Trends Technol 4(7)

11. Navaraja P (2014) Crack detection system for railway track using ultrasonic and PIR sensor. Int J Eng Trends Technol 5
12. Mahalakshmi V (2013) GPS based railway track survey system. Int J Comput Sci Secur
13. Ramesh S (2011) Detection of cracks and railway collision avoidance system. Int J Electron Electr Eng 4(3):321–327

Study of Network-Based DoS Attacks

Sonali B. Wankhede

Abstract The denial of service attack exploits the weakness in the TCP/IP implementation. In a privacy attack, the attacker tries to get access to the resources to which he is not authorized. The DoS attack blocks the authorized users from accessing the resources. The affected systems crash or disconnect from the Internet. In some cases, they are not much harmful. In other cases, they can be harmful, especially when you run a corporate network or ISP. This paper presents the denial of service attacks, and also the difference between DoS and DDoS is mentioned.

Keywords DoS · DDoS · Network-based attacks · Methods of attack

1 Introduction

In denial of service attack, the server is flooded with the traffic affecting the system resources and it becomes impossible for the authenticated user to use them [1]. Such an attack which floods the network or servers is difficult to recover. Various strategies have been recommended by the experts to protect the system from denial of service attack. If a DoS attack is predicted, the ISP should be contacted to check the overall scenario. This will help to reduce the attack [2]. The malicious traffic will be re-routed and using the load balancers will reduce the effects.

Some of the detection tools, intrusion detection systems and firewalls, can detect a DoS attack. There are many instances where attackers have demanded payment from the victim to stop the attacks [3, 4].

Distributed attacks are sophisticated DoS attacks in which the means traffic is directed from multiple attack systems. DoS attacks that originate from a single source are easier to reduce, by blocking the network traffic [3]. But DoS attacks that originate from multiple sources are difficult to detect. Since it is difficult to

S. B. Wankhede (✉)
Computer Engineering Department, Veermata Jijabai Technological Institute (VJTI), Mumbai, India
e-mail: sonaliw270@gmail.com

V. Nath and J. K. Mandal (eds.), *Nanoelectronics, Circuits and Communication Systems*, Lecture Notes in Electrical Engineering 511,
https://doi.org/10.1007/978-981-13-0776-8_58

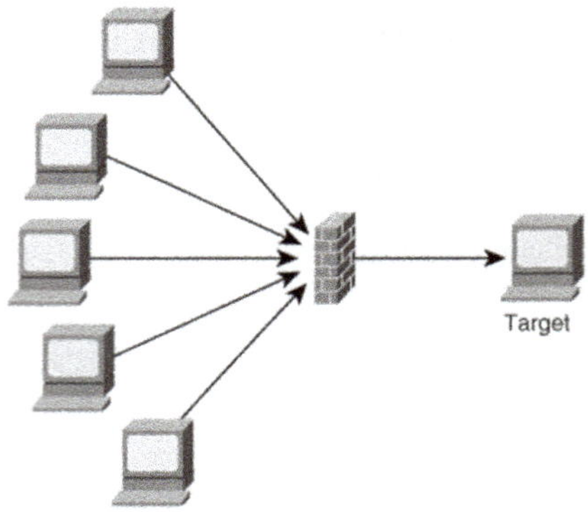

Fig. 1 Denial of service attack

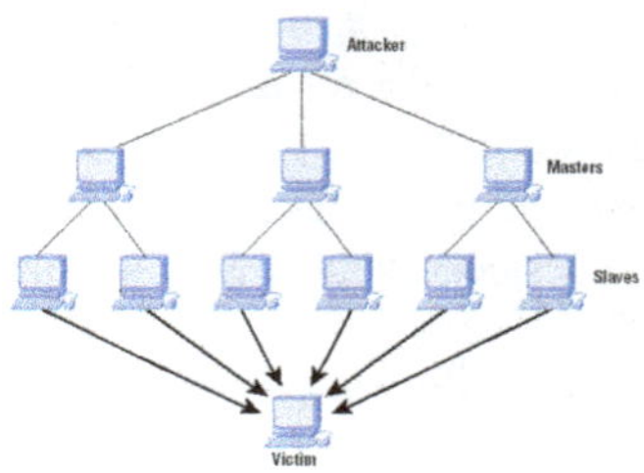

Fig. 2 Distributed DDoS attack

determine legitimate traffic and malicious traffic, in distributed denial of service attack, a service is made unavailable by blocking it with the traffic from multiple sources [5]. Here, the victims are the end target systems (Fig. 1).

In distributed denial of service attack, the incoming traffic is originated from many different sources and to stop the attack by just blocking a single IP address is not enough (Figure 2).

Difference between DOS and DDOS

An attack focused to deny a service to its legitimate users can be called a DoS attack. DDoS generates more attack traffic. Hence, it is very difficult to block attacks. Similarly, it is very difficult to find the actual attacker who initiated the attack

Both DoS and DDoS use the same technique of sending a large number of connection requests to the victim. The concept of DDoS is used in Botnets. Each system participating in the attack is the Botnet. Both the attacks are difficult to control. In recent studies, it is clear that many advanced techniques to bypass the firewall are being used [6].

Some of the techniques include attacking a victim by various network layers using different vectors by sending ICMP messages and later UDP, TCP, etc.

2 Network-Based Attacks

Network-based attacks are the simple attacks. It is easy to consume the resources on target system and lead the services to fail.

(1) *Ping of death*: *ICMP echo variation that could cause a system crash.th*: Sends excessive ICMP datagrams to the victim host. The ping commands are used to check whether the host is alive and the ICMP requests, and reply messages are used.
(2) *Teardrop attack*: The data sent is fragmented at the source and re-assembled at the destination.
(3) *SYN flood attack*: Various packets with an invalid source IP address are sent to the target. The target systems receive the packets and respond with SYN/ACK packets.
(4) *UDP flood attacks*: The echo and chargen services are used for network testing. The DOS services can be started by connecting the chargen to echo ports.
(5) *Land attack*: The packet includes the IP address of the target. As a result, a cycle is created in the target, which leads the victim system to hang.
(6) *Smurf attack*: It creates traffic to the victim's path by broadcasting the ICMP packets with the spoofed IP, which leads to all the host on the network to reply to the ICMP requests.

2.1 Ping of Death

The attacker deliberately sends IP packets of more than 65,536 bytes. Many operating systems were unable to find out a solution after receiving an oversized packet, so it leads to crashing or rebooting the system [7, 8].

Ping of death is basically a DoS attack that was widespread in legacy systems. Ping of death is conducted by using the ping command.

It is written as ping 150.24.35.46 or ping www.acme.net.

Most of the operating systems are vulnerable to the ICMP packets. Hence, to specify large number of packets cause an overflow in the system's internals that can result in system crash. The syntax varies depending on the operating system you were using. Given are the examples of Windows and Sun Solaris (Fig. 3).

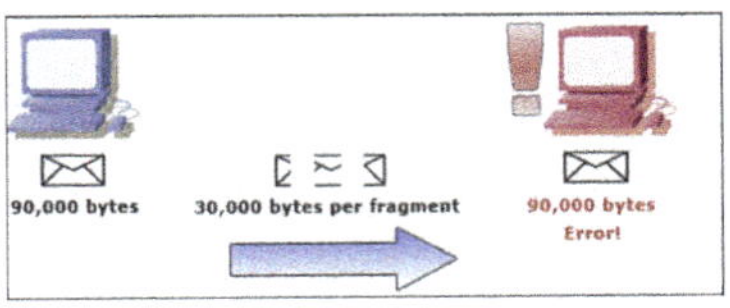

Fig. 3 Fragmented packets in ping of death attack

- Windows: ping-165527-s 1 hostname, and
- Solaris: ping-s hostname 65527.

2.2 *SYN Attack*

The normal TCP three-way handshake is exploited by the SYN flood attack. The resources on the targeted server re-consumed and are made unresponsive.

In SYN flood DDoS, the TCP connection requests are rapidly transmitted by the attacker, which causes network saturation [8].

When a normal TCP three-way handshake connection is established,

1. The client sends an SYN (Synchronize) message for requesting a connection to the server.
2. The message is acknowledged by the server and an SYN-ACK (synchronize–acknowledge) message is sent to the client.
3. A response with ACK (acknowledge) message is given by the client and a connection is established.

The attacker sends a huge amount of SYN packets to every port on the targeted server. In this case, a fake IP address can also be used. The server receives multiple requests that appear to be legitimate. It responds to every request with an SYN-ACK packet.

2.3 *Teardrop Attack*

It is a denial of service attack in which the fragmented packets are sent to the target machine. The machine cannot reassemble the packets due to bug in TCP/IP fragmentation reassembly.

Hence, the packets overlap one another which leads to the target network device to crash. This happens on older versions of operating systems like Windows NT, Windows 3.1x, Windows 95 and different versions of the Linux kernel prior to 2.1.63.

The fragment offset field indicates the start position or offset of data. If the sum of offset and the size of one packet differ from the next fragmented packet, the packets overlap. This results in a denial of service condition (Fig. 4).

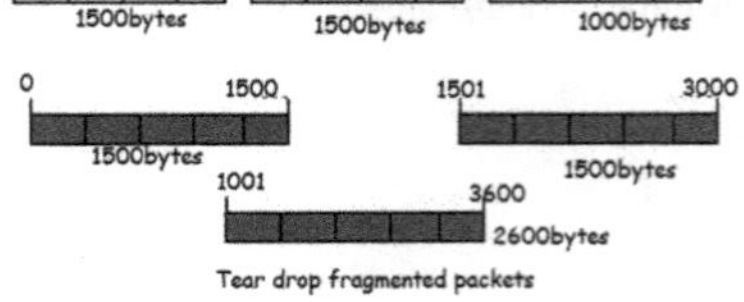

Fig. 4 Teardrop attack

2.4 *Smurf Attack*

A smurf attack exploits the IP protocol broadcast addressing to create a denial of service. The ICMP is used by the network nodes to exchange information.

The Smurf program builds a packet that is originated from another address. The packet contains an ICMP ping message [7]. The echo replies to the messages are sent to the victim's address, and several ping and echo messages can flood the network.

How a smurf attack works,

(1) The IP address victim is identified.
(2) The ICMP packets are transmitted at the broadcast address of intermediary sites.
(3) The broadcast is delivered by the intermediary sites at layer 2 to all the hosts on their subnet.
(4) Host replies to the victim network.

Suppose, the hacker has a cable modem and sends a 1-Mbps spoofed ICMP traffic at the intermediary sites. If the intermediary sites have 150 hosts that are responding, this results in a 150-Mbps attack traveling from the "amplifiers" toward the victim

An SYN attack with IP spoofing results in a land attack.

3 Comparison of Network-Based Attacks

Parameters	Ping of death	Smurf	SYN flood	Teardrop
Causes	Flaw in TCP/IP system	Spoofing an IP address	Incomplete TCP three-way handshake	Fragments are sent in improper
Transmission	Excessive IP Datagrams are sent	Creates traffic to victim's path	Packets with invalid IP address are sent	Creates bug in the TCP/IP fragmentation
Crashes	Oversized packets lead to crash the system	Several pings and echo messages flood the network	The increasing amount of TCP requests leads to network saturation	Accumulation of fragments leads to system crash
Solutions	Blocking the fragmented Ping requests	Disabling the broadcast feature on routers	Proper filtering of firewalls and Linux Kernels	Proper upgrade of hardware and software

4 Conclusion

In this paper, various denial of service and distributed denial of service attacks are mentioned. The different methods for performing the SYN flood attacks are also presented. By reviewing various DoS attack techniques and classifying them, this paper highlights the characteristics of DoS attacks.

References

1. https://www.incapsula.com/ddos/ddos-attacks/denial-of-service.html
2. Caelli WJ, Raghavan SV, Bhaskar, SM, Georgiades J (2011) Policy and law: denial of service threat
3. Mohay G, Ahmed E, Bhatia S, Nadarajan Ravindran AB, Tickle AB, Vijayasarathy R (2011) Detection and mitigation of high-rate flooding attacks
4. Boyd C, Gonzalez-Nieto J, Kuppusamy L, Narasimhan H, Pandu Rangan C, Rangasamy J, Smith J, Stebila D, Varadarajan V (2011) Cryptographic approaches to denial-of-service resistance
5. Semerci M, Yamaç M, Cemgil AT, Sankur B, Coşar DS (2016) Detection of (D)DOS attacks based on online change point analysis. In: 2016 24th Signal processing and communication application conference (SIU)
6. Ettiane R, Elkouch R, Chaoub A (2016) Protection mechanisms for signaling DoS attacks on 3G mobile networks: comparative study and future perspectives. In: 2016 4th IEEE International colloquium on information science and technology (CiSt)
7. http://www.cs.utexas.edu/users/chuang/dos.html
8. http://itsecurity.telelink.com/udp-flood
9. Si W, Starobinski D, Laifenfeld M (2016) Protocol-compliant DoS attacks on CAN: demonstration and mitigation. In: 2016 IEEE 84th vehicular technology conference (VTC-Fall)
10. Aleroud A, Alsmadi I (2016) Identifying DoS attacks on software defined networks: a relation context approach. In: NOMS 2016 IEEE/IFIP network operations and management symposium
11. http://www.cisco.com/c/en/us/about/press/internet-protocol-journal/back-issues/table-contents-34/syn-flooding-attacks.html

Author Index

A
Abinash, M. J., 185
Agarawal, Sakshi, 535
Agarwal, Sparsh, 599
Aleem, Asfia, 105
Anand, Satvika, 409
Ankita, 269
Anwar, Shamama, 365
Ashank, Vishal, 365

B
Balakannan, S. P., 313
Banerjee, Arindam, 139
Baral, Manish Mohan, 13
Baro, Raj Kamal Roye, 549
Begum, Amreen, 105
Bhattacharya, Suman, 435
Bhattacharyya, Swapan, 139

C
Chandra, Bibhas, `
Chandra, Mahesh, 25, 457
Chaudhary, Shantanu, 507
Chithaluru, Premkumar, 89

D
Dabhade, Sandeep, 239
Damayanti, K., 127
Das, A. K., 281
Dehuri, Satchidananda, 219, 229
Deyasi, Arpan, 139

E
Eleswarapu, Lalitha, 573

F
Fatima, Ajaz, 105
Fatma, Sabiha, 383

G
Ganapathy, S., 1
Ganesh Madhan, M., 337
Ghosh, Momojit, 565
Ghosh, Trisha, 483
Goel, Vidushi, 557, 605
Gupta, Hari Shanker, 353
Gupta, Neha, 527

H
Hari Charan, E. V. V., 549

J
Jyoti, 493

K
Kandar, Debdatta, 115
Kannan, A., 1
Karuppasamy, M., 313
Kaunoujiya, Suraj, 591
Kavitha, C., 299
Khalkho, Ayushman, 599
Khan, Amer Ali, 209
Kulothungan, K., 1
Kumar, Abhinav, 397
Kumar, Adesh, 493
Kumar, Akshay, 195
Kumar, Ankush, 591
Kumar, Anupam, 177
Kumar, Avanish, 255

V. Nath and J. K. Mandal (eds.), *Nanoelectronics, Circuits and Communication Systems*, Lecture Notes in Electrical Engineering 511,
https://doi.org/10.1007/978-981-13-0776-8

Kumar, Chandan, 415
Kumari, Dipti, 515
Kumari, Pinki, 557
Kumar, Manoj, 177
Kumar, Shishir, 239
Kumar, Shubham, 605
Kumar, Sumit, 239

M
Madhan, M. Ganesh, 299
Mahato, Shalini, 323
Majumdar, Sananaya, 535
Mandal, M. K., 281
Markham, Naveen, 605
Maurya, Divyesh Kumar, 591
Mishra, Annapurna, 229
Mishra, Punit, 353
Mohan, Rohit, 535
Mohan, Utkarsh, 195
Mohapatra, Nihar, 353
Mohapatra, Satyajit, 353
Mondal, Pratik, 425
Muralidharan, Aditya, 605
Murali Krishna, K., 337

N
Nancy Noella, R. S., 53
Nath, Vijay, 373, 383, 397, 409, 415, 451, 469, 475, 507, 535, 549, 557, 565, 573, 591, 599, 605
Neema, Vaibhav, 527
Nidhi, Neha, 451, 469, 475, 507

P
Pal, Indrajit, 549
Pallavi, Smita, 195
Padlia, Minal, 161
Parihar, Priyanka, 527
Parui, Susanta Kumar, 425
Paul, Babu Sena, 115
Paul, Partha, 77
Paul, Sanchita, 323
Prakash, Ravi, 89
Prakash, Suraj, 599
Prasad, Brij Mohan, 65
Prasad, Deepak, 451, 469, 557, 591, 599, 605
Prava, Arka, 507
Priya, Annu, 149
Priyadarshini, J., 53
Priyadarshi, Rahul, 289

Q
Qadeer, Shaik, 105

R
Rajnish, Kumar, 515
Raju, Deril, 573
Raj, Utkarsh, 475
Ranjan, Prabhat Kumar, 245
Razi, Qaiser, 373
Reddy, B. M., 127
Riya, 557

S
Sahana, Sudip Kumar, 149, 269
Sahay, J., 483
Sahoo, G., 457
Saiv, Rohith, 573
Sajid, Md, 209
Sangwan, Anil, 493
Sanjay Kumar, S., 599
Saraiyan, Satyam, 565
Saranya, M. S., 1
Sarkar, Angsuman, 139
Selvi, M., 1
Sharma, Ashwani, 435
Sharma, Deepak, 89
Sharma, Jankiballabh, 161
Sharma, Prashant, 289
Sharma, Shalini, 435
Shikha, Prachi, 557
Shipra, 25
Singh, B. P., 209
Singh, Inder, 435
Singh, Jatindeep, 353
Singh, Nalini, 219
Singh, Praveen, 527
Sinha, Akash, 549
Sinha, Keshav, 77
Sinha, Paritosh Kumar, 565
Sinha, Subham Kumar, 365
Siva Kumar Reddy, Bathula, 97
Sivasubramanian, K. B., 239
Soni, Surender Kumar, 289
Srikanth, P., 89
Srinivasulu, Avireni, 127
Srivastava, Praveen, 37
Srivastava, Sanjiv Kumar, 37
Srivastava, Sumit, 457
Sunil Reddy, T., 127
Suraj, Akash Keneth, 535

Suryakalavathi, M., 209

T
Tanushree, 557
Thakura, P. R., 65, 245, 255
Thangaramya, K., 1
Tiwari, Sneha, 483

V
Vasudevan, V., 185
Verma, Amitabh, 13
Vivek, Snehanshu, 365

W
Wankhede, Sonali B., 611

CPSIA information can be obtained
at www.ICGtesting.com
Printed in the USA
LVHW02*2148190818
587448LV00001B/156/P

9 789811 307751